Economic Geography

This book series (now indexed in SCOPUS) serves as a broad platform for scientific contributions in the field of Economic Geography and its sub-disciplines. *Economic Geography* wants to explore theoretical approaches and new perspectives and developments in the field of contemporary economic geography. The series welcomes proposals on the geography of economic systems and spaces, geographies of transnational investments and trade, globalization, urban economic geography, development geography, climate and environmental economic geography and other forms of spatial organization and distribution of economic activities or assets.

Some topics covered by the series are:

- Geography of innovation, knowledge and learning
- Geographies of retailing and consumption spaces
- Geographies of finance and money
- Neoliberal transformation, urban poverty and labor geography
- Value chain and global production networks
- Agro-food systems and food geographies
- Globalization, crisis and regional inequalities
- Regional growth and competitiveness
- Social and human capital, regional entrepreneurship
- Local and regional economic development, practice and policy
- New service economy and changing economic structures of metropolitan city regions
- Industrial clustering and agglomeration economies in manufacturing industry
- Geography of resources and goods
- Leisure and tourism geography.

Publishing a broad portfolio of peer-reviewed scientific books *Economic Geography* contains research monographs, edited volumes, advanced and undergraduate level textbooks, as well as conference proceedings. The books can range from theoretical approaches to empirical studies and contain interdisciplinary approaches, case studies and best-practice assessments. Comparative studies between regions of all spatial scales are also welcome in this series. Economic Geography appeals to scientists, practitioners and students in the field.

If you are interested in contributing to this book series, please contact the Publisher.

Pawel Churski · Czesław Adamiak ·
Anna Dubownik · Maciej Pietrzykowski ·
Barbara Szyda

Inner Peripheries in the New Conditions of Cohesion Post-SARS-CoV-2

Recommendations for European Cohesion Policy

Springer

Pawel Churski
Faculty of Human Geography and Planning
Adam Mickiewicz University
Poznań, Poland

Anna Dubownik
Faculty of Earth Sciences and Spatial Management
Nicolaus Copernicus University
Toruń, Poland

Barbara Szyda
Faculty of Earth Sciences and Spatial Management
Nicolaus Copernicus University
Toruń, Poland

Czesław Adamiak
Faculty of Earth Sciences and Spatial Management
Nicolaus Copernicus University
Toruń, Poland

Maciej Pietrzykowski
Institute of International Business and Economics
Poznan University of Economics and Business
Poznań, Poland

ISSN 2520-1417 ISSN 2520-1425 (electronic)
Economic Geography
ISBN 978-3-032-10935-4 ISBN 978-3-032-10936-1 (eBook)
https://doi.org/10.1007/978-3-032-10936-1

This Springer imprint is published by the registered company Springer Nature Switzerland AG
The registered company address is: Gewerbestrasse 11, 6330 Cham, Switzerland

Acknowledgements The research on which this book is based was funded by the National Science Centre in Poland under the project "The Trajectories and Challenges of the Development of Inner Peripheries in the New Conditions of Cohesion Post-SARS-CoV-2" no. 2020/37/B/HS4/01034 (http://tiperico.web.amu.edu.pl/en/).

Competing Interests The authors have no competing interests to declare that are relevant to the content of this manuscript.

Contents

Chapter 1
Introduction

Abstract In this chapter, we present the contemporary challenges of Cohesion Policy in the context of increasing spatial disparities in socio-economic development, particularly under the growing pressure of developmental shocks. Against this backdrop, we define our understanding of inner peripheries—an understanding that motivated our research endeavor and guided the approach we took and how we organized the findings. We define and systematize the research objective, further specifying it as hypotheses that are verified throughout the various stages of the analysis. We also present our methodological approach, outlining the underlying assumptions, justifying the research structure, and specifying the research methods. Finally, to facilitate the reader's understanding and enable quick access to the most relevant content, we provide a concise overview of the individual chapters of this monograph.

Keywords Spatial disparities of development · Inner peripheries · Aim and scope of the analysis · Structure and content of the chapters · Research approach and applied methods

1.1 Inner Peripheries in the Process of Development in Turbulent Times: Understanding, Motivation, Aims of Research

Inherent to socio-economic development is its spatial differentiation. This results, on the one hand, from the heterogeneity of space in terms of the quantity and quality of resources and, on the other hand, from the influence of agglomeration and scale effects, which serve as natural catalysts for the spatial polarization of development processes. The impact of these effects evolves under the influence of megatrends (Rodríguez-Pose et al. 2024). It nevertheless does not lead to the outcomes anticipated in discussions about the "death of distance," the "end of geography," or the projected dominance of the "weightless economy" (Cairncross 1997, O'Brien 1992, Quah 1999), as emphasized in the "flat world" concept proposed by Thomas

P. Churski et al., *Inner Peripheries in the New Conditions of Cohesion Post-SARS-CoV-2*, Economic Geography,
https://doi.org/10.1007/978-3-032-10936-1_1

Friedman (Friedman 2005). This underlines the need to consider the differentiation of places and their inhabitants when formulating development policy assumptions, as this is a prerequisite for effectively enhancing economic, social, and territorial cohesion (McCann 2008; Rodríguez-Pose and Crescenzi 2008). In designing development policies, particularly Cohesion Policy, the goal should not be the idealistic and unattainable equalization of spatial development disparities. Instead, it is important to acknowledge the inevitability of spatial differences in development, which when maintained within socially acceptable limits (Faludi and Rocco 2022) constitute a fundamental element of the functioning of a capitalist economy (Harvey 2005).

Assuming that a certain level of spatial development inequalities is inherent and unavoidable, it is important to recognize that the persistence of those inequalities (particularly when accompanied by a tendency to deepen) leads to entrenchment, resulting in growing scale and costs that are increasingly difficult to ignore (Rodríguez-Pose et al. 2024). The OECD's 2023 report, tellingly titled The Long-standing Geography of Inequalities, highlights three key categories of these costs (OECD 2023): (1) Economic costs stem from the fact that lagging regions and/or those trapped in vicious cycles of long-term stagnation contribute a significant share of economic activity across all countries and represent untapped potential for driving growth and socio-economic development. This has led to a situation in which "left-behind places" have become a central theme in contemporary discussions on geographic inequality and the challenges of Cohesion Policy (Farole et al. 2018; Pike et al. 2023). (2) Social costs are associated with the failure to ensure adequate access to essential public services and infrastructure. This problem is characteristic of both rapidly growing areas (e.g., high housing prices, congestion) and marginalized regions (e.g., poor quality and limited availability of social services). These challenges were starkly exposed by the social impacts of the SARS-CoV-2 pandemic and further exacerbated by the economic consequences of Russia's aggression against Ukraine (Ballantyne et al. 2023; Huang et al. 2022; Mbah and Wasum 2022; Zhai et al. 2021). (3) Political costs are linked to rising discontent and social withdrawal, which in turn contribute to the growth of populism, which poses a serious threat to the foundations of democracy in Europe (Dijkstra et al. 2020; Hadjimichalis 2021; Rodríguez-Pose 2018).

The persistence and, in some EU Member States, the expansion of the group of "left-behind" places constitute one of the clearest indicators of the dysfunctions of Cohesion Policy. This issue is powerfully illustrated by the fact that within the territory of the European Union, there are areas particularly affected by economic stagnation, population decline, low productivity, high unemployment, a limited number of highly skilled jobs, and a lack of knowledge and innovation centers (MacKinnon et al. 2022; Rodríguez-Pose 2018). These conditions lead to the impoverishment of local populations, declining real wages, and increasing social tensions (MacLeod and Jones 2020). "Left-behind" places are often characterized by limited connectivity. They are poorly linked to regional and national growth centers, which contributes to a sense of detachment and isolation from the rest of the country (Davenport and Zaranko 2020;

Mattinson 2020; Tomaney et al. 2021). This form of isolation is reflected in the insufficient or absent provision of essential basic services, including transport and communication infrastructure. These areas also lack adequate social infrastructure such as schools, healthcare facilities, and public services, many of which are closed or in poor condition due to population decline and financial constraints (Davenport and Zaranko 2020; MacLeod and Jones 2020). These deficits severely limit the capacity of such regions to implement the European Union's current development objectives related to green and digital transitions. These transitions pose significant development challenges for lagging areas and risk deepening spatial inequalities, disproportionately affecting economically weaker regions (Rodríguez-Pose et al. 2024). This vulnerability has become particularly evident in the wake of recent global crises, which have underscored the low resilience of left-behind areas (Floerkemeier et al. 2021). On the other hand, the social and economic consequences of recent global crises, including the SARS-CoV-2 pandemic, have also led to changes in social behavior and organizational practices. These changes may lay the groundwork for new development opportunities in left-behind regions, calling for a rethinking of development drivers and their contemporary influence on development processes—especially regarding their spatial differentiation (European Commission 2024a; Hertrich and Brenner 2024; OECD 2023).

Nevertheless, the fact remains that the scale of spatial development inequalities, particularly at the subregional level, is increasing (European Commission 2024b), which has intensified the debate on the underlying causes of "left-behind places" and "abandoned areas" (Pike et al. 2023). An important yet still insufficiently studied category within this group is inner peripheries (Copus et al. 2017; Servillo et al. 2016). These are areas characterized more by a lack of relational linkages with growth centers than by significant physical distance due to geographical location. Identifying the causes of developmental dysfunctions in these areas, as well as defining effective and efficient development interventions tailored to this specific category of "left-behind places," represents a significant research gap—one we aim to address through the findings presented in this book. Inspired by the findings of the ESPON PROFECY project "Inner Peripheries: National territories facing challenges of access to basic services of general interest" (ESPON PROFECY 2017, 2018), we seek to deepen the analysis of the root causes behind the occurrence of disadvantageous processes, characteristics, and evolutionary dynamics in these areas factors that constrain their development potential. In our view, understanding these dynamics is crucial for designing effective and efficient Cohesion Policy interventions targeted specifically at this category of regions, in line with the territorial, place-based policy approach.

We understand inner peripheries as areas whose development potentials are dormant or lost due to their social or economic peripherality. This peripherality primarily results from low transport accessibility and relatively long travel times to economic centers, a limited range of functional linkages, and a lack of capacity or persistent difficulties in establishing and maintaining such connections. These areas are characterized by lower efficiency in the functioning of the territorial socio-economic system, reduced access to public goods and services, and, consequently, an overall lower quality of life (well-being). As a result of these conditions, such areas experience declining competitive potential and persistent marginalization. We argue that expanding the normative basis of inner peripheries requires incorporating the concept of territorial justice into the assumptions of territorial cohesion. Spatial justice calls for the explicit recognition of space as a factor of social inequality, reproduced through socio-economic mechanisms that structure society spatially. In parallel, the concept of the foundational economy should be integrated, as it advocates for the protection ("sheltering") of those sectors of the economy that provide essential goods and services, such as utilities and basic public services. We assume that inner peripheries are not internally homogeneous, which leads to the emergence of central areas and surrounding zones within their internal structure, which is analogous to the general dichotomy of socio-economic space. Local "poles" within inner peripheries exhibit similar limitations to their surrounding areas, with their relative position stemming from internal variations in the scale of problems and in the concentration of economic activities and population.

The main objective of the research presented in this monograph is to identify the patterns of socio-economic development changes in selected inner peripheries, with particular emphasis on their vulnerability and resilience to the consequences of the SARS-CoV-2 crisis. The accompanying aim is to derive recommendations for adjusting the orientation and implementation of interventions under the European Cohesion Policy. The research was conducted using the case of Poland, which is the largest beneficiary of EU Cohesion Policy development interventions and serves as a kind of laboratory for testing its effectiveness and efficiency. The main objective can be further specified into three groups of specific goals: empirical, methodological, and applicative. The empirical goals of the study are:

- EG1: Delimitation and typology of spatial development disparities within the framework of Functional Urban Areas (FUAs), leading to the identification of inner peripheries
- EG2: Determination of the developmental trajectories of inner periphery areas in relation to more developed areas, with particular emphasis on their vulnerability and resilience to the SARS-CoV-2 crisis and its socio-economic consequences
- EG3: Determination of the intensity of the diffusion of developmental processes in spatial terms between inner periphery areas and urban centers serving as subregional hubs of functional areas, as well as the factors facilitating and hindering these processes
- EG4: Determination of the susceptibility of individual industries and economic sectors to the shocks of the SARS-CoV-2 crisis, along with an assessment of the observed and anticipated economic and social consequences
- EG5: Identification of developmental factors and barriers shaping socio-economic transformations that influence the living standards and conditions of residents in selected inner peripheries, considering the consequences of the SARS-CoV-2 crisis shock

The methodological goals of the research whose results are presented in this monograph are:

- MG1: Preparation of a model indicator system enabling the identification of inner periphery areas and monitoring the developmental processes occurring in these areas
- MG2: Development of principles and criteria for the delimitation and typology of inner periphery areas
- MG3: Development of principles and criteria for identifying the developmental trajectories of inner periphery areas
- MG4: Preparation of assumptions and testing of a monitoring system for socio-economic changes in inner periphery areas

The application-oriented goal of the study (AG1) is the formulation of recommendations regarding the orientation and implementation methods (including the selection of instruments) of Cohesion Policy developmental interventions in inner periphery areas. Within this goal, particular emphasis is placed on the role of their urban centers in creating conditions for networking and the diffusion of developmental processes.

The scientific aim of the research, specified through detailed empirical, methodological, and applicative goals, is further concretized by formulating six research hypotheses.

- H1: The inner peripheries identified within the system of functional areas are characterized by varying sizes of both core cities and their surrounding areas.
- H2: The development trajectories of inner peripheries differ from those of more developed areas and exhibit greater sensitivity to the socio-economic consequences of the SARS-CoV-2 crisis.
- H3: The spatial spread and diffusion of development processes between inner peripheral areas and urban centers serving as subregional hubs of functional areas are constrained by factors unfavorable to these processes, which have gained prominence as a result of the SARS-CoV-2 crisis.
- H4: Inner peripheries demonstrate a specific vulnerability of individual industries and economic sectors to the shocks of the SARS-CoV-2 crisis, resulting in distinct economic and social consequences compared to more developed areas.
- H5: Social and institutional capital is the most critical factor shaping the socio-economic transformations that determine the living standards and conditions of residents in inner peripheral areas. Its importance increases under the impact of the consequences of the SARS-CoV-2 crisis shock.
- H6: Ensuring the effectiveness and efficiency of Cohesion Policy interventions targeted at inner peripheral areas requires a greater focus on changing implementation instruments rather than on changing the principles of programming this policy.

The research was conducted within two spatial frameworks. The first comprises FUAs of medium-sized and small towns in Poland, delineated using an original methodology based on migration and commuting matrices. These FUAs include both the towns themselves and the surrounding municipalities. It was assumed that the identification of inner peripheries should be grounded in a system of functional areas that reflect the scope of socio-economic linkages developing at a supra-local level. The second spatial framework consists of purposefully selected FUAs representing different types of inner peripheries.

The temporal scope of the analysis covers the years 2011–23, with particular emphasis on the period of the SARS-CoV-2 crisis and its consequences (2020–23).

1.2 Research Approach

Given the complex and multidimensional nature of inner peripherality, we designed literature-conceptual and empirical research divided into several stages and carried it out using data and research method triangulation. In the first stage, based on a literature review, we formulated a theoretical-conceptual model illustrating the interconnection between theoretical approaches, spatial structures, and analytical frameworks that we applied in planning the empirical study. We developed our definition of inner peripheries and determined their significance in contemporary socio-economic processes. We described the impact of asymmetric development shocks on spatial socio-economic inequalities and the responses of European Cohesion Policy. To refine our approach, we drew on theoretical frameworks explaining spatial socio-economic disparities and regional policy responses, including the core–periphery theory, geographical and organized proximity, Modern World Systems, foundational economy, spatial justice, and place-based policy. Based on a review of the significance of functional areas in identifying development disparities and systematizing development factors, we developed assumptions that we applied in the empirical analysis.

Considering the theoretical and conceptual findings, we planned two stages of empirical research. In the first stage (national level), we examined inner peripheries at the country scale using secondary statistical data and quantitative analysis methods. In the second stage (case study), where we focused on four selected case study areas, we supplemented secondary sources with primary field research, including quantitative methods (survey) and qualitative methods (interviews, document analysis). The sequence of stages was not accidental. The first stage, typological in nature, enabled us to select research areas representing different types of inner peripheries for the second stage (idiographic research). The empirical study was then divided into a sequence of six research steps, as illustrated in Fig. 1.1. The final step—a summary phase—focused on formulating synthetic conclusions and policy recommendations, completing the research framework.

When defining the time frame for our analysis, we were constrained on the one hand by the research topic, which focuses on post-pandemic transformations, and on the other hand by the availability of secondary data, which at the beginning of the project covered only the early period of the pandemic. Therefore, in the first stage of our nationwide study, we primarily used data from the most recent available years before the SARS-CoV-2 pandemic. In contrast, to investigate the case study areas, we relied mainly on primary data. We collected the primary data mostly in 2023, after

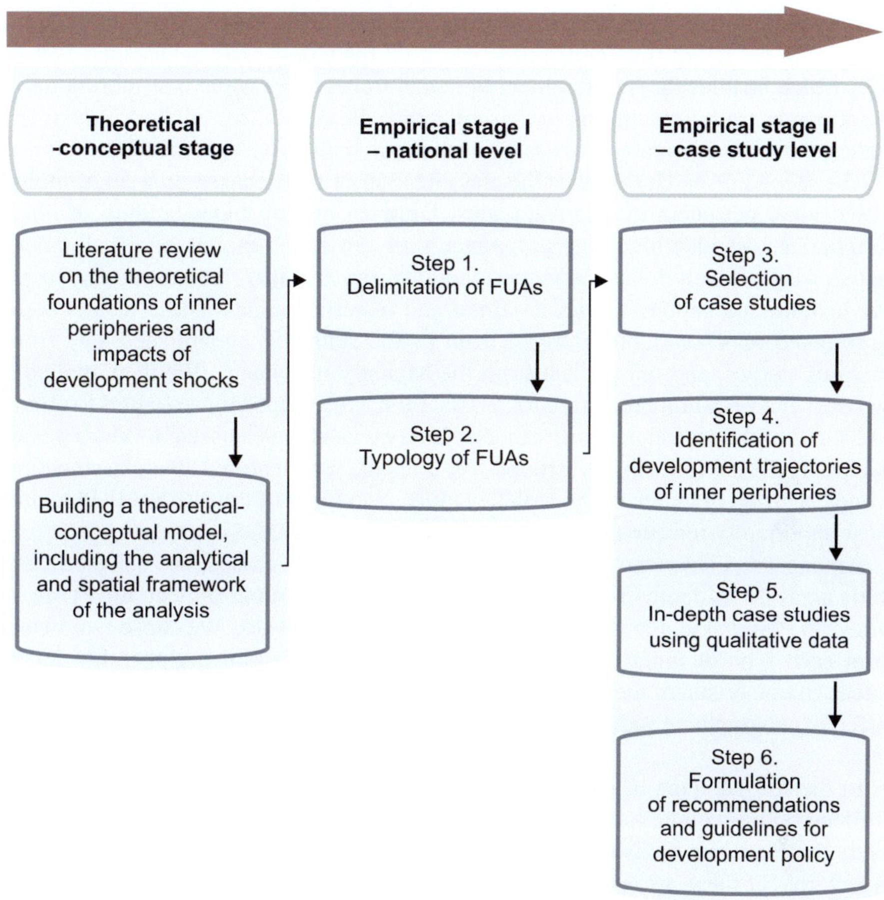

Fig. 1.1 Study procedure; *Source* Own elaboration

the pandemic had ended. This approach allowed the transformations that occurred following this health crisis to be examined retrospectively.

The territorial scope of our study covers Poland. When selecting the spatial resolution for analysis, instead of relying on predefined administrative units (which correspond to statistical territorial units), we concluded that it was necessary to design new spatial analysis units that would capture functional linkages between cities (including small ones) and their surrounding rural areas. Thus, our first research task was to delineate FUAs (which ultimately were to total 413 in number), which became the spatial framework for our nationwide analysis. We defined FUAs as territorially contiguous areas that include a central city and at least one adjacent municipality functionally linked to it through population flows (commuting to work and bidirectional migration) and located within an isochrone of 1 h (or for the largest cities, 1.5 h) of travel time to the central city. To delineate FUAs, we relied on inter-municipal migration and commuting matrices provided by Statistics Poland. To account for travel-time constraints, we used the Google Maps Distance Matrix API, which enables the measurement of actual travel times while considering daily variations in traffic conditions. After collecting these data, we applied a modified highest-flow linkage method to assign each municipality to a specific FUA.

The second research task was the classification of FUAs based on a set of multi-dimensional peripherality characteristics. Drawing on our understanding of inner peripheries, we identified five components of territorial capital of development factors whose deficit defines inner peripherality: accessibility, human capital, social and institutional capital, financial capital and material capital. These factors were represented by 47 variables derived from public statistics, unpublished data from previous studies, and online data from the Ministry of Finance. We then applied a two-step classification procedure for FUAs. First, we grouped the variables to eliminate those highly correlated with others. Next, we used the Gaussian mixture model to classify FUAs into similar groups. As a result, we identified five classification groups: a group of major FUAs with key cities, two types of transitional FUAs, and, most importantly for our study, two types of peripheral FUAs.

Among the FUAs classified into the two peripheral groups, we selected case study areas for in-depth examination in the second stage of our project: the FUAs of Drawsko Pomorskie, Przasnysz, Sandomierz, and Włocławek. We chose two FUAs from each type of inner periphery. These four FUAs exhibit peripherality index values characteristic of the two inner periphery groups. They represent the country's different geographical and historical regions and vary in the sizes of FUAs and central cities.

In the second stage of the empirical study, we focused on the case study areas, utilizing both secondary and primary data and qualitative and quantitative research methods. First, we analyzed official statistical data to examine the development trajectories of these areas in comparison to the FUA groups and the entire country within a short period from pre-pandemic to post-pandemic years (2017–23). For the dynamic analysis, we considered only 26 of the 47 variables used in the FUA typology.

The primary information we collected in the case study areas included both qualitative and quantitative data. On the qualitative side, we gathered data from regional and local planning and policy documents as well as group and individual interviews.

In our document analysis, we examined: development strategies at the municipal, county, and regional levels (larger administrative units); spatial development plans; and sectoral strategic documents—including revitalization plans, sustainable mobility plans, low-emission economy plans, innovation strategies, social problem mitigation strategies, and environmental protection plans. These documents underwent substantive analysis, firstly to compare the development challenges associated with peripherality against the issues we identified as characteristic of inner peripheries and secondly to identify local responses to these challenges.

To gain firsthand insights into the diagnoses and experiences of local leaders and stakeholders, we conducted both group and individual interviews. Our interviewees included local politicians and administrative staff, entrepreneurs, and representative of local non-governmental organizations (NGOs). We asked them about their perceptions of quality of life, development challenges, and the changes that occurred as a result of the SARS-CoV-2 pandemic. The interviews were conducted in person by researchers or a contracted research agency. All interviews were recorded and transcribed verbatim. We then analyzed the content of the responses, focusing on the various dimensions of territorial capital identified in the earlier stages of our study and the changes resulting from the SARS-CoV-2 pandemic.

While the interviews allowed us to understand the perspectives of local leaders in the studied areas, we used a survey to delve deeper into the opinions and behaviors of residents. The survey was divided into four thematic sections, in addition to socio-demographic information about the respondent: changes in the face of crisis (referring to perceptions of social and economic changes due to the pandemic), changes in the labor market (individual changes in professional situations and experiences), identity and security (history and attachment to the current place of residence and changes in perceptions of security), and communication (mobility behaviors, their changes, and perceptions of transportation opportunities). We designed the survey questionnaire, and we entrusted the sampling procedure and survey administration to a social research company. A geographic and demographic stratified sampling procedure was used to select respondents living in different environments (both in the local central city and surrounding rural areas, within each municipality forming the FUA) and representing various age and gender groups. Residents were contacted by phone—or in person, when phone contact with respondents from specific strata was impossible. After excluding incomplete or incorrect responses, the analysis included 1,557 survey participants (ranging from 314 to 470 in each FUA). In analyzing the results, we used descriptive statistics and cross-tabulation tables to identify typical structures and diversity of outcomes across different areas.

Throughout the project, we gathered both primary and secondary empirical data. We utilized quantitative and qualitative analysis methods to ensure that the theory and knowledge accumulated by other researchers and the insights from qualitative research would be integrated into the quantitative studies with descriptive, explanatory, and validating orientation. For the analysis of quantitative data, we used IBM SPSS software and the R programming language. The presentation of spatial data also involved the use of ESRI ArcGIS software.

1.3 Structure of Book

The book is divided into seven chapters, including the introduction, a theoretical section, an empirical section, and the conclusion. To facilitate the reader's navigation and help them find the most relevant content, a concise overview of each chapter's content has been provided.

Chapter One, serving as an introduction, presents the contemporary challenges of Cohesion Policy in the context of spatial differentiation trends in socio-economic development processes under the conditions of intensifying development shocks.

Against this background, we define the study's adopted concept of inner peripheries, which forms the basis for the research rationale and the approach to conducting and structuring the empirical investigation we present. We also define and systematize the main research objective, specifying it through a set of hypotheses that are subject to verification throughout the successive stages of the analysis. Furthermore, we outline the adopted methodological approach by presenting the underlying assumptions, justifying the structure of the research process, and describing the methods applied. Finally, to facilitate comprehension and help readers quickly locate the content most relevant to their interests, we provide a concise overview of each chapter.

Chapters 2, 3, and 4 constitute the theoretical part of the monograph. In Chap. 2, we present the origins and definition of inner peripheries. Our discussion is preceded by an overview of current transformation trends in EU Cohesion Policy, which involve a reorientation of its objectives, a shift in the understanding of proximity, and the growing importance of the subregional level in identifying and addressing spatial disparities in socio-economic development—an issue of particular relevance to inner peripheries. In this chapter, we refine our conceptualization of inner peripheries. We elaborate on the theoretical foundations of the inner peripheries concept, based on a triangulated theoretical-methodological-narrative framework that draws on the notion of geographical peripherality, the Modern World System theory, the concept of organized proximity, the foundational economy approach, and the theory of spatial justice. We also justify the significance of FUAs in the delimitation of inner peripheries, which serve as the spatial framework for the research design adopted. Additionally, we propose an analytical framework for the study of inner peripheries, grounded in the concept of territorial capital and the systematization of development factors.

In Chap. 3, we analyze the SARS-CoV-2 pandemic as an example of a global development shock, highlighting its multidimensional impact on the economy and society from a spatial perspective. The discussion begins with a classification of economic shocks and their consequences, as well as the role of territorial resilience in responding to crises. Particular attention is devoted to the spatial asymmetry of the pandemic's effects and the role of territorial capital as an interpretative framework for understanding the differentiated adaptive capacities of core and peripheral areas. The chapter demonstrates how the pandemic deepened existing socio-economic and infrastructural inequalities and how the availability of various components of capital—human, social, institutional, financial, and material—influences the ability of regions to respond to crises and recover in their aftermath. It also examines the spatial variation in the course of the pandemic, including an analysis of virus transmission at national and regional levels both in Poland and across Europe while considering demographic, infrastructural, institutional, and climatic conditions. The importance of access to public services, transport, and digital infrastructure is emphasized, as these factors proved essential in mitigating the pandemic's negative effects. The chapter concludes with a systematization of the socio-economic consequences of the pandemic in the context of inner peripheries, highlighting their particular vulnerability to crises and the need for targeted Cohesion Policy interventions. This chapter provides the theoretical and empirical foundation for the subsequent analyses presented in the study, which focus on identifying and characterizing the functioning of inner peripheries under the impact of a development shock.

In Chap. 4, we present an original synthesis of the theoretical and conceptual findings outlined in the previous chapters, which serves as the foundation for our operational approach to identifying and analyzing the functioning of inner peripheries, as applied in the empirical investigation presented in this study. The starting point for this synthesis is the formulation of the theoretical-methodological-narrative triangulation framework we adopted to define and explain the dynamics of inner peripheries. Building on this theoretical foundation and our understanding of inner peripheries, we present the operationalization of these assumptions, which guided our research process. This includes the spatial framework for delimiting inner peripheries, the analytical framework used to characterize their condition and transformations, and the role we attribute to inner peripheries in addressing the challenges of development policy programming in response to the consequences of development shocks. These considerations enable us to present a theoretical-conceptual model along with the assumptions underlying its empirical operationalization, which together form the basis of the research approach adopted in this study. This approach supports the identification, typology, and characterization of the functioning of inner peripheries under the impact of the development shock triggered by the SARS-CoV-2 pandemic.

Chapters 5 and 6 present the results of the empirical research conducted based on the adopted theoretical assumptions, which served to verify the formulated research hypotheses. Chapter Five presents the first three steps of the empirical analysis: the delimitation of Functional Urban Areas and their typology in order to identify the inner peripheries in Poland, as well as the choice of four case study areas for in-depth investigation. In this chapter, we first briefly present the state of knowledge on the regional development disparities in the country. After that, we describe how we divided the territory of Poland into 413 FUAs based on the inter-municipal movement of population: commuting and migration. Following a discussion of the resulting division of Poland into FUAs and confronting it with the administrative division of the country, we use the FUAs as the building blocks for the typology of territories based on their multidimensional core-peripherality characteristics, which, in turn, is used to identify territories that we label as inner peripheries. The classification of FUAs and the identification of two types of inner peripheries are described in the third section of the chapter and then confronted with previous studies on regional development disparities in Poland. The final part of the chapter outlines the rationale for the selection and briefly describes the characteristics of the four case study FUAs, which are the focus of the next chapter.

Chapter 6 focuses on the empirical analysis of development trajectories in inner peripheries and presents the findings of qualitative research conducted in four FUAs classified as inner peripheries and selected for further investigation as case studies. The chapter begins with a comparative analysis of the dynamics of territorial capital formation in Poland during the SARS-CoV-2 pandemic, highlighting differences among core, intermediate, and peripheral areas. This dynamic analysis enabled the identification of development trajectories specific to inner peripheries in relation to more developed areas of the country and provided the basis for selecting the specific cases for in-depth qualitative research. The following section discusses in detail the research methods used in the case studies, including content analysis, resident surveys, and both in-depth and focus-group interviews with local leaders. The next part presents the findings related to transport accessibility and its changes—considered a key development factor differentiating inner peripheries from central areas. The results concerning human, social, and institutional capital are presented in an integrated manner, which allowed for the identification of interdependencies and overlapping deficits that influence the development potential of the analyzed peripheral areas. The analysis of financial and material capital in the studied FUAs concludes the presentation of the in-depth research findings. The final section of the chapter explores the significance of the identified factors in shaping the relational development environment of inner peripheries. It highlights the interconnections between various dimensions of territorial capital, reveals relational disconnection from development cores, and identifies discrepancies between strategic planning and the actual needs of local communities. The chapter also emphasizes the limited resilience of internal peripheries to development shocks.

In the concluding Chap. 7, we present the final conclusions and operational recommendations based on the research findings discussed throughout this book. We begin by drawing the reader's attention to the key findings that underpin the verification of the research hypotheses formulated in this study. We highlight how the achievement of the empirical, methodological, and applicative objectives outlined in the book led to the formulation of these findings, which in turn served as the basis for confirming or falsifying the hypotheses. We also offer our perspectives on how the identified facts and patterns may influence the inclusion of inner peripheries in development processes aligned with the priority directions of contemporary Cohesion Policy transformations. Within the operational recommendations, we formulate both general and specific guidelines. In the general section, we propose priority actions that, in our view, should be considered when shaping conditions for the development of inner peripheries. These are structured according to the main contemporary shifts in Cohesion Policy, addressing the challenges of the post-2027 EU Cohesion Policy framework. In the specific section, we put forward two proposals for operational changes regarding the principles of programming and implementing development interventions financed under Cohesion Policy. The first concerns the use of functional areas as the primary actors and beneficiaries of such interventions. The second stresses the need for a stronger emphasis on territorial social justice in the programming and implementation of EU Cohesion Policy interventions—through the full application of integrated development planning approaches and the use of an operational definition of a spatial social minimum.

***

This monograph provides a synthetic presentation and summary of the key findings from the research project TIPERICO (http://www.tiperico.web.amu.edu.pl), funded by the National Science Center of Poland (OPUS 19–2020/37/B/HS4/01034): Trajectories and Development Challenges of Inner Peripheries under the New Cohesion Conditions Post SARS-CoV-2.

References

Ballantyne P, Singleton A, Dolega L, Macdonald J (2023) Integrating the who, what, and where of U.S. retail center geographies. Annal Am Assoc Geogr 113(2):488–510. https://doi.org/10.1080/24694452.2022.2098087

Cairncross F (1997) The death of distance: how the communications revolution will change our lives (Issues C20–21). Harvard Business School

Copus A, Mantino F, Noguera J (2017) Inner peripheries XE "Inner peripheries": an oxymoron or a real challenge for territorial cohesion XE "territorial cohesion" ? Ital J Plan Pract 7(1):24–49

Davenport A, Zaranko B (2020) Levelling up: where and how? London

Dijkstra L, Poelman H, Rodríguez-Pose A (2020) The geography of EU discontent. Reg Stud 54(6):737–753. https://doi.org/10.1080/00343404.2019.1654603

ESPON (2017) PROFECY Inner peripheries: national territories facing challenges of access to basic services of general interest. https://archive.espon.eu/inner-peripheries

ESPON (2018) PROFECY Inner peripheries in Europe: possible development strategies to overcome their marginalising effects. https://archive.espon.eu/inner-peripheries

European Commission: Directorate-General for Regional and Urban Policy (2024a) Forging a sustainable future together—cohesion for a competitive and inclusive Europe—report of the high-level group on the future of Cohesion Policy. Publications Office of the European Union. https://data.europa.eu/doi/; https://doi.org/10.2776/974536

European Commission (2024b) Ninth report on economic, social and territorial cohesion. Directorate-General for Regional and Urban Policy. https://ec.europa.eu/regional_policy/information-sources/cohesion-report_en

Faludi A, Rocco R (2022) Faludi blogging: chasing territorialism. TU Delft Open Publishing, Short essays by Andreas Faludi

Farole T, Goga S, Ionescu-Heroiu M (2018) Rethinking lagging regions: using Cohesion Policy to deliver on the potential of Europe's regions. World Bank

Floerkemeier MH, Spatafora MN, Venables A (2021) Regional disparities, growth, and inclusiveness. International Monetary Fund

Friedman TL (2005) The world is flat: A brief history of the twenty-first century. Macmillan

Hadjimichalis C (2021) An uncertain future for the post-Brexit, post-COVID-19 European Union. European Urb Reg Stud 28(1):8–13. https://doi.org/10.1177/0969776420968961

Harvey D (2005) Spaces of neoliberalization: towards a theory of uneven geographical development. Steiner

Hertrich TJ, Brenner T (2024) When the past becomes the future: the problem of policies in 'left behind' places (Working Papers on Innovation and Space No. 02/2024). Philipps Universität Marburg

Huang X, Lu J, Gao S, Wang S, Liu Z, Wei H (2022) Staying at home is a privilege: evidence from fine-grained mobile phone location data in the United States during the COVID-19 pandemic. Ann Am Assoc Geogr 112(1):286–305. https://doi.org/10.1080/24694452.2021.1904819

MacKinnon D, Kempton L, O'Brien P, Ormerod E, Pike A, Tomaney J (2022) Reframing urban and regional 'development' for 'left behind' places. Camb J Reg Econ Soc 15:39–56

MacLeod G, Jones M (2020) Explaining 'Brexit capital': uneven development and the austerity state. In: Boyle M, Paddison R, Shirlow P (eds) Brexit geographies. Routledge, pp 1–26

Mattinson D (2020) Beyond the red wall: why labour lost, how the conservatives won and what will happen next? Biteback Publishing

Mbah RE, Wasum DF (2022) Russian-Ukraine 2022 war: a review of the economic impact of Russian-Ukraine crisis on the USA, UK, Canada, and Europe. Adv Soc Sci Res J 9(3):144–153

McCann P (2008) Globalisation and economic geography: the world is curved, not flat. Camb J Reg Econ Soc 1(3):351–370

O'Brien R (1992) Global financial integration: the end of geography. Council on Foreign Relations

OECD (2023) OECD regional outlook 2023: the longstanding geography of inequalities. OECD Publishing

Pike A, Béal V, Cauchi-Duval N, Franklin R, Kinossian N, Lang T, Leibert T, MacKinnon D, Rousseau M, Royer J, Servillo L, Tomaney J, Velthuis S (2023) Left behind' places: a geographical etymology. Reg Stud. https://doi.org/10.1080/00343404.2023.2167972
Quah DT (1999) The weightless economy in growth. Bus Econ 30:40–53
Rodríguez-Pose A (2018) The revenge of the places that don't matter (and what to do about it). Camb J Reg Econ Soc 11(1):189–209. https://doi.org/10.1093/cjres/rsx024
Rodríguez-Pose A, Bartalucci F, Lozano-Gracia N, Davalos M (2024) Overcoming left-behindedness: moving beyond the efficiency versus equity debate in territorial development (Policy Research Working Paper No. 10734). World Bank Group. http://documents1.worldbank.org/curated/en/099224303262426681/pdf/IDU17dc414d91a08a142ad1a8da1e2c5bf4ade41.pdf
Rodríguez-Pose A, Crescenzi R (2008) Mountains in a flat world: why proximity still matters for the location of economic activity. Camb J Reg Econ Soc 1(3):371–388
Servillo L, Russo AP, Barbera F (2016) Inner peripheries: towards an EU place-based agenda on territorial peripherality. Ital J Plan Pract 6(1):[page range]
Tomaney J, Pike A, Natarajan L (2021) Land-use planning, inequality and the problem of 'left-behind places'—a provocation for the UK2070 Commission (The UK2070 Papers). https://discovery.ucl.ac.uk/id/eprint/10125613/7/Natarajan_UK2070Papers%20Series3_extracted.pdf
Zhai W, Liu M, Fu X, Peng ZR (2021) American inequality meets COVID-19: uneven spread of the disease across communities. Ann Am Assoc Geogr 111(7):2023–2043. https://doi.org/10.1080/24694452.2020.1866489

Chapter 2
Inner Peripheries in Contemporary Spatial Development Differences

Abstract In this chapter, we present the origins and definition of inner peripheries. Our discussion begins by highlighting contemporary trends in the transformation of the EU Cohesion Policy, which involve the reorientation of its objectives, a shift in the understanding of proximity, and the increasing importance of the subregional level for identifying and addressing spatial disparities in socio-economic development. These factors are seen as particularly relevant to the concept of inner peripheries. We refine our understanding of inner peripheries by elaborating on the theoretical foundations of the concept, which are based on a triangulated theoretical-methodological-narrative approach. This approach integrates the concept of geographical peripherality, the Modern World System theory, the foundational economy framework, and theories of spatial justice. Additionally, we justify the significance of Functional Urban Areas (FUAs) in the delimitation of inner peripheries, which serve as the foundation for the spatial framework adopted in our research. Finally, we propose an original analytical framework for inner peripheries, drawing on the concept of territorial capital and a systematic analysis of development factors.

Keywords Lagging areas · Inner peripheries · Theoretical foundations of the concept of inner peripheries · Functional areas · Development factors

2.1 Inner Periphery: Genesis and Understanding

The genesis and definition of how inner peripheries are understood in the literature require, at the outset, attention to their connection with the changes in the goals of the EU Cohesion Policy, the increasing importance of the subregional level in identifying spatial development disparities, as well as the evolving interpretation of the concept of proximity.

Spatial disparities in the level of development, which, as we have already noted in the introduction to this book, are a characteristic feature of this process, exhibit a highly varied range and scale. They are certainly not limited to the quantitative dimension of the development process, drawing attention to the fundamental differences

P. Churski et al., *Inner Peripheries in the New Conditions of Cohesion Post-SARS-CoV-2*, Economic Geography,
https://doi.org/10.1007/978-3-032-10936-1_2

between economic growth and development. The traditional approach to development, which confines this process solely to quantitative changes measured by gross domestic product, is certainly insufficient for identifying spatial development disparities. This is especially true in times when there is an increasing emphasis on the need to focus on post-growth concepts, which highlight new development goals within alternative economic streams (Lange et al. 2021). Additionally, this stems from the challenges of future EU Cohesion Policy emphasizing the need to shift focus toward reducing disparities in well-being based on distributive and procedural justice, rather than succumbing to the pressures of growth and competitiveness (Fritsch et al. 2023).

The necessity for a fundamental change in the approach to identifying development disparities and directing public interventions aimed at reducing them arises, on the one hand, from the expanding qualitative interpretation of the socio-economic development process, and on the other hand, from the increasing need to reorient the goals of the European Cohesion Policy. There is a growing recognition that the objective of intervention activities should not be to equalize outcomes but rather to ensure development opportunities for all residents of the Community (European Commission: Directorate-General for Regional and Urban Policy 2024). This aligns with the core principles of the future EU Cohesion Policy, which emphasize the need to create development opportunities for everyone, counteracting exclusion and economic poverty (European Commission 2024). Every resident should be provided with opportunities for development and access to general services and basic infrastructure that enable their development. Unfortunately, in sparsely populated or economically declining areas, including inner peripheries, meeting this requirement is becoming increasingly costly and consequently less efficient, presenting a growing challenge for social justice in the spatial dimension (Fritsch et al. 2023). Overcoming this challenge requires the implementation of unconventional solutions in shaping the relationships between core areas and their surrounding regions. These solutions must utilize not only product innovations—such as demand-responsive transport—but also fully rely on organizational innovations such as the organizational and tariff integration of public transport, as well as social innovations such as enhanced participation in seeking socially acceptable solutions for lagging regions, including peripheral areas (OECD 2023). Ensuring development opportunities for all residents of the Community theoretically represents the desired end state that regional policy intervention should strive for. However, an open question remains regarding the scope of public services that guarantee equal opportunities, also referred to as the "spatial social minimum" (Dworkin 1981; Rawls 1971; Smith 1994, 2000), and, most importantly, the method of ensuring these services in every spatial unit (Domański 2021), including inner peripheries. This sheds new light on the functioning and mutual relationships between areas where population and economic activity are concentrated and their surrounding areas, which are explained by theories such as the growth poles theory and the core-periphery theory (Boudeville 1964, 1972; Friedmann 1967; Friedmann and Alonso 1964; Hägerstrand 1951, 1952, 1967; Isard 1960; Paelinck 1965). They form the foundation of the theory of uneven development, which highlights the dynamics of polarization and diffusion. It is worth emphasizing that their core principles are universal in nature and remain relevant (Churski et al. 2021).

The operationalization of actions related to the aforementioned changes in the practice of EU policies, particularly the Cohesion Policy, is not feasible at the NUTS-2 regional level. While the importance of this spatial level in the allocation of Cohesion Policy funds, through the application of the Berlin algorithm, seems politically secure, the programming and implementation of developmental interventions are now officially perceived as ineffective and inefficient (European Commission 2024). The EU is achieving developmental convergence at the national level, but it is debatable whether the same is true at the regional (NUTS-2). However, there is undeniable evidence of increasing developmental disparities at the subregional level (European Commission: Directorate-General for Regional and Urban Policy 2024). Peripheralization and marginalization of many territories, as we have noted, on one hand, lead to increasing costs—not only economic but also social—posing a real threat to European democracy (Rodríguez-Pose et al. 2024). On the other hand, these processes result in a situation where "left-behind places" become a central theme in contemporary discussions on geographical inequalities and the challenges of the Cohesion Policy (Farole et al. 2018; Pike et al. 2023). In this context, new approaches to programming and implementing developmental interventions are being sought, exemplified by the widespread adoption of a place-based policy approach. Attempts to adapt this approach at level of the NUTS-2 regions have not, due to their significant inner disparities, yielded the expected results. This confirms Roberta Capello's (2011, 2014) highlighted need for greater consideration of microscale factors in regional development research, including social behaviors and other intangible aspects of developmental processes. It also points to the necessity of applying the place-based policy approach at a lower, subregional spatial level, while leaving coordination competencies with regional governments (McCann 2023). This requires an objective identification of intraregional development disparities and the preparation of interventions tailored to the needs of each territory (Toptsidou et al. 2024). In our opinion, the foundation of these actions should be an analysis of functional linkages leading to the real strengthening of the relationships between cities and their surrounding areas, which can help overcome developmental barriers in areas trapped in path-dependent development, including inner peripheries.

When identifying peripheral areas, especially when attempting to address contemporary spatial disparities in development, it becomes essential to consider the evolving significance of the concept of proximity, which holds particular importance in geographical research. The role of proximity gains prominence with the increasing adoption of the space-of-flows approach (Castells 2004; Hudson 2004), which emphasizes the primacy of relationships over static conditions in the modern economy. The literature reveals a clear evolution in the concept of proximity (Balland et al. 2014; Boschma 2005; Capello 2007, 2020; Lalrindiki and O'Gorman 2021; Rallet and Torre 1999; Salazar and Claudel 2021; Wilke and Pyka 2024). This evolution expands the conventional understanding of the geographical characteristics of space (Balland et al. 2014), enabling the analysis of spatial development disparities to account not only for physical distance but also for other dimensions of proximity. Proximity is a crucial category in describing the agglomeration economy, as Capello (2007, p. 18) argues that economies of scale depend on production size and are not necessarily spatially determined. According to Capello (2014), proximity evolves in regional studies to encompass not only geographical proximity but also socio-cultural, relational, organizational, institutional, and cognitive proximities. However, as noted in Chap. 1, this does not signify the "end of geography" (Cairncross 1997), but rather underscores the necessity of a broader understanding of proximity in spatial analyses. It is important to recognize that each category of proximity is always reinforced by geographical proximity, which acts as a catalyst for establishing relationships and influences their strength and scope. Furthermore, these categories of proximity mutually enhance one another, creating a more favorable environment for socio-economic development processes (Delgadillo et al. 2021). A multidimensional perception of proximity is critical for the accurate identification of inner peripheries. These areas, as we have noted, are characterized more by a lack of relational connections to growth centers than by physical distance arising from their geographical location. Identifying the type and extent of proximity deficits is therefore key not only for determining spatial development disparities, particularly in lagging areas, but also for establishing the principles for programming and implementing development interventions tailored to these territories.

Studying peripherality, particularly inner peripherality, is a complex task. This is well-illustrated by Gould's (1969, p. 37) assertion that peripherality is "…a slippery notion…one of those common terms everyone uses until faced with the problem of defining and measuring it." The challenge is compounded by the frequent interchangeable use of the terms "peripherality" and "marginalization," despite their distinct meanings.

Traditionally, peripheral areas have been defined by their geographical remoteness. However, in the case of inner peripheries, this interpretation is not accurate. Peripherality in this context is based on the evaluation of interactions and arises as a consequence of the loss of or inability to establish connections, rather than being determined by a region's location. Marginalization, on the other hand, is associated more with low levels of socio-economic growth and development, regardless of the underlying causes or geographical location (Danson et al. 2012; Jussila et al. 1998). This raises the question of whether the identification of inner peripheries implies the existence of external peripheries. In the literature, these terms are sometimes used to distinguish peripheries along external national borders from those along internal administrative boundaries (Bajerski 2008; Czapiewski and Janc 2009; Máliková et al. 2016). Other authors suggest that these terms can also reflect the perception of peripheral areas by their inhabitants and outsiders (Schmidt 1998; Woods 2012). In all these cases, peripheries are traditionally viewed through the lens of their geographical location and how that location is perceived. We propose that the basis for distinguishing between these types of peripherality lies in the difference between traditional and contemporary perspectives on the concept. External peripheries can be understood as areas where peripherality is primarily physical, determined by geographical remoteness from core areas. Socio-economic marginalization in these areas largely results from this remoteness, which can act as an objective barrier. Addressing such marginalization often requires significant investment in transport infrastructure, posing challenges that are difficult to overcome and, in some cases, economically unjustifiable for weakly populated and economically underdeveloped areas. In contrast, inner peripheries are areas where peripherality is relational, not necessarily tied to geographical location but instead defined by the degree of disconnection from core areas and, consequently, from major value and supply chains. Socio-economic marginalization in these areas primarily stems from the lack of or difficulties in establishing such connections. This marginalization is more easily mitigated than that of external peripheries and does not always require substantial public investment. It often does not involve overcoming the objective barrier of geographical location. It is important to recognize that peripherality is multifaceted. Its geographical, economic, and social dimensions may not coexist, presenting a significant challenge for its conceptualization and the delimitation of peripheral areas. This leads to the conclusion that both the concept of peripherality and the methodologies for its delimitation require a degree of revitalization (Bajerski 2008). Expanding this discussion to include a new perspective on inner peripheries, which emphasizes the need for a broader understanding of developmental processes beyond the simple core–periphery dichotomy, aligns with this recommendation.

The traditional approach to peripherality in regional studies has primarily focused on the geographical location of these areas. Peripheral regions were considered to be those situated along national borders, where the socio-economic situation was also influenced by political factors. The marginalization of these areas was largely due to their treatment as heavily guarded military buffer zones (Servillo et al. 2016). As a result, these areas were overlooked by large-scale industrial and service investments, with their economic profile being primarily based on agricultural and forestry production. The worsening socio-economic situation of these areas led to the launch of the special cross-border cooperation program INTERREG, which continues to operate today, delivering tangible results despite relatively very limited funding (Chilla and Lambracht 2022; Perkmann 2007). A similar approach was applied to areas situated along inner borders, sometimes referred to, as mentioned earlier, as "inner peripheries." The marginalization of these areas was not merely a consequence of the influence of national-level administrative factors but was more strongly determined by the impact of socio-economic and cultural boundaries (Servillo et al. 2014).

The issue of inner peripheries gained significance as a consequence of the financial crisis (2007–2009), which highlighted the stark contrasts in the state and dynamics of development between western and eastern countries, as well as the clear disparities between rural areas and urban/metropolitan regions (Servillo et al. 2016). This was a result of the increasing fragmentation of the territorial structure of the European Union, driven by the multifaceted disparities between territories and their ability to function under crisis conditions, as well as the growing competition (Russo et al. 2012). The emergence of pronounced centralist trends in response to the crisis, both in terms of competencies and access to resources, which led to the rapid privileging of core areas, was also significant (Moisio and Paasi 2013). To mitigate the consequences of these changes, the EU began promoting a place-based policy approach in 2008, aiming to reduce activities based on the "one size fits all" assumption (Morgan 2016). Over time, this approach became the foundation of the new Cohesion Policy paradigm (Barca 2009; OECD 2009a, b). This shift was accompanied by an expansion of the conceptual foundations of territorial cohesion and efforts to reduce spatial development disparities, incorporating the principles of spatial justice (Soja 2010a, b) and the foundational economy (Bowman et al. 2014).

The concept of inner peripheries was formally introduced for the first time in a new, broader context, emphasizing the relational deficits of these areas rather than the consequences of their geographical location, in a document prepared for the informal meeting of the ministers of the member states responsible for spatial planning and territorial development, held in Godollo, Hungary, in 2011 (Ministry of National Development and Vàti Nonprofit Ltd. 2011, p. 57).

Inner peripheries are unique types of rural peripheries in European terms. The vast majority of these areas are located in Central and Eastern Europe and south-eastern Europe, and most of them have serious problems. Their peripherality comes primarily from their poor accessibility and paucity of real urban centers where central functions can be concentrated. These problems derive from the historical underdevelopment of these territories and are often compounded by specific features of the settlement network or social characteristics. The main problems of these areas are their weak and vulnerable regional economies and their lack of appropriate job opportunities. In these circumstances, negative demographic processes, notably out-migration and aging of the population, are getting stronger and stronger. These trends create the conditions for social exclusion and even territorial exclusion from mainstream socio-economic processes and opportunities. While rural ghettos are mainly a result of social factors, ethnic segregation can worsen difficult situations. This is the case, for example, in the rural peripheries of Slovakia, Hungary, and Romania, where there are areas with high proportions of Roma population.

This meeting was dedicated to adopting the Territorial Agenda of the European Union 2020 (2011), which defined the common territorial development goals of the European Union relating to the implementation of the Europe 2020 Strategy (2010), concretizing actions in the EU's 2014–2020 multiannual financial framework. However, it should be noted that the concept of inner peripheries did not make it into the final text of the Territorial Agenda 2020. It is also difficult to find in other documents related to regional policy or in the planning practices of member states (ESPON GEOSPECS 2012). Moreover, it has not been widely disseminated in the scientific discourse on left-behind areas (Pike et al. 2023). This leads to the conclusion that it is a relatively new concept, both in theoretical and conceptual terms and in operational and practical applications, requiring further elaboration, broader discussion, and greater utilization in Cohesion Policy practice. This is particularly important in the context of expanding the interpretation of socio-economic development, the increasing importance of the subregional level in the programming and implementation of the Cohesion Policy, and the evolution of the concept of proximity, which, as mentioned, are characteristic features of contemporary changes in development conditions.

This is confirmed by the findings of Noguera and Copus (2016), who state that the literature on inner peripheries is scarce. Initially, this term was used incidentally to describe patterns associated with the functioning of the global core-periphery system, as described by Wallerstein (1976, 1991). Examples of such studies include research on Appalachia (Hanna 1995; Walls 1978) and Lesotho (Weisfelder 1992). In more contemporary studies, the term has been used, among others, to describe sparsely populated areas in Moravia and the Alps (Vaishar and Zapletalová 2008) or rural areas relatively close to economic activity centers (Kaganskii 2013). An important contribution comes from the research by Musil and Müller (2008), whose definition of inner peripheries emphasizes a territorial approach and the internal differentiation of these areas, highlighting that they are always characterized by the presence of a local node. Notably, in their approach to the delimitation of inner peripheries, which they conducted for the Czech Republic, the authors adopted the assumption of the spatial continuity of these areas. According to them, these areas consist of subregional units distinguished by lower values of selected demographic, economic, and living standard indicators compared to the threshold values they identified as typical for inner peripheries. Additionally, the authors employed both a narrow and broad definition of inner peripheries. The narrow definition (hard core peripheries) corresponds, in their view, to sub-regional units that have been repeatedly assessed as peripheral based on the values of various criteria confirmed by indicators. The broader definition includes subregional units assessed as peripheral only when specific criteria related to selected indicators are applied. In other studies related to the Czech Republic conducted by Bernard and Šimon (2017), the term "inner peripheries" refers to areas with a concentration of various forms of social disadvantage. The typology of peripheries proposed by these authors is based on three dimensions: demographic (long-term consequences of selective migration leading to high proportions of the oldest age cohorts and low natural population growth), labor market (job availability, job quality), and social (poverty, dependency on social benefits, household amenities, education). The Italian political initiative supporting "inner areas," whose delimitation was based on very similar assumptions (Lucatelli et al. 2013; Mantino and Fano 2015), also played a significant role in promoting the concept of inner peripheries. This became the foundation for interesting studies on the potential use of the inner peripheries concept in operationalizing place-based Cohesion Policy actions (Servillo et al. 2016). Important findings regarding the need to popularize the concept of inner peripheries also emerged from the results of the aforementioned ESPON GEOSPECS project—Geographic Specificities and Development Potentials in Europe (ESPON GEOSPECS 2012), which was implemented as part of the ESPON 2013 program aimed at identifying spatial development disparities within the EU. On the one hand, its findings identified deficits in the use of the inner peripheries concept in the practice of regional policy in selected EU member states (Belgium, the Netherlands, and Germany), while on the other hand, they highlighted the need for in-depth research on this category of neglected areas.

Undoubtedly, the most significant contributions to the existing research on inner peripheries come from the findings of the ESPON PROFECY project—"Inner Peripheries: National territories facing challenges of access to basic services of general interest" (ESPON PROFECY 2017, 2018). This project was a response to a call under the ESPON 2020 program, which this time was specifically dedicated to inner peripheries. The project results indicate that inner peripherality is a complex, multidimensional phenomenon, consisting of the effects of various socio-economic processes that lead to disconnection from external territories and networks. A characteristic feature of inner peripheries is their degree of "disconnection" rather than their geographical location relative to "core areas." A common trait of inner peripheries is their generally lower performance, lower level of development, reduced access to services, and lower quality of life compared to neighboring territories. However, it is important to note that these areas are not homogeneous. They are highly individual hybrids, each shaped by a unique history, multiple factors/processes, and territorially rooted contextual elements. Inner peripherality is determined by the ability of a territory to "connect" with its surroundings. This is influenced not only by "geography," although geography is not insignificant, but primarily by relational connections—that is, non-spatial factors and processes (Copus et al. 2017; Servillo et al. 2016). Thus, in defining and identifying inner peripheries, the role of social conditions and participatory approaches is increasing (Máliková et al. 2016; Wójcik et al. 2018). Relational connections generate synergies, networks, and other types of linkages that enable a territory to make appropriate decisions that consider the needs of all actors. The absence of these connections and a limited ability to create them result in peripherality, independent of the geographical location of the territories. The way local actors interact, the level of integration into relevant networks, and the ability of local institutions, organizations, and businesses to establish contacts with other entities in neighboring and distant territories illustrate the importance of "being connected." Well-connected territories offer greater and better development opportunities, improved access to services of general interest (SGI), and a more dynamic labor market capable of retaining skilled residents. All these features create conditions for establishing new contacts and, through synergies, strengthening the relational foundations for sustainable development processes. Unfortunately, these dynamics also work in the opposite direction. "Disconnected" areas, regardless of the combination of causal processes and factors, tend to reproduce and perpetuate their unfavorable situation over time due to the evolutionary nature of "disconnection" and its feedback effects. The findings of the ESPON PROFECY project (2017, 2018) propose three main concepts of inner peripherality, highlighting different specific characteristics that form the basis for their delimitation and indicate diverse needs for development intervention:

- In enclaves with low economic potential (IP1), there is a connectivity gap related to long travel times to regional centers (longer than in neighboring NUTS-3 units), leading to low "population and economic potential" (lower population and GDP accessibility potential compared to neighboring NUTS-3 units, relative to the regional or national average). Development intervention for this category of inner peripheries should focus on improving connections to major transportation networks through conventional infrastructure improvements, logistical systems, or travel cost reductions.
- In areas with poor access to services of general interest (SGI) (IP2), there is a connectivity gap related to long travel times to regional centers (longer than in neighboring NUTS-3 units), resulting in poor access to services of general interest

(SGI) (worse conditions for accessing public services compared to the average in surrounding NUTS-3 areas and/or the NUTS-2 region). Development intervention for this category of inner peripheries should be focused on improving intraregional service provision/access, including innovative organizational solutions and the use of information technology. Changes in administrative structures aimed at achieving economies of scale may create conditions for the emergence of this category of inner peripheries or exacerbate their unfavorable situation.
- In areas lacking relational proximity (IP3), there is a gap in "organized proximity," leading to rising unemployment, population decline, and a decrease in GDP per capita. Development intervention in this category of inner peripheries should emphasize strengthening non-spatial factors and processes, whose absence or low levels result in poor socio-economic outcomes. The intervention should focus on enhancing interactions between local actors.

The classification of NUTS-3 units in EU member states conducted within the framework of the project, using the above-mentioned concepts of inner peripherality, leads to the following results (ESPON PROFECY 2017, 2018):

- Areas where the primary factor of peripherality is poor economic conditions or lack of relational proximity (IP1 and IP3) account for 46% of all areas identified as inner peripheries.
- Areas where the main driver of peripherality is poor access to services and/or regional centers (IP2) have a very similar share, making up 45% of all units classified as inner peripheries.
- Areas showing signs of influence from all factors, recognized by the authors as areas of deep inner peripherality (IP1, IP2, and IP3), constitute only 9% of all areas identified as inner peripheries.

The spatial distribution of inner peripheries in Europe reveals certain patterns. As many as 80% of inner peripheries with low economic potential (IP1) or poor access to SGI services (IP2) are located in non-urban regions, with an equal share between rural areas and intermediate areas. Additionally, around half of the inner peripheries with poor access to SGI services (IP2) are found in mountainous regions. Areas lacking relational proximity (IP3) are more frequently found in urban (32.2%) and metropolitan areas (43%) compared to areas with low economic potential (IP1) and poor access to SGI services (IP2). This highlights the fact that peripheralization processes caused by poor socio-economic conditions resulting from a loss of relational proximity can also affect "enclaves" in areas stereotypically perceived as highly developed core regions. However, it is important to note that there is a high degree of overlap between the coverage of inner peripheries and underdeveloped areas as defined by the EU Cohesion Policy. In this context, inner peripheries with low economic potential (IP1) and those lacking relational proximity (IP3) show a greater alignment with the designated underdeveloped areas, suggesting that in these cases, the causes and effects of peripherality are more directly linked to economic outcomes. It is also noteworthy that geographically peripheral areas significantly overlap with the extent of inner peripheries, including border areas (as opposed to frontier areas), both national and regional. The findings suggest that inner peripheries face diverse development challenges stemming from the conditions of their peripherality. A common characteristic among them is an unfavorable demographic situation, manifested in depopulation, an aging population, and declining birth rates. However, the most significant challenge in this regard is the migration of young and skilled residents. These areas are characterized by relatively poorer economic performance (GDP, GVA), with varying conditions in terms of entrepreneurship levels and access to public services (SGI). The labor market in these areas does not generally exhibit particularly negative traits, and in some cases, even shows above-average indicator values.

The research conducted by the ESPON PROFECY project team has inspired further in-depth studies, both by members of this team—such as in relation to the metropolitan area of Valencia and the Central Belt area in Scotland (Noguera and Copus 2016)—as well as by other authors, for example, in highlighting the need to address the development challenges of inner peripheries in the programming and implementation of rural development policies (De Toni et al. 2021). The findings of our TIPERICO project, presented in this book, contribute to expanding this discussion by proposing new assumptions that better explain the functioning of inner peripheries, as well as new recommendations to enhance the effectiveness and efficiency of development interventions planned and implemented in these areas.

2.2 Theoretical Foundations of the Concept of Inner Peripheries

The theoretical foundations of the contemporary concept of inner peripheries, which has emerged in the current discussion on spatial disparities in socio-economic development and the challenges of regional policy, according to Copus et al. (2017), have two theoretical roots that originated in the 1970s and 1980s. The first is associated with positivist spatial analysis based on patterns identified in the literature as Geographical Peripherality. The second stems from the structuralist school and is based on the assumptions of the "Modern World System" theory, adapted to the regional level. These two theoretical currents developed completely independently. Their current convergence is, on the one hand, a confirmation of the expanding research pluralism observed in geographical studies, manifested in the eclectic borrowing from various scientific concepts (Wójcik and Suliborski 2021). On the other hand, it underscores the need for theoretical-methodological-narrative triangulation, which offers a more comprehensive and balanced explanation and understanding of the existing geographical disparities in socio-economic development (Churski et al. 2021). Fully agreeing with these assertions and considering the contemporary, growing challenges of regional policy exacerbated by global crises related to reducing its inefficacy and inefficiency in diminishing spatial disparities in socio-economic development, we advocate for expanding the theoretical foundations of inner peripheries under consideration with two additional theoretical concepts: the Foundational Economy and Spatial Justice. This is essential, as Servillo et al. (2016) state, to strengthen the normative dimension of inner peripheries in the pursuit of territorial cohesion—a state that should be socially accepted.

Based on this theoretical foundation, which incorporates our proposals for defining, spatial structuring, and analytical frameworks for inner peripheries presented in this chapter, as well as contemporary challenges in development policy programming responding to the consequences of global developmental shocks (Chap. 3), we formulate a theoretical-conceptual model. The model, presented in Chap. 4, serves as the foundation for identifying and explaining the functioning patterns of inner peripheries in this study.

2.2.1 Geographical Peripherality

Geographical peripherality is one of many types of peripherality considered in various relative approaches and across different spatial scales (Pugh and Dubois 2021). Its systematization is directly related to the evolution of the concept of proximity, which we presented in Table 2.1 in Sect. 2.1. This is because proximity, or rather its absence, determines the accessibility of any given territory, and this serves as the basis for identifying various types of peripherality, including but not limited to geographical, socio-cultural, relational, organizational, institutional, and cognitive peripherality (Capello 2014).

Geographical peripherality is associated with the relationships between cores and peripheries. It originates from one of the fundamental paradigms of contemporary economic geography, which emphasizes the spatial differentiation of the socio-economic development process and the significance of physical distance in shaping it (Fujita and Thisse 2009). It is synonymous with the geographical remoteness of a given area from growth centers (Kühn 2015), referred to as geographical proximity (Capello 2014). Geographical peripherality remains crucial for a comprehensive understanding of the territorial component of development, which currently determines the programming and implementation of development policy interventions (Churski et al. 2021). This leads to the conclusion that it is impossible to identify and analyze socio-economic development processes occurring in inner peripheries without considering these fundamental assumptions about spatial differentiation of development, which are based on broad theoretical foundations in economics and economic geography. The decreasing impact of geographical distance on shaping socio-economic processes in the context of improved temporal and economic transportation accessibility does not change this fact. However, as previously mentioned, it does not lead to the "end of geography," which would justify ignoring physical distance in shaping spatial disparities in socio-economic development (Cairncross, 1997; Friedman 2005; O'Brien 1992; Quah 1999). Although inner peripheries do not necessarily have to be physically distant areas, and their formation and functioning fundamentally go beyond the geographical location determinants typical of

Table 2.1 Evolution of the concept of proximity

Type of proximity	Selected authors	Understanding
Geographical proximity	Marshall (1890), Arrow (1962), Romer (1986)	Physical proximity facilitating knowledge transfer that is easier under monopoly conditions
	Porter (1990)	Physical proximity facilitating knowledge transfer within industries and types of activities driven by local competition, enabling the generation of innovation
	Jacobs (1969)	Physical proximity facilitating the transfer and diffusion of knowledge regardless of industry or type of activity, increasing competition and stimulating innovation
Socio-cultural proximity (and geographical proximity)	Becattini (1979), Granovetter (1985), Coenen et al. (2010)	Collaboration and productive capacities are easier to achieve when embedded in social, formal, and informal relationships based on trust, as well as shared values and principles applied in interactions between residents, local governments, and businesses
Relational proximity (and geographical proximity)	Camagni (1991)	New knowledge is easier to acquire through the learning process within the framework of collaboration among residents, local governments, and economic entities (a local multilateral network), along with the local mobility of specialists and skilled workers
Organized proximity (and geographical proximity)	Rallet (1993), Rallet and Torre (1995), Hansen (2014), Ievoli et al. (2019)	Socio-economic development is stimulated by knowledge sharing between residents, local governments, and businesses, which is easier when interactions occur face to face (though it can also take place through ICT, especially within organizational networks)
Institutional proximity (and geographical proximity)	Ludnvall (1992), Cooke and Morgan (1994), Asheim (1996), Malmberg and Maskell (2002), Dubois (2019), Lopolito et al. (2022)	The socio-economic development strengthened by the learning process (both formal and informal) occurring through interactions between residents, local government, businesses, and research institutions, all originating from the same social and institutional system and applying the same standards, codes, and rules of conduct

(continued)

Table 2.1 (continued)

Type of proximity	Selected authors	Understanding
Cognitive proximity (possible with geographical proximity)	Boschma (2005), Capello (2009), Janssen and Abbasiharofteh (2022)	Economic development based on the acquisition of new knowledge, which serves as the foundation for innovation, through the collaboration of actors possessing a certain level of foundational knowledge and sufficient cognitive abilities to communicate, understand, and utilize new knowledge. This complementary nature stimulates cooperation

Source Own elaboration based on Capello (2014), Kopczewska et al. (2017), and Wilke and Pyka (2024)

classic peripheries, their formation allows for a different perspective on the distinction between peripheries and peripheralization proposed by Kühn (2015). In this context, peripheries are no longer merely areas physically distant from cores; they can emerge in relatively close geographical proximity. Peripheralization is not limited to the consequences of deficits in basic social relationships determined by geographical location. Instead, it accounts for the growing impact of an expanding range of multidimensional relational deficits on the process of creating peripheries. These trends do not change the fact that the significance of physical distance in shaping the regularities of socio-economic processes contributing to the creation and development within inner peripheries remains highly important and cannot be ignored.

The origins and theoretical foundations of geographical peripherality should be sought in the theoretical and methodological contributions of both economics and economic geography, which explain the relationships between cores and peripheries. In this regard, location theories provide important insights into the significance of physical geographical distance in shaping the regularities of developmental processes. The precursor of location theory is Thünen, who in 1826 proposed a model for the distribution of different types of agricultural activities around a single, centrally located market, which was the city. In this simple model, physical distance from the city—the market—was one of the key factors influencing transportation costs and determining the type of activity conducted. Despite its many limitations, this model became an inspiration for other authors, and the regularities identified by Thünen regarding the relationship between the center and its surroundings, which were significantly shaped by physical distance and its impact on transportation costs, became an important element in explaining spatial disparities in socio-economic development (O'Kelly and Bryan 1996). Thünen's location theories were further developed by Weber, who sought the optimal location for a firm. In his proposed model, Weber also emphasized the importance of physical distance, assuming that minimizing transportation costs could be achieved by locating the firm where the combined costs of raw material supply and finished product transport are the lowest. It is worth noting that Weber's theory includes an important observation regarding peripheral areas: agglomeration benefits (low transportation costs) decrease as one moves away from the center toward the periphery; however, labor costs, which are highest in central areas and lowest in peripheral areas, also decline concurrently (McCann 1998). Lösch, a successor to the works of Thünen and Weber, developed the central place theory to explain the causes of socio-economic development disparities between different regions. Building on the assumptions of the settlement system organization outlined in Christaller's central place theory, Lösch created a model in which he assumed that the size of a firm's market area, where it sells its products, is determined by the interaction between economies of scale and transportation costs, both of which are influenced by physical distance. According to the assumptions of the model, if economies of scale outweigh transportation costs, all production should take place in a single facility. However, if transportation costs are high compared to economies of scale, firms should be dispersed across the region. In such conditions, market areas of different sizes emerge: common goods are produced in many locations (small market areas), while rare goods are produced in fewer locations (large market areas). Central places, in turn, emerge where the market areas for different products overlap (Ponsard 1983). The examples cited confirm that location theories point to fundamental factors for identifying geographical peripherality, which include high transportation costs and the lack of agglomeration effects (Copus 2001). It is also important to note that location theories, while highlighting the significance of physical distance, emphasize the spatial dichotomy in terms of the course and effects of socio-economic development processes, distinguishing between central areas and peripheries within that space. They view the formation of these areas as a natural characteristic of the differentiation of socio-economic processes in space. Importantly, location theories do not treat peripheries as problem areas or areas in need of support. They simply view them as areas with different characteristics compared to central areas, whose formation and functioning depend largely on their geographical location (Grosel 2023).

The fundamental significance for explaining geographical peripherality in the process of shaping spatial development disparities lies in the concepts of growth poles and core-periphery models, which form the foundation of the theory of uneven development, highlighting the regularities of its polarization and diffusion (Boudeville 1964, 1972; Friedmann 1967; Friedmann and Alonso 1964; Hägerstrand 1951, 1952, 1967; Isard 1960; Paelinck 1965). The theory of growth poles originates from the French school and has its roots in the concept of growth poles proposed in 1949 by Perroux (1955). This non-spatial sectoral proposition was later modified toward a spatial approach in Boudeville's concept of the polarized region and Paelinck's theory of polarized growth, which is regarded in the literature as the foundation of all contemporary polarization theories (Grzeszczak 1978, 1999, 2007). The core-periphery theory, of American origin, was proposed by Friedmann (1967; Friedmann and Alonso 1964). It is widely considered the most mature attempt to create a theory of polarized development in the sense of regional development theory. Despite differences between these concepts (Churski 2014), they are useful for explaining the regularities associated with the spread of development processes in economic space, taking into account the influence of physical distance. This usefulness arises, among other things, from the fact that the common feature of the theory of growth poles and the core-periphery concept is the perception of the development process through the lens of its spatial polarization, based on agglomeration effects (as understood by Marshall-Arrow-Romer, Porter, or Jacobs) in growth areas, and its diffusion based on spillover effects to surrounding areas. However, it is important to emphasize that this does not only refer to the aforementioned classical spatial diffusion known from Hägerstrand's works (1951, 1952, 1967), which is based on the spread of specific Schumpeterian innovations (Schumpeter 1911) in economic space. The impact of growth poles on economically weaker areas is considered in a broader context of the spillover effects of development and relates to the liberal doctrine in economics. This doctrine assumes that leveling development disparities can occur freely without the need for state intervention, which might otherwise lead to dependencies of areas surrounding cores on social support (Lambooy 2005; Richardson 1973; Williamson 1975). The condition for initiating this process is the opening of economically weaker areas to exogenous factors. This can lead to a more developed area having positive influences on its surroundings, which can occur at any level of the settlement system. However, it always follows the principle that the size of the center, which influences the scope and strength of its functions, determines the geographical extent of its impact. This impact does not necessarily involve the diffusion of innovations but can, for example, lead to improvements in infrastructure, thereby enhancing access to employment and higher-order services in areas surrounding the cores and improving the quality of life for their inhabitants (Knowles et al. 2008). However, the influence of growth poles on surrounding areas can also lead to unfavorable consequences related to economic dependency and resource depletion, which exacerbates the process of peripheralization and, over time, socio-economic marginalization of these areas. To counter these tendencies, it is necessary to strengthen surrounding areas by utilizing their endogenous resources and preparing them to absorb influences from core areas. In this way, the theory of growth poles and the core-periphery theory explain the relationships associated with the socio-economic spatial dichotomy, in which core and peripheral areas spontaneously emerge, highlighting the persistent importance of physical distance in shaping these regularities. Physical distance determines transportation costs, which are one of the fundamental factors considered in these theoretical frameworks. The location of geographical peripheries generates high transportation costs, which disadvantages them in the process of socio-economic development according to the models developed within these theoretical approaches.

Evidence of the persistent significance of physical distance in shaping spatial disparities in socio-economic development is found in the propositions of New Economic Geography. Its author, Krugman (1991a, b, 1995), placed even greater emphasis on the spatial aspects of inequality. In his model, relationships between regions depend on transportation costs (including distance) and agglomeration benefits. When transportation costs are very high, production occurs everywhere; when transportation costs are zero, production shifts to the peripheries; and at intermediate levels of transportation costs, production moves to the core (Barry 1996). It is worth noting that due to the assumption of mobility of production factors in the model, the roles can change over time—central areas can become peripheral, and vice versa (Grosel 2023).

The reduction of geographical peripherality, which manifests in spatial distance as defined by transport accessibility and concretized through various theoretical approaches that use the Newtonian analogy of gravity to study changes in economic potential across space, is one of the priorities of the EU Cohesion Policy's intervention measures. This understanding of peripherality has been reinforced in the political discourse surrounding the orientation of regional policy intervention streams in the EU, particularly following the accession of Spain and Portugal (1986) and Sweden and Finland (1995) to the organization. This had a tangible impact, exemplified by the creation of Objective 6, dedicated to peripheral areas even in wealthy countries such as Sweden. Today, geographical peripherality is increasingly interpreted through the lens of social and economic well-being and the role of accessibility to services of general interest (SGI), which shifts attention away from economic potential. As a result, a new formulation of peripherality emerges, although still calibrated in terms of geographical distance, focusing on areas poorly served by SGI. Since the key parameter is the distance to the nearest service center, this definition of peripherality tends to identify "enclaves" within regions that, from the perspective of conventional "economic potential" models, were considered part of the core (Copus et al. 2017).

The identified regularities fully justify the need to consider geographical peripherality in explaining the emergence and functioning of inner peripheries, which we aim to propose in our conceptual model, serving as the basis for the empirical analysis conducted in this study.

2.2.2 The Modern World System and "Organized Proximity"

The development of road infrastructure, the decrease in transportation service costs, the advancement of telecommunications, and the effects of the very dynamic expansion of the Internet, as well as structural transformations in the economy that are leading to an increased share of activities less sensitive to transportation costs, have resulted in the diminished significance of physical distance in shaping spatial disparities in socio-economic development. However, this does not lead, as already indicated, to a radical change in the perception of traditional patterns aligned with Friedman's (2005) concept of a "flat world." Instead, it forms the basis for the necessity to consider other factors beyond geographic distance in explaining the differentiation of economic space and the emergence of core areas and peripheries within its boundaries. A particularly significant approach in this regard is the network perspective, which emphasizes the importance of relationships in shaping contemporary social and economic processes (Grosel 2023). This approach highlights global transformations corresponding to the new economy of flows at the turn of the twenty-first century, which shape contemporary development processes (Castells 2004; Todaro 1999). Consequently, the traditional understanding of peripherality is changing, and aspatial models of peripherality are gaining popularity. These models take into account characteristics such as the quality of information technology infrastructure, the quality of human capital and its ability to utilize ICT technologies, the quality of local economic and social networks, the quality of institutions, and local embeddedness in delineating peripheral areas (Copus 2001). Proposals in this area are met with alternative concepts that emphasize that, despite ongoing transformations, the contemporary knowledge-based economy also exhibits tendencies toward spatial polarization, favoring core areas and struggling to develop in surrounding regions (Kühn 2015). This leads to the conclusion that explaining the causes of the formation and functioning of peripheries requires considering not only geographic proximity but also other factors that influence the differentiation of socio-economic development processes across space. Among these, organized proximity holds particular importance. Such an approach to identifying peripheries is key to the concept of inner peripheries, which seeks to combine both geographical proximity and other dimensions of proximity, recognizing the fundamental significance of the former and the dominant role of the latter in contemporary socio-economic processes. To understand contemporary transformations in the definition of peripheries, the assumptions of the Modern World System theory are particularly useful. This theory highlights the concept of inner peripheries and organized proximity, which redefines the understanding of proximity. These ideas expand the theoretical foundations for describing the functioning patterns of inner peripheries (Copus et al. 2017).

The concept of the Modern World System by Wallerstein (1976, 1991) is regarded as one of the key innovations in the social sciences since the 1970s. It highlights the growing globalization of the capitalist world economy, spreading from Western Europe and North America, whose geographical structure is based on the exploitation of peripheral areas for the benefit of the capitalist class in the core. Wallerstein developed his world systems theory by utilizing dependency theory and the ideas of Marx and the Annales School. Alongside economic imperialism (Galtung 1971) and *dependencia* (Palma 1978; Prebisch 1959), it is considered one of the theories explaining the formation of core-periphery relations on a global scale. All these theoretical concepts focus on the dualism between the global core and the global periphery to explain the magnitude and persistence of global social and economic inequalities, arguing that the economic development of peripheral regions is always limited to a greater or lesser extent by the core (Glückler et al. 2023). The assumptions of Wallerstein's concept quickly gained popularity among radical geographers. This was due to its global scope, historicity, critique of state-centered approaches, and spatial focus on different labor processes related to various geographical zones (Agnew 2021). Without delving into the historical and political assumptions and consequences of Wallerstein's world system proposition, it is important to focus on the fundamental elements of his concept, particularly those concerning the division of the world. These elements were subsequently adapted to explain patterns at the regional level, including the formation of peripheral areas with the consideration of aspatial factors (Hanna 1995; Vaishair and Zapletalova 2008; Walls 1978; Weisfelder 1992). By organizing historical and geographical conditions on a global scale, Wallerstein proposes a division of the world's countries based on the consequences of social relations rather than geographic distance. According to this framework, wealthy, capitalist societies in the core succeed by exploiting poorer peripheral societies. Between them lie semi-peripheral societies, forming an uncertain global middle class. This leads to the division of the world into three types of countries or regions: the core, the periphery, and the semi-periphery, marking the first use of the concept of inner peripheries in the literature. Core countries are dominant, capitalist states that exploit peripheral countries for their labor and resources. Wallerstein argues that poor, peripheral nations become increasingly impoverished as the developed core countries exploit their resources to become wealthier. It is important to emphasize that this typology is based on the distribution of power and the differentiated processes of capitalist exploitation, where the economic development of the core relies on cheap sources of raw materials and labor from the periphery. Semi-peripheral countries share in the exploitation of the periphery and aspire to become part of the core, but they lack the freedom of action and dominance associated with the latter. In Wallerstein's approach, the periphery does not necessarily refer to geographically distant areas. Rather, it refers to social systems shaped by complex processes of economic, demographic, and political changes that influence decision-making within the prevailing system of norms and socio-cultural values (Naumann et al. 2013). This allows for a much broader understanding of peripheralization, which is not solely determined by geographical location but is significantly shaped in socially constructed spaces and networks, exhibiting considerable flexibility in terms of context and scale. This applies to all spatial levels: countries, regions, cities, and local neighborhoods (Kühn 2015). The primary factor differentiating cores and peripheries is the extent of power, which influences the ability to organize social and economic processes and their agency, significantly impacting the development processes within their boundaries. This expanded understanding of peripheries and peripheralization must be taken into account when defining and explaining the functioning of inner peripheries (Copus et al. 2017).

The evolution of the interpretation of the concept of proximity, highlighted at the beginning of this chapter, is associated with a process ongoing since the 1970s of incorporating social aspects (including political, institutional, and cultural dimensions) into considerations regarding the determinants of spatial differentiation in socio-economic development. Consequently, alongside geographical proximity (which dominated until the late 1960s), other categories of proximity emerged, such as socio-cultural and relational proximity (the 1970s and 1980s), organizational and institutional proximity (the 1990s), and finally, cognitive proximity, whose significance was identified after 2000 (Table 2.1) (Kopczewska et al. 2017). This has resulted, among other outcomes, in the expanding understanding of peripheries, and the process of peripheralization (Dawkins 2003) has led to increased attention on aspatial indicators of peripherality. Among these, network peripherality has been deemed particularly significant, being associated with the consequences of exclusion from networks of social, political, and economic connections (Herrschel 2011). This results in the emergence of peripheries as enclaves located outside the network, not necessarily distant from core areas, which is typical for inner peripheries.

Network and aspatial approaches to defining peripheralization are based on the theory of organized proximity proposed by Torre and Rallet (2005). They distinguish between geographical proximity, which is based on physical distance and inherently geographical in nature, and organized proximity, which is relational and lacks a geographical character. Geographical proximity, expressed in terms of measurable physical distance (e.g., kilometers), is monotonic (something can be near or far), physical (quantifiable), and doubly relative (weighted by time and/or transportation costs and influenced by subjective perceptions of distance, time, and cost, shaped by factors like age, health, or financial status). In contrast, organized proximity is relational rather than geographical. Torre and Rallet (2005, p. 4) define organized proximity as "the capacity of an organization to facilitate interaction among its members. The organization facilitates interactions within itself, or at least makes them a priori easier than with individuals outside the organization …" Organized proximity is shaped by two logics: the logic of belonging and the logic of similarity. The logic of belonging pertains to the effects of individuals operating within an organization. Membership in an organization fosters interactions among its members, which are, as evolutionary language suggests, embedded in its "genes" (routines). Two members of the same organization are considered close because they interact, and their interactions are facilitated by (explicit or implicit) rules and behavioral routines they adhere to. The logic of similarity relates to the system of representation, beliefs, or knowledge prevailing in the organization, which facilitates collaboration. This social relationship is primarily implicit. Two members of an organization are deemed close because they are "similar," sharing the same way of perceiving, storing, and encoding information, which enhances their ability to interact. It is important to note that these two logics are partly complementary and partly substitutive. They are complementary because shared beliefs reduce the potential for divergent interpretations of formal rules. They are substitutive because, in informal communities (i.e., organizations without strong, explicit rules), weak cooperation among members can be offset by strong behavioral cohesion, which establishes implicit interaction norms (Boschma 2005; Torre and Rallet 2005). Applying this framework to the analysis of spatial disparities in socio-economic development provides a unique analytical grid that incorporates both geographical and organized proximity. This approach has been operationalized to identify and explain the functioning of inner peripheries (Copus et al. 2017).

The factors of spatial differentiation in socio-economic development, as indicated by Wallerstein in the concept of the Modern World System and by Torre and Rallet in the theory of organized proximity, highlight the fact that this process is not solely conditioned by physical distance expressed through geographical proximity. The distribution of power, decision-making processes, prevailing rules and norms, and the conditions shaping social relations, including business relationships, fundamentally alter the definition of peripheries and the explanation of the peripheralization process. They emphasize that peripheries may not necessarily emerge from geographical remoteness. Instead, areas predisposed to becoming peripheral, which may be located close to core regions, are often territories lacking elites, exhibiting deficits in their functioning due to limited competencies, and characterized by entrenched signs of provincialism, frequently conditioned by historical factors. Incorporating these new dimensions into the identification and explanation of the functioning of inner peripheries is not an easy task. This is confirmed by our experiences in conducting the research process, the results of which we present in this book. However, this does not change the fact that considering aspatial factors is essential for properly studying inner peripheries. Our proposals in this regard are presented in a conceptual model that forms the basis of the conducted research process.

2.2.3 Foundational Economy

The evolving approach to socio-economic development processes, expanding their scope and redefining goals toward improving the living standards and conditions of residents rather than striving for the highest possible level and rate of economic growth at all costs, has created the conditions for the formulation of alternative concepts (Earle et al. 2018). Among these are ideas such as the everyday economy (Reeves 2018), the foundational economy (Foundational Economy Collective 2020), and universal basic services (Cote 2020), which advocate for adjustments in regional policy to focus directly on enhancing the well-being of all citizens, rather than intensifying the activities of industries heavily reliant on research and development (Martynovich et al. 2023). This signifies a shift in attention toward "the part of the economy that creates and distributes goods and services consumed by everyone (regardless of income and status) because they support daily life" (Bentham et al. 2013, p. 7). This is linked to the assumption that everyone should have access to the resources and services necessary to meet their basic life needs as a right, not a privilege (Cote 2020). The economy associated with these activities relies on essential material infrastructure that provides public utilities and local transportation organization and encompasses care services that guarantee basic health protection and basic education (Foundational Economy Collective 2020). These proposals are particularly attractive to lagging areas, including inner peripheries (Gough 2020; Moore and Collins 2020). On one hand, in many cases, these areas lack the territorial capital resources that would allow them to enter a development path based on knowledge and innovation, even with substantial development support. Strengthening the foundational economy in these areas, however, ensures an improvement in the living standards and conditions of residents, contributing to an increase in the settlement attractiveness of these regions. This occurs both by offering employment opportunities in these activities and ensuring access to the satisfaction of residents' basic needs. On the other hand, in some cases, the very limited resources of some of these areas do not economically justify development interventions beyond the level that guarantees the minimum satisfaction of residents' basic needs. Attempts to program and implement growth-oriented interventions in these areas have proven ineffective and inefficient, becoming a basis for total criticism of Cohesion Policy actions. For these reasons, the incorporation of foundational economy principles becomes necessary in identifying and analyzing the functioning of inner peripheries and in programming intervention measures dedicated to these areas.

The foundational economy should be seen as one of several progressive socio-economic development concepts based on a participatory approach, rooted in the needs of residents, that have emerged in the literature in recent years. These include community wealth building, shared prosperity, and universal public services (Brown and Jones 2021; Coote and Percy 2020; Moore and Woodcraft 2019). These concepts are undoubtedly interrelated, and their application in the context of intensifying neo-endogenous development creates favorable conditions for explaining new trends and operationalizing actions directed toward evolving goals that account for new developmental processes (MacKinnon et al. 2022). The term "foundational economy" was introduced by Bentham et al. (2013) to describe the part of the economy that provides goods and services to meet citizens' basic needs and ensures the infrastructure for daily life. It is subject to everyday consumption, distinguishing it from other types of economies, whose products and services are consumed occasionally or incidentally (Hall and Schafran 2017). The foundational economy consists of two parts: material and care-based (Martynovich et al. 2023). The material foundational economy connects households with daily needs and includes public utilities (electricity, gas, water), transport and telecommunications infrastructure, food production and distribution, as well as private banking services. The care-based foundational economy includes a set of activities related to the provision of social services (education, healthcare, and care services) and income maintenance systems, including the controversial universal basic income (UBI), which many authors debate as a potential "solution" to poverty and inequality by promoting unconditional cash payments for all (Cote 2020). A common feature of these two types of foundational economy is their inherent social value and the fact that they are collectively delivered through branches and networks (Froud et al. 2020). Furthermore, the foundational economy creates a significant number of local jobs, which are relatively resilient to the socio-economic impacts of recurring crises (Sissons et al. 2021). Employment in the foundational economy cannot fall below a certain threshold, making this type of economy particularly attractive to lagging areas, including inner peripheries. These characteristics of the foundational economy have formed the basis of increasingly raised demands for a radical reorientation of regional policy "toward developing the foundational economy, rather than pursuing the chimera of competitiveness" (Engelen et al. 2017, p. 408). This leads to a shift in development goals from increasing efficiency to setting limits on the sufficiency of resources allocated to providing goods and services that improve the well-being of all residents (Foundational Economy Collective 2018).

The significance of the foundational economy for the development of lagging regions has been emphasized by many authors (Gough 2020; Moore and Collins 2020). This also applies to inner peripheries (Servillo et al. 2016). A key argument in this discussion is the fact that these regions have been most severely affected by deindustrialization, while having very limited capacity to generate employment opportunities in competitive economic sectors (Rodríguez-Pose 2018). As a result, the jobs created in the foundational economy in these regions constitute a larger share of the demand side of their local labor markets, highlighting their importance in shaping development processes based on endogenous resources (Earle et al. 2018; Heslop et al. 2019; Morgan 2019). This can be particularly relevant during crises, when "the hardest impacts are felt in regions already disadvantaged" (Tomaney et al. 2010, p. 773). Research, including studies by Martynovich et al. (2023), indicates that the foundational economy can help improve economic conditions in these areas during and after crises. This is due to its unique nature, characterized by limited exposure to international and interregional competition and a stable demand for foundational goods and services (Calafati et al. 2019). These traits allow the foundational economy to be viewed as an important element of "protected regional economies" (Fratesi and Rodríguez-Pose 2016), which become evident when non-export-oriented activities dominate a region's economic structure. However, it is important to note that comparisons between the foundational economy and protected economies should be limited to their similarities in terms of exposure to international/interregional competition (Froud et al. 2020). The functioning of the foundational economy in times of crisis in lagging regions, including inner peripheries, and its impact on the local socio-economic system, shows significant variability over time. The foundational economy can serve as a stabilizing force for the economy of a given region during periods of contraction (Engelen et al. 2017). Areas with higher levels of employment in the foundational economy may, at least in the early stages of a crisis, benefit from its stabilizing effect, which limits the scale of job losses (Źenka et al. 2019). Thus, in the short term, the foundational economy enhances a region's ability to mitigate job destruction and/or recover pre-crisis employment levels. This phenomenon is referred to as "engineering resilience" (Boschma 2015; Fröhlich and Hassink 2018; Simmie and Martin 2010), which denotes the ability to "return to the state we were in" (Davoudi 2012, p. 302). This can also be partially linked to the concept of "ecological resilience" from the literature, which describes a socio-economic system's response to disruptions, aiming to return not to its pre-crisis state, but to a new stable equilibrium (Martynovich et al. 2023). Unfortunately, in the long term, the foundational economy lacks the capacity to stimulate stronger economic recovery, meaning it cannot create conditions for "evolutionary resilience," which refers to a region's ability to maintain long-term development through the adaptation and reconfiguration of industrial, technological, and institutional structures within an evolving economic system (Boschma 2015). In this context, the positive role of the foundational economy is diminished, and it could even be argued that a high share of employment in foundational activities leaves a region "ill-prepared for adaptation to changes through processes of creative destruction" (Frates and Rodríguez-Pose 2016, p. 38). These patterns highlight the fact that the development of the foundational economy should not be regarded as a universal solution to the development challenges of inner peripheries. However, its presence in the economic structure of these regions should certainly be considered due to its protective role and its contribution to mitigating the social consequences of spatial development inequalities. It should also be noted that in the challenging economic conditions of inner peripheries, the development of foundational economy activities faces serious obstacles due to a lack of budgetary resources and high costs stemming from low and declining populations (MacKinnon et al. 2022). This calls for the exploration of alternative ways to foster its development, which may provide a basis for its preservation in these areas. Recognizing the importance of these relationships for accurately explaining spatial socio-economic development processes, we aim to incorporate them into the conceptual framework of our research and use them to interpret the results obtained at various stages of the investigation.

2.2.4 Spatial Justice

Incorporating the concept of spatial justice into the theoretical foundations has become essential in the context of contemporary discussions on broadly understood inequalities. Spatial justice emphasizes the need to explicitly recognize space as a factor of social inequality, reproduced by the socio-economic mechanisms that organize society within spatial frameworks (Watt 2009). As Servillo and colleagues (2016) highlight, after the 2008 financial crisis, the expansion of the normative foundation for inner peripheries is fully justified. It should be noted that the socio-economic impacts of subsequent global crises—such as those resulting from the SARS-CoV-2 pandemic and Russia's aggression against Ukraine, both of which demonstrate significant spatial differentiation—lead to the conclusion that, under current conditions, this is not only necessary but essential. Utilizing the concept of spatial justice, alongside insights from previously mentioned theories such as Geographical Peripherality, the Modern World System and Organized Proximity, and the Foundational Economy, enables a rethinking of how to design and implement development interventions. This approach also allows for better management of the socio-economic processes occurring in inner peripheries.

The concept of justice is widely employed within the social sciences, although this does not imply a universal agreement on how to define and apply it in explaining socio-economic realities. Its origins can be traced back to the ancient works of Plato and Aristotle, through which social justice became a central element of many concepts forming the foundations of ethics, political philosophy, and law. In the context of addressing inequalities, within the broader concept of social justice, two main types are distinguished, both of which dominate the contemporary discourse on this issue (Rawls 1971; Sen 1997, 2009): distributive justice and procedural (commutative) justice. Distributive justice (also known as distributional or allocative justice) ensures that all members of a community share in the common good, though this share is not equal and corresponds to each member's contribution and status. Procedural or commutative justice (equitable or reciprocal) is concerned with the pursuit of justice through a well-structured legal system, particularly focusing on the process of gathering information, exchanging arguments, making decisions, and compensating deficiencies for those who possess them for various reasons. This approach, in a sense, involves leveling the distribution of goods and services. The first type of social justice is future-oriented, focusing on the creation of future opportunities, while the second type is oriented toward the past, emphasizing the importance of addressing the consequences of previous actions or inactions that determine the current situation of each community member.

The concept of territorial justice is rooted in decades of academic discussions on what Pirie (1983, p. 470) refers to as "the social justice of distribution in space ..." Its origins can be traced back to the radical geographers of the 1960s, who emphasized the need to explain the relationship between society and space within urban areas, highlighting the necessity of ensuring fair access to resources and minimizing disparities in this regard (Rauhut 2017). The term "territorial justice" was first introduced in academic literature in the context of spatial consideration of social justice at the regional or territorial scale by Davies in his book *Social Needs and Resources in Local Services: A Study of Variations in Provision of Social Services between Local Authority Areas* published in 1968 (Davies 1968). An important contribution to the discussion was made by Harvey in his work *Social Justice and the City* (1973), where he pointed out the lack of social justice in urban spaces and the challenges of its precise definition. Contemporary geographic discourse on territorial justice heavily draws on the views of Soja, who argues that it stems from "a deliberate and focused emphasis on the spatial aspects of justice and injustice" (Soja 2009, p. 4). At its core is the equitable distribution of socially valuable resources and opportunities to utilize them in space. The concept of territorial justice focuses on the patterns of spatial differentiation of resources and the ways they are geographically allocated in relation to either enhancing or limiting people's opportunities to participate in development processes and benefit from their outcomes. It addresses the political and socio-economic organization of space and its influence on creating just and unjust geographic disparities (Soja 2010a; Weck et al. 2021). Thus, territorial justice should be viewed as a form of distributive justice, with an added geographical dimension. This leads to the definition of territorial justice as "the fair distribution of valuable resources in space and opportunities to utilize them" (Soja 2010b, p. 62). It is worth emphasizing that Soja's understanding of territorial justice is based on a socio-spatial dialectic, which assumes that "two systems of relations (i.e., social and spatial relations) are only seemingly homologous because they originate from the same sources in the mode of production, but are also dialectically intertwined and inseparable" (Soja 1980, p. 209). Therefore, it can be argued that the spatiality of justice affects society and social life just as social processes shape spatiality or the specific geography of justice, while perfect spatial justice remains unattainable (Papadopoulos 2019). This way of defining territorial justice can be seen as an attempt to spatialize Rawls' (1971) concept of justice. Consequently, territorial justice allows for an integrated understanding of justice that overcomes the dichotomy between procedural and distributive justice (Madanipour et al. 2021). Overcoming this dichotomy is possible through the use of the relational concept of space, which does not separate spatial systems from the processes that create them (Lefebvre 1974). This requires shifting the perception of space from merely a location for distribution ("container") to an element of the broader process of distribution ("process") (Papadopoulos 2019, p. 8; Pirie 1983; Soja 1980). This approach replaces the static pattern of resource and opportunity distribution with a continuous and democratic process that creates fairer and more equal opportunities, regardless of location. Territorial justice, therefore, entails striving to equalize opportunities for society across space, while ensuring the fulfillment of procedural aspects of social justice (Weck et al. 2021). It is based on the connection between the distribution of resources and opportunities (distributive justice) and fair and transparent decision-making regarding resource management (procedural justice) (Davoudi and Brooks 2014; Israel and Frenkel 2017; Pirie 1983).

The concept of territorial justice, in relation to inner peripheries, offers a framework for addressing the dilemmas associated with the economic efficiency of development interventions targeted at lagging areas. To achieve this, it is both possible and necessary to apply the principles of social justice in a territorial dimension, which are based on the spatial adaptation of Rawls' social justice and Sen's capability deprivation. Rawls (1971) grounds his concept on the assumption that society is obliged to ensure a fair distribution of opportunities and resources among its citizens. This highlights the need for the state to undertake corrective actions aimed at providing fair and equal opportunities for people in both peripheral and central areas (Kukathas and Pettit 1990), ensuring that such opportunities are not directly determined by place of birth (Arneson 2008). Sen (1997) also underscores the necessity of public intervention in cases of market failure, particularly focusing on areas that cannot effectively reduce inequalities and their consequences on their own. The fundamental premise of territorial justice—that an individual's location should not affect their opportunities or lead to the deprivation of their capabilities—has deep roots in Sen's reasoning. In seeking to operationalize these assumptions, one might refer to the proposal by Magel (2017), who, in the context of balancing living conditions in Bavaria, put forward a model of territorial justice based on four dimensions. This model could be applied in the programming of development interventions aimed at inner peripheries: (1) equality of opportunity, ensuring that every newborn child has access to education and more; (2) distributive justice, which assumes the availability of basic infrastructure necessary for living and working; (3) procedural justice, guaranteeing equal rights for all; and (4) intergenerational justice, which highlights the need for sustainable development to ensure that future generations are not worse off at the outset (Churski 2022). However, the operational implementation of this model presents a significant challenge, particularly in terms of maintaining economic efficiency at the scale of the entire socio-economic system (Rawls 1971). Development interventions should not be provided at all costs, and their scope must be conditioned by the economic efficiency of public resource allocation. This leads to the conclusion that ensuring equality of opportunity for residents of lagging areas, including inner peripheries, requires the establishment of a minimum level of access to basic services, such as education and healthcare, that must be guaranteed for every resident. To this end, it is necessary to develop and implement a specific spatial social minimum based on the foundational economy (Sect. 2.2.3), which on the one hand would define a catalog of essential services (SGI) that must be accessible in every location within a defined time-access isochrone and on the other hand would justify limiting economically inefficient actions in areas lacking broader development opportunities (Gorzelak 1985; Halamska 2018). Our proposals regarding inner peripheries in this context are formulated in the final recommendations (Chap. 6).

2.3 Inner Peripheries in Spatial Differences of Development

Expanding the understanding and interpretation of the socio-economic development process increases the multifaceted nature of the spatial disparities in this process. Consequently, this leads to numerous typologies of spatial units in terms of their level of development, with delimitation and classification criteria based on a very broad range of indicators (Pike et al. 2023). These indicators describe selected aspects and dimensions of the socio-economic process, which are most often organized according to development factors. While the categories of development factors show significant stability, the way they are interpreted changes, reflecting shifts in their scope of influence. These changes are the result of the increasing impact of global conditions, leading to more frequent and deeper crises that act as catalysts for reinforcing spatial development disparities (Churski et al. 2021). Alongside the multifaceted nature of contemporary development disparities, the spatial level at which these disparities are considered is particularly important from the perspective of Cohesion Policy practice. Existing experiences suggest that the NUTS-2 regional level is certainly not sufficient for programming and implementing effective and efficient development interventions. Consequently, the literature continues to debate the optimal spatial level for, on the one hand, objectively identifying and, on the other, best addressing spatial development disparities. It appears that the subregional level, utilizing the distribution of functional areas, should become a greater focus of attention in this regard (European Commission: Directorate-General for Regional and Urban Policy 2024). A common feature of these typologies and classifications is the identification of the state and dynamics of spatial development disparities, as well as their causes. The goal is to find ways to reduce their scale, which poses an increasing challenge for the evolving EU Cohesion Policy. It is important to remember that, in this context, inner peripheries represent only one type of "left-behind" area. However, in our view, this is a category that concentrates spatially on negative socio-economic phenomena and their consequences, which should become a particular focus of the EU Cohesion Policy, contributing to its much-needed reorientation.

2.3.1 Spatial Developmental Differences and Their Typologies

In analyzing the position of inner peripheries within the highly diverse and rich set of typological spatial differentiations of development, we aim to highlight their similarities and differences in relation to other categories of "left-behind areas," as well as to demonstrate their relationship to the spatial classifications of development disparities used at the level of programming and evaluation of the EU Cohesion Policy.

The literature on the subject contains numerous proposals for typologies of spatial development disparities. These typologies arise from different approaches to analyzing the patterns of development differentiation across space. Most are based on a geographical approach that refers to the classical physical interpretation of the core-periphery relationship (Copus 2001; Keeble et al. 1988). Others, taking into account the expanding interpretation of the concept of socio-economic development, are based on a sociological approach that distinguishes between peripheries and semi-peripheries (Connell 2005), or an anthropological approach that introduces the concept of remote areas. All of these typologies are characterized by different assumptions regarding the criteria used for delimitation, different spatial levels at which delimitation and classification procedures are carried out, and additionally, either a static or dynamic approach to analyzing the level of socio-economic development (Fig. 2.1). It should be emphasized that the scope and level of detail of these typologies are largely dependent on the availability of data. The limitations in this regard significantly reduce the substantive and empirical value of the research conducted, particularly when they involve longer time series or lower spatial levels or include international comparisons.

A significant increase in interest in the analysis of spatial disparities in development, particularly in identifying spatial concentrations of developmental deficits, has been noted since the 2008 financial crisis. In response, several initiatives were launched at the EU level to both objectively identify spatially differentiated needs and develop targeted actions to address them. Among these was the initiative aimed at lagging regions, whose level of development was significantly below the EU average. Today, this initiative is referred to as the "catching-up regions initiative." It

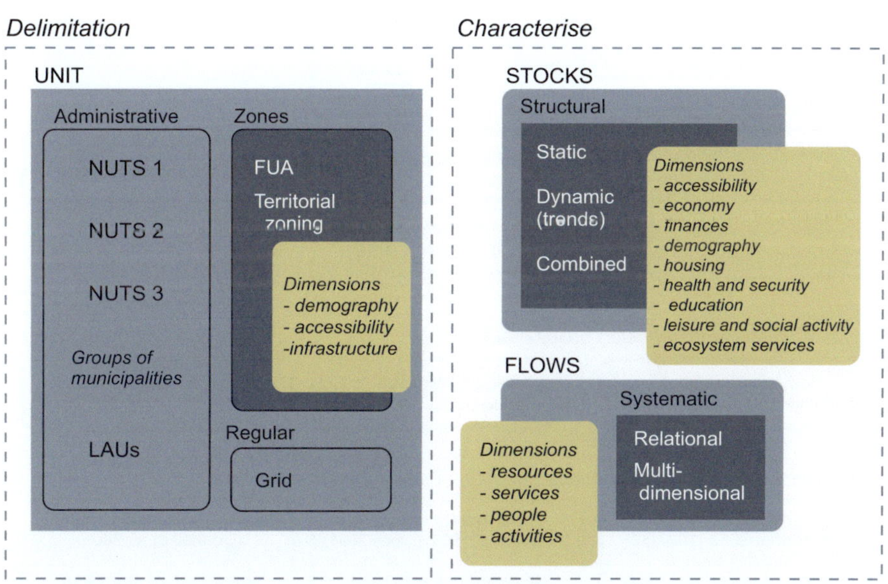

Fig. 2.1 Main approaches to the identification and typology of spatial socio-economic development disparitie. *Source* Own elaboration based on Stjernberg et al. (2023)

identified two categories of such areas: low-growth regions and low-income regions (European Commission 2017). In line with the growing focus on this issue, the concept of "left-behind areas" gained popularity (Pike et al. 2023). This term has become widely and universally used to describe the challenging situation of former industrial centers and rural areas adversely affected by austerity policies, globalization, economic changes, and technological advancements (Görmar et al. 2019; Hendrickson et al. 2018; McCann 2019). Some scholars have further emphasized that the issue is not only about left-behind areas but also the left-behind residents of these areas (Dijkstra et al. 2020). This perspective arises from the fact that, in response to the adverse effects of these processes, residents increasingly expressed social dissatisfaction and rejection of their living conditions. As a result, left-behind areas became a focal point of analysis in studies on the geography of discontent and the rise of economic nationalism and populism (Florida 2021; Martin 2021; Rodríguez-Pose 2018; Rodríguez-Pose et al. 2023). The widespread use of the term "left-behind areas" has led to its frequent replacement of terminology previously used in geographical research, such as "inequalities," "divisions," "gaps," "disparities," and "imbalances" (Dunford 2010). This concept has become a universal keyword encompassing various categories of less-developed areas. As Pike et al. (2023) argue, it reflects diverse perspectives and ways of using the term across different methodological, geographical, and temporal contexts. Within this conceptual category, the notion of inner peripheries has also been distinguished (Servillo et al. 2016) (Fig. 2.2).

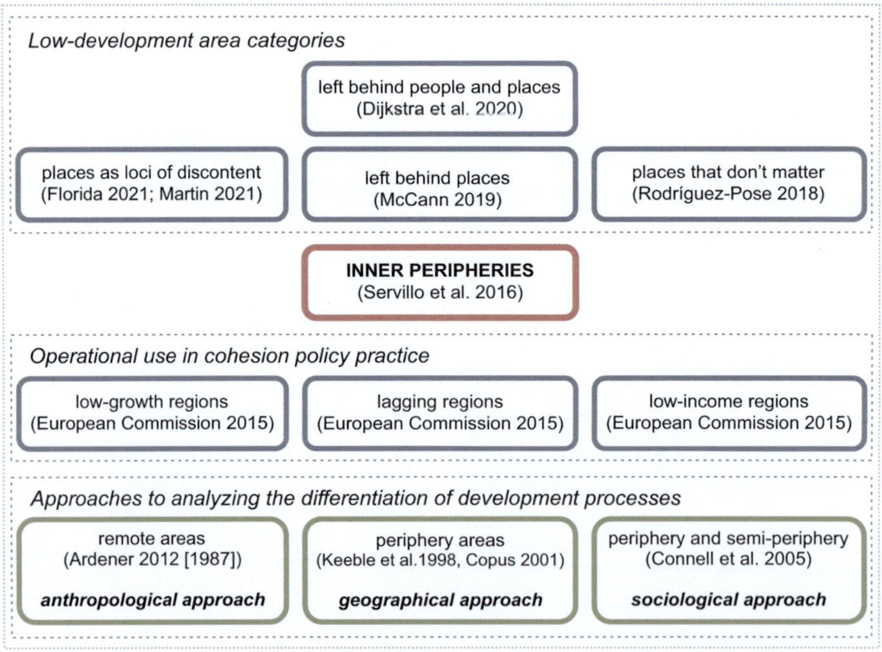

Fig. 2.2 Selected conceptual categories of less-developed areas. *Source* Own elaboration

Efforts to clarify the concept of "left-behind areas" require that two fundamental questions be addressed (Pike et al. 2023): (1) Who and what is being left behind—and behind what? (2) Where and when is something being left behind?

The answer to the first question refers to a broad discussion, well-documented in the literature that seeks to resolve the dilemma between inequalities of places and inequalities of people—the residents of those places. This discussion also extends to the focus of development policies, particularly the confrontation between people-based policy and place-based policy, which forms the foundation of the contemporary paradigm of the EU Cohesion Policy. The former approach is used in the literature to describe development actions primarily focused on residents and their development opportunities, regardless of their location (Bentley and Pugalis 2014). However, it is important to note that this concept was initially used by other authors to describe actions that, while focusing on the residents of specific territories, paid more attention to their embedded human and social capital and their inherent development opportunities rather than just their material resources and economic potential (place-based people strategies) (Crane and Manville 2008). This ambiguity has led to unjustified debates in the context of regional policy, treating place-based policy in opposition to people-based policy (Garcilazo et al. 2010). Assuming, in line with Roberto Camagni's concept of territorial capital (2008), which emphasizes that territory is a space relationally shaped by its residents, this dilemma becomes less relevant. This leads to the conclusion that "left-behind areas" are territories that, due to deficits determined by the state of their resources, are unable to independently shape competitive territorial capital. These deficits are multidimensional (Table 2.2), and their scope and structure form the basis for distinguishing various categories of areas and their residents who are left behind compared to areas and residents in relatively better situations regarding the considered aspects of their territorial capital. It is important to remember that not every dimension is visible in every left-behind area. The characteristics of these areas vary over time and space, requiring consideration of their causes, geography, and dynamics of change (Martin et al. 2021). This also applies to inner peripheries, where, as we have already noted, the key issues are the barriers related to limited relationships and their resulting consequences.

Considering the answer to the second question—where and when are areas left behind?—we again encounter a highly diverse situation in the literature regarding the application of this concept to the examination of spatial development inequalities at virtually every spatial scale. These include communities, neighborhoods, localities, towns, cities, metropolitan regions, and regions (Bolton et al. 2020; Jennings and Stoker 2019; Kemeny and Storper 2020). The use of this term in relation to specific geographical or temporal dimensions sometimes leads to the use of various names such as "frostbelt," "rustbelt," "Trumpland," or "Brexitland," which emphasize the common developmental deficits occurring in a given area. This is relevant to practically every country, although the term "left-behind areas" is not always explicitly used (Daniel and Whalan 2021; McQuarrie 2017; Sawers and Tabb 1984; Sobolewska and Ford 2020). The situation is further complicated by the fact that in the identification and analysis of "left-behind areas" found in the literature, these areas are treated both as a temporary and incidental state, and more commonly, as a

Table 2.2 Potential dimensions of left-behind places

Dimensions	Examples
Economic	Falling real wages Limited and/or poor-quality opportunities for training and/or employment Limited wealth
Social	Lack of social and/or spatial mobility Limited bridging social capital (but high levels of bonding social capital) High levels of attachment and belonging to place Low levels of civic participation
Environmental	Degraded physical environments Poor air quality
Political	Neglect by mainstream political parties and politicians Disengagement from representative democracy Populist, nativist and/or nationalist beliefs and views
Institutional and Governmental	Absent or weak local leadership Limited decentralized powers, resources, and/or capabilities Lack of political voice
Cultural	Inferior, subordinate, common, proletarian, or plebeian worldviews Disconnection from and/or rejection of dominant attitudes/values Backward-looking, behind-the-*zeitgeist* outlooks
Infrastructural	Lack of public investment Limited and/or unequal access to public services Unequal provision of infrastructure systems and services

Source Pike et al. (2023, p. 6)

long-term, entrenched condition that has developed over many years (Dijkstra et al. 2020; Martin et al. 2021). In this context, it should be noted that in the case of inner peripheries, their identification typically occurs at the subregional level. This level often corresponds to NUTS3 units, or, as we recommend in this book, to clusters of NUTS 5 units that form functional areas corresponding to the extents of local labor markets and local service markets. These are certainly areas that have developed over an extended period, with long-standing and entrenched developmental deficits and barriers, making them difficult to overcome and change.

Delimitations and typologies of spatial development disparities also have operational applications for the functioning of community policies. To ensure the conduct of international comparisons and in the context of the needs of implementing and evaluating Cohesion Policy interventions, EUROSTAT launched a legislative initiative called "Tercet" aimed at integrating the most important typologies with the NUTS regulation. This initiative resulted in the adoption of *Regulation (EU) 2017/2391 of the European Parliament and of the Council of 12 December 2017, amending Regulation (EC) No 1059/2003 as regards territorial typologies (Tercet)* (OJ L 350, 29.12.2017). This regulation introduced the following typologies:

- A typology based on a grid with a resolution level of 1 km^2 (kilometer grid) that is used to delineate urban centers, urban clusters, and grid cells of rural areas.
- Typologies established at the level of Local Administrative Units (LAU) that are used to determine the degree of urbanization (DEGURBA), Functional Urban Areas (FUA), and coastal areas.
- Typologies established at the NUTS 3 level that form the basis for urban–rural typologies, metropolitan typologies, and coastal area typologies.

Unfortunately, despite the well-developed methodology for delineating these areas (EUROSTAT 2019), there is not a sufficiently rich data set available to allow for the measurement of various dimensions of socio-economic development disparities in these spatial frameworks. This leads to significant compromises and forces the analysis of development disparities at higher spatial levels than those at which these disparities actually occur to the greatest extent. This situation highlights the institutional inefficiencies that limit our understanding of the causes, processes, consequences, and potential interventions for changing contemporary spatial development disparities. This also applies to attempts to identify and typologize inner peripheries, as was the case in the research conducted for this study.

Initiatives attempting to overcome this unfavorable situation include projects carried out under the ESPON Program. Unfortunately, even in these cases, limited access to data at lower spatial levels forces the conduct of more generalized analyses. An example is the EDORA project (European Development Opportunities for Rural Areas) carried out under the ESPON 2013 program, which developed a typology of intermediate and predominantly rural areas at the NUTS3 level in 27 EU member states (ESPON EDORA 2011). Another example is the DEMIFER project (Demographic and Migratory Flows affecting European Regions and Cities), which resulted in the analysis of demographic, labor market, and migration trends at the NUTS2 level (ESPON DEMIFER 2010). This group also includes the previously mentioned PROFECY project (Inner Peripheries: National territories facing challenges of access to basic services of general interest) carried out under the ESPON 2020 program, in which the identification of inner peripheries in the EU was conducted at the NUTS3 level (ESPON PROFECY 2017, 2018).

As a result, EU initiatives aimed at less-developed areas are also not targeted at lower spatial levels. An example is the Catching-up Regions Initiative, introduced in 2015 to identify areas characterized by developmental deficits. The NUTS2 level adopted for analysis allowed for the identification of very general patterns, which led to the distinction of "low-growth regions," mainly in Southern Europe, and "low-income regions," mainly in Central and Eastern Europe. Actions dedicated to these areas were subsequently prepared, with a greater focus on the latter category. However, their effectiveness was limited, prompting the search for more effective and efficient actions (Brown et al. 2017; European Commission 2017; Farole et al. 2018; Rodríguez-Pose and Ketterer 2020). The need for these searches was fully confirmed by the consequences of subsequent crises, drawing attention to the fact that spatial development disparities primarily form at the subregional level. It is at this spatial level that their causes should be identified and that development interventions aimed at reducing them should be programmed and implemented. The effects of the crisis also highlighted new challenges for lagging regions, which can pose both threats and opportunities related to the ongoing redefinition of development factors. This requires rethinking and changing current approaches, as well as programming and implementing interventions that provide a territorially targeted response to new opportunities and threats, to prevent them from exacerbating spatial socio-economic development disparities (ESPON ESCAPE 2020; Pilati and Hunter 2020). To this end, it is essential to expand research on the delimitation and typology of "left-behind areas," including inner peripheries at the subregional level, lower than NUTS3, taking into account new development conditions and the need to build long-term resilience to crisis phenomena.

2.3.2 Functional Areas and Their Importance in Identifying and Interpreting Spatial Developmental Differences

The assertion about the significance of cities for regional development is axiomatic, yet understanding the specifics of this influence is a crucial task in theoretical and analytical dimensions within regional studies. Cities themselves and the processes occurring within them are significant factors of economic growth; they are the driving forces behind social and economic activities in the region. However, the outcomes of their changing roles and functions are highly uncertain and require continuous investigation (Clark et al. 2018; Zdražil and Pernica 2021). Therefore, interpreting the developmental processes of cities and their roles within the region is not possible solely from the perspective of the urban center (Shuang and Long 2020). It is important to analyze the city's relationship with its surroundings; as it develops, the complexity of these relationships increases, and they become interdependent. Cities develop and spread developmental processes within their influence zones, but they also cease to be self-sufficient, especially in the context of the availability of human resources, services, etc. (Śleszyński 2013). On one hand, cities constitute substantial labor markets and provide educational services and a range of other public services meeting the needs existing within their areas of influence. On the other hand, the surrounding areas meet the demand for a qualified workforce largely employed in the cities, and also fulfill supply functions and take over activities previously occurring in the cities. It is important to note that these patterns directly relate to the perception of the development process through the lens of its spatial polarization, based on the effects of agglomeration benefits (as understood by Marshall-Arrow-Romer, Porter, or Jacobs) in growth areas (cities) and its diffusion driven by spillover effects to surrounding areas (peri-urban areas). This was previously highlighted in the discussion of the concept of geographical peripherality (Sect. 2.2.1).

The extent of a city's (core) impact seems to depend on fundamental socio-economic development indicators and the size of the human capital mobility zone

(Castells-Quintana et al. 2020; Lewin et al. 2013). Hence, it is widely believed that developmental impulses are usually sent by large or medium-sized cities, whose economic and social potential significantly translates into observable developmental processes occurring in their surroundings. Research increasingly provides evidence of the spread of developmental processes from core cities to areas where small towns serve as local growth poles (Rio Fernandes and Seixas 2018). They can play an important regional economic role as driving forces of their surroundings—both as markets for goods and services and as knowledge centers (Nilsen et al. 2023). Small towns provide access to services and the labor market, as well as national administration services. Often, microregion identity develops around them. However, what distinguishes these units negatively is low transport accessibility and weak relations with higher-order centers—regional ones. Despite this, these same small towns remain definite centers of their hinterlands, usually due to a lack of competition (Vaishar and Zapletalová 2008). Thus, the position of an urban unit undoubtedly depends on the spatial scale of reference. Whereas on a macro scale a given city may be part of the periphery, on a local level it is recognized as the core of a nodal region (Komárek and Chromý 2020).

The uneven impact of urban centers on surrounding spaces corresponds to an imbalanced level of economic development. The explanation for these disparities is rooted in theories of regional development polarization, based on the concepts of growth poles and core-periphery dynamics, which form the foundation of the uneven development theory, identifying patterns of spatial differentiation. The utility of both concepts for explaining developmental changes in economic space is discussed in Sect. 2.2.1. Their practical application is realized through the polarization-diffusion model, which is employed in the planning and implementation of developmental interventions as an alternative to the equalization model. This model highlights the possibility that areas with low levels of development within the analyzed socio-economic territorial systems are not necessarily doomed to regression. Centrifugal tendencies, such as the so-called "trickle-down effect" (Dyjach 2013), may also occur. Cores, by influencing surrounding areas, can stimulate their development, although this growth remains dependent, with the core maintaining its dominant position. As a result of the diffusion of economic development from the center to the periphery, these areas often continue to serve a subordinate role, both economically and socially. However, in line with the core-periphery concept, they can be viewed as "ascending peripheries," experiencing improvements in socio-economic conditions, as opposed to "draining areas" undergoing progressive marginalization. The latter are referred to by Friedmann (1967) as "descending peripheries."

The outlined patterns provide a better understanding of the concept of Functional Urban Areas (FUAs), which explains the formation and operation of areas influenced by the relational interactions of cities. This concept has been widely applied in many countries at both the diagnostic and programming stages of socio-economic development processes. Its utility stems from its consideration of the circulatory nature of socio-economic linkages between the main city (urban core) and its surroundings, which is particularly relevant in the context of the contemporary increase in population mobility (Churski et al. 2023). The FUA research paradigm originally emerged in the past century, grounded in the concept of the nodal region, understood as a type of analytical region serving as a tool for spatial analysis (Whittlesey 1954). What distinguishes a nodal region is its homogeneous internal structure or organization, characterized by the presence of a node and the surrounding area, delineated by its connections and the extent of the node's influence.

In a broad sense, Functional Urban Areas have been a subject of interest in geographical studies since the first half of the twentieth century. The development of theories and empirical research on city-environment relationships was accompanied by the popularization of a number of concepts. As early as the first half of the last century, the concept of metropolitan areas was formulated (McKenzie 1933), followed by economic, market, commercial areas, urban fields, spheres of urban influence, and others (Berry 1967; Gerard 1958; Huff 1963; Smart 1974), urban fields (Friedmann and Miller 1965), urban spheres of influence (Huff 1973), urban daytime system (Berry 1973), and urban functional regions (Hay and Hall 1977). The contemporary understanding of a Functional Urban Area derives mostly from Berry's concept of daily urban systems (1973) and Hall's concept of functional urban regions (Hay and Hall 1977). The advantage of these approaches was the consideration of the circulatory nature of socio-economic linkages between the main city (urban core) and its surroundings, particularly significant under conditions of increasing mobility.

An FUA is defined as a geographical area encompassing a central city (with the possibility of more than one core) and its surrounding rural and urban areas, which constitute separate administrative units but have common economic, social, and infrastructural linkages (Busłowska and Marcinkiewicz 2023; Dijkstra et al. 2019; Gross et al. 2024; Zdražil and Pernica 2021). Within the boundaries of an FUA, there is a flow of materials, capital, and information, and these areas provide space for living, production, recreation, and economic development (Busłowska and Marcinkiewicz 2023).

The delimitation of FUAs should be based on the results of objective diagnoses identifying the extent of spatial relationships, but potential collaboration opportunities and the effective cooperation of the units forming them—both vertically and horizontally—should also be considered. The boundaries of FUAs are determined differently depending on the research topic, usually based on one criterion or one diagnostic feature. Depending on the approach taken, the extent of the FUA delineated for the same city will vary. These discrepancies can also result from the purposes for which the delimitation work is undertaken. Objectives such as cognitive, statistical, or planning purposes determine not only the selection of criteria but also the spatial reference systems.

In general, the identification of FUAs is the result of applying various criteria (Busłowska 2024; Domenech et al. 2012; Dijkstra et al. 2019; Kurek et al. 2020; Śleszyński 2013):

- Administrative: defined as an arbitrarily determined group of administrative units.
- Morphological: identified as areas meeting criteria for building characteristics, population density, and settlement density.
- Functional: defined as areas of interaction between the core (or cores) and the surrounding areas, based on commuting patterns and other economic interactions.
- Socio-economic: identified as areas meeting labor market criteria. A network-based approach is recommended (Domenech et al. 2012; ESPON 2007), where an FUA is determined by a multifaceted network of interactions and connections.

The delimitation of FUAs based on the described approaches is fraught with potential problems; relying on "traditional" static data from incomplete and infrequently updated databases, presented at different levels of spatial aggregation, portrays FUAs as static units, which contradicts the dynamic nature of these areas.

In recent years, new possibilities have emerged for measuring population flows and land-use structures, guaranteeing high spatial and temporal precision through the use of Big Data generated by users. In this way, the delineated extents of FUAs constitute evolving, self-adjusting urban systems (Chen et al. 2017). Based on Big Data, attempts have been made to determine the spatial extents of FUAs using data generated by ride-hailing applications to capture the directions and intensities of commuting flows (Shuang and Long 2020), as well as taxi trajectory data and the distribution of points of interest (POI) (Liu et al. 2023). Additionally, Li et al. (2022) proposed a method for classifying areas by combining the analysis of morphological features of land development and Land Surface Temperature (LST) obtained through thermal remote sensing. This method can be effectively used in the delimitation stage of FUAs. The results of delimitations based on Big Data and innovative data acquisition methods allow for a view of the city's influence on its surroundings that is realistic and fully independent from administrative factors (level of data aggregation).

The growing interest in Functional Urban Areas and their delineation is a consequence of the evolving model of regional policy adopted by the European Union, which is focused on a place-based approach. This model aims at the comprehensive planning and implementation of actions in functionally defined territories rather than administratively defined ones (Nowak 2018). This new perspective on development policy and its territorial dimension necessitates a shift from the traditional paradigm, where the functional approach increasingly supersedes the administrative approach, which has limited the scope of developmental activities (Ładysz 2021). One of the main attributes of the place-based approach is the spatial focus of actions aimed at identifying and fully incorporating the specifics of individual functionally defined territories (Barca 2009). As emphasized by Dijkstra and others (2019, p. 4):

FUA are a powerful tool to compare socio-economic and spatial trends in cities and to design urban development policy. FUA are better suited than administrative areas to capture agglomeration economies and they encompass the full extent of the city's labor market.

In 2011, the European Commission and the OECD developed an operational definition of a Functional Urban Area (FUA). As highlighted in its rationale, the FUA concept enables the integration of densely populated urban areas with their closer and more distant surroundings, accounting for the extent of daily socio-economic relationships, expressed, among other ways, through commuting patterns. The advantage of the functional approach lies in its comprehensive capture of local labor and housing markets without fragmenting them across multiple municipalities (local administrative units, LAU2). It also avoids merging numerous distinct urban systems into a single spatial unit, as is often the case with many NUTS3 regions. Additionally, FUAs exhibit less variation in size compared to those observed within sets of municipalities (LAU2) or subregions (NUTS3). It is worth noting that the idea of functional areas is also relevant to the ongoing discussion about the long-term vision for rural area development in the EU (Dijkstra and Jacobs-Crisioni 2023). In this context, attention is drawn to the necessity of considering the potential to shape the reach of functional relationships determined by travel time to the nearest small town or large village. These form "daily rural systems" that often extend beyond the administrative boundaries of local units (LAU2).

The increased operational application of the FUA concept in the European Cohesion Policy was recommended in the conclusions of the 8th Cohesion Report (European Commission: Directorate-General for Regional and Urban Policy 2022). The report highlighted the need to strengthen the management of functional areas to ensure better coordination of services, infrastructure development, and investment in these areas. This would help mitigate the scale of issues revealed during the SARS-CoV-2 pandemic. Recognizing growing intraregional development disparities, the report also emphasized the importance of enhancing urban–rural linkages and strengthening the role of smaller towns and cities in supporting rural areas. Actions in this regard should aim to improve the situation of core areas while expanding their influence on surrounding regions. This would optimize the functioning of FUAs and strengthen the process of multifunctional rural development, thereby reducing spatial disparities in development levels. However, these recommendations have not received sufficient implementation, resulting in continued growth of intraregional development disparities within the EU. Consequently, these recommendations have been reiterated in recent reports, including the High-Level Group Report on challenges for future Cohesion Policy (European Commission: Directorate-General for Regional and Urban Policy 2024) and the 9th Cohesion Report (European Commission 2024).

Thus far, the operational application of FUAs has been realized through the instrument of Integrated Territorial Investments (ITI) (European Commission 2015). The implementation of the Cohesion Policy through ITIs is intended to enhance efficiency and facilitate the execution of multidimensional and cross-sectoral interventions (Ferry 2019). Despite the numerous advantages of this instrument, we hold the position that the use of FUAs in the programming and implementation of the EU Cohesion Policy, aimed at effectively reducing spatial development disparities, remains insufficient.

The utilization of the proposals presented in this book, which advocate for basing the delimitation of inner peripheries on the FUA framework and recommend the operational application of FUAs in implementing development interventions tailored to the spatially differentiated needs of specific territories, has the potential to change this situation. This approach enables the creation of a comprehensive framework based on functional connections and provides a relational perspective on space, transcending the boundaries of individual administrative units. The extents of individual FUAs correspond to the varying functional influence zones of specific cities (cores), determined by their size. In contrast, administrative divisions (LAUs) typically cover compact territories whose spatial boundaries result from political decisions and compromises. These units often fail to encompass "functional wholes," within which interlinked socio-economic activities occur, leading to a low correspondence with actual conditions and limited utility (Churski et al. 2023). Understanding FUAs as networks of connections enhances the operational dimension of development policy, improving its effectiveness. Only through this lens can functional relationships be fully identified, their spatial extents determined, and the territorial capital of areas characterized, ultimately allowing for the proper design of development interventions. This creates opportunities to design collaborative efforts among local government units, strengthening the dynamics of development processes. It enables the initiation and implementation of actions tailored to units that are genuinely interconnected, thereby extending beyond the boundaries of administrative units—a principle directly reflected in the latest recommendations concerning the future of EU Cohesion Policy (European Commission: Directorate-General for Regional and Urban Policy 2024, European Commission 2024).

2.3.3 Redefinition of Development Factors

The socio-economic development of inner peripheries, as previously defined, aligns with the assumptions of the body of theories on uneven development. Identifying the unique factors and specific barriers to development in inner peripheries requires tracing the evolution of theoretical approaches to identifying development factors. Initially, economists defined factors of economic growth based on the assumption of a homogeneous space and focused on external factors such as labor, land, and capital (understood as financial resources), drawing on the classical concepts of Smith and Ricardo. These were later adapted for the needs of neoclassical development models, expanded to include factors related to technological progress (Solow 1956), while maintaining their external nature. A breakthrough occurred when these models were applied at the regional level by Richardson (Churski et al. 2021; Richardson 1973), initiating an important shift in the perception of development factors as not only exogenous but also endogenous. This responded to the need to explain the functioning of economies in new conditions shaped by global economic crises. These changes marked a turn from dominant macroeconomic models to microeconomic and regional perspectives, allowing for a better consideration of territorial specificity in explaining the causes of spatial disparities in socio-economic development. This approach was further deepened by the work of Lucas and Romer, who emphasized the importance of endogenizing development processes. Their concepts highlighted not only the role of human, social, and institutional capital as internal drivers of development but also the heterogeneity of space, reshaping the view of territories as differentiated development units rather than homogeneous areas (Capello et al. 2009, 2015; Leon-Ledesma and Thirlwall 2002; Romer 1990). This perspective enabled a more comprehensive approach to socio-economic development, emphasizing the importance of internal resources and potentials within a given territory. A key aspect of this was the focus on smaller-scale spatial disparities, highlighting the significance of the local level. More attention was given to the qualitative nature of development factors, identifying and quantifying social factors. The significance of spatial differentiation in development, shaped by factors such as economies of scale and agglomeration benefits, as well as the diffusion not only of innovations, was recognized. These influences remain relevant despite the increasing pressures of globalization and international economic integration, applying the assumptions of the New Economic Geography (Fujita and Thisse 2009; Krugman 1991b). The New Economic Geography underscored the importance of local conditions and interregional interactions, opposing the neoclassical assumptions of spatial homogeneity. The significance of agglomerations, which can enhance the innovation and efficiency that directly impact regional development, was emphasized (Rahman and Dimand 2020). The development of new institutional economics (Amin and Thrift 1994; Hayter 2004; Martin 2000; North 1990) brought formal and informal social institutions functioning in a given area into the analysis of development factors. Contemporary approaches to identifying and interpreting development factors have highlighted the importance of dynamics, innovation, and networks in regional development. Evolutionary economics and geography emphasized the need to consider the historical dynamics of change to justify the selection of various development pathways (Boschma and Frenken 2011; Martin and Sunley 2006). Relational geography offered a multidimensional view of space, stressing the importance of spatial and social relationships in shaping regional development factors (Bathelt et al. 2004). Behavioral economics pointed to how, influenced by emotions and social context, the bounded rationality of individuals affects economic decisions, increasing the importance of social factors in development processes (Akerlof and Kranton 2010; Storper 1997). Finally, highly prevalent contemporary resilience theories have focused on the capacities of regional systems to adapt to changes and return to equilibrium after various disruptions, aiming to identify the development factors that most strongly shape these patterns (Simmie and Martin 2010) (Fig. 2.3).

Considering the above, the systematization of development factors typically distinguishes five fundamental groups: economic, geographical, socio-cultural, institutional, and technological. This catalog of factors is generally stable; however, researchers, in response to global and local (territorial) conditions of the development process, refine their names (Stimson et al. 2006), redefine their conceptual scope, and reassess their influence on development processes (Churski et al. 2021). In line with the assumptions of the territorially oriented approach, the classical groups of factors began to be treated as new categories of capital (Camagni 2008; Camagni and Capello 2013). In addition to the long-standing financial and material capital, human and social capital were also identified (Romer 1986, 1990, 1994), along with institutional capital (Amin 1999; Williamson 1981).

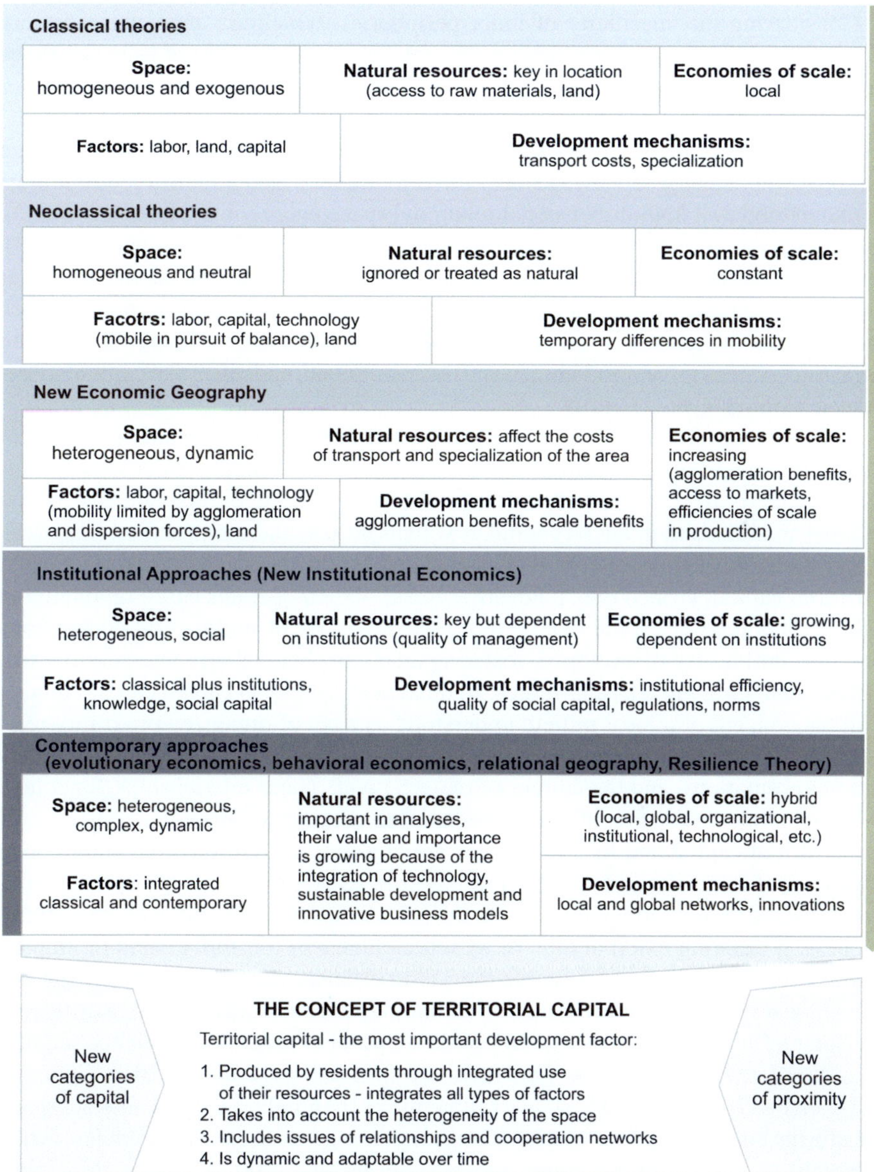

Fig. 2.3 Evolution of research on socio-economic development factors. *Source* Own elaboration

Considering the specificity of inner peripheries, which are characterized more by a lack of relational connections with growth centers than by physical distance determined by geographical location, identifying accurate development factors for these areas becomes a significant challenge. In our view, it is crucial to account for the reinterpretation of the concept of proximity, as discussed in Sect. 2.1. Recognizing that proximity cannot be viewed solely through a spatial-geographical lens but must be understood as a multidimensional relational space undergoing constant evolution determines the necessary actions in this regard (Godart 2012; Nowakowska 2023). Therefore, we believe that any attempt to identify development factors for inner peripheries should be based on Camagni's (2008) concept of territorial capital, which encompasses the full spectrum of relationally connected material (natural, cultural, infrastructural resources) and immaterial (human, social, and relational capital) assets located within a given territory.

These assets can be considered both universal metrics and relevant development factors (Camagni 2017). Camagni and Capello (2013) identify traditional development factors of territorial capital, referred to as the "traditional square," which includes material and financial capital as well as human and social capital (Fig. 2.4). A key element of this concept is the so-called innovative development factors of territorial capital, termed the "innovative cross," which operates based on relational capital. This type of capital can be interpreted as a set of multilateral connections that local entities have developed, fostering an atmosphere of easy interaction, trust, shared behavioral models, and values. In this sense, relational capital can be equated with the concept of a local milieu, understood as a set of organized proximity relationships that connect and integrate the local production system, the system of actors and representations, and the culture of management. These relationships generate a localized dynamic process of collective learning (Camagni 1991).

Territorial capital and its components are closely linked to both the competitive advantage of territories and cooperation-based advantage (Morretta 2021). The material elements of territorial capital form the foundation of a territory's overall attractiveness, generating external effects. Mixed elements of territorial capital ("impure public goods," "club goods") enable long-term efficiency in the economic exploitation of local resources while enhancing technology transfer and knowledge diffusion. The intangible elements of territorial capital support learning and experience acquisition, stimulate knowledge flows, and facilitate joint actions (Camagni 2016; Camagni and Capello 2013; Tóth 2015, 2017). Camagni (2016) emphasized that abstract space transforms into a "place" or "territory" encompassing localized competencies, skills, externalities, relationships, values, representations, and a set of beliefs, principles, and practices. According to the concept of territorial capital, a specific set of territorial principles and assets will induce higher economic outcomes. This approach also suggests that territorial characteristics cannot be replaced or replicated, as each territory has unique conditions for integrating immobile or hard-to-replicate resources while leveraging endogenous and exogenous network effects (Jewtuchowicz 2013; Nijkamp 2016; Nowakowska 2018, 2023; Tóth 2023). Territorial capital is thus socially produced based on the resources of a given area through functional human and institutional relationships. The capital of a given territory is always created by

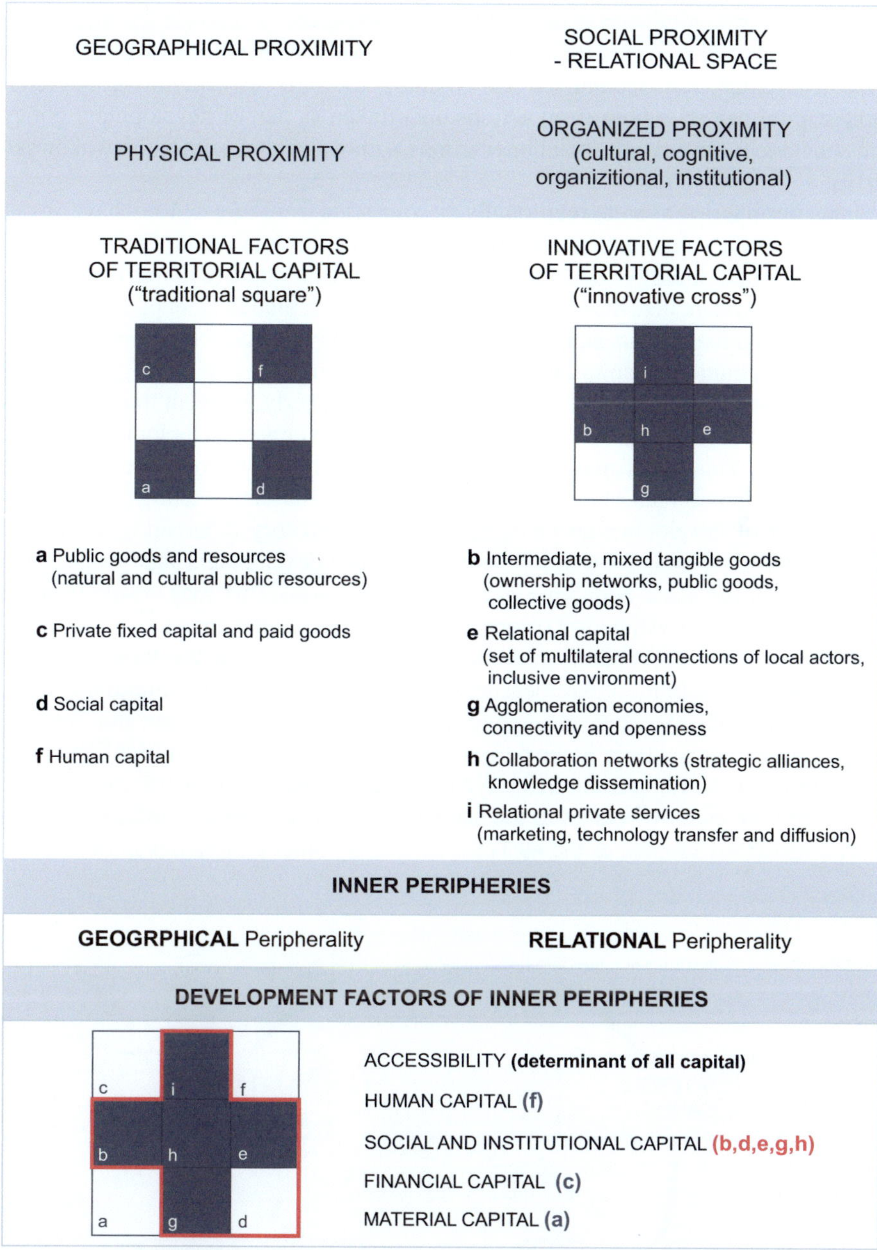

Fig. 2.4 Adaptation of the concept of territorial capital in identifying development factors for inner peripheries. *Source* Own elaboration based on Camagni (2008)

a specific community with a defined culture, resources of knowledge and skills, shaped by historical events that influence its network connections (Morretta et al. 2020; Tóth 2023; van Geenhuizen and Nijkamp 2000). It is worth noting that territorial capital can arise both from actions undertaken as part of development policy and spontaneously through social interactions within a given territory (Nowakowska 2019).

Inner peripheries, as areas relationally disconnected from core regions, have, in our opinion, specific development conditions, which justifies our attempt to modify the well-established catalog of development factors in the literature (Churski et al. 2021). In seeking to systematize the components of territorial capital in inner peripheries, we have identified five factors shaping their development: accessibility, human capital, social and institutional capital, financial capital, and material capital (Fig. 2.5). By a "development factor," we understand an element, property, or event that influences changes in the state of a given territory, that is, exerts an impact on its socio-economic development. This factor may either be located within the territory, constituting an endogenous determinant of development, or in its surroundings, acting as an exogenous driver of developmental changes within the territory (Churski et al. 2021). Consequently, we define a development barrier as a development factor that negatively affects the socio-economic development process. This may result from its absence (element), dysfunction (property), or negative impact (event).

In the subsequent part of this subsection, we characterize each development factor, starting with its definition and measurement method based on the literature. Next, we identify subfactors that, in our view, are specific to each factor and describe their characteristics within inner periphery areas. This explanation highlights why these subfactors are important in analyzing the development of such areas. We also considered the patterns related to the state and influence of these subfactors within FUAs, which we adopted as the basis for delimiting inner peripheries in our study.

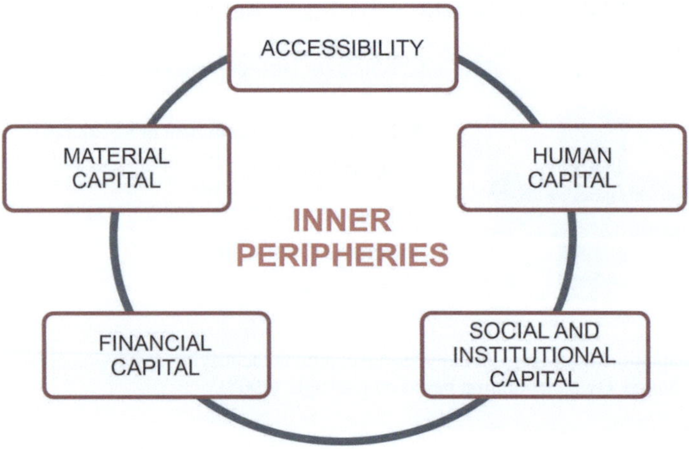

Fig. 2.5 Development factors of inner peripheries. *Source* Own elaboration

Accessibility

Considering the nature of inner peripheries, characterized by their disconnection from core areas—not always in a geographical sense but more often in a relational context—accessibility must be regarded as a key development factor for these areas. Accessibility, present in the concepts of both geographical peripherality and relational peripherality (Sects. 2.2.1 and 2.2.2), is a fundamental concept not only in socio-economic geography and spatial planning but also in the broader social sciences. However, its precise definition is not straightforward and remains a subject of extensive discussion in the literature. As Gould (1969) states, "accessibility" is one of those commonly used terms that everyone employs but not everyone can accurately define or measure. A widely accepted definition in the literature is Hansen's (1959), which defines accessibility as the potential for interaction opportunities. It is important to understand these interactions broadly, encompassing both economic and social dimensions (Handy and Niemeier 1997). They are influenced by various factors, including infrastructure, traffic organization, time, and the personal characteristics of participants (Geurs and van Wee 2004). In the most general sense, accessibility can be defined as the ability to establish (or enable) relationships between more than one element within a set, conditioned by two characteristics (Komornicki et al. 2009):

1. The existence of at least two elements in socio-economic space (absolute or geodetic), which can be unilaterally or mutually accessible and are thus theoretically capable of interacting with each other. These elements constitute the source and target components of accessibility.
2. The presence of a carrier for this relationship, typically identified as a means of communication, most commonly associated with a mode of transport.

It should be emphasized that the realization of these relationships in the real world is hindered by numerous barriers of a physical, political, social, and economic nature.

The diverse ways of defining accessibility result in a varied set of methods for its measurement (Dong et al. 2006; El-Geneidy and Levinson 2006; Geurs and van Wee 2004; Helden 2002; van Wee et al. 2001). The methods of measuring accessibility discussed in the literature primarily focus on transport accessibility. However, it is important to note that contemporary accessibility is also shaped by the degree of infrastructure development in information and communication technologies (ICT) and broadband Internet access, the limitations of which constitute one of the dimensions determining the level of digital exclusion (Ilnicki and Janc 2009; Janc 2019; Warf 2013, 2018). These methods can be organized into four categories most commonly used in research (Komornicki et al. 2009; Rosik 2012):

- Infrastructure-based accessibility measure: Accessibility measured by infrastructural features, using indicators that quantify the amount and quality of infrastructure, such as road and rail network density or the number and type of transport stops (Carruthers et al. 2005; Geurs and van Eck 2001; Spiekermann and Neubauer 2002). It can also include parameters of infrastructure supporting broadband Internet access (Benda et al. 2020; Gould and Shierholz 2020).

- Distance-based accessibility measure: Accessibility measured by physical distance—either straight-line (Euclidean) distance, actual physical distance via transportation networks, travel time, or economic cost to a destination or set of destinations. For example, the average travel time between the origin and other points of interest in the network, such as cities serving as core centers in Functional Urban Areas (e.g., Gutiérrez 2001; Spiekermann and Neubauer 2002).
- Cumulative isochronic-based accessibility measure: Accessibility assessed by estimating the number of destinations reachable within a specific time or cost threshold. This involves analyzing travel-time-based isochrones for destinations reachable within 15, 30, 45, 60, or 90 min from the origin (e.g., Keeble et al. 1982; Spiekermann and Neubauer 2002; Śleszyński et al. 2015).
- Potential-based accessibility measure: Accessibility measured by the potential for interaction between an origin and a set of destinations, using potential models. For example, accessibility is calculated using the Weighted Mean Distance Time (WMDT) index (e.g., Hansen 1959; Komornicki et al. 2008, 2010, 2014; Spiekermann and Schürmann 2007).

The definitional scope of accessibility as a factor in the development of inner peripheries, as understood in this context, encompasses three subfactors: external transport accessibility, internal transport accessibility, and e-accessibility.

A characteristic feature of inner peripheries is their variability in external accessibility. In some cases, this status is relatively favorable, while in others, it exhibits significant deficiencies. On one hand, deficits in this dimension of accessibility—particularly in terms of transport accessibility to FUAs, which structure the identification of inner peripheries, and the relatively long travel times to economic centers—result in a limited range of functional linkages between these areas and the poles of socio-economic growth and development, as well as challenges to, or the impossibility of establishing sustainable connections. This leads to the social and economic marginalization of these areas, potentially stifling or even eroding their developmental potential. On the other hand, overly rapid infrastructure investments to improve external accessibility can lead to the relocation of businesses and workforce to core areas, further deepening the marginalization of peripheral regions (Rodriguez-Pose et al. 2024). These observed dynamics align directly with the dependencies outlined in the concept of geographical peripherality, as discussed in Sect. 2.2.1.

Internal accessibility shapes the internal cohesion of inner peripheries, determined by the degree of heterogeneity within their spaces and the opportunities for developing relationships between the central node of a functional area and its surroundings. A common feature of internal accessibility in these peripheries is its low level. This is influenced not only by the condition of the transport network infrastructure but also, to a greater extent, by how it is organized to facilitate public transport services. Equally important is the capacity to foster relationships among stakeholders in developmental processes, emphasizing the importance of not only geographical peripherality but also organized proximity (Sect. 2.2.2).

E-accessibility, or accessibility in terms of electronic communication, is determined by access to broadband Internet. In this respect, inner peripheries exhibit significant deficiencies. This affects both their central hubs and, particularly, their outer areas within the boundaries of FUAs. E-accessibility in inner peripheries can enhance external accessibility by providing residents with opportunities for remote work in companies located in geographically distant regions. Local entrepreneurs may engage in cooperative networks with an expanded range of contacts. Furthermore, e-accessibility can also improve internal communication efficiency, enabling residents outside the urban core of a functional area to access e-services and e-administration. It creates opportunities to overcome the limitations imposed by geographical peripherality (Sect. 2.2.1) by fostering conditions for the development of organized proximity (Sect. 2.2.2).

Human Capital

The most important development factor shaping the territorial capital of any areas is its residents. Their characteristics, which can be systematized as human capital, play a key role in creating spaces that serve as relational developmental environments. Deficits in this development factor, particularly in the current era dominated by a knowledge-based economy and the strong pressures of globalization and economic integration, significantly influence the formation of spatial disparities in development, contributing to the emergence of inner peripheries. Human capital is also critically important for shaping territorial capital. It should be noted that residents, by utilizing the resources of a territory, create a relational developmental space that determines the quality of territorial capital (Jewtuchowicz 2013; Nijkamp 2016; Nowakowska 2018, 2023; Tóth 2023). Human capital is understood as the resource inherent in individuals and is associated with their knowledge, creativity, skills, abilities, and health (Becker 2009; Ployhart et al. 2014). It is one of the traditionally recognized drivers of socio-economic development (Becker 1990, 1993; Coleman 1988; Faggian and McCann 2009; Fukuyama 1997; Mahroum 2000; McDonald and Roberts 2002). Residents who exhibit knowledge, skills, and experience have a significant impact on forming unique functional linkages, which is particularly important for the developmental processes occurring within FUA, where inner peripheries are identified. Their collective resources can contribute to grassroots developmental processes (Marginson 2019). The state and structure of human capital are influenced by demographic factors, while its quality is conditioned by the educational level of the population. Additionally, the utilization of human capital is tied to the equilibrium of the labor market, which indicates the extent of human capital's engagement in economic processes (Churski 2008; Crescenzi et al. 2016).

The measurement of the level and quality of human capital is typically conducted using the following methods:

- **Cost-based and income-based methods**: Cost-based methods focus on evaluating the investments made in the development of human capital, assuming that the value of human capital equals the total costs incurred for its development. Income-based methods, on the other hand, posit that human capital is a source of future income, and its value corresponds to the discounted value of future earnings (Becker 1993; Mincer 1974; Schultz 1961; Trinh et al. 2005).

- **Educational indicators**: These are the most widely used methods for estimating human capital. They typically measure the number of years of education, the percentage of the population with higher education, and the number of graduates employed in innovative sectors. These indicators are commonly applied by institutions such as the OECD and the World Bank (OECD 2018, 2023; World Bank 2020a).
- **Health indices**: These include metrics such as life expectancy, health status, and access to healthcare services (World Bank 2020a). The quality of human capital in the health domain—and consequently, the economic efficiency of a community—is influenced by factors such as demographic structure and the youth potential of the labor force (William and Lewis 2009).

The definitional scope of human capital as a factor in the development of inner peripheries, as understood in this context, encompasses four subfactors: demographic situation, health and safety, education, and the labor market, which allows for evaluating the utilization of human capital.

Demographic situation strongly influences contemporary developmental processes. In inner peripheries, there is a concentration of negative demographic trends. Rapid population decline is accompanied by a deterioration in the biological structure. Peripheral areas experience an outflow of population, particularly among younger age groups, leading to disruptions in demographic structure and challenges in the local labor market. Depopulation occurs concurrently with population aging. These phenomena affect the outer areas within the boundaries of FUAs more significantly than their cores. Unfortunately, in the case of inner peripheries, these challenges also often pose development issues for many central hubs in these areas.

The quality of human capital, in line with the concept of foundational economy (Sect. 2.2.3) and the concept of social justice in a territorial dimension (Sect. 2.2.4), is determined by the population's sense of security. This is understood in various dimensions, including economic (sufficient financial resources for living), health-related (threats to life), and property-related (natural disasters, crime) (Hastings 2009). In inner peripheries, worsening access to healthcare is observed. Trends in population decline reduce the tax base for local governments, increasing the per-unit costs of delivering public services and ensuring public safety. These negative phenomena affect both core and outer areas within FUAs identified as inner peripheries, but their intensity is generally greater in outer areas than in central areas.

Access to education shapes the educational attainment of the population. The skills and professional competencies of residents form the foundation for economic development and directly impact wages and worker productivity (Stiglitz et al. 2009). Better-educated individuals are less likely to face unemployment; they enjoy better health, have stronger social connections, and exhibit greater civic engagement. In inner peripheries, students cannot access the same quality and diversity of educational offerings as those available in developmental cores. A similar disparity exists for higher education and lifelong learning opportunities. These disparities affect both inner relationships within the boundaries of inner peripheries structured as FUAs and the relationships between relatively strong and weak FUAs. Access to education is a fundamental condition for shaping residents' developmental opportunities, which, at a basic level, must be accessible to all, as emphasized by the foundational economy (Sect. 2.2.3) and territorial justice concepts, which argue that such access should be ensured at a comparable level for everyone (Sect. 2.2.4).

Access to educational services and other characteristics of human capital directly impact the labor market, which serves as an indicator of the extent to which labor resources are engaged in economic processes. A mismatch between qualifications structure and employer demand leads to increased unemployment, causing numerous socio-economic challenges. Unfortunately, relatively high unemployment rates are typical of inner peripheries. However, there are noticeable differences in this regard. The situation in central areas of FUAs within inner peripheries is better compared to outer areas within FUA boundaries. This is a direct consequence of the concentration of jobs in the central hubs of these areas.

Social and Institutional Capital

A particularly significant factor in the development of any territory, especially when viewed through the lens of territorial capital that emphasizes the importance of the relational developmental environment created by its residents, is social capital. Its role in the development of inner peripheries is exceptional due to its importance in shaping relational connections, which are the foundation of building organized proximity (Sect. 2.2.2). A deficit or dysfunction in these connections is one of the causes of the emergence of inner peripheries. Social capital is understood as a network of relationships between people that enables the effective functioning of the entire community, fundamentally relying on the utilization and development of organized proximity. When analyzing social capital, it is essential to distinguish between two types: bonding social capital and bridging social capital. Bonding social capital refers to strong connections between like-minded individuals. These are relatively insular communities that are internally supportive and cohesive but may lack external adaptability. Such strong ties can hinder the ability of these communities to adapt to changing external developmental conditions (Cortinovis et al. 2017; Knack and Keefer 1997). Bridging social capital, on the other hand, relates to inclusive social structures that build trust between heterogeneous groups and facilitate the diffusion of knowledge. This type of capital can increase the likelihood of developing new specializations and support socio-economic development processes (Cortinovis et al. 2017).

Social capital is perceived as a collection of informal social institutions (Bourdieu 1986; Coleman 1988; Putnam 1993). Alongside it, we distinguish institutional capital, which refers to formal institutions (Hall and Soskice 2001; Knight 1992; North 1990; Ostrom 1990). Institutions are defined, following Hodgson (2006), as "durable systems of socially embedded, established, and sustained social rules that structure social interactions." When institutions—sets of norms, rules, and patterns of behavior—emerge spontaneously in the course of social relations, they form the core of an area's social capital. However, when institutions are consciously constructed with the intent of coordinating actions (e.g., through legal regulations) to achieve specific goals, they are referred to as "organizations," which may have economic and/or public functions (Górniak 2013; North 1991). These organizations form the foundation of an area's institutional capital. We consider social capital and institutional capital to be inherently interconnected, leading us to integrate them in our approach. Social capital is recognized as a beneficial characteristic of social organizations, rooted in mutual trust, shared values, and the development of networks of cooperation based on solidarity and social engagement (Hellerstein and Neumark 2020; Kwiatkowski 2023). Such networked relationships can facilitate the mobilization of local social and economic resources and contribute to the transfer of knowledge, which is more efficiently disseminated when stakeholders are embedded within a system of social relationships (Cortinovis et al. 2017; Echebarria and Barrutia 2013; Putnam et al. 1993). Meanwhile, institutional capital refers to formal social institutions, which may be public or private. The quality of public institutions in managing developmental processes reflects the quality of institutional capital. Rodriguez-Pose et al. (2024) demonstrate that weak institutions, particularly at the local government level, are a major cause of persistent developmental inequalities, including inner peripherality. The quality and efficiency of local governments directly influence local development by stimulating interactions among economic actors, potentially leading to increased productivity, employment, and labor efficiency, as well as reduced transaction costs (Ganau and Rodriguez-Pose 2019; Kwiatkowski 2023). Effective local public institutions promote innovation, which is crucial for the development of areas with weak growth rates (Rodriguez-Pose and Ketterer 2020). They also foster a social climate of entrepreneurship in the region, which is as important as the availability of infrastructure (Audretsch and Belitski 2017; Martin 2000). High-quality formal institutions also act as intermediaries for local and regional public investments, securing and distributing funds under the EU Cohesion Policy. They influence the effectiveness of interventions, particularly in developing local human capital, promoting and diffusing innovation and knowledge, and creating a social atmosphere of trust in institutions and within the community (Rodriguez-Pose et al. 2024). Formal social institutions include public entities such as local governments and private organizations, namely businesses aimed at achieving specific profit-oriented goals (Gitling 2013). Their proper functioning relies on being embedded in economic networks. Economic development requires a network of institutions cooperating with businesses. Business-supporting institutions include public, private, or hybrid entities that create favorable conditions for entrepreneurship. In peripheral areas, the activities of business-supporting institutions are often weaker or heavily dependent on external support. Denser clusters of such institutions often correlate with higher levels of territorial production organization, leading to synergistic effects within the local economy. Ahlerup et al. (2009) argue that, during the development process, a substitution effect can be observed between formal and informal institutions, with social capital being particularly important when formal institutions are weak. Similarly, Rodriguez-Pose (2013) asserts that, when effective regulatory frameworks are introduced in a region, the economic demand for informal institutions diminishes significantly.

Social and institutional capital are multidimensional constructs operating on various interconnected levels that influence and interact with one another. Measuring social capital depends on quantifying elements of this capital that are challenging or, in some cases, impossible to quantify in a specific social context (Putnam 2000). A key dilemma is whether to measure access to resources or their utilization; whether to measure potentially existing social capital that is accumulated and mobilized when needed or the capital that is actively used (Snijders 2002). As a consequence:

- Social capital metrics focus on social activity and engagement in local initiatives (Putnam 2000), analysis of the structure and frequency of social interactions (network intensity) (Bourdieu 1986), or direct studies of social trust in neighbors, coworkers, or local authorities (World Bank 2019).
- Institutional capital metrics pertain to efficiency, understood as cross-sectoral cooperation and the effectiveness of public institution expenditures (North 1990). Other indicators include legislative stability and public trust in the legal system (World Bank 2019).

In our opinion, the assessment of the significance of social and institutional capital as a factor in the development of inner peripheries should consider four subfactors: social activity, use of leisure time, business infrastructure, and economic structure and activity. Due to the lack of comprehensive data, their identification must often rely on the use of indirect indicators for measuring social and institutional capital.

Social activity is understood as grassroots initiatives driven by trust and residents' need to organize and collaborate. Its level is lower in inner peripheries than in development centers. Within the framework of FUAs, it is notable that their bonding social capital often remains high, while bridging social capital is frequently lacking. This reinforces the insularity of these communities, limiting their openness to the diffusion of various innovations. Consequently, this significantly hampers the development of organized proximity (Sect. 2.2.2) in an external dimension, which, in turn, impacts the developmental potential of the entire FUA classified as inner peripheries.

An analysis of how leisure time is spent within a community also provides insights into the social relationships in the studied area, helping to evaluate the quality of its social capital. This is reflected in the limited involvement of local communities in creating and participating in non-governmental organizations, a typical characteristic of inner peripheries. Deficiencies in this area hinder the strengthening of the social and institutional capital necessary to develop relationships that mitigate the barriers caused by geographical peripherality (Sect. 2.2.1).

The quantification of the quality of social and institutional capital should also, in our view, include an evaluation of the functioning of business support institutions, the presence of firms from the creative sector, and the economic activity of the population. In inner peripheries, both existing and newly registered businesses are primarily engaged in basic production and service activities, which are vital for the functioning of the foundational economy (Sect. 2.2.3). Micro, small, and medium-sized enterprises dominate the business structure, with a noticeable lack of large companies and firms from the creative sector. Under these conditions, the insufficient availability of business support services—which are crucial for fostering economic activity among residents—further exacerbates the developmental challenges. It is important to emphasize that these patterns exhibit internal variations within FUAs categorized as inner peripheries. Central areas within these regions tend to show a concentration of economic activity, while outer areas within FUAs face significant deficits. Economic activity in these peripheral zones largely relies on individual agriculture. Additionally, the rate of new business formation in inner peripheries is much lower than in more developed FUAs. Overcoming these challenges is made more difficult by the previously mentioned limitations related to the characteristics of bridging social capital.

Financial Capital

We consider financial capital as a traditional factor of development (Richardson 1973; Solow 1956) that builds the resources of territorial capital in any area, including in inner peripheries. The spatial differentiation in its availability significantly shapes the conditions for achieving territorial justice, which aims to guarantee similar developmental opportunities for all residents (Sect. 2.2.4). Financial capital is understood as financial resources available to public entities, local governments, as well as private enterprises and individual residents. It encompasses both equity financing (including own and external sources, such as foreign direct investment or public policies of the European Union [Churski 2008]) and debt financing, which can be allocated toward acquiring assets, expanding material capital, or supporting private consumption. As such, financial capital constitutes a critical resource underpinning every aspect of development processes, influencing the mitigation of geographical peripherality (Sect. 2.2.1), facilitating the establishment of organized proximity (Sect. 2.2.2), and creating conditions conducive to the functioning of the foundational economy (Sect. 2.2.3).

The conceptualization of financial capital in the literature is most commonly based on the analysis of financial situation (Schinasi 2004) and financial condition (Siddik and Kabiraj 2018). Financial situation is defined as the ability of an entity (public or private) to manage its finances in a way that ensures the fulfillment of current and future financial obligations. It is a key indicator of economic stability and a fundamental component of credit risk assessment and financial planning (Schinasi 2004, 2005). Financial condition, on the other hand, refers to the systemic capacity of entities to meet their financial obligations in a stable and efficient manner (Siddik and Kabiraj 2018). The main aspects related to financial condition include (Maher et al. 2020; Stanny and Strzelczyk 2018): the ability to meet financial obligations on time, an adequate level of taxation, budgetary balance, and a limited role of debt in financing activities. The primary criterion distinguishing the analysis of financial situation from financial condition is the time perspective: short-term and operational for financial situation, versus long-term and strategic for financial condition. In our research, for the purposes of delimitation, we utilized the assessment of financial situation, referring to the state of finances at a specific point in time. Elements of financial condition were analyzed by examining the trajectories of identified and delimited Functional Urban Areas (FUAs). This approach allowed us to evaluate trends in the development of individual indicators across FUA groups, including the inner peripheries, which were the primary focus of our study.

The analysis of financial capital in the literature typically addresses the financial situation and condition of local governments, businesses, and residents. In local governments, financial capital is used to finance public services, implement infrastructure projects, and improve the quality of life for residents. The ability of local authorities to effectively utilize financial capital directly impacts the extent to which local community needs are met, the organization of spatial development, and the execution of development projects that can enhance residents' quality of life in the long term (Satoła 2015; Sobko et al. 2021). Local government units are also responsible for a significant portion of public investments, often of supra-local importance. Consequently, their activities, shaped by their financial capabilities, have a direct effect on the country's economic situation (Swianiewicz and Łukomska 2020). For businesses, financial capital refers to the resources available for investments, operations, and growth. This also encompasses the ability to generate profits, manage costs, and maintain financial liquidity. Financial capital is crucial for business innovation and growth, and its effective management enhances enterprise competitiveness (Dziekański and Prus 2020). The financial stability of businesses is essential for the fiscal health of local governments, as it influences the level of tax revenues, which in turn fund local budgets. These funds are used to improve residents' quality of life and strengthen investment attractiveness (Feld et al. 2013). Businesses with strong financial positions have greater investment capacity, which contributes to an increase in local production value-added, employment growth, and improvements in residents' living standards (Gunnlaugsson 2017; Yakymova and Kuz 2019). Businesses also create jobs, and high levels of employment contribute to higher household incomes and may increase demand for locally produced goods and services (Delfmann and Koster 2016). The financial capital of residents refers to their income, savings, and access to credit. A high level of financial capital among residents is essential for meeting basic needs and investing in education, health, and housing, which are crucial for personal and societal development. Limited access to financial capital can lead to poverty and social marginalization (Pawlik and Karpińska 2020; Gunnlaugsson, 2017). Conditions in this domain influence the potential for foundational economic development (Sect. 2.2.3) and the achievement of territorial equity (Sect. 2.2.4).

Indicators that can be used to assess the financial situation include measures describing the liquidity, profitability, and credit/debt policies of individual entities and organizations (Coss-Mandiola et al. 2022; Dziekański and Prus 2020; Filipiak 2016; Lukáč et al. 2021; Standar and Kozera 2019). These categories may encompass the following:

- **Financial resources and income:** Own revenues, current revenues, property revenues, EU funds, other non-repayable funds, budget surplus, annual revenue value, net profitability, share of equity in financing activities, average household income, and savings levels;

- **Access to financing:** Number of projects co-financed from external sources, value of grants obtained, number of venture capital projects, average size of investment or consumer loans, and average creditworthiness;
- **Investments and development:** Share of investment expenditures in the budget, investment spending per capita, value of completed R&D projects, average expenditures on technology development, and spending on education, training, and skills enhancement.

In our understanding, the analysis of financial capital as a factor in the development of inner peripheries encompasses three subfactors: the financial situation of local government units, the financial situation of enterprises, and the financial situation of residents.

Inner peripheral areas are characterized by limited financial resources and flows across all groups of entities operating within these regions. When examining inner peripheries within the framework of Functional Urban Areas (FUAs), it is evident that internal differentiation exists. The financial situation and condition in their central areas are better compared to the outer zones within the boundaries of such FUAs. Nevertheless, this does not alter their overall weaker position in terms of financial capital when compared to more developed FUAs.

The financial situation of local governments in inner peripheral areas is weaker due to lower own-source revenues, a high dependence on external financing, and limited creditworthiness. Low general subsidies for local government units, combined with their restricted credit capacity, hinder their ability to generate the required co-financing for securing grants from external sources. Consequently, these local governments are unable to independently fund both operational and investment expenditures. The challenging financial situation of municipalities within FUAs can impede natural cooperation processes, such as co-financing investments or jointly applying for funding. As a result, local governments in peripheral areas often struggle to provide an adequate level of public services or to implement developmental projects, reducing both the investment attractiveness of the area and the quality of life for its residents.

The financial situation of enterprises is a reflection of their market position and their ability to effectively achieve competitive advantages. In inner peripheries, this situation exhibits internal differentiation, consistent with the previously identified patterns. Enterprises concentrated in the cores of FUAs are in relatively better financial condition, while those that are dispersed and economically weaker, located in the outer zones within FUAs classified as inner peripheries, face relatively poorer financial conditions. A weak financial position also affects the ability of businesses to undertake investment challenges and secure credit commitments, which are often essential for carrying out production tasks or expanding service activities. The inability to make investments financed by own capital, coupled with limited creditworthiness, becomes a barrier to business development, leading to stagnation or even the liquidation of enterprises. This results in fewer job opportunities, which directly impacts the financial situation of residents. The negative financial situation of enterprises not only limits the opportunities available to residents but also directly affects the revenues of local governments through their share in local taxes and fees. Enterprises with limited financial capital have a minimal impact on regional economic growth, further entrenching the marginalization of peripheral areas. It is worth noting that the financial condition of enterprises significantly affects the level of tax revenues available to local governments and, consequently, their ability to perform public tasks.

The financial situation of residents is typically weaker in peripheral areas. Wages in peripheral regions are generally lower than in economic centers, limiting both consumption and savings capacity. Banks and financial institutions show less interest in financing activities in areas with low economic activity. The lack of financial capital hinders investments in education, healthcare, and housing. The financial situation of residents is closely tied to that of businesses and local governments. Taxpayers with low purchasing power, caused by low wages and incomes from economic activities, do not form a robust or effective tax base for local governments. Insufficient financial resources among residents constrain consumption levels, which negatively impacts the revenues of local entrepreneurs. The lack of access to better-paying jobs within commuting distance—both within and between FUAs—restricts opportunities for economic improvement, weakening the reach and strength of functional connections. This situation adversely affects efforts to reduce geographical peripherality (Sect. 2.2.1) and limits opportunities to strengthen organized proximity (Sect. 2.2.2).

The outlined patterns highlight significant feedback loops among the identified subfactors of financial capital, which critically shape developmental opportunities. These opportunities pertain to, among other aspects, reducing geographical peripherality (Sect. 2.2.1), fostering organized proximity (Sect. 2.2.2), creating conditions for the functioning of the foundational economy (Sect. 2.2.3), and, ultimately, establishing the foundations for achieving territorial equity (Sect. 2.2.4).

Material Capital

Material capital is a traditional type of capital and should be regarded as the most classical factor in the development of any area (Dunford 1988; Keynes 1936; Schumpeter 1934; Solow 1956). Its impact on the spatial differentiation of development processes has undergone significant transformations (Churski et al. 2021). In highly developed areas, where material capital is most abundant, environmental components of material capital are gaining increasing importance. However, in less-developed regions, including inner peripheries, deficits and the low quality of its infrastructural components remain the most critical factors (Pike et al. 2023). Material capital is understood as the collection of all material goods of natural and anthropic origin that humans use in production or activities aimed at improving the quality of life for residents (Churski et al. 2018; Hines 2000; Pike et al. 2016). The quality of material capital thus influences both the mitigation of geographical peripherality (Sect. 2.2.1), the functioning of the foundational economy (Sect. 2.2.3), and the ability to achieve territorial justice (Sect. 2.2.4).

The state and dynamics of material capital are under significant pressure from the contemporary process of settlement deconcentration, driven by centrifugal forces (influenced by greater availability and economic attractiveness of areas surrounding core regions) counteracted by centripetal forces that encourage settlement in core areas (Colby 1933). In terms of anthropogenic material goods, extensive land-use forms are significant, as they increase spatial use costs and reduce access to market and public services (Bontje 2004; Couch et al. 2005). Deconcentrated development necessitates the construction of new transportation routes and the organization of public transport and poses challenges such as pollution, road safety issues, and losses associated with greater travel distances. Deconcentration may also weaken social ties. These trends hinder the effective functioning of the foundational economy (Sect. 2.2.3) and weaken the ability to achieve territorial justice (Sect. 2.2.4). Urbanized areas with concentrated development, however, benefit from relatively lower unit costs for infrastructure construction due to economies of scale. They also have a higher density of housing stock, greater population density, higher market absorption, stronger consumer purchasing power, and better access to the labor market. This creates conditions for more rational land management. Nevertheless, these areas are not without development challenges, such as the rapidly growing issue of congestion. The state and quality of material capital significantly shape living conditions, which, according to the concept of territorial justice (Sect. 2.2.4), greatly influence both the material and immaterial aspects of quality of life. Material conditions are relatively easy to measure objectively and are crucial for assessing social progress and quality of life. They have been recognized by the Stiglitz Commission as the first of eight dimensions of well-being (Stiglitz et al. 2009). Their condition influences the achievement of spatial justice (Sect. 2.2.4). A crucial component of material capital is natural material goods, which determine environmental quality. Their importance stems not only from the need to pursue sustainable development but also from their direct impact on quality of life (Churski and Perdał 2021; Rudolf et al. 2002). In line with the principles of spatial justice (Sect. 2.2.4), residents should have equitable access to environmental services, such as clean water, air, forests, and recreational areas. Green spaces are essential for health-promoting, recreational, and sports activities, particularly in urban areas, and for maintaining biodiversity in natural environments. Protected areas enhance opportunities for contact with nature, positively affecting mental health—a critical factor in urbanized contexts with rising stress levels (Liotta et al. 2020). A healthy environment increases life expectancy, reduces morbidity, and improves subjective well-being (ESPON QoL 2021).

The measurement of material capital is conducted using indicators related to:

- **Investment in infrastructure**: This includes expenditures on technical, social, and housing infrastructure (OECD 2018).
- **Availability and quality of infrastructure**: Indicators such as the density of transportation networks, energy capacity and efficiency, access to high-speed Internet, the number of hospital beds, and housing standards, which impact the availability of public services (World Bank 2020a, b).

- **State of the natural environment and its resources**: This includes air quality, water quality and availability, soil quality, the share of green spaces, biodiversity indicators, and carbon footprint (United Nations 2021; Global Footprint Network 2019).

Urban development often creates pressure for green spaces to be repurposed for construction, leading to potential conflicts of interest. Local authorities may prioritize development to increase tax revenues and attract investment, sometimes at the expense of environmental protection, thereby lowering ecological standards (Xu et al. 2022). However, exploitative human interference in environmental resources results in global and local changes. Global climate change leads to natural disasters that damage property and threaten lives at a local scale. The increasing scale and frequency of these events significantly influence decisions about where people choose to live and conduct business (Eurostat 2017; Knopman et al. 2015; Pukelienė and Starkauskienė 2015).

The definitional scope of material capital as a factor in the development of inner peripheries, as understood here, encompasses three subfactors: quality of land use, housing conditions, and ecosystem goods and services.

Inner peripheries are areas lacking the economic foundations necessary to initiate developmental processes due to deficiencies in technical and social infrastructure. On one hand, these areas exhibit multidimensional deficits in the state and quality of material capital compared to more developed regions. On the other hand, within FUAs, inner peripheries display notable internal disparities. Infrastructural components of material capital tend to be concentrated and of relatively higher quality in core areas, whereas outer areas within the boundaries of FUAs categorized as inner peripheries exhibit significant deficiencies, greater dispersion, and lower quality. Conversely, environmental components of material capital often display an opposite distribution. Their condition and quality are relatively poorer in the core areas of inner periphery FUAs, whereas outer areas within these FUAs are more attractive in this regard. These outer areas often possess more abundant and diverse environmental resources in relatively better condition, allowing for greater utilization in strengthening ecosystem services that improve quality of life for residents. Overall, inner peripheries demonstrate a relative advantage in environmental components of material capital compared to more developed areas. This competitive advantage became particularly pronounced during the SARS-CoV-2 pandemic and could present a significant development opportunity for these areas.

We consider quality of land use to be a key subfactor of material capital. Its importance has increased in the context of observed suburbanization processes, which, in the absence of adequate legal regulations, can lead to increasing spatial chaos described in the literature as "urban sprawl" (Kudłacz and Karwińska 2021; Lityński 2023; Rosati 2025). This phenomenon has significant economic and social costs, and it poses a real threat to the environmental components of material capital. In inner peripheries within FUAs, land-use quality is typically low. This is due to degraded central area buildings with low levels of revitalization and chaotic construction in outer areas, which lack spatial order and threaten natural resources. These trends significantly influence the proper functioning of the foundational economy (Sect. 2.2.3) and spatial justice (Sect. 2.2.4).

Housing conditions, a critical subfactor of material capital, directly determine residents' quality of life. Addressing housing needs is a fundamental requirement for achieving territorial justice (Sect. 2.2.4). Unfortunately, inner peripheries are characterized by inadequate housing stock and low rates of growth in both quantity and quality. This deficit is exacerbated by the poor condition of existing housing, with many units, even in the central areas of inner periphery FUAs, lacking access to heating and sewer systems. In core areas, the average living space per person is often significantly smaller than in the cores of more developed FUAs. In outer areas of inner periphery FUAs, residential construction tends to be more dispersed than in the outer areas of economically stronger FUAs, increasing economic and social costs associated with spatial chaos.

Changes in material capital as a development factor emphasize the importance of including ecosystem goods and services as a subfactor. This is particularly relevant for inner peripheries adapting to post-SARS-CoV-2 conditions. Many inner peripheries offer residents access to extensive, often forested green spaces. While such areas improve air quality, they may also limit economic activity, affecting the local labor market and property management freedom. Not all inner peripheries exhibit excellent environmental quality. Many are impacted by the effects of low emissions due to residential development patterns. In central areas of inner periphery FUAs, individual coal-fired heating systems are common, while in dispersed outer areas, a variety of fuels, including waste, are used for heating. The low level of resident activity in obtaining funding for home insulation and modernization exacerbates these issues.

The characterization of development factors, their current state, and their impact on inner peripheral areas identified within the framework of FUAs allows us to refine the catalog of factors by specifying subfactors. This approach forms an analytical framework applicable to the characterization of development processes in inner peripheries within our study (Fig. 2.6). We utilize this framework both during the delimitation and typology stages of inner peripheries and during field research conducted in case study areas. The aim is to identify and explain the development processes of these areas, taking into account the socio-economic impacts of the SARS-CoV-2 pandemic.

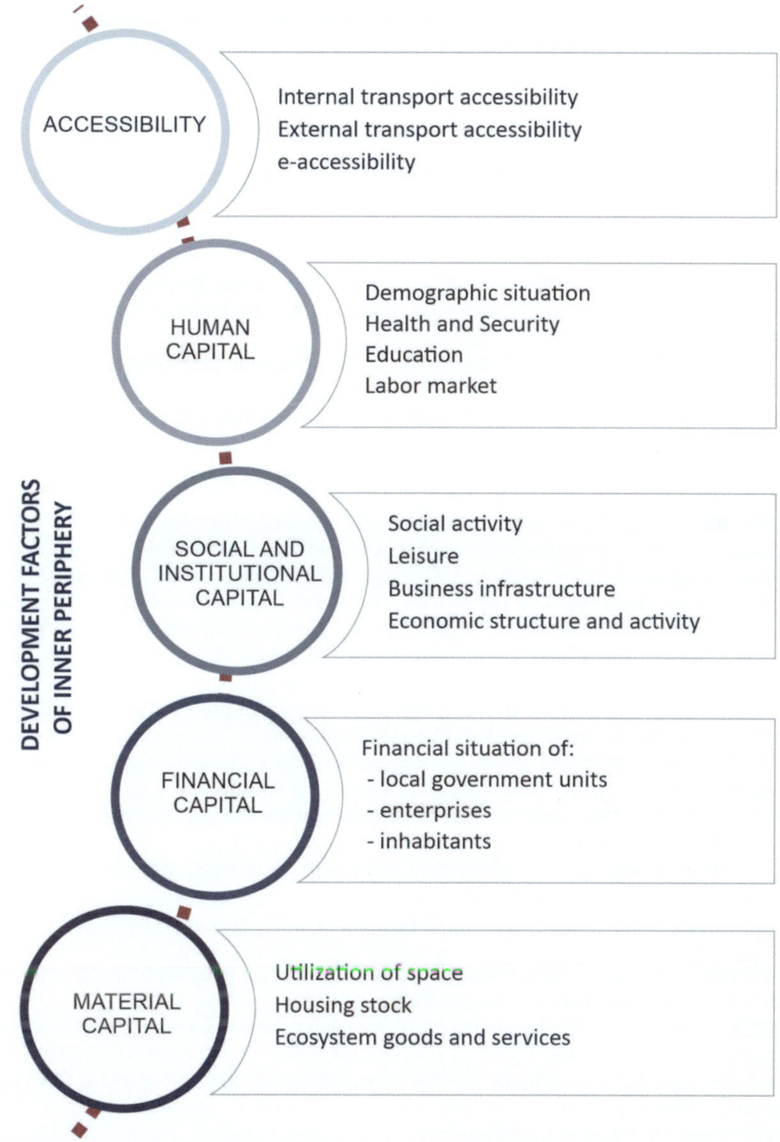

Fig. 2.6 Factors and subfactors of development in inner peripheries. *Source* Own elaboration

References

Agnew J (2021) Immanuel Wallerstein, the "modern world-system," and radical human geography. Hum Geogr 14(1):17–30. https://doi.org/10.1177/1942778620974056

Ahlerup P, Olsson O, Yanagizawa D (2009) Social capital vs institutions in the growth process. Eur J Polit Econ 25:1–14

Akerlof GA, Kranton RE (2010) Identity economics: how our identities shape our work, wages, and well-being. Princeton University Press
Amin A (1999) An institutionalist perspective on regional economic development. Int J Urban Reg Res 23:365–378. https://doi.org/10.1111/1468-2427.00201
Amin A, Thrift N (1994) Globalization, institutions, and regional development in Europe. Oxford University Press
Ardener E (2012) Remote areas: some theoretical considerations. J Ethnogr Theory 2(1):519–533
Arneson R (2008) Rawls, responsibility, and distributive justice. In: Fleurbaey M, Weymarkin JA (eds) Justice, political liberalism, and utilitarianism: themes from Harsanyi and Rawls. Cambridge University Press, pp 80–107
Arrow K (1962) Economic welfare and the allocation of resources for invention. In: The rate and direction of inventive activity: economic and social factors. Princeton University Press, pp 609–626
Asheim B (1996) Industrial districts as learning regions: a condition for prosperity? Eur Plan Stud 4:379–400
Audretsch T, Belitski M (2017) Entrepreneurial ecosystems in cities: establishing the framework conditions. J Technol Transf 42(5):1030–1051. https://doi.org/10.1007/s10961-016-9473-8
Bajerski A (2008) Problemy wydzielania peryferii społeczno-gospodarczych. Ruch Prawniczy Ekonomiczny i Socjologiczny Rok LXX, Z. 2:159–167
Balland PA, Boschma R, Frenken K (2014) Proximity and innovation: from statics to dynamics. Reg Stud 49(6):907–920. https://doi.org/10.1080/00343404.2014.883598
Barca F (2009) An agenda for a reformed Cohesion Policy: a place-based approach to meeting European Union challenges and expectations. Independent report prepared at the request of Danuta Hübner, Commissioner for Regional Policy. European Commission. https://www.europarl.europa.eu/meetdocs/2009_2014/documents/regi/dv/barca_report_/barca_report_en.pdf
Barro RJ, Sala-i-Martin X (1995) Economic growth. McGraw-Hill
Barry F (1996) Peripherality in economic geography and modern growth theory: evidence from Ireland's adjustment to free trade. World Econ 9(3):345–365
Bathelt H, Malmberg A, Maskell P (2004) Clusters and knowledge: local buzz, global pipelines and the process of knowledge creation. Prog Hum Geogr 28(1):31–56
Becattini G (1979) Dal settore industriale al distretto industriale. Alcune considerazioni sull'unità di indagine dell'economia industriale. Il mulino
Becker GS (1993) Human capital: a theoretical and empirical analysis, with special reference to education. University of Chicago Press
Becker GS, Murphy KM, Tamura R (1990) Human capital, fertility, and economic growth. J Polit Econ 98(5, Part 2):S12–S37
Becker GS (2009) Human capital: a theoretical and empirical analysis, with special reference to education, 3rd ed. University of Chicago Press
Benda NC, Veinot TC, Sieck CJ, Ancker JS (2020) Broadband internet access is a social determinant of health. Am J Public Health 110(8):1123–1125. https://doi.org/10.2105/AJPH.2020.305784
Bentham J, Bowman A, De La Cuesta M, Engelen E, Ertűrk I, Folkman P, Froud J, Johal S, Law J, Leaver A, Moran M, Williams K (2013) Manifesto for the foundational economy (CRESC Working Paper 131). Centre for Research on Socio-Cultural Change, University of Manchester
Bentley G, Pugalis L (2014) People-centred models: active regional development, space-blind policies and place-based approaches. Local Econ 29(4–5):283–294
Bernard J, Šimon M (2017) Vnitřní periferie v Česku: Multidimenzionalita sociálního vyloučení ve venkovských oblastech. Czech Sociol Rev 53(1):1–25
Berry BJL (1973) Growth centres in the American urban system. Ballinger
Berry BJL (1967) Functional economic areas and consolidated urban regions of the US. Social Sciences Research Council
Bolton M, Day R, Leach M (2020) England's overlooked neighbourhoods: defining, understanding and regenerating 'left behind' communities. J Urb Regen Renew 12:116–123. https://www.ingentaconnect.com/content/hsp/jurr/2019/00000013/00000002/art00002

Bontje M (2004) From suburbia to post-suburbia in the Netherlands: potentials and threats for sustainable regional development. J Housing Built Environ 19(1):25–47
Boschma R (2005) Proximity and innovation: a critical assessment. Reg Stud 39(1):61–74. https://doi.org/10.1080/0034340052000320887
Boschma R (2015) Towards an evolutionary perspective on regional resilience. Reg Stud 49:733–751
Boschma R, Frenken K (2011) Technological relatedness and regional branching. In: Cooke P (ed) Handbook of regional innovation and growth. Edward Elgar Publishing, pp 187–201
Boudeville JR (1964) Note sur l'intégration des espaces économiques. Cahiers de l'I.S.E.A. [Institut de Science Economique Appliquée], 153(14):5–74
Boudeville JR (1972) Aménagement du territorie et polarization. Génin M-Th (ed)
Bourdieu P (1986) The forms of capital. In: Richardson J (ed) Handbook of theory and research for the sociology of education. Greenwood, pp 241–258
Bowman A, Ertürk I, Froud J, Johal S, Law J, Leaver A, Moran M, Williams K (2014) The end of the experiment? Oxford University Press, From competition to the foundational economy
Brown M, Jones RE (2021) Paint your town red: how Preston took back control and your town can too. Repeater Books
Brown A, Gardiner B, Römisch R, Stenning J (2017) Economic challenges of lagging regions I: Fiscal and macroeconomic environment (WIIW Research Report No. 421). The Vienna Institute for International Economic Studies
Busłowska A, Marcinkiewicz J (2023) Social cohesion of Functional Urban Areas (example of Eastern Poland). Soc Indic Res 167:451–473. https://doi.org/10.1007/s11205-023-03119-4
Busłowska A (2024) Sustainable development of Functional Urban Areas of voivodship's capital cities in Poland. Econ Environ 2(89):1–16. https://doi.org/10.34659/eis.2024.89.2.720
Cairncross F (1997) The death of distance: how the communications revolution will change our lives. Harvard Business School Press
Calafati L, Froud J, Johal S, Williams K (2019) Building foundational Britain: from paradigm shift to new political practice. Renewal 27:13–23
Camagni R (1991) Innovation networks. John Wiley and Sons
Camagni R (2008) Regional competitiveness: towards a concept of territorial capital. In: Capello R, Camagni R, Chizzolini B, Fratesi U (eds) Modelling regional scenarios for the enlarged Europe: European competitiveness and global strategies. Springer, pp 33–48
Camagni R (2016) Towards creativity-oriented innovation policies based on a hermeneutic approach to the knowledge-space nexus. In: Cusinato A, Philippopoulos-Mihalopoulos A (eds) Knowledge-creating milieus in Europe. Springer, pp 341–358
Camagni R, Capello R (2013) Regional competitiveness and territorial capital: a conceptual approach and empirical evidence from the European Union. Reg Stud 47(9):1383–1402
Camagni R (2017) Territorial capital, competitiveness and regional development. In Huggins R, Thompson P (eds) Handbook of regions and competitiveness: contemporary theories and perspectives on economic development. Edward Elgar, pp 232–244. https://doi.org/10.4337/9781783475018.00016
Capello R (2007) Regional economics. Routledge
Capello R (2009) Spatial spillovers and regional growth: a cognitive approach. Eur Plan Stud 17(5):639–658. https://doi.org/10.1080/09654310902778045
Capello R, Caragliu A, Fratesi U (2015) Advances in regional growth forecasting models: conceptual challenges and methodological responses. Int Reg Sci Rev 40(1):3–11
Capello R, Caragliu A, Nijkamp P (2009) Growth: Increasing returns in cognitive knowledge use, Tinbergen Institute Discussion Paper No. 2009–059/3), Tinbergen Institute
Capello R (2011) Location, regional growth and local development theories. Aestimum 58:1–25. https://doi.org/10.13128/Aestimum-9559
Capello R (2014) Regional economics, 2nd ed. Routledge
Capello R (2020) Proximity and regional competitiveness. Scienze Regionali Ital J Reg Sci 3:373–394. https://doi.org/10.14650/98284

Carruthers R, Dick M, Anuja S (2005) Affordability of public transport in developing countries. The World Bank Group

Castells M (2004) Space of flows, space of places: materials for a theory of urbanism in the information age. In: Graham S (ed) The cybercities reader. Routledge, pp 82–93

Castells-Quintana D, Royuela Veneri P (2020) Inequality and city size: an analysis for OECD Functional Urban Areas. Reg Sci 99(4):1045–1064. https://doi.org/10.1111/pirs.12520

Chen Y, Liu X, Li X, Liu X, Yao Y, Hu G, Xu X, Pei F (2017) Delineating urban functional areas with building-level social media data: a dynamic time warping (DTW) distance based k-medoids method. Landsc Urban Plan 160:48–60. https://doi.org/10.1016/j.landurbplan.2016.12.001

Chilla T, Lambracht M (2022) Institutional mapping of cross-border cooperation: INTERREG programme analyses with KEEP data. Eur Plan Stud 31(4):700–718. https://doi.org/10.1080/09654313.2022.2058321

Churski P (2022) Miasto a sprawiedliwość społeczna w ujęciu terytorialnym. In: Nowak MJ (ed) Prawo do miasta a wyzwania polityki miejskiej w Polsce. Wydawnictwo Naukowe Scholar, pp 129–177

Churski P, Perdał R (2021) Geographical differences in the quality of life in Poland: challenges of regional policy. Soc Indic Res 164(1):31–54. https://doi.org/10.1007/s11205-021-02829-x

Churski P, Perdał R, Konecka-Szydłowska B, Herodowicz T (2018) Redefinicja czynników rozwoju regionalnego w świetle megatrendów społeczno-gospodarczych. Studia Regionalne i Lokalne 3(73):70–98

Churski P, Adamiak C, Szyda B, Dubownik A, Pietrzykowski M, Śleszyński P (2023) Nowa delimitacja miejskich obszarów funkcjonalnych w Polsce i jej zastosowanie w praktyce zintegrowanego podejścia terytorialnego (place based approach). Przegląd Geograficzny 95(1):29–55

Churski P, Perdał R, Konecka-Szydłowska B, Herodowicz T (2021) European regional development: contemporary regional and local perspectives of socio-economic and socio-political changes. Springer

Churski P (2008) Czynniki rozwoju regionalnego i polityka regionalna w Polsce w okresie integracji z Unią Europejską, vol 79. Wydawnictwo Naukowe UAM

Churski P (2014) The polarization-diffusion model in the changes to the Cohesion Policy-the consequences to the direction of the growth policy. In: Churski P (ed) The social and economic growth vs. the emergence of economic growth and stagnation areas. Bogucki Wydawnictwo Naukowe, pp 13–27

Clark J, Harrison J, Miguelez E (2018) Connecting cities, revitalizing regions: the centrality of cities to regional development. Reg Stud 52(8):1025–1028. https://doi.org/10.1080/00343404.2018.1453691

Coenen L, Raven R, Verbong G (2010) Local niche experimentation in energy transitions: a theoretical and empirical exploration of proximity advantages and disadvantages. Technol Soc 32(4):295–302. https://doi.org/10.1016/j.techsoc.2010.10.006

Colby CC (1933) Centrifugal and centripetal forces in urban geography. Ann Assoc Am Geogr 23(1):1–20

Coleman JS (1988) Social capital in the creation of human capital. Am J Sociol 94(1):S95–S120. http://www.jstor.org/stable/2780243

Collective FE (2018) Foundational economy: the infrastructure of everyday life. (No Title)

Commission E (2010) EUROPE 2020. A strategy for smart, sustainable and inclusive growth. Communication from the Commission, Brussels 3(3):2010

Connell RW (2005) Masculinities. University of California Press

Cooke P, Morgan K (1994) The creative milieu: a regional perspective on innovation. In: Dodgson M, Rothwell R (eds) The handbook of industrial innovation. Edward Elgar, pp 25–32

Coote A, Percy A (2020) The case for universal basic services. Polity Press

Copus A (2001) From core-periphery to polycentric development: concepts of spatial and aspatial peripherality. Eur Plan Stud 9(4):539–552. https://doi.org/10.1080/713666491

Copus A, Mantino F, Noguera J (2017) Inner peripheries: an oxymoron or a real challenge for territorial cohesion? Ital J Plan Pract 7(1):24–49

Cortinovis N, Xiao J, Boschma R, van Oort FG (2017) Quality of government and social capital as drivers of regional diversification in Europe. J Econ Geogr 17(6):1179–1208. https://doi.org/10.1093/jeg/lbx001

Coss-Mandiola J, López JV, Rojas A, Carrasco R, Dubó P, Campillay MC (2022) Characterization of communes with quality accredited primary healthcare centers in Chile. Int J Environ Res Public Health 19(15):9189. https://doi.org/10.3390/ijerph19159189

Cote A (2020) Universal basic services and the foundational economy (Foundational Economy Working Paper 7). Foundational Economy Collective

Couch C, Karecha J, Nuissl H, Rink D (2005) Decline and sprawl: an evolving type of urban development—observed in Liverpool and Leipzig. Eur Plan Stud 13(1):117–136

Crane R, Manville M (2008) People or place? Revisiting the who versus the where of urban development. Land Lines 1–5. Lincoln Institute of Land Policy

Crescenzi R, Di Cataldo M, Rodríguez-Pose A (2016) Government quality and the economic returns of transport infrastructure investment in European regions. J Reg Sci 56(4):555–582

Czapiewski K, Janc K (2009) Internal peripheries of socio-economic development in Poland. In: Kovács AD (ed) Old and new borderlines/frontiers/margins. Centre for Regional Studies of the Hungarian Academy of Sciences, pp 109–121

Daniel Z, Whalan R (2021) *Greetings* from Trumpland: how an unprecedented presidency changed everything. HarperCollins

Danson M, De Souza P (eds) (2012) Regional development in Northern Europe: peripherality, marginality and border issues. Routledge

Davies BP (1968) Social needs and resources in local services: a study of variations in provision of social services between local authority areas. Joseph Rowntree Foundation

Davoudi S (2012) Resilience: a bridging concept or a dead end? Plan Theory Pract 13:299–333

Davoudi S, Brooks E (2014) When does unequal become unfair? Judging claims of environmental injustice. Environ Plan A 46(1):2686–2702. https://doi.org/10.1068/a130346p

Dawkins CJ (2003) Regional development theory: conceptual foundations, classic works, and recent developments. J Plan Lit 18(2):131–172

De Toni A, Vizzarri M, Di Febbraro M, Lasserre B, Nouguera J, Di Martino P (2021) Aligning inner peripheries with rural development in Italy: territorial evidence to support policy contextualization. Land Use Policy 100:104899. https://doi.org/10.1016/j.landusepol.2020.104899

Delfmann H, Koster S (2016) The effect of new business creation on employment growth in regions facing population decline. Ann Reg Sci 56:33–54. https://doi.org/10.1007/s00168-015-0738-1

Delgadillo E, Reyes T, Baumgartner RJ (2021) Towards territorial product-service systems: a framework linking resources, networks and value creation. Sustain Prod Consum 28:1297–1313. https://doi.org/10.1016/j.spc.2021.08.003

Dijkstra L, Poelman H, Rodríguez-Pose A (2020) The geography of EU discontent. Reg Stud 54(6):737–753. https://doi.org/10.1080/00343404.2019.1654603

Dijkstra L, Jacobs-Crisioni C (2023) Developing a definition of functional rural areas in the EU (JRC135599). European Commission

Dijkstra L, Poelman H, Veneri P (2019) The EU-OECD definition of a Functional Urban Area (OECD Regional Development Working Papers No. 2019/11). OECD Publishing. https://doi.org/10.1787/d58cb34d-en

Domański B (2021) Sprawiedliwość społeczna a przestrzenne zróżnicowanie rozwoju społeczno-gospodarczego Polski—kilka refleksji dyskusyjnych. Studia Regionalne i Lokalne 2(84):109–116

Domenech RB, Veneri P, Almenar V, Hernández F (2012) Polycentric metropolitan areas in Europe: towards a unified proposal of delimitation. Paper presented at the 51st Congress of the European Regional Science Association, Barcelona, Spain

Dong X, Ben-Akiva M, Bowman JL, Walker JL (2006) Moving from trip-based to activity-based measures of accessibility. Transp Res Part A 40:163–180

Dubois A (2019) Translocal practices and proximities in short quality food chains at the periphery: the case of North Swedish farmers. Agric Hum Values 36(4):763–778. https://doi.org/10.1007/s10460-019-09953-y
Dunford M (1988) Capital, the state, and regional development, vol 1. Pion
Dunford M (2010) Regional development models. University of Sussex. https://www.sussex.ac.uk/webteam/gateway/file.php?name=modelsrd.pdf&site=2
Dworkin R (1981) What is equality? Part I: equality of welfare; Part II: equality of resources. Philos Public Aff 10:185–246. https://doi.org/10.2307/2264894
Dyjach K (2013) Teorie rozwoju regionalnego wobec zróżnicowań międzyregionalnych. Annales Universitatis Mariae Curie-Skłodowska. Sectio H, Oeconomia 47(1):49–59
Dziekański P, Prus P (2020) Financial diversity and the development process: case study of rural communes of Eastern Poland in 2009–2018. Sustainability 12(16):1–25. https://doi.org/10.3390/su12166446
Earle J, Froud J, Johal S, Williams K (2018) Foundational economy and foundational politics. Welsh Econ Rev 26:38–45
Echebarria C, Barrutia J (2013) Limits of social capital as a driver of innovation: an empirical analysis in the context of European regions. Reg Stud 47:1001–1017
El-Geneidy A, Levinson D (2006) Access to destinations: development of accessibility measures. Minnesota Department of Transportation
Engelen E, Froud J, Johal S, Salento A, Williams K (2017) The grounded city: from competitivity to the foundational economy. Camb J Reg Econ Soc 10:407–423
ESPON DEMIFER (2010) Demographic and migratory flows affecting European regions and cities. https://archive.espon.eu/demifer
ESPON EDORA (2011) European development opportunities in rural areas. https://archive.espon.eu/programme/projects/espon-2013/applied-research/edora-european-development-opportunities-rural-areas
ESPON ESCAPE (2020) European shrinking rural areas: challenges, actions and perspectives for territorial governance. https://archive.espon.eu/escape
ESPON GEOSPECS (2012) European perspective on specific types of territories. https://archive.espon.eu/sites/default/files/attachments/GEOSPECS_Final_Report_inner_peripheries_v14.pdf
ESPON PROFECY (2017) Inner peripheries: national territories facing challenges of access to basic services of general interest. https://archive.espon.eu/inner-peripheries
ESPON PROFECY (2018) Inner peripheries in Europe: possible development strategies to overcome their marginalising effects. https://archive.espon.eu/inner-peripheries
ESPON (2007) Study on urban functions (ESPON Project 1.4.3. Final Report). ESPON Coordination Unit. https://www.espon.eu/programme/projects/espon-2006/studies-and-scientific-support-projects/study-urban-functions
ESPON QoL (2021) Quality of life measurements and methodology (Final Report). https://archive.espon.eu/sites/default/files/attachments/ESPON%20QoL%20-%20Final%20Report%20-%2001292021-FINAL.pdf
EUROSTAT (2017) *F*inal report of the expert group on quality of life indicators. https://doi.org/10.2785/021270
EUROSTAT (2019) Methodological manual on territorial typologies. Publications Office of the European Union
European Commission, Directorate-General for Regional and Urban Policy (2022) Cohesion in Europe towards 2050: eighth report on economic, social and territorial cohesion. Publications Office of the European Union
European Commission (2015) Scenarios for integrated territorial investments. Directorate-General for Regional and Urban Policy. https://ec.europa.eu/regional_policy/en/information/publications/reports/2015/scenarios-for-integrated-territorial-investments
European Commission (2017) Competitiveness in low-income and low-growth regions: the lagging regions report (Commission Staff Working Document SWD (2017)

132 final). https://ec.europa.eu/regional_policy/sources/docoffic/official/reports/lagging_regions2017/lagging_regions_2017_en.pdf
European Commission (2024) Ninth report on economic, social and territorial cohesion. Directorate-General for Regional and Urban Policy. https://ec.europa.eu/regional_policy/information-sources/cohesion-report_en
European Commission: Directorate-General for Regional and Urban Policy (2024) Forging a sustainable future together—cohesion for a competitive and inclusive Europe—report of the high-level group on the future of Cohesion Policy. Publications Office of the European Union. https://data.europa.eu/doi/; https://doi.org/10.2776/974536
Faggian A, McCann P (2009) Universities, agglomerations and graduate human capital mobility. Tijdschr Econ Soc Geogr 100(2):210–223
Farole T, Goga S, Ionescu-Heroiu M (2018) Rethinking lagging regions: using Cohesion Policy to deliver on the potential of Europe's regions. World Bank
Feld LP, Heckemeyer JH, Overesch M (2013) Capital structure choice and company taxation: a meta-study. J Bank Finance 37(8):2850–2866
Ferry M (2019) Integrated territorial investments as an effective tool of the Cohesion Policy (Budgetary Affairs. In-depth Analysis). European Parliament's Committee on Budgetary Control. https://www.europarl.europa.eu/cmsdata/162823/25032019_CONT_Briefing_ITI_Final.pdf
Filipiak BZ (2016) Dylematy pomiaru potencjału finansowego jednostek samorządu terytorialnego—dobór czynników i ich pomiar. Prace Naukowe Uniwersytetu Ekonomicznego We Wrocławiu 451:75–88
Florida R (2021) Discontent and its geographies. Camb J Reg Econ Soc 14:619–624. https://doi.org/10.1093/cjres/rsab014
Foundational Economy Collective (2020) The foundational approach. Foundational Economy Collective
Fratesi U, Rodríguez-Pose A (2016) The crisis and regional employment in Europe: what role for sheltered economies? Camb J Reg Econ Soc 9:33–57
Friedman TL (2005) The world is flat: a brief history of the twenty-first century. Macmillan
Friedmann J, Alonso W (1964) Introduction. In: Friedmann J, Alonso W (eds) Regional development and planning: a reader. MIT Press, pp 1–13
Friedmann J, Miller J (1965) The urban field. J Am Inst Plann 31(4):312–320
Friedmann J (1967) A general theory of polarized development. Ford Foundation
Fritsch M, Kahila P, Nemeth S, Scott JW (eds) (2023) Spatial justice and cohesion. Routledge
Fröhlich K, Hassink R (2018) Regional resilience: a stretched concept? Eur Plan Stud 26:1763–1778
Froud J, Haslam C, Johal S, Williams K (2020) (How) does productivity matter in the foundational economy? Local Econ 35:316–336
Fujita M, Thisse JF (2009) New economic geography: an appraisal on the occasion of Paul Krugman's 2008 Nobel prize in economic sciences. Reg Sci Urban Econ 39(2):109–119
Fukuyama F (1997) Trust: the social virtues and the creation of prosperity. Simon and Schuster
Galtung J (1971) A structural theory of imperialism. J Peace Res 8:81–117
Ganau R, Rodríguez-Pose A (2019) Do high-quality local institutions shape labour productivity in Western European manufacturing firms? Pap Reg Sci 98(4):1633–1666
Garcilazo JE, Martins JO, Tompson W (2010) Why policies may need to be place-based in order to be people-centred. http://voxeu.org/article/why-policies-may-need-be-place-based-order-be-people-centred
Gerard R (1958) Commuting and the labour market area. J Reg Sci 1:124–130
Geurs KT, van Eck R (2001) Accessibility measures: review and applications (RIVM Report 408505 006). National Institute for Public Health and the Environment
Geurs K, van Wee B (2004) Accessibility evaluation of land-use and transport strategies: review and research directions. J Transp Geogr 12:127–140
Gitling M (2013) Człowiek w organizacji: Ludzie, struktury, organizacje. Difin
Global Footprint Network (2019) National footprint and biocapacity accounts

Glückler J, Shearmur R, Martinus K (2023) Liability or opportunity? Reconceptualizing the periphery and its role in innovation. J Econ Geogr 23(1):231–249. https://doi.org/10.1093/jeg/lbac028

Godart FC (2012) Trend networks: multidimensional proximity and the formation of aesthetic choices in the creative economy. Reg Stud 49(6):973–984

Görmar F, Lang T, Nagy E, Raagmaa G (2019) Re-thinking regional and local policies in times of polarisation: an introduction. In: Lang T, Görmar F (eds) Regional and local development in times of polarisation: re-thinking spatial policies in Europe. Palgrave Macmillan, pp 1–24

Górniak J (2013) Instytucje, wiedza, działanie i kapitał instytucjonalny. Zarządzanie Publiczne 2–3:24–25

Gorzelak G (1985) Przestrzenne aspekty jakości życia i sprawiedliwości społecznej. In: Kukliński A (ed) Gospodarka przestrzenna Polski wczoraj, dziś i jutro II. Wydział Pracy Posłów i Radnych Stowarzyszenia PAX, pp 69–83

Gough I (2020) The case for universal basic services. LSE Public Policy Rev 1:1–9

Gould E, Shierholz H (2020) Not everybody can work from home: black and Hispanic workers are much less likely to be able to telework (Working Economics Blog—19 March 2020). Economic Policy Institute. https://www.epi.org/blog/black-and-hispanic-workers-are-much-less-likely-to-be-able-to-work-from-home/. Accessed 25 Nov 2021

Gould P (1969) Spatial diffusion. Association of American Geographers

Granovetter M (1985) Economic action and social structure: the problem of embeddedness. Am J Sociol 91(3):481–510. https://doi.org/10.1086/228311

Grosel M (2023) Ku nowej definicji peryferyjności. Rozwój Regionalny i Polityka Regionalna 63:73–87. https://doi.org/10.14746/rrpr.2023.63.05

Gross M, Dudzińska M, Dawidowicz A, Wolny-Kucińska A (2024) Transportation management in urban functional areas. Real Estate Manag Valuat 32(3). https://doi.org/10.2478/remav-2024-0021

Grzeszczak J (2007) Teoria biegunów wzrostu w warunkach polskich: przeszłość i teraźniejszość. In: Lach J, Borowice M, Rachwał T (eds) Procesy transformacji społeczno-ekonomicznych i przyrodniczych struktur przestrzennych. Instytut Geografii Akademii Pedagogicznej, pp 120–129

Grzeszczak J (1978) Teoria biegunów wzrostu. Przegląd Zagranicznej Literatury Geograficznej. IGiPZ PAN

Grzeszczak J (1999) Bieguny wzrostu a formy przestrzeni spolaryzowanej. Prace Geograficzne 173. IGiPZ PAN

Gunnlaugsson SB (2017) Credit risk of Icelandic municipalities. Oradea J Bus Econ 2(2):7–15. https://doi.org/10.47535/1991ojbe022

Gutiérrez J (2001) Location, economic potential and daily accessibility: an analysis of the accessibility impact of the high-speed line Madrid–Barcelona–French border. J Transp Geogr 9:229–242

Hägerstrand T (1967) Innovation diffusion as a spatial process. University of Chicago Press

Hägerstrand T (1951) Migration and growth of cultural regions. Lund Studies in Geography

Hägerstrand T (1952) The propagation of innovation waves. Lund Studies in Geography

Halamska M (2018) Wspierać czy zalesiać? Dylematy rozwoju wiejskich obszarów problemowych. Wieś i Rolnictwo 3(180):69–91

Hall S, Schafran A (2017) From foundational economics and the grounded city to foundational urban systems (Foundational Economy Working Paper 3). Foundational Economy Collective

Hall PA, Soskice D (2001) Varieties of capitalism: the institutional foundations of comparative advantage. Oxford University Press

Handy SL, Niemeier DA (1997) Measuring accessibility: an exploration of issues and alternatives. Environ Plan A 29:1175–1194

Hanna S (1995) Finding a place in the world-economy: core-periphery relations, the nation-state and the underdevelopment of Garrett County. Polit Geogr 14(5):451–472

Hansen WG (1959) How accessibility shapes land-use. J Am Inst Plann 25:73–76

Hansen T (2014) Juggling with proximity and distance: collaborative innovation projects in the Danish cleantech industry. Econ Geogr 90(4):375–402. https://doi.org/10.1111/ecge.12057
Harvey D (1973) Social justice and the city. Johns Hopkins University Press
Hastings DA, UN ESCAP (2009) From human development to human security: a prototype human security index. https://hdl.handle.net/20.500.12870/1201
Hay D, Hall P (1977) Urban regionalization of Great Britain: 1971. University of Reading
Hayter R (2004) Economic geography as dissenting institutionalism: the embeddedness, evolution and differentiation of regions. Geografiska Annaler Ser b Hum Geogr 86(2):95–115
Helden D (2002) Using accessibility measures to integrate land use and transport policy in Edinburgh and the Lothians. Transp Policy 9:313–324
Hellerstein JK, Neumark D (2020) Social capital, networks, and economic wellbeing (IZA DP No. 13413). Institute of Labor Economics
Hendrickson C, Muro M, Galston WA (2018) Countering the geography of discontent: strategies for left behind places. Brookings Institution
Herrschel T (2011) Regional development, peripheralization and marginalization—and the role of governance. In: Herrschel T, Tallberg P (eds) The role of regions? Networks, scale territory. Sveriges Kommuner och Landsting, pp 45–66
Heslop J, Morgan K, Tomaney J (2019) Debating the foundational economy. Renew A J Labour Polit 27:5–12
Hines C (2000) Localization: a global manifesto. Earthscan
Hodgson G (2006) What are institutions? J Econ Issues 40(1):1–24
Hudson R (2004) Conceptualizing economies and their geographies: spaces, flows and circuits. Prog Hum Geogr 28(4):447–471
Huff DL (1963) Defining and estimating a trade area. J Mark 28:34–38
Huff DL (1973) The delineation of a national system of planning regions on the basis of urban spheres of influence. Reg Stud 7:323–329
Ievoli C, Belliggiano A, Marandola D, Milone P, Ventura F (2019) Information and communication infrastructures and new business models in rural areas: the case of Molise region in Italy. Eur Countryside 11(4):475–496. https://doi.org/10.2478/euco-2019-0027
Ilnicki D, Janc K (2009) Topology, nodality and space of internet flows. Acta Universitatis Palackianae Olomucensis—Geographica 40(2):15–26
Isard W (1960) Methods of regional analysis: an introduction to regional science. John Wiley & Sons
Israel E, Frenkel A (2017) Social justice and spatial inequality: toward a conceptual framework. Prog Hum Geogr 42(5):647–665. https://doi.org/10.1177/0309132517702969
Jacobs J (1969) The economy of cities. Random House
Janc K (2019) Przestrzeń cyfrowa i internet jako przedmiot zainteresowań w badaniach geograficznych. Przegląd Geograficzny 91(2):21–37. https://doi.org/10.7163/PrzG.2019.2.2
Janssen MJ, Abbasiharofteh M (2022) Boundary spanning R&D collaboration: key enabling technologies and missions as alleviators of proximity effects? Technol Forecast Soc Chang 180:121689. https://doi.org/10.1016/j.techfore.2022.121689
Jennings W, Stoker G (2019) The divergent dynamics of cities and towns: geographical polarisation and Brexit. Polit Quart 90(S2):155–166. https://doi.org/10.1111/1467-923X.12612
Jewtuchowicz A (2013) Terytorium i współczesne dylematy jego rozwoju. Wydawnictwo Uniwersytetu Łódzkiego
Jussila H, Leimgruber W (eds) (1998) Perceptions of marginality: theoretical issues and regional perceptions of marginality in geographical space. Ashgate
Kaganskii V (2013) Inner periphery is a new growing zone of Russia's cultural landscape. Reg Res Russ 3(1):21–31
Keeble D, Owens PL, Thompson C (1982) Regional accessibility and economic potential in the European Community. Reg Stud 16:419–432
Keeble D, Offord J, Walker S (1988) Peripheral regions in a community of twelve member states. Commission of the European Community

Kemeny T, Storper M (2020) Superstar cities and left behind places: disruptive innovation, labour demand and inter-regional inequality (Working Paper No. 41). International Inequalities Institute, LSE

Keynes JM (1936) The general theory of employment, interest and money. Macmillan

Knack S, Keefer P (1997) Does social capital have an economic payoff? A cross-country investigation. Quart J Econ 112(4):1251–1288. http://www.jstor.org/stable/2951271

Knight J (1992) Institutions and social conflict. Cambridge University Press

Knopman D, Zmud J, Ecola L, Zhimin M, Crane K (2015) Quality of life indicators and policy strategies to advance sustainability in the Pearl River Delta. RAND Corporation

Knowles R, Shaw J, Docherty I (2008) Transport geographies: mobilities, flows and spaces. Blackwell

Komárek M, Chromý P (2020) The institutional thickness of an inner periphery in the cross-border region between Central Bohemia and Eastern Bohemia. Geografie 125(4):423–446. https://doi.org/10.37040/geografie2020125040423

Komornicki T, Śleszyński P, Pomianowski W, Rosik P, Siłka P, Stępniak M (2008, 2010, 2014) Opracowanie metodologii liczenia wskaźnika międzygałęziowej dostępności transportowej terytorium Polski oraz jego oszacowanie; (+ Aktualizacja). Instytut Geografii i Przestrzennego Zagospodarowania PAN

Komornicki T, Śleszyński P, Rosik P, Pomianowski W (2009) Dostępność przestrzenna jako przesłanka kształtowania polskiej polityki transportowej. Biuletyn KPZK PAN 241

Kopczewska K, Churski P, Ochojski A, Polko A (2017) Measuring regional specialisation: a new approach. Palgrave Macmillan. https://doi.org/10.1007/978-3-319-51505-2

Krugman P (1991a) Increasing returns and economic geography. J Polit Econ 99(3):483–499

Krugman P (1991b) Geography and trade. MIT Press

Krugman P (1995) Development, geography, and economic theory. MIT Press

Kudłacz M, Karwińska A (2021) The specificity of urban sprawl in Poland: the spatial, social, and economic perspectives. Public Governance 1(55):50–67. https://doi.org/10.15678/ZP.2021.55.1.05

Kühn M (2015) Peripheralisation: theoretical concepts explaining socio-spatial inequalities. Eur Plan Stud 23(2):367–378. https://doi.org/10.1080/09654313.2013.862518

Kukathas C, Pettit P (1990) Rawls—a theory of justice and its critics. Polity Press

Kurek S, Wójtowicz M, Gałka J (2020) Functional Urban Areas in Poland: demographic trends and migration patterns. Springer. https://doi.org/10.1007/978-3-030-31527-6

Kwiatkowski T (2023) Kapitał społeczny i wyposażenie instytucjonalne jako czynniki sukcesu regionalnych polityk klastrowych. Instytut Geografii i Gospodarki Przestrzennej UJ

Ładysz J (2021) Delimitacja transgranicznych obszarów funkcjonalnych. Wydawnictwo Uniwersytetu Ekonomicznego

Lalrindiki M, O'Gorman B (2021) The role of proximity in developing an inter-regional innovation system. Triple Helix 8(3):534–577. https://doi.org/10.1163/21971927-bja10024

Lambooy J (2005) Innovation and knowledge: theory and regional policy. Eur Plan Stud 13(8):1137–1152

Lange B, Hülz M, Schmid B, Schulz C (eds) (2021) Post-growth geographies: spatial relations of diverse and alternative economies. Columbia University Press

Lefebvre H (1974) The production of space. Blackwell

León-Ledesma MA, Thirlwall AP (2002) The endogeneity of the natural rate of growth. Camb J Econ 26(4):441–459. https://doi.org/10.1093/cje/26.4.441

Lewin P, Weber B, Holland D (2013) Core-periphery dynamics in the Portland, Oregon, region: 1982–2006. Ann Reg Sci 51:411–433. https://doi.org/10.1007/s00168-013-0552-6

Li B, Liu Y, Xing H, Meng Y, Yang G, Liu X, Zhao Y (2022) Integrating urban morphology and land surface temperature characteristics for urban functional area classification. Geo-Spat Inf Sci 25(2):337–352. https://doi.org/10.1080/10095020.2021.2021786

Liotta C, Kervinio Y, Levrel H, Tardieu L (2020) Planning for environmental justice—reducing well-being inequalities through urban greening. Environ Sci Policy 112:47–60. https://doi.org/10.1016/j.envsci.2020.03.017
Lityński P (2023) Living in sprawling areas: a cost–benefit analysis in Poland. J Housing Built Environ 38:1069–1096. https://doi.org/10.1007/s10901-022-09986-6
Liu T, Cheng G, Yang J (2023) Multi-scale recursive identification of urban functional areas based on multi-source data. Sustainability 15:13870. https://doi.org/10.3390/su151813870
Lopolito A, Falcone PM, Sica E (2022) The role of proximity in sustainability transitions: a technological niche evolution analysis. Res Policy 51(3):104464. https://doi.org/10.1016/j.respol.2021.104464
Lucatelli S, Carlucci C, Guerrizio M (2013) A strategy for the inner areas of Italy. Paper presented at the European Rural Futures Conference, Asti
Lukáč J, Teplická K, Čulková K, Hrehová D (2021) Evaluation of the financial performance of the municipalities in Slovakia in the context of multidimensional statistics. J Risk Financ Manag 14(12):570. https://doi.org/10.3390/jrfm14120570
Lundvall BA (1992) National systems of innovation: towards a theory of innovation and interactive learning, vol 242. Pinter, London
MacKinnon D, Kempton L, O'Brien P, Ormerod E, Pike A, Tomaney J (2022) Reframing urban and regional 'development' for 'left behind' places. Camb J Reg Econ Soc 15:39–56. https://doi.org/10.1093/cjres/rsab034
Madanipour A, Shucksmith M, Brooks E (2021) The concept of spatial justice and the European Union's territorial cohesion. Eur Plan Stud 30(5):1–18. https://doi.org/10.1080/09654313.2021.1928040
Magel H (2017) Territorial justice for urban and rural regions? About the responsibility and role of the Bavarian Academy for Rural Areas. Geomat Land Manag Landsc 1:65–79
Maher CS, Ebdon C, Bartle JR (2020) Financial condition analysis: a key tool in the MPA curriculum. J Public Aff Educ 26(1):4–10. https://doi.org/10.1080/15236803.2020.1717321
Mahroum S (2000) Highly skilled globetrotters: mapping the international migration of human capital. R&D Manag 30:23–32. https://doi.org/10.1111/1467-9310.00154
Máliková L, Farrell M, McDonagh J (2016) Perception of marginality and peripherality in an Irish rural context. Quaes Geogr 35(4):93–105. https://doi.org/10.1515/quageo-2016-0037
Malmberg A, Maskell P (2002) The elusive concept of localization economies: towards a knowledge-based theory of spatial clustering. Environ Plan A 34(3):429–449
Mantino F, Fano G (2015) New concepts for territorial rural development in Europe: the case of most remote rural areas in Italy. Paper presented at the XXVI Congress of the European Society of Rural Sociologists, Aberdeen
Marginson S (2019) Limitations of human capital theory. Stud High Educ 44(2):287–301. https://doi.org/10.1080/03075079.2017.1359823
Marshall A (1890) Principles of political economy. Macmillan
Martin R (2000) Institutional approaches in economic geography. In: Sheppard E, Barnes TJ (eds) A companion to economic geography. Blackwell, pp 77–94
Martin R (2021) Rebuilding the economy from the Covid crisis: time to rethink regional studies? Reg Stud Reg Sci 8:143–161. https://doi.org/10.1080/21681376.2021.1919191
Martin R, Sunley P (2006) Path dependence and regional economic evolution. J Econ Geogr 6(4):395–437
Martin R, Gardiner B, Pike A, Sunley P, Tyler P (2021) Levelling up left behind places: the scale and nature of the policy challenge. Regional Studies Association
Martynovich M, Hansen T, Lundquist K-J (2023) Can foundational economy save regions in crisis? J Econ Geogr 23(3):577–599. https://doi.org/10.1093/jeg/lbac027
McCann P (2019) Perceptions of regional inequality and the geography of discontent: insights from the UK. Reg Stud 54:256–257
McCann P (1998) The location of the firm in theory. In: The economics of industrial location. Springer, pp 27–44. https://doi.org/10.1007/978-3-662-03702-7_3

McCann P (2023) How have place-based policies evolved to date and what are they for now? Background paper for the OECD-EC High-Level Expert Workshop Series on “Place-Based Policies for the Future”. https://www.oecd.org/cfe/regionaldevelopment/place-based-policies-for-the-future.htm
McDonald S, Roberts J (2002) Multiple forms of human capital in an augmented Solow model: a panel data investigation. Econ Lett 74(2):271–276
McKenzie RD (1933) The metropolitan community. McGraw-Hill
McQuarrie M (2017) The revolt of the rustbelt: place and politics in the age of anger. Br J Sociol 68(S1):120–152. https://doi.org/10.1111/1468-4446.12328
Mincer J (1974) Schooling, experience, and earnings. National Bureau of Economic Research
Ministry of National Development and Váti Nonprofit Ltd (2011) The territorial state and perspectives of the European Union. 2011 update. Background document for the Territorial Agenda of the European Union 2020
Moisio S, Paasi A (2013) From geopolitical to geoeconomic? The changing political rationalities of state space. Geopolitics 18(2):267–283. https://doi.org/10.1080/14650045.2012.723287
Moore HL, Collins H (2020) Towards prosperity: reinvigorating local economies through universal basic services (Working Paper 01–2020/04). Institute for Global Prosperity, UCL
Moore HL, Woodcraft S (2019) Understanding prosperity in East London: local meanings and ‘sticky’ measures of the good life. City Soc 31:275–298
Morgan K (2016) Nurturing novelty: regional innovation policy in the age of smart specialization. Eviron Plann C Gov Policy 34(4):1–15. https://doi.org/10.1177/0263774X16645106
Morgan K (2019) The future of place-based innovation policy (as if ‘lagging regions’ really mattered). In: Barzotto M et al (ed) Revitalising lagging regions: smart specialisation and industry 4.0. Taylor and Francis, pp 15–32
Morretta V (2021) Territorial capital in local economic endogenous development. Reg Sci Policy Pract 13:103–119. https://doi.org/10.1111/rsp3.12317
Morretta V, Syrett S, Ramirez LS (2020) Territorial capital as a source of firm competitive advantage: evidence from the North and South of Italy. Eur Plan Stud 28(1):1–19. https://doi.org/10.1080/09654313.2020.1722067
Musil J, Müller J (2008) Vnitřní periferie v České republice jako mechanismus sociální exkluze. Sociologický Časopis/czech Sociol Rev 44(2):321–348
Naumann M, Fischer-Tahir A (eds) (2013) Peripheralisation: the making of spatial dependencies and social injustice. Springer
Nijkamp P (2016) The “resourceful region”: a new conceptualization of regional development strategies. J Reg Res 36:191–214
Nilsen T, Grillitsch M, Hauge A (2023) Varieties of periphery and local agency in regional development. Reg Stud 57(4):749–762. https://doi.org/10.1080/00343404.2022.2106364
Noguera J, Copus A (2016) Inner peripheries: what are they? What policies do they need? Agriregionieuropa 12(45). https://agriregionieuropa.univpm.it/it/content/article/31/45/inner-peripheries-what-are-they-what-policies-do-they-need
North DC (1990) Institutions, institutional change and economic performance. Cambridge University Press
North DC (1991) Institutions. J Econ Perspect 5(1):97–112
Nowak A (2018) Miejsce miejskich obszarów funkcjonalnych w procesie rozwoju regionalnego. Rozwój Regionalny i Polityka Regionalna 41:51–69
Nowakowska A (2018) Od regionu do terytorium—reinterpretacja znaczenia przestrzeni w procesach rozwoju gospodarczego. Gospodarka Narodowa 3:223–236
Nowakowska A (2023) Reinterpretacja pojęcia bliskości i jej roli w rozwoju terytorialnym. In: Rynio D, Zakrzewska-Półtorak A (eds) Przestrzeń i regiony w nowoczesnej gospodarce. Wydawnictwo Uniwersytetu Ekonomicznego we Wrocławiu, pp 164–174
Nowakowska A (2019) Kapitał terytorialny—źródło przewagi konkurencyjnej. In Danielewicz J, Sikora Fernandez D (eds) Zarządzanie rozwojem współczesnych miast. Wydawnictwo Uniwersytetu Łódzkiego, pp 165–174

O'Brien R (1992) Global financial integration: the end of geography. Council on Foreign Relations
OECD (2009a) How regions grow: trends and analysis. OECD Publishing. https://doi.org/10.1787/9789264039469-en
OECD (2009b) Regions matter: economic recovery, innovation and sustainable growth. OECD Publishing. https://doi.org/10.1787/9789264076525-en
OECD (2023) OECD economic outlook, Volume 2023 Issue 1: a long unwinding road. OECD Publishing. https://doi.org/10.1787/ce188438-en
OECD (2018) Regions and cities at a glance 2018. OECD Publishing
O'Kelly M, Bryan D (1996) Agricultural location theory: von Thunen's contribution to economic geography. Prog Hum Geogr 20(4):457–475. https://doi.org/10.1177/030913259602000402
Ostrom E (1990) Governing the commons: the evolution of institutions for collective action. Cambridge University Press
Paelinck J (1965) La théorie du développement régional polarisé. Cahiers de l'I.S.E.A. 159(15):5–47
Palma G (1978) Dependency: a formal theory of underdevelopment or a methodology for the analysis of concrete situations of underdevelopment? World Dev 6:881–924
Papadopoulos AG (2019) Editorial: spatial justice in Europe: territoriality, mobility and peripherality. Europa XXI 37:5–21. https://doi.org/10.7163/Eu21.2019.37.1
Pawlik A, Karpinska U (2019) The financial situation as a stimulant of the competitive position of rural communes in the Świętokrzyskie Voivodeship. Baltic J Econ Stud 5(5). https://ideas.repec.org/a/bal/journl/2256-07422017551.html
Perkmann M (2007) Policy entrepreneurship and multilevel governance: a comparative study of European cross-border regions. Eviron Plann C Gov Policy 25(6):861–879. https://doi.org/10.1068/c60m
Perroux F (1955) Note sur la notion de "pôle de croissance.". Econ Appl 8(1–2):307–320
Pike A, Rodriguez-Pose A, Tomaney J (2016) Local and regional development, 2nd ed. Routledge. https://doi.org/10.4324/9781315767673
Pike A, Béal V, Cauchi-Duval N, Franklin R, Kinossian N, Lang T, Leibert T, MacKinnon D, Rousseau M, Royer J, Servillo L, Tomaney J, Velthuis S (2023) Left behind places: a geographical etymology. Reg Stud 58(6). https://doi.org/10.1080/00343404.2023.2167972
Pilati M, Hunter A (2020) Research for REGI Committee—EU lagging regions: state of play and future challenges. European Parliament
Pirie G (1983) On spatial justice. Environ Plan A 15(4):465–473. https://doi.org/10.1068/a150465
Ployhart RE, Nyberg AJ, Reilly G, Maltarich MA (2014) Human capital is dead; long live human capital resources! J Manag 40(2):371–398. https://doi.org/10.1177/0149206313512152
Ponsard C (1983) August Losch. In: History of spatial economic theory. Springer, pp 143–158. https://doi.org/10.1007/978-3-642-82125-7_9
Porter ME (1990) The competitive advantage of nations. Harv Bus Rev 68(2):73–93
Prebisch R (1959) Commercial policy in the underdeveloped countries. Am Econ Rev 49:251–273
Pugh R, Dubois A (2021) Peripheries within economic geography: four "problems" and the road ahead of us. J Rural Stud 87:267–275. https://doi.org/10.1016/j.jrurstud.2021.09.007
Pukelienė V, Starkauskienė V (2015) Assessment of changes in the quality of life of emerging economies in the context of developed economies of the European Union. Organ Mark Emerg Econ 6(2):103–120
Putnam R, Leonardi R, Nanetti R (1993) Making democracy work. Princeton University Press
Putnam RD (2000) Bowling alone: the collapse and revival of American community. Simon & Schuster
Quah DT (1999) The weightless economy in growth. Bus Econ 30:40–53
Rahman J, Dimand RW (2020) The emergence of geographical economics: at the contested boundaries of economics, geography and regional science. https://doi.org/10.31219/osf.io/xg5m9
Rallet A (1993) Choix de proximité et processus d'innovation technologique. Revue D'economie Régionale Et Urbaine 3:365–386

Rallet A, Torre A (1999) Is geographical proximity necessary in the innovation networks in the era of global economy? GeoJournal 49:373–380. https://doi.org/10.1023/A:1007140329027
Rallet A, Torre A (eds) (1995) Economie industrielle et économie spatiale. Economica
Rauhut D (2017) Polycentricity—one concept or many? Eur Plan Stud 25(2):332–348
Rawls J (1971) A theory of justice. Oxford University Press
Reeves R (2018) The everyday economy: a reply to Neil McInroy. Polit Quart 89:618–620
Regulation (EU) 2017/2391 of the European Parliament and of the Council of 12 December 2017 amending Regulation (EC) No 1059/2003 as regards the territorial typologies (Tercet). Off J Eur Union L 350:1–6
Richardson HW (1973) Regional growth theory. Macmillan
Rio Fernandes JA, Seixas J (2018) Cities and urbanisation in democratic Portugal. Méditerranée 130.https://doi.org/10.4000/mediterranee.10698
Rodríguez-Pose A (2013) Do institutions matter for regional development? Reg Stud 47(7):1034–1047
Rodríguez-Pose A (2018) The revenge of the places that don't matter (and what to do about it). Camb J Reg Econ Soc 11:189–209
Rodríguez-Pose A, Ketterer T (2020) Institutional change and the development of lagging regions in Europe. Reg Stud 54(7):974–986. https://doi.org/10.1080/00343404.2019.1608356
Rodríguez-Pose A, Dijkstra L, Poleman H (2023) The geography of EU discontent and the regional development trap (Working Papers 03/2023). European Commission
Rodríguez-Pose A, Bartalucci F, Lozano-Gracia N, Davalos M (2024) Overcoming left-behindedness: moving beyond the efficiency versus equity debate in territorial development (Policy Research Working Paper 10734). World Bank Group. http://documents1.worldbank.org/curated/en/099224303262426681/pdf/IDU17dc414d91a08a142ad1a8da1e2c5bf4ade41.pdf
Romer PM (1986) Increasing returns and long-run growth. J Polit Econ 94:1002–1037
Romer PM (1994) The origins of endogenous growth. J Econ Perspect 8(1):3–22
Romer PM (1990) Endogenous technological change. J Polit Econ 98(5, Part 2):S71–S102
Rosati RM (2025) Urban sprawl and routing: a comparative study on 156 European cities. Landsc Urban Plan 253:105205. https://doi.org/10.1016/j.landurbplan.2024.105205
Rosik P (2012) Dostępność lądowa przestrzeni Polski w wymiarze europejskim. Prace Geograficzne 223. IGiPZ PAN
Rudolf H, Watts J (2002) Quality of life in substance abuse and dependency. Int Rev Psychiatry 14(3):190–197. https://doi.org/10.1080/09540260220144975
Russo A, Smith I, Atkinson R, Servillo L, Madsen B, Van Den Borg J (2012) The attractiveness of European regions and cities for residents and visitors (Final Report). ESPON. https://archive.espon.eu/programme/projects/espon-2013/applied-research/attreg-attractiveness-european-regions-and-cities
Salazar MA, Claudel M (2021) Spatial proximity matters: a study on collaboration. PLoS ONE 16(12):e0259965. https://doi.org/10.1371/journal.pone.0259965
Satoła Ł (2015) Kondycja finansowa gmin w warunkach zmiennej koniunktury gospodarczej. J Agribus Rural Dev 1(35):115–123
Sawers L, Tabb WK (eds) (1984) Sunbelt/Snowbelt: urban development and regional restructuring. Oxford University Press
Schinasi GJ (2004) Defining financial stability. IMF Work Pap 4(187):1. https://doi.org/10.5089/9781451859546.001
Schinasi GJ (2005) Defining financial stability. In: Safeguarding financial stability. International Monetary Fund, pp 73–96. https://doi.org/10.5089/9781589064409.071.ch005
Schmidt MH (1998) An integrated systemic approach to marginal regions: from definition to development policies. In: Jussila H, Leimgruber W, Majoral R (eds) Perceptions of marginality: theoretical issues and regional perceptions of marginality. Ashgate, pp 45–66
Schultz TW (1961) Investment in human capital. Am Econ Rev 51(1):1–17
Schumpeter JA (1911) The theory of economic development. Harvard University Press

Schumpeter JA (1934) The theory of economic development, 2nd ed. Harvard University Press
Sen A (1997) On economic inequality. Clarendon Press
Sen A (2009) The idea of justice. Harvard University Press
Servillo L, Russo AP, Barbera F (2016) Inner peripheries: towards an EU place-based agenda on territorial peripherality. Ita J Plan Pract 6(1):42–75
Servillo L, Atkinson R, Smith I, Russo A, Sýkora L, Demazière C, Hamdouch A (2014) Small and medium sized towns in their functional territorial context (Final Report). ESPON
Shuang M, Ying L (2020) Functional Urban Area delineations of cities on the Chinese mainland using massive Didi ride-hailing records. Cities 97:102532. https://doi.org/10.1016/j.cities.2019.102532
Siddik Md, Kabiraj S (2018) Does financial inclusion induce financial stability? Evidence from cross-country analysis. Australas Account Bus Financ J 12(1):34–46. https://doi.org/10.14453/aabfj.v12i1.3
Simmie J, Martin R (2010) The economic resilience of regions: towards an evolutionary approach. Camb J Reg Econ Soc 3:27–43
Sissons P, Tiwasing P, Ferreira J (2021) Working in the foundational economy during Covid-19 (Foundational Economy Working Paper 9). Foundational Economy Collective
Śleszyński P (2013) Delimitation of the Functional Urban Areas around Poland's voivodship capital cities. Przegląd Geograficzny 85(2):173–197. https://doi.org/10.7163/PrzG.2013.2.2
Śleszyński P (2015) Problemy delimitacji miejskich obszarów funkcjonalnych w Polsce. Rozwój Regionalny i Polityka Regionalna 29:37–53. https://doi.org/10.14746/rrpr.2015.29.04
Smart MW (1974) Labour market areas: uses and definition. Prog Plan 2:238–353
Smith DM (1994) Geography and social justice. Blackwell
Smith DM (2000) Moral geographies: ethics in a world of difference. Edinburgh University Press
Snijders T (2002) An approach to the measurement of individual social capital
Sobko R, Janiszewska D, Ossowska L, Oklevik O (2021) The role of the financial condition in the development of coastal municipalities in Poland. Eur Res Stud J 24(3B):688–703. https://doi.org/10.35808/ersj/2495
Sobolewska M, Ford R (2020) Brexitland: identity, diversity and the reshaping of British politics. Cambridge University Press
Soja EW (1980) The socio-spatial dialectic. Ann Assoc Am Geogr 70(2):207–225. https://doi.org/10.1111/j.1467-8306.1980.tb01308.x
Soja EW (2009) The city and spatial justice. Justice Spat/Spat Justice 1(1):1–5
Soja EW (2010a) Seeking spatial justice. University of Minnesota Press
Soja EW (2010b) The city and spatial justice. In: Bret B, Gervais-Lambony P, Hancock C, Landy F (eds) Justice et injustices spatiales. Presses Universitaires de Paris Nanterre, pp 56–73
Solow RM (1956) A contribution to the theory of economic growth. Quart J Econ 70(1):65–94
Spiekermann K, Neubauer J (2002) European accessibility and peripherality: concepts, models and indicators (Nordregio Working Paper)
Spiekermann K, Schürmann C (2007) Update of selected potential accessibility indicators (Final Report). Spiekermann & Wegener
Standar A, Kozera A (2019) The role of local finance in overcoming socioeconomic inequalities in Polish rural areas. Sustainability 11(20):5848. https://doi.org/10.3390/su11205848
Stanny M, Strzelczyk W (2018) Kondycja finansowa samorządów lokalnych a rozwój społeczno-gospodarczy obszarów wiejskich, Ujęcie przestrzenne. IRWIR PAN
Stiglitz JE, Sen A, Fitoussi JP (2009) The measurement of economic performance and social progress revisited. OFCE
Stimson R, Stough R, Roberts B (2006) Regional economic development: analysis and planning strategy. Springer
Stjernberg M, Norlén G, Vasilevskaya A, Tapia C, Berchoux T (2023) Scoping report on European rural typologies (GRANULAR Project). Mediterranean Agronomic Institute of Montpellier
Storper M (1997) The regional world: territorial development in a global economy. Guilford Press
Swianiewicz P, Łukomska J (2020) Finanse samorządu terytorialnego w dobie pandemii

Todaro MP (1999) Economic development, 7th ed. Longman
Tomaney J, Pike A, Rodríguez-Pose A (2010) Local and regional development in times of crisis. Environ Plan A 42:771–779
Toptsidou M, Böhme K, Aalbu S, Schuh B, Dallhammer E, Gorny H, Gaugitsch R (2024) Cohesion Policy and the single market: the cost of non-cohesion. European Committee of the Regions
Torre A, Rallet A (2005) Proximity and localization. Reg Stud 39(1):47–59. https://doi.org/10.1080/0034340052000320842
Tóth BI (2015) Territorial capital: theory, empirics and critical remarks. Eur Plan Stud 23(7):1327–1344
Tóth BI (2017) Territorial capital—a fuzzy policy-driven concept: context, issues, and perspectives. EUROPA XXI 33:5–19. https://doi.org/10.7163/Eu21.2017.33.1
Tóth BI (2023) Territorial capital in the European Union: Measuring the territorial endowments of the EU-28 NUTS-2 regions over the 2010s. Reg Stat 13(1):3–35. https://doi.org/10.15196/RS130101
Territorial Agenda of the European Union 2020 (2011) Towards an inclusive, smart and sustainable Europe of diverse regions
Trinh LTV, Gibson L, Oxley L (2005) Measuring the stock of human capital in New Zealand. Math Comput Simul 68(5–6):485–498
United Nations (2021) Managing infrastructure assets for sustainable development: a handbook for local and national governments. https://www.un.org/development/desa/financing/document/un-handbook-infrastructure-asset-management
Vaishar A, Zapletalová J (2008) Small towns as centres of rural microregions. Eur Countryside 1(2):70–81. https://doi.org/10.2478/v10091-009-0006-4
van Geenhuizen M, Nijkamp P (2000) The learning capabilities of regions: conceptual policies and patterns. In: Boekema F, Morgan K, Bakkers S, Rutten R (eds) Knowledge, innovation and economic growth. Edward Elgar, pp 38–56
Van Wee B, Maat K (2001) Land use and transport: a review and discussion of Dutch research. Paper presented at the International Conference on Land Use and Modal Choice, Amsterdam
Wallerstein I (1976) The modern world system I: capitalist agriculture and the origins of the European world-economy in the 16th century. Academic Press
Wallerstein I (1991) Geopolitics and geoculture: essays on the changing world system. Cambridge University Press
Walls D (1978) Internal colony or internal periphery? A critique of current models and an alternative formulation. In: Lewis MH (ed) Colonialism in modern America: the Appalachian case. Appalachian Consortium Press, pp 319–349
Warf B (2013) Global geographies of the Internet. Springer
Warf B (2018) Digital divides in the twenty-first century United States. In: Felgenhauer T, Gäbler K (eds) Geographies of digital culture. Routledge, pp 115–130
Watt P (2009) Housing stock transfers, regeneration and state-led gentrification in London. Urban Policy Res 27(3):229–242. https://doi.org/10.1080/08111140903154147
Weck S, Madanipour A, Schmitt P (2021) Place-based development and spatial justice. Eur Plan Stud 30(5):1–18. https://doi.org/10.1080/09654313.2021.1928040
Weisfelder R (1992) Lesotho and the inner periphery in the new South Africa. J Mod Afr Stud 30(4):643–668
Whittlesey D (1954) The regional concept and the regional method. In: James PE, Jones CF (eds) American geography: inventory and prospect. Syracuse University Press, pp 19–68
Wilke U, Pyka A (2024) Sustainable innovations, knowledge and the role of proximity: a systematic literature review. J Econ Surv 1–26.https://doi.org/10.1111/joes.12617
William J, Lewis M (2009) Health investments and economic growth: macroeconomic evidence and microeconomic foundations, World Bank Policy Research Working Paper
Williamson OE (1981) The economics of organization: the transaction cost approach. Am J Sociol 87(3):548–577. https://doi.org/10.1086/227496
Williamson OE (1975) Markets and hierarchies: analysis and antitrust implications. Free Press

Wójcik M, Dmochowska-Dudek K, Jeziorska-Biel P, Tobiasz-Lis P (2018) Understanding strategies for overcoming peripherality: a polish experience of transition. Bull Geogr Soc Econ Ser 40(40):173–192. https://doi.org/10.2478/bog-2018-0022
Wójcik M, Suliborski A (2021) Poznawcze i praktyczne dylematy w geograficznych badaniach funkcjonalnych wsi. Rozwój Regionalny i Polityka Regionalna 55:9–20. https://doi.org/10.14746/rrpr.2021.55.03
Woods M (2012) New directions in rural studies? J Rural Stud 28(1):1–4. https://doi.org/10.1016/j.jrurstud.2011.12.002
World Bank (2019) Governance indicators database
World Bank (2020a) World development indicators
World Bank (2020b) Human capital index 2020 update: human capital in the time of COVID-19
Xu Y, Ge W, Liu G, Su X, Zhu J, Yang C, Ran Q (2022) The impact of local government competition and green technology innovation on economic low-carbon transition: new insights from China. Environ Sci Pollut Res 30(9):23714–23735. https://doi.org/10.1007/s11356-022-23857-1
Yakymova L, Kuz v (2019) The use of discriminant analysis in the assessment of municipal company's financial health. Econ Sociol 12(2):64–78. https://doi.org/10.14254/2071-789x.2019/12-2/4
Zdražil P, Pernica B (2021) Stimuli contributing to local property taxation—with the focus on spatial effects of industry and Functional Urban Areas. Central Eur J Public Policy 15(1):55–66. https://doi.org/10.2478/cejpp-2021-0004
Źenka J, Slach O, Pavlĭk A (2019) Economic resilience of metropolitan, old industrial, and rural regions in two subsequent recessionary shocks. Eur Plan Stud 27(12):2288–2311

Chapter 3
Navigating Shocks: The Role of European Cohesion Policy During the SARS-CoV-2 Pandemic

Abstract In this chapter, we analyze the SARS-CoV-2 pandemic as an example of a global development shock, focusing on its impact on the economy, spatial inequalities, and EU Cohesion Policy. We present the origins and classification of economic shocks, highlighting their multidimensional nature and diverse socio-economic consequences. We demonstrate how the pandemic has exposed asymmetries in adaptive capacities between core and peripheral areas. Our analysis is structured around the concept of territorial capital, considering both tangible and intangible resources, such as transport and digital accessibility, human, social, and institutional capital, as well as financial and material assets. We examine how the pandemic has deepened inequalities in access to public services, the labor market, education, and infrastructure, particularly in peripheral areas. We also emphasize the role of EU Pohesion Policy as a tool for mitigating the effects of the pandemic and promoting sustainable regional development. Finally, we highlight the need for more integrated actions that would enhance regional resilience, reduce inequalities, and support spatial integration, preparing regions for future challenges.

Keywords Global shock · Resilience · Adaptation · Territorial capital · Cohesion Policy

3.1 The SARS-CoV-2 Pandemic as an Example of a Global Development Shock

The concept of shock, originally rooted in medicine, has gained universal significance across various fields such as economics, psychology, and sociology, where it describes both sudden crises and intense emotional reactions. In contemporary usage, it has expanded to encompass both negative and positive experiences, reflecting the complexity of human emotions and social phenomena (Plichta et al. 2022). Economic shocks, which are at the core of our analysis, are associated with sudden, usually external, and unpredictable events that have a significant impact on the economy, making them economically relevant (Bernanke 1986; Ramey 2016). These events

P. Churski et al., *Inner Peripheries in the New Conditions of Cohesion Post-SARS-CoV-2*, Economic Geography,
https://doi.org/10.1007/978-3-032-10936-1_3

can affect socio-economic relations both positively and negatively, and they vary in scale and intensity. Research on the origins, dynamics, and consequences of economic shocks dates to the early twentieth century, with roots in the work of Keynes, who in the 1930s highlighted the crucial role of aggregate demand and expectations in shaping economic responses to shocks. His theories also emphasized the importance of expectations and uncertainty in amplifying the effects of shocks and pointed to the need for government intervention, including fiscal and monetary policies, to counteract the negative effects of recessions (Kurmann and Otrok 2013). These ideas became the foundation of economic shock analysis and an inspiration for subsequent generations of economists. The oil shocks of the 1970s and business cycle analyses accelerated research on economic shocks (Kar and Bhattacharya 2011), leading to the development of the first classifications and typologies. The emergence of analytical tools such as VAR and DSGE models significantly contributed to a deeper understanding of the dynamics of economic shocks and their propagation in different economic systems (Collard and Fève 2008; Giacomini 2013). A key development in this research was the introduction of the concept of regional resilience to economic shocks. Martin's (2011) studies on economic resilience and recessionary shocks underscored the importance of regional differences in responses to economic disturbances. Modern research on economic shocks increasingly considers complex interdependencies and global challenges. These analyses focus on the interaction between shocks and business cycles, emphasizing structural differences between economies and the role of institutions in building economic resilience. Concepts such as hysteresis have been introduced, indicating that the effects of certain shocks can have a lasting impact on economic potential and employment levels, highlighting the need for appropriate monetary and fiscal policies to support long-term economic stability (Kienzler and Schmid 2014). The development of analytical tools such as SVAR (Structural Vector Autoregression) models and research on nonlinear economic relationships has enabled more precise forecasting and assessment of shock effects (Ramey 2016). At the same time, increasing interest in post-Keynesian concepts, which integrate elements of uncertainty, historical time, and institutions, has broadened the research perspective, allowing for a deeper understanding of how shocks influence modern economies. As a result, the analysis of economic shocks has evolved from simple supply-and-demand models to multidimensional frameworks that consider both local and global conditions.

Economic shocks can be classified in various ways according to different criteria (Drop 2001; Lettau and Ludvigson 2011; Reed 2020). A basic classification of shocks, along with a brief description, is presented in Table 3.1.

Economic shocks can belong to multiple categories simultaneously, as these classifications are not mutually exclusive. They can also be cumulative, meaning that one shock may trigger another, and they may overlap. According to this classification, the shock caused by SARS-CoV-2 was a global, multidimensional supply-and-demand shock of large scale with lasting consequences (Table 3.2).

In the context of our research, the analysis of economic shocks and their consequences must also address the crucial issue of the spatial asymmetry in responses to

Table 3.1 Characteristics of economic shocks

Criterion	Type of shock	Description
Source of impact	Demand-side	Arise from changes in demand for products/ services in a country, region, or sector. These may result from shifts in consumer preferences, changes in public spending and transfers, or the introduction of trade policy instruments domestically or by a trading partner
	Supply-side (including technological)	Related to changes in the availability of production factors, which may stem from technological advancements (technological shocks), natural disasters, or restrictions on resource utilization
	Financial	Associated with disruptions in financial markets, leading to changes in the availability or cost of money and financial system stability, which ultimately affects the real economy
	Political	Resulting from deliberate or accidental actions of public authorities in areas such as fiscal, monetary, or trade policy
Area of impact, geographical scope	International (global)	Affecting an area beyond the borders of a single country
	National	Limited to a single country or a specific jurisdiction
	Regional	Affecting a specific region, sometimes spanning multiple countries but usually on a smaller geographical scale
	Local	Occurring within a very limited geographical area, usually with minor consequences for the overall economy
	Sectoral	Related to one or more specific sectors, independent of geographical location
Shock origin (source)	Internal	Originating from domestic conditions, including fiscal or monetary policy changes within the country
	External	Stemming from external factors such as international price fluctuations, global economic crises, changes in trade policies, etc
Duration of impact	Short term (temporary)	Having a brief impact on the economy, resulting from transient and short-lived disruptions
	Long term (persistent)	Having a long-lasting and significant impact on the economy, often associated with fundamental structural changes

(continued)

Table 3.1 (continued)

Criterion	Type of shock	Description
Intensity and scale	Large-scale shocks	Typically arising from global processes affecting multiple sectors and countries, such as the global financial crisis or the SARS-CoV-2 pandemic
	Small-scale shocks	Affecting specific sectors, regions, or single countries, usually caused by local factors, such as the collapse of a major company in a particular location
Shock effects	Single-dimensional	Effects are limited in scope and impact only a specific aspect of the territorial socio-economic system
	Multidimensional	Effects extend across multiple dimensions, affecting various aspects of the territorial socio-economic system (e.g., economic, social, cultural impacts)

Source Own elaboration

these shocks and the related concept of economic resilience across different territories. The relationship between the asymmetry of economic shocks and spatial development is a complex, multidimensional issue that has attracted significant attention in socio-economic geography and regional studies.

One of the key aspects of understanding economic shocks is recognizing that their effects are not evenly distributed across space and largely depend on the resilience of a given area. While the shock itself may be uniform, its impact can vary based on the socio-economic context and territorial capital, leading to asymmetries. Areas with stronger economic foundations and diversified industrial sectors are generally better equipped to withstand shocks and adapt more quickly to changing economic conditions compared to more vulnerable regions. Martin and Sunley (2014) discuss the concept of regional economic resilience, emphasizing that different regions respond to economic shocks based on their structural characteristics and governance systems. For example, regions with diversified economies are better prepared to absorb shocks compared to those reliant on a single industry. This resilience plays a crucial role in long-term socio-economic growth patterns, as regions that frequently experience significant economic disruptions may face persistent disparities in economic well-being (Martin 2011).

Cities play a key role in shaping spatial development patterns, particularly in the context of their Functional Urban Areas (FUAs), which include both the urban core and its surrounding influence zones. The fact that resilience during crises varies significantly depending on structural characteristics is confirmed by research from Sutton and Arku (2022). Their analysis of Canadian cities during the 2020 economic crisis revealed that larger cities demonstrated lower initial resilience but a higher capacity for recovery compared to smaller cities, which faced greater challenges in this regard. This finding suggests that the size and structure of a city, as well as

Table 3.2 Classification of the SARS-CoV-2 economic shock

Type of shock	Characteristic
Supply-side	SARS-CoV-2 caused disruptions in the availability of production factors due to breaks in global supply chains, factory closures, and restrictions on worker mobility. It represented a clear supply shock, resulting from the inability to utilize production resources
Demand-side	The pandemic led to a decline in demand for many goods and services, especially in sectors such as tourism, hospitality, and transportation, due to mobility restrictions and shifts in consumer preferences toward savings and essential goods
Political	The introduction of lockdowns, sanitary restrictions, and government relief programs was a response to the pandemic, but these measures themselves constituted a political shock, impacting economies
Global	The crisis affected nearly all economies worldwide, disrupting international trade, capital flows, and the functioning of global labor markets
External	The pandemic originated from external factors beyond the control of individual economies, extending beyond local or national conditions, as SARS-CoV-2 spread on a global scale
Long term (persistent)	The pandemic accelerated lasting changes, such as the growth of remote work, digitalization, and e-commerce, fundamentally altering economic structures in the long run
Short term (temporary)	In some areas, such as the decline in demand for luxury goods or production disruptions, the effects of the pandemic were temporary and subsided once restrictions were eased
Large-scale	The pandemic impacted multiple sectors and countries simultaneously, leading to a global recession and necessitating international cooperation to combat the crisis
Multidimensional	The effects of the pandemic extended beyond the economic sphere, affecting public health, education, culture, and social relations. It forced changes in work models, lifestyles, and the functioning of public institutions

Source Own elaboration

its FUA, can significantly influence its ability to cope with economic shocks and recovery processes. The complexity of this relationship further complicates the link between shock asymmetry and spatial development. Another aspect highlighted by Fochezatto et al. (2023) is the spatial differentiation in how affected areas absorb and retain income in response to economic shocks. Studies conducted in Brazil and Chile showed that economically weaker regions demonstrated a greater ability to retain income during supply shocks, suggesting that such shocks can contribute to reducing spatial inequalities. In contrast, research in China (Li et al. 2022) indicated varying levels of resilience in border regions during the pandemic, with trade and consumer behavior being particularly sensitive to shocks.

It can therefore be concluded that different areas respond to shocks in varying ways due to their unique economic structures, governance systems, and internal resilience. Understanding the dynamics of these processes, which are shaped by the diverse territorial capital of each area, is crucial for Cohesion Policy aimed at promoting spatially balanced development. It is also essential for policymakers who use economic and social policy tools to mitigate the negative effects of shocks and alleviate the social tensions they generate.

3.1.1 Global Shocks as Challenges for Development Processes

Crises, shocks, and economic slowdowns are inherent aspects of economies worldwide. Regardless of their origin or scope, they often impact the fundamental structures of the economy, shaping development opportunities and dynamics while frequently exacerbating spatial inequalities between regions. Analyzing the effects of crises from a spatial perspective allows for a deeper understanding of disparities in resource availability, their flows, and adaptive capacities between central and surrounding (including peripheral) areas. This perspective is crucial both for the functioning of the foundational economy and for achieving territorial justice. From the standpoint of the foundational economy, spatial analysis helps identify areas particularly vulnerable to the effects of crises, enabling targeted interventions to strengthen local markets and ensure continuous access to essential services. Improving the functioning of the foundational economy is also a prerequisite for enhancing local resilience and mitigating the long-term consequences of crises. Proper support can contribute to reducing disparities in access to goods and public services, advancing the goal of territorial justice while simultaneously reinforcing the foundational economy as a pillar of local development. To better understand how less-developed areas, including inner peripheries, cope with such challenges, we have adopted Camagni's approach based on the concept of territorial capital. This framework emphasizes that developmental factors are closely tied to the local context, shaped by the relational development environment (relational space) created by residents. This approach considers both tangible and intangible capital, including factors such as accessibility, human, social, institutional, financial, and material capital. Such a perspective allows for a more comprehensive reflection of a given territory's specific resources and their influence on competitive advantages and resilience to shocks. The theoretical and historical foundations of uneven development that guide our analysis suggest that a thorough understanding of the development processes of internal peripheries requires that both exogenous (external) and endogenous (internal) factors be considered. The evolution of economic approaches from classical and neoclassical theories to modern concepts related to New Economic Geography, new institutional economics, evolutionary economics, resilience theories, and behavioral economics, demonstrates the necessity of this broad analytical framework. The approach we have adopted and adapted from Camagni is particularly relevant for internal peripheries, which suffer more from relational isolation than from geographical remoteness. Each of the resources and factors considered in our study plays a specific role in enhancing a territory's ability to adapt and minimize the negative effects of shocks. Recognizing "accessibility" as a key element of this analysis allows for a better understanding of challenges related to limited infrastructural connections, as well as access to digital and institutional resources, which influence the ability to withstand crises and adapt to new economic conditions. Human, social, institutional, and material capital support adaptive processes, but their effectiveness depends on local conditions and how efficiently they are utilized, both individually and in combination. Applying the concept of territorial capital not only helps identify spatial inequalities in resources and adaptive potential but also provides insight into how the interconnections among these components shape regions' ability to respond to crises and recover over the long term. The stability and flexibility of the set of factors (further detailed in our analysis through subfactors), aligned with the principles of Camagni's territorial capital framework, enable the adaptation of the analysis to change local and global conditions. This approach facilitates the systematic examination of the relationships between local resources and development processes, while also creating flexible analytical frameworks that consider both material and intangible aspects crucial for sustainable regional development under crisis conditions. The more diverse the resources and potential of a given area, the greater its resilience to external disruptions and its capacity to return to a growth trajectory. Conversely, the greater the deficits in the state and transformation of territorial capital, the more severe the social and economic consequences, leading to peripheralization and marginalization of the region.

Thanks to such a comprehensive analysis, we will be able to better understand how various development factors respond to crises and identify the most effective strategies for mitigating them. This, in turn, will enable the formulation of more effective development policies tailored to the specific characteristics and needs of different territories.

Accessibility

Economic and social crises, both local and global, highlight the critical importance of accessibility as a foundation for the functioning of modern societies and economies. Accessibility, understood in both transport (internal and external) and digital dimensions, forms the backbone of regional integration, labor mobility, and the efficient operation of markets (Sect. 2.3.3). It also plays a key role in maintaining territorial cohesion and justice, particularly in regulating access to essential goods and public services across different spatial dimensions. Analyzing accessibility provides deeper insights into the mechanisms responsible for differences in resilience across regions and their vulnerability to crises. The ability of residents to mzove within a functional area (internal accessibility) is determined by the state of transport infrastructure networks (material capital) and the availability of transportation services (financial, social, and institutional capital). Peripheral areas, characterized by lower population density and less-developed transport infrastructure, experience significantly lower accessibility compared to core areas. Crises further expose accessibility deficits, disproportionately affecting peripheral areas and limiting their ability to respond to economic and social shocks (Cao et al. 2023). During crises, inadequate transport infrastructure in peripheral areas hinders access to essential public services and markets. Challenges also arise in external trade and access to specialized services, weakening the competitiveness of local businesses and leading to job losses. Accessibility, defined by the number and quality of transport connections, is also crucial for local and regional labor markets. Labor mobility plays a central role in economic adaptation to crises and shocks. In core areas, greater workforce mobility, better opportunities for retraining, and higher demand for diverse skills enable a more flexible response to changing economic conditions (Arpaia et al. 2016). Workers can relocate quickly to where they are needed, reducing pressure on local labor markets and mitigating the effects of economic disruptions. In contrast, peripheral areas face numerous employment and mobility constraints, which typically exacerbate the negative consequences of economic shocks (Ray et al. 2016). As a result, peripheries experience prolonged unemployment and economic stagnation. Crises can also lead to drastic reductions in transportation services (Borkowski et al. 2021). Economic downturns and declining public revenues can reduce funding for public transportation, leading to service cuts and fare increases. This disproportionately affects low-income individuals who rely heavily on public transport (Efthymiou and Antoniou 2017). Furthermore, the economic consequences of crises may result in decreased investment in transport infrastructure, further limiting accessibility in the future (Borca et al. 2021). Inadequate public transport efficiency leads to increased reliance on private cars as the primary means of transportation (Kołsut and Stryjakiewicz 2023; Komornicki 2011), which raises mobility costs and excludes individuals who do not have access to personal vehicles. Residents of peripheral areas face greater challenges due to longer commuting distances and fewer public transport options, making them more dependent on private vehicles or informal transport solutions. Core areas, typically characterized by denser transport networks and more frequent connections, can respond to crises more effectively and maintain functional cohesion. A similar dynamic applies to an area's ability to establish connections with higher-order central hubs (external accessibility). The quality and capacity of transport infrastructure, such as highways, rail lines, and regional bus connections, play a crucial role in minimizing the impact of shocks. During crises, peripheral areas often lose critical transport links, weakening their integration with central hubs. This, in turn, reduces access to a wide range of goods and services, such as specialized medical or educational facilities, while also limiting trade opportunities and economic exchange. From the perspective of local labor markets, low external accessibility restricts labor mobility, exacerbating the effects of economic shocks and reducing the ability of peripheral areas to adapt to economic changes (Ray et al. 2016).

New opportunities for improving accessibility arise from e-accessibility, which leverages digital technologies as an important alternative to traditional methods of enhancing physical accessibility. In the context of global transformations, digital transformation and technological progress are playing an increasingly significant role in shaping labor markets. The introduction of new work models, such as remote work, the gig economy, freelancing, and digital platforms, is changing labor market dynamics, particularly in core areas that have access to modern digital infrastructure (Ma 2024; Zhu et al. 2023). Digitalization enhances labor market flexibility, allowing workers to adapt more quickly to changing economic conditions. Technology also creates new employment opportunities in high-tech sectors, strengthening economic resilience to shocks. However, peripheral areas often lag in this process. Limited access to technology, a lack of investment in digital infrastructure, and low digital literacy among workers prevent these regions from fully utilizing modern technological solutions, even though such advancements could significantly improve their accessibility (Wang 2024; Zhou et al. 2023). In this context, digital resilience is gaining increasing importance (Copestake et al. 2024; Knuepling et al. 2024). Solutions such as remote work tools, digital education platforms, and systems supporting local entrepreneurship can not only mitigate the economic impact of crises but also reduce inequalities between core and peripheral areas. Implementing digital technologies allows peripheral regions to better adapt to evolving economic conditions, strengthening their ability to respond to shocks and enhancing spatial cohesion. However, this also deepens the digital divide between core and peripheral areas. The latter not only struggle to adapt to modern work models but also remain more vulnerable to the long-term effects of economic crises. E-accessibility also includes remote access to public services, which becomes particularly crucial during periods of economic shocks. Underfunded telecommunication infrastructure in peripheral areas limits service accessibility, leading to digital exclusion, exacerbating development inequalities, slowing economic growth, and further marginalizing these regions while concentrating resources and investments in core areas.

In times of crisis, accessibility responds unevenly, largely reflecting existing infrastructural and organizational disparities. Core areas, with their well-developed transport and digital infrastructure, can adapt more quickly to changing conditions, maintaining access to key resources and markets. In contrast, peripheral areas, which face deficiencies in both physical and digital accessibility, experience long-term crisis effects, deepening their dependence on central regions and reinforcing developmental inequalities. The decline in public transport use and the rise of individual mobility during crises highlight the need for a balance between public and private transport. Improving transport infrastructure enhances connectivity within Functional Areas (FAs), enabling residents' easier access to essential social services such as education, healthcare, and administrative support. Investments in transport infrastructure and improved accessibility can drive significant economic growth and generate momentum for further development, especially in resource-limited regions (Badiru et al. 2023; Luz et al. 2016). To mitigate the negative effects of crises and strengthen resilience, an integrated approach to transport and digital accessibility is essential. Addressing existing deficits and promoting equal access across all areas requires investments in physical infrastructure, such as expanding road and rail networks and optimizing their use, which must go hand in hand with digital infrastructure development, ensuring a balance between physical and digital accessibility. A key priority is improving the quality of public transport, particularly in peripheral regions, to increase internal mobility and reduce spatial inequalities. At the same time, digitalizing public services, implementing modern technologies, and enhancing residents' digital skills are critical components of a long-term strategy to reinforce the socio-economic resilience of different areas. The combination of these measures can not only reduce inequalities but also establish the foundations for more sustainable and adaptable development in the face of future shocks. Tackling these challenges requires targeted policies that prioritize equity and inclusivity in planning efforts, ensuring that all communities have access to essential services and opportunities, regardless of their geographic location or socio-economic status.

Human Capital

Crises significantly deepen spatial inequalities and pose major developmental challenges for peripheral areas, impacting human capital and its key subfactors: demographic trends, health and safety, education, and the labor market (Sect. 2.3.3). Analyzing these subfactors provides a better understanding of the mechanisms responsible for differences in regional resilience, highlighting how inequalities in these areas shape the ability to respond to crises and adapt to new conditions. In terms of demographic trends, the effects of crises can manifest in population dynamics, migration patterns, birth rates, and demographic resilience, often exacerbating socio-economic disparities. Core areas typically exhibit a greater capacity for demographic adaptation during crises. Economic shocks frequently trigger urbanization and gentrification processes, leading to increased population density in city centers and attracting young, highly skilled labor. This, in turn, fosters demographic renewal, stimulating and reinforcing regional development processes (Carlucci et al. 2018). Conversely, peripheral areas experience population decline, leading to a deterioration in demographic structure and a reduction in development potential due to shrinkage in the tax base and labor force needed to sustain local economies (Colantoni et al. 2020; Hălbac-Cotoară-Zamfir et al. 2021). A relatively higher level of economic security and better access to healthcare and childcare infrastructure (e.g., nurseries) in core areas also support procreation trends, resulting in less severe demographic changes during crises. In peripheral areas, families may prioritize financial stability, discouraging family expansion (Benassi and Salvati 2019). This further exacerbates labor shortages and restricts the ability of peripheral regions to overcome economic difficulties in the future.

One of the most visible effects of crises and shocks is the shift in migration patterns. Core areas, with their well-developed economic and social infrastructure, attract individuals seeking economic stability and better career prospects (Benassi and Salvati 2019). In contrast, peripheral areas experience an ongoing "brain drain" and depopulation (Salvia et al. 2020). Over the long term, these trends hinder the implementation of development policies in such regions, as local human resources become insufficient to support economic recovery efforts. Additionally, internal migration reinforces demographic disparities between core and peripheral areas, further marginalizing the latter. As a result, core areas have greater demographic resilience. Developed urban centers can adapt more effectively to demographic changes due to economic specialization and the availability of public services (Colantoni et al. 2020). The lack of access to resources and public services in peripheral areas limits their ability to attract and retain residents, leading to further weakening of their demographic potential. Alonso et al. (2023) highlight that the limited economic opportunities in peripheral regions significantly hinder their ability to recover after crises, perpetuating marginalization and spatial inequalities. The consequences of negative demographic processes include economic decline, which leads to additional social tensions. This impact is particularly evident in access to education, healthcare, and employment, contributing to cycles of social exclusion and a continued deterioration in quality of life (Price et al. 2013).

Crises also have a varied impact on public health and perceptions of health security, resulting in different experiences and consequences across regions. Core areas, due to their better healthcare infrastructure, are generally more prepared to manage crises and respond effectively to health threats (Ženka et al. 2017). In contrast, healthcare systems in peripheral areas are often underfunded, which limits access to medical facilities and personnel; under crisis conditions, these limitations can further exacerbate health inequalities and result in higher rates of illness (Ferguson et al. 2020; Guzal-Dec and Zwolińska-Ligaj 2023; Marmet et al. 2020; Yang et al. 2022). Differences also exist in the availability of psychological support, potentially resulting in increased anxiety, depression, and other mental health issues during crises (Brown and Cowling 2021).

Another important factor is public safety. Economic downturns can lead to increased crime, particularly in areas where unemployment rises and social cohesion weakens, making peripheral regions especially vulnerable. Economic isolation and stress in these areas can further deepen mental health challenges and contribute to a sense of insecurity, creating a difficult cycle of unfavorable conditions (Brown and Cowling 2021; Kaneko et al. 2021). In core areas, while crime rates may also increase, stronger law enforcement presence and social programs help mitigate these effects (Annoni et al. 2019). The interconnection between economic stability and health outcomes underscores the need to address regional inequalities in health and security, particularly during times of crisis.

Crises affect access to education and learning outcomes in different ways across regions. Financial constraints resulting from economic downturns tend to impact peripheral areas more severely, as these regions rely more heavily on public funding sources (Fernández-Albertos et al. 2013). Budget cuts can lead to larger class sizes, fewer resources, and even school closures, making it more difficult for students in peripheral areas to access quality education (Halikiopoulou and Vasilopoulou 2016; Kholyavin 2017). The challenges in these areas are further exacerbated by insufficient infrastructure, outdated materials, and a shortage of qualified teachers (Oravský et al. 2020), which is particularly concerning in an era of digital transformation. Digital skills and access to modern technologies and educational resources are essential for ensuring equal learning opportunities and leveraging e-accessibility for remote education. In contrast, core regions typically have more resources available to invest in teacher training and educational technologies, allowing them to maintain higher educational standards even during crises (Antonová and Zapletalová 2014).

Economic downturns can also lead to lower academic performance. The relatively poorer financial situation in peripheral areas may force students to drop out of school to contribute to family income, with long-term consequences for their future employment prospects (Demirci et al. 2019). Wider availability of public support programs, scholarship systems, and social assistance in core regions can help mitigate these effects (Boomgaarden et al. 2011). This dichotomy contributes to a cycle of poverty in peripheral areas, where lower educational attainment limits economic mobility and opportunities for future generations. Mental health and student well-being also play a crucial role in educational outcomes. Shocks and crises tend to intensify anxiety and stress, particularly in peripheral areas where, as previously mentioned, access to mental health support is limited (Aslan and Çinibulak 2022; Srinivasan et al. 2014). The long-term effects of economic crises on education are especially severe in peripheral regions, where interruptions in learning can lead to significant difficulties in the labor market. As Amate-Fortes et al. (2017) point out, individuals who experience disruptions in education are more likely to face unemployment and lower wages, creating a cycle of disadvantage that is difficult to break.

A lower level of education restricts access to better job opportunities and career development, reinforcing socio-economic inequalities (MacKenzie et al. 2014). In contrast, residents of core regions, with better access to educational resources and support systems, have greater opportunities to overcome economic hardships, leading to stronger labor market outcomes (Marangos and Triarchi, 2022). It is therefore essential to recognize these disparities and implement targeted strategies to support education in peripheral areas, ensuring equal access to quality education and resources during crises (Huikari et al. 2019; Leite 2019).

Economic slowdowns and crises can also lead to significant changes in employment patterns, job availability, and overall labor market dynamics. During economic turbulence, employment rates typically decline, but in core areas, a diversified economic structure, greater flexibility, and labor market resilience can provide a certain degree of stability (Stan 2022). Peripheral areas, with a less diversified structure and lower economic resilience, experience sharper fluctuations in unemployment, higher job loss rates, and longer periods of unemployment during crises (Lavesson 2016; Partridge et al. 2010).

Core and peripheral areas also differ in terms of job quality. Core regions often have a higher concentration of jobs requiring qualifications and offering better working conditions, which can help mitigate the impact of economic shocks (Ženka et al. 2017). In contrast, peripheral regions may experience a decline in job quality, as many workers are forced into low-paid, unstable forms of employment as economic conditions deteriorate (Gaag and Beer 2014). Formal employment opportunities shrink, while the number of workers in the informal sector increases (Treviño and Cuellar 2021). These workers often operate outside regulated legal and social frameworks, making their employment unstable and poorly compensated. Since informal jobs do not provide social security or legal protections, these regions become especially vulnerable to rising instability during crises. As a result, unemployment and income insecurity increase, exacerbating existing issues and delaying economic recovery. In the long run, the lack of stable income and limited access to social insurance restrict workers' ability to save and invest, ultimately harming peripheral economies. By contrast, core regions tend to recover more quickly from economic shocks due to their more diverse employment opportunities. Core economies typically encompass multiple sectors, such as financial services, technology, and advanced industries, allowing for better risk distribution during crises (Bisello et al. 2020). A diversified economy and well-developed social institutions enable core areas to absorb shocks in one sector while others continue to function, facilitating faster recovery.

Migration is another key factor influencing labor markets. During crises, workers migrate in search of employment, increasing labor supply in core regions while decreasing it in peripheral areas, which further hampers economic recovery in the latter (Giannakis et al. 2018; Tomao et al. 2021; Yin et al. 2014). Access to training programs also plays a critical role in labor market outcomes. Core areas typically have more resources to invest in workforce development and skills programs, helping workers adapt to changing labor market demands (Pavel et al. 2020). Job opportunities can also improve through public support programs for businesses (Giannakis et al. 2018). However, peripheral regions often have limited access to such resources, making it difficult for workers to acquire new skills or transition to other industries during crises (Gailhard and Kataria 2014). This inequality can lead to long-term negative effects on employment prospects and economic stability in peripheral areas. Individuals who experience job loss during economic downturns often struggle to re-enter the labor market, leading to a rise in long-term unemployment (Dong 2023; Rost et al. 2010; Xie 2023).

The consequences of crises clearly affect human capital in different ways. Core areas, due to better infrastructure and greater resources, demonstrate higher resilience to shocks, maintaining demographic stability, higher educational standards, and better health outcomes (Li and Wang 2022). Additionally, these regions invest in education systems and scientific research, which supports further human capital development and builds a strong foundation for long-term competitiveness (Martin and Sunley 2014; Martin et al. 2016). The adaptive capacities of core areas enable them to sustain high economic dynamics despite external shocks, strengthening their competitive advantage and attracting skilled labor.

In contrast, peripheral regions face limitations in access to public services, leading to population decline, labor force depletion, and cycles of social exclusion that reinforce their marginalization. Prolonged crises can result in declining health and education outcomes, as well as reduced opportunities for professional development (Frankenberg and Thomas 2017; Nübler et al. 2020). The erosion of human capital due to such shocks leads to a lasting deterioration in quality of life and the entrenchment of poverty, limiting opportunities for future generations (Harnani et al. 2022). To reduce these disparities, investments in local education systems, healthcare infrastructure, and support for microenterprises and training programs in peripheral regions are essential. Improving access to technology, developing digital skills, and ensuring equal opportunities in mobile healthcare and education services can enhance the adaptability of these areas, contributing to the reduction of inequalities and the construction of long-term socio-economic resilience.

Social and Institutional Capital

Social capital, both bonding and bridging, along with institutional capital, plays a crucial role in a region's ability to respond to crises and recover from shocks. The key subfactors of this development component include social activity, leisure time utilization, business infrastructure, and economic structure and activity (Sect. 2.3.3). These are resources that support social cohesion, institutional stability, and the capacity for innovation and dynamic development within local communities. Economic crises often highlight the differences in levels of social and institutional capital between core and peripheral areas, deepening existing inequalities and leading to the reconfiguration of social networks and relationships, as well as changes in the economic structure of regions. Understanding these mechanisms provides valuable insight into why some regions cope better with crises while others become more vulnerable to their long-term effects.

During periods of economic shocks, there is often an increase in the role of bonding social capital (Hörisch and Obert 2020), which helps people cope with stress more effectively (Collaborative Research 2013; Rückert and Labonté 2014). Regions with strong bridging social capital, on the other hand, are better able to mobilize resources and adapt to the consequences of crises (De et al. 2021; Hannan 2017; Martínez and Wissink 2022; Mohammad and Khan 2021). Peripheral areas exhibit a different dynamic in terms of both types of social capital during crises. Bonding social capital tends to be relatively stronger than in core areas due to its highly integrated nature (Ijeoma 2017; Johansson 2011). In times of crisis, residents of peripheral areas, particularly rural communities, often rely on family and social networks for support, which can strengthen the resilience of the entire community (Lange 2014). However, reliance on bonding social capital can also lead to isolation, limiting exposure to innovative ideas and the creation of new resources that bridging social capital could provide (Drakaki and Tzionas 2017; Giacomini et al. 2023). Bridging social capital is particularly at risk in peripheral areas. Communities in these regions may struggle to maintain connections with broader external networks, which can make it difficult to access information, resources, and support from outside the community (Akgün and Türkoğlu 2023; Blažek et al. 2018). In such cases, geographical distance further reinforces relational distance. Additionally, competency-based exclusion may become a significant barrier, limiting residents' ability to use available digital and educational resources, which could otherwise compensate for limited physical accessibility. This isolation can exacerbate the effects of crises, as these areas often lack institutional support and infrastructure available in core urban regions (Moudud-Ul-Huq et al. 2021; Murphy and Mercille 2019).

Public policy also plays an important role in this regard. In core areas, public interventions often focus on strengthening social cohesion and supporting social networks, which can reinforce both bonding and bridging social capital (Benink 2020; Garbelli 2018). Initiatives aimed at promoting social engagement and cooperation strengthen social ties and facilitate resource-sharing during crises. In contrast, peripheral areas may experience a lack of dedicated support, leading to the weakening of social capital (Macartney 2011; Waldron 2018). The difference in intervention levels can significantly impact the resilience of social capital in different geographical contexts.

Social and institutional capital also responds differently to crises in core and surrounding areas in terms of long-term consequences. Communities in core areas that successfully adapt to crisis conditions may emerge stronger, with reinforced networks, reconfigured relationships, and a more effective system of mutual support (McQuade and Schmitz 2016; Miković 2022). This can contribute to greater resilience against future crises. In contrast, in peripheral areas, the erosion of social capital, which limits access to external resources and support, can reinforce vulnerability to marginalization (Chowdhury and Żuk 2018; Yesin 2015). The mechanisms described above influence social activity and the frequency of interactions. The extensive social networks of core areas facilitate interactions during crises, promote the rapid mobilization of resources and support systems, and enhance crisis response effectiveness. This flexibility and speed of reaction are supported by social service infrastructure, such as social media platforms and networks of organizations, which contribute to an efficient crisis response (Drakaki and Tzionas 2017). In peripheral areas, lower population density, weaker social networks, and poorer social service infrastructure hinder broader social activity and effective responses to shocks (Adekola and Clelland 2019; Morrison et al. 2017).

At the same time, the ability of communities to learn and innovate in the face of challenges can play a crucial role in adapting to the consequences of crises. The process of collective learning and cooperation among different stakeholders during crises facilitates the implementation of social innovations, which help communities adjust to new conditions (Rhouiri et al. 2023). This is particularly important in peripheral areas, where local initiatives can lead to sustainable practices and community-driven solutions that enhance resilience (Campos et al. 2021; Imperiale and Vanclay 2016).

Social relationships and activity are directly influenced by economic conditions. Economic stability is a key component of community resilience, as it directly affects the ability to recover from crises. Core areas, benefiting from a broader diversity of economic opportunities, are better able to shield themselves from the impacts of crises (Pavel et al. 2020; Song and Huang 2021). The role of local businesses in building resilience is particularly evident in peripheral areas, where firms can act as anchors for community support and resource mobilization (Steiner and Atterton 2014, 2015).

Crises significantly alter how people spend their free time. Economic uncertainty during a crisis typically leads to reduced spending on leisure activities, as households prioritize essential needs over discretionary consumption (Glückler and Panitz 2023; Kaicker et al. 2023). Core areas generally have better infrastructure and a wider range of leisure facilities, which can be used during crises to promote physical and mental well-being (Glückler and Panitz 2023; Halikiopoulou and Vasilopoulou 2016). However, economic downturns can lead to service reductions, shortened operating hours, or even the closure of leisure facilities, further limiting access to recreational activities (Glückler and Panitz 2023; Kaicker et al. 2023). As a result, leisure activities often shift toward digital services (Tušl et al. 2021). Peripheral areas, lacking a diverse range of leisure options, may rely more on community-based activities such as local festivals, gatherings, and team sports, which can strengthen social cohesion and resilience during difficult times (Glückler and Panitz 2023; Tilmann et al. 2021). A strong sense of belonging and social support in rural areas can help mitigate some of the negative effects of crises on mental health and engagement in leisure activities (Glückler and Panitz 2023; Tilmann et al. 2021). E-accessibility plays a crucial role in this context. In areas with well-developed digital infrastructure, access to online services and platforms allows residents to adapt their leisure activities to crisis-related restrictions. Streaming services, virtual events, and social media become essential tools for maintaining social connections and engaging in recreational activities despite physical distancing (Tušl et al. 2021). However, peripheral areas may struggle with digital access, particularly in terms of high-speed internet availability, limiting their ability to participate in digital leisure activities (Glückler and Panitz 2023; Halikiopoulou and Vasilopoulou 2016). This digital divide can exacerbate existing inequalities in leisure participation between core and peripheral populations during crises.

Crises can influence leisure activities not only in the short term but also over the long term. While some communities temporarily shift activities to digital and home-based spaces, others adopt more lasting changes, focusing on more sustainable and community-oriented forms of recreation. These long-term shifts can strengthen local social ties, encourage appreciation of local resources, and ultimately enhance community resilience against future crises (Tilmann et al. 2021).

The consequences of crises also have different implications for institutional capital. Stronger institutional frameworks and generally more efficient public offices, by benefiting from economies of scale in core areas, are better equipped to handle the effects of crises (Hardy et al. 2017; Pizzuto 2020). Urban enterprises typically have better access to resources, including financial capital, which allows them to flexibly adjust their operations and sustain economic activity during difficult times (He et al. 2022; Warren and Grigaliūnaitė 2023). A larger tax base can act as a buffer against financial shocks (Ramírez et al. 2022; Totouom 2023).

In contrast, peripheral areas face significant challenges in maintaining economic activity and institutional capital during crises. Peripheral economies are often reliant on a limited number of industries, making them more vulnerable to economic shocks (Lee et al. 2022; Stengel et al. 2022). A lack of diversification can lead to higher business closure rates and prolonged economic stagnation during crises. Peripheral enterprises frequently struggle to secure financing, which hampers their ability to adapt and recover (Giannakis and Mamuneas 2022; Guzal-Dec and Zwolińska-Ligaj 2023). The absence of institutional support can further exacerbate the negative effects of crises on economic activity in peripheral areas, reducing their overall resilience.

Peripheral areas lack a well-developed network of business-supporting infrastructure, which leads to isolation and limited capacity for collective action during economic slowdowns. The absence of a supportive business ecosystem can further exacerbate the negative effects that crises have on economic activity in these regions (Yu and Kim 2021; Zarębski et al. 2022). In contrast, core areas benefit from a high density and concentration of businesses and non-governmental organizations, which foster collaboration and innovation. This environment allows firms to share resources and knowledge, creating synergies that enhance resilience (Brown and Cowling 2021; Ramos-Vielba and Woolley 2018). The agglomeration effect strengthens the ability of businesses to navigate crises more effectively by leveraging existing networks and partnerships.

There are also notable spatial differences in institutional support and economic policy. In core areas, government interventions frequently focus on stimulating economic activity through financial aid, infrastructure investments, and policy reforms designed to support businesses (Hu and Yang 2019; Hu et al. 2022). These measures contribute to building long-term resilience at both the industry and macroeconomic levels. In contrast, peripheral areas receive less targeted support, which results in a slower return to normal economic conditions and an extended recovery period following economic downturns (Arbolino et al. 2019; Uyarra and Valdaliso 2021). The variation in public intervention strategies can have a significant impact on the resilience of institutional capital in different geographical contexts, further deepening disparities in economic development and crisis recovery between core and peripheral regions.

Institutional efficiency is also reflected in the quality of governance and the ability to manage adversity. The greater flexibility and adaptability of institutions in core areas, particularly in maintaining essential services and providing support to residents during crises, highlight the importance of institutional frameworks in building long-term resilience (Mia et al. 2021; Taylor et al. 2021). This adaptability extends to long-term processes as well. Institutions that successfully adjust during crises can emerge stronger, with diversified funding sources, increased operational efficiency, and overall improvements in competitiveness (Matharu and Juneja 2021; Medina-García et al. 2021; Zhou et al. 2023). This transformation can contribute to sustained economic growth and improved service provision. Institutions in core areas have a greater likelihood of experiencing such improvements due to better access to resources, support networks, and government interventions. In contrast, peripheral areas face greater challenges in recovering from crises, as difficulties in institutional and business adaptation can lead to declining local economic activity and rising unemployment (Hermann 2014; Musil 2014). Structural weaknesses in peripheral economies can hinder their ability to rebuild after crises, reinforcing their marginalization and limiting long-term development opportunities.

The impact of crises on institutional capital can also be analyzed through the lens of social capital. Strong social networks in core areas facilitate cooperation and resource-sharing among businesses, improving their ability to respond to crises. This social capital serves as a crucial asset during difficult times, enabling businesses to leverage community support and collective resources. In contrast, peripheral areas may experience a decline in social capital during crises, as economic hardships put pressure on social ties and limit collaboration. The erosion of social capital can further weaken the institutional frameworks necessary for economic resilience in these regions.

Social and institutional capital are inherently interconnected in crisis response, and their condition significantly influences a region's ability to cope with difficulties and adapt to changing conditions. The impact of crises exposes vulnerabilities in these forms of capital, particularly in peripheral areas, where limited resources and weaker cooperation networks make it harder to mobilize communities and support non-governmental organizations. As a result, the absence of established mechanisms for cross-sector collaboration further complicates effective crisis management. The long-term consequences of these processes highlight the need to develop sustainable cooperation structures and promote social and institutional innovation. Strengthening collaboration between public institutions, non-governmental organizations, and the private sector is essential, alongside fostering local initiatives and integrating social and institutional actions at all levels. Special attention should be given to peripheral areas, where external support, appropriate policy frameworks, and supportive infrastructure can enhance social and institutional resilience to future shocks. Building the capacity for effective crisis response requires a holistic approach that takes into account local resources, social activity, and the needs of institutions supporting economic development. By addressing these factors, it is possible to create a more resilient and adaptive system that ensures both economic stability and social cohesion in times of crisis.

Financial Capital

Financial stability is the foundation of economic resilience, but, during crises, the differences between core and peripheral areas become particularly pronounced. Analyzing the impact of crises on financial capital across local governments, businesses, and residents reveals key distinctions in how these areas mobilize resources, cope with income losses, and manage economic risk. Examining these aspects provides a better understanding of the mechanisms responsible for financial resilience disparities and helps identify intervention strategies that could reduce inequalities between regions.

Crises typically lead to an immediate reduction in public revenue, which often results in increased debt and subsequent cuts in public services and infrastructure investments. These fiscal constraints can have long-term implications for community well-being and economic growth (Barrett et al. 2018; McDonald 2018). Core areas, with their broader tax base and diversified revenue streams, are generally more capable of absorbing financial shocks (Kim and Warner 2017; Klein and Linnemann 2018). They also tend to have greater access to external financing, which strengthens their resilience and allows for quicker adaptation to the consequences of a crisis (Gorina et al. 2017; Hinkley and Weber 2020; Mitchell et al. 2020; Park et al. 2018). In contrast, peripheral municipalities rely more on property taxes and external transfers, making them significantly more vulnerable to revenue fluctuations (Anundsen and Nymoen 2015; Kass et al. 2022). Rural peripheral municipalities, particularly those dependent on resource-based sectors, often generate revenue from low-yield agricultural and forestry taxes. The difficulty in diversifying income sources limits their ability to respond effectively during crises (Buerger 2023; Ford and Ihrke 2016). Dependence on a narrow tax base exacerbates financial difficulties, leading to deeper cuts in services and infrastructure investments compared to core regions. Moreover, peripheral local governments often lack the administrative capacity and resources needed to manage fiscal challenges effectively, which increases the severity of the economic shocks they experience (Giuliani 2020; Mandelkern 2015). However, municipalities with strong local leadership and active community engagement are better prepared to navigate financial crises, as these factors enhance their ability to mobilize resources and respond effectively to shocks (Araújo et al. 2020; Hendrick and Wang 2018). The level of financial support from higher government tiers, whether regional or national, also plays a crucial role in shaping financial resilience. Core areas generally receive more proportionate financial aid during crises than peripheral areas (Chernick et al. 2011; Picchio and Santolini 2020; Prescott and Gjerde 2022). The way regions overcome difficulties and adapt to crises often determines their long-term financial resilience. Core areas, which tend to adapt more successfully during economic downturns, often emerge with strengthened financial structures, diversified revenue sources, and improved financial management practices (Dellepiane-Avellaneda 2014; Park 2023). This can lay the groundwork for sustained economic growth and enhanced service provision. On the other hand, peripheral areas face significant challenges in post-crisis recovery, as the erosion of financial capital can restrict their ability to invest in critical infrastructure and services, further reinforcing their vulnerability to marginalization (Jun and Bryer 2016; Overton and Bland 2014). Without targeted intervention, these disparities in financial resilience risk becoming deeply entrenched, limiting the capacity of peripheral areas to achieve long-term economic stability.

Just as with municipal capital, businesses in core areas are generally better prepared to adapt to the effects of crises. More diversified revenue sources, greater resilience to declines in demand, and a broader customer base enable these businesses to absorb shocks more effectively, emphasizing the importance of a solid financial structure (Kapelko and Stefanou 2017; Kestens et al. 2011). Such firms also typically have better access to financial and credit institutions, making it easier for them to secure funding during crises (Binnashir-Alketbi 2023; Smith et al. 2020). In contrast, businesses that rely primarily on local markets may lack the financial resources necessary to withstand economic downturns (Badenhorst 2017; Curran and Zignago 2010). These firms often face difficulties obtaining financing, which hinders their ability to invest in essential changes or innovations during crises (Kang and Na 2018; Suder et al. 2022). This, in turn, leads to stagnation in economic activity, forcing businesses to scale back or even close, which negatively affects municipal finances.

Variations also exist in terms of ownership structure and business size. Small- and medium-sized enterprises, which are more commonly located in peripheral areas, have smaller financial buffers and struggle to adapt to new conditions in financial markets. Over time, shifts in financial markets often lead to consolidation, where stronger firms acquire those facing financial difficulties. During prolonged crises, this process can result in the bankruptcy of entire sectors or the collapse of large companies. Another crucial factor, as previously emphasized in discussions on social and institutional capital, is the concentration of businesses and business support institutions in core areas. Strong social networks and well-developed infrastructure foster innovation, knowledge diffusion, and resource-sharing (Chen 2023; Fadhil and Burhan 2021; Kampe et al. 2021). The agglomeration effect strengthens long-term resilience, as businesses can leverage their networks to navigate crises more effectively. In contrast, peripheral areas lack the developed networks needed to share knowledge efficiently often resulting in a decline in social capital, increased isolation, and a reduced capacity for collective action during economic downturns (Feldkircher 2014; Kazana-McCarthy 2021; Motoc 2022; Pongsakornrungsilp et al. 2021). The erosion of social capital further weakens the financial frameworks necessary for economic resilience in peripheral areas, making them more vulnerable to long-term economic instability.

Undoubtedly, in the case of residents' financial capital, crises have a less severe impact on core areas. Residents of these regions tend to have a more diversified income structure, making them better prepared to absorb the effects of economic shocks (Allanjawi et al. 2023; Eleuteri and Arduino 2014). They also typically have easier access to external financing, enabling them to secure loans and financial assistance more quickly during crises (Poměnková et al. 2019; Zavras 2020). As a result, the decline in demand caused by reduced personal incomes is not as abrupt, which has broader implications for public finances and business stability.

Residents of peripheral areas, on the other hand, are more dependent on often monofunctional local economies and face greater difficulties in coping with economic slowdowns (Mansor et al. 2022; Simona-Moussa and Ravazzini 2018). Those living in peripheral rural areas receive significantly less support compared to core residents (Lusardi et al. 2011; Waldron and Redmond 2015) and face more difficulties in obtaining credit, which hinders their ability to adapt to the consequences of crises (Dolls et al. 2022; Volkos and Symvoulakis 2021).

Social networks and institutional capital also play a crucial role in financial resilience. Core areas offer significantly greater economic and social opportunities, supported by a broader financial ecosystem that facilitates effective adaptation to changing conditions (Abdullah et al. 2022; Lombe et al. 2013). Strong social networks help residents cooperate and share resources, improving their ability to respond to crises (Zhao et al. 2022; Chita et al. 2020). In contrast, peripheral areas often lack such opportunities, leading to isolation and limited economic adaptability during downturns (Crespo and Faytong-Haro 2022; Waldron and Redmond 2016). The absence of well-developed financial institutions, coupled with weaker social networks, further restricts residents' ability to navigate crises, increasing their long-term vulnerability.

The existing differences in the absorption of crisis impacts between core and peripheral areas prompt the search for measures that strengthen financial stability and build the resilience of peripheral regions. Creating local financial support mechanisms, developing credit institutions, and promoting financial innovations can serve as the foundation for effective crisis management and more sustainable development, reducing the gap between regions and preparing them for future shocks.

Material Capital

Crises, regardless of their origin, have a profound impact on material capital, revealing both its weaknesses and potential in the development of core and peripheral areas. Material capital, which includes the quality and use of space, housing conditions, and environmental resources, shapes residents' quality of life and the growth potential of local economies. Analyzing the impact of crises on material capital in the context of core and peripheral regions provides a deeper understanding of the mechanisms responsible for differences in resilience between these areas. At the same time, it allows for the identification of actions that could counteract growing inequalities and promote territorial justice.

Vulnerability to shocks and natural crises largely depends on the quality of infrastructure, which is a key factor differentiating core and peripheral areas. Core areas are characterized by better maintained and modern infrastructure, making them more prepared for shocks. These regions often have better access to resources, including financing and advisory services, which enables them to respond quickly to infrastructure challenges such as improving healthcare and educational facilities or enhancing digital connectivity (Clark-Ginsberg et al. 2020; Naffa and Dudás 2023). Moreover, the presence of strong governance structures in core areas facilitates coordinated crisis responses, allowing for the efficient allocation of resources and infrastructure management (Knodt et al. 2021). With a solid economic base and stronger financial standing, these areas can afford investments in climate-resilient infrastructure, which is crucial for adapting to climate change and other crises (Climate-resilient infrastructure 2018; Fisk et al. 2023). In contrast, peripheral areas, due to weaker infrastructure development, are more vulnerable to the negative consequences of shocks (Mostafavi 2017; Sathurshan et al. 2022). Outdated infrastructure that often fails to meet modern standards leads to a decline in service quality in critical sectors such as healthcare and education, particularly in times of crisis (Bakchan et al. 2021; Sanne et al. 2021). Additionally, limited access to social infrastructure, including the absence of effective social organizations and support networks, exacerbates problems for residents who may lack access to essential services and assistance during emergencies (Arvanitis 2023; Gloger et al. 2021). The weaker structural and financial foundations of peripheral areas make it more challenging to implement necessary infrastructural improvements, which further deepens their vulnerability to future crises.

The physical characteristics of space also play a significant role in how crises affect its use. Core areas often benefit from well-planned urban solutions that support diverse applications and activities, allowing for flexible responses to crisis situations (Fathi et al. 2020; Mostafavi 2017). However, the lack of space, dysfunctions in its development, and poor environmental conditions can lead to spatial conflicts, making it difficult to manage resources effectively and ensure their availability during crises. In contrast, peripheral areas may be characterized by rigid spatial configurations, which limit their flexibility and lead to a decline in the quality of space utilization during crises (Bakchan et al. 2021; Sanne et al. 2021). At the same time, these areas have relatively greater spatial availability, both geographically and economically, which can help reduce spatial tensions and promote more sustainable resource use in times of threats and shocks.

The way space is utilized and managed plays a crucial role in shaping crisis responses. In core areas, efficient institutions and active governance can lead to the implementation of policies prioritizing infrastructure resilience and sustainable development (Dimelli 2023; Holgersen 2013). In contrast, governance structures in peripheral areas may be less effective and reactive, resulting in delays in implementing necessary adaptations during crises and, consequently, lower material capital quality (Bakchan et al. 2021; Sanne et al. 2021). This disparity can create a feedback loop, where poor infrastructure quality and flawed spatial management in peripheral areas, despite relatively better spatial availability, further weaken their economic prospects and social cohesion (Bakchan et al. 2021; Liu et al. 2022a,b).

A critical element of spatial management, particularly in the context of residents' quality of life and well-being, is housing conditions. Housing markets are sensitive to both positive and negative shocks, leading to varied responses across spatial configurations. Even temporary shocks can have long-term consequences. As a result, adaptation in core areas, which have stronger economic foundations, may be faster than in peripheral areas, where structural constraints hinder recovery. The dynamics of housing market adaptation are further complicated by the spatial spillover effect of crises, particularly in interconnected territorial socio-economic systems. This interdependence increases the risk of shock transmission from central to peripheral areas, deepening disparities in housing stability and economic resilience.

Housing markets are also directly linked to financial markets, particularly through interest rates, credit availability, and the ability of residents to generate equity. These economic dependencies make it easier for core areas to cushion shocks due to their significantly greater financial resource-generation capacity, more diversified economies, and higher labor market flexibility. On the other hand, housing demand in core areas is more elastic, leading to higher price fluctuations (Rubio and Carrasco-Gallego 2016), especially during periods of employment growth and economic booms (Mohan et al. 2019). Peripheral areas do not experience such sudden demand shifts, resulting in a more stable or slower price increase (Ökkola and Brunelle 2018). However, challenges arise in terms of residents' purchasing power, lack of financial resources, and institutional support, which often prevent effective responses to housing needs. During economic crises, peripheral areas often experience an increase in evictions and property losses, which can lead to homelessness and housing instability (Brill and Raco 2021). Limited housing policies in these regions may exacerbate the effects of crises, leaving vulnerable social groups without adequate support systems (Coşkun 2022). Property price fluctuations can also drive speculative behavior, more likely in core areas (Pani 2022), as well as gentrification (Boterman and Gent 2014), a process that displaces lower-income residents, forcing them to migrate to city outskirts or peripheral regions. This, in turn, exacerbates inequalities in housing accessibility and quality of life. Meanwhile, peripheral areas often struggle with abandoned and neglected housing stock, particularly as young people migrate to more developed core regions. The lack of investment in housing infrastructure and limited access to mortgage financing prevent many residents from improving their living conditions. As a result, real estate in peripheral areas loses value, reducing investment attractiveness and limiting development opportunities. Additionally, weak employment prospects and poor access to modern services encourage further outmigration, exacerbating demographic challenges and leading to further degradation of residential spaces. In contrast, core areas, supported by investments in housing quality and sustainable development, are better equipped to withstand future crises (Grazioli 2022).

Crisis phenomena shape the state of ecosystem goods and services in core and peripheral areas in different ways. In core areas, the demand for ecosystem goods and services, such as urban parks, clean air, and access to green spaces, is very high. As a result, these resources are often overburdened, leading to numerous compromises between social and ecological needs, as well as disruptions to ecosystem balance (Raudsepp-Hearne et al. 2010). Ongoing urbanization is typically associated with additional loss of green spaces and natural habitats, reducing biodiversity and the quality of ecosystem services such as air purification, carbon sequestration, and recreational opportunities (Chen and Huang 2021; Furberg et al. 2019). Shocks, such as a sudden influx of population or economic expansion, can lead to excessive exploitation of green areas and natural resources (Tong et al. 2022). Balancing intensive urbanization with the preservation of green spaces is a challenge, especially during crises, when attracting investments often takes priority over environmental protection. Peripheral areas, on the other hand, possess richer natural resources and more extensive land available for ecosystem services, but their potential is often underutilized. In the face of economic shocks and disruptions, these areas may be forced to exploit natural resources more intensively to compensate for revenue losses in other sectors. Such overexploitation can lead to environmental degradation and a decline in the sustainability and effectiveness of ecosystem services, ultimately impacting the quality of life in these areas. The loss of agricultural land due to urban sprawl reduces ecosystem services such as food production and soil fertility maintenance (Ba et al. 2023; Reinhart et al. 2023).

The quality of ecosystem services is also influenced by governance structures during crises. Core areas, thanks to more proactive management, are more likely to implement policies that promote sustainable development and the protection of ecosystem services (Vindigni et al. 2021; Wu et al. 2019). One example is the integration of green infrastructure in urban planning, which enhances ecosystem services such as stormwater management and urban cooling, facilitating adaptation to climate change (Minixhofer and Stangl 2021). In contrast, in peripheral areas, less effective governance leads to insufficient responses to environmental challenges and a decline in the quality of ecosystem services (Zhang et al. 2023). The lack of resources for effective environmental management often results in deforestation, habitat loss, and decreased biodiversity (Hu et al. 2023).

The consequences of crises for ecosystem goods and services highlight the importance of targeted policy interventions that promote equal access to environmental resources and support resilience both in core areas and on the periphery.

An analysis of the impact of crises on material capital in core and peripheral areas reveals key differences in their resilience and adaptive capacity. While core areas, thanks to their well-developed infrastructure, stronger governance, and greater financial resources, are better equipped to handle challenges, peripheral areas struggle with limited access to healthcare, housing, and digital infrastructure, as well as underfunded green spaces. These disparities deepen social and economic inequalities, hindering sustainable development and territorial cohesion.

To counteract these effects, material capital must be seen as a foundation of social resilience, requiring the integration of various aspects of infrastructure and services. Emphasis should be placed on the development of local healthcare facilities, revitalization of housing infrastructure, and improvement of access to digital technology and recreational spaces. Only through targeted and evenly distributed investments will it be possible to mitigate the effects of crises and build more resilient and equitable communities.

3.1.2 SARS-CoV-2: Genesis, Process, Variability

SARS-CoV-2 was first identified in China, near the city of Wuhan, in December 2019. On January 30, 2020, the World Health Organization (WHO) declared the virus a Public Health Emergency of International Concern (PHEIC), and on March 11, 2020, it was officially classified as a global pandemic (Cucinotta and Vanelli 2020). The first case in Europe was recorded in January 2020 in France. The pandemic quickly spread worldwide, reaching its peak at the beginning of 2022 (Fig. 3.1).

By April 2024 (when most countries stopped reporting coronavirus-related data), nearly 705 million COVID-19 cases had been recorded worldwide, with just over 7 million deaths and nearly 676 million recoveries. An important indicator illustrating the scale of the global health crisis was the number of confirmed deaths due to SARS-CoV-2 per 100,000 inhabitants. In Europe, Bulgaria recorded the highest rate at 567.122, while globally, Peru had the highest rate at 660.111. The global average stood at 88.428. Poland, with a rate of 314.679, was slightly above the EU (27) average of 281.644 and slightly below Italy's rate of 331.340 (The Economist 2024) (Fig. 3.2).

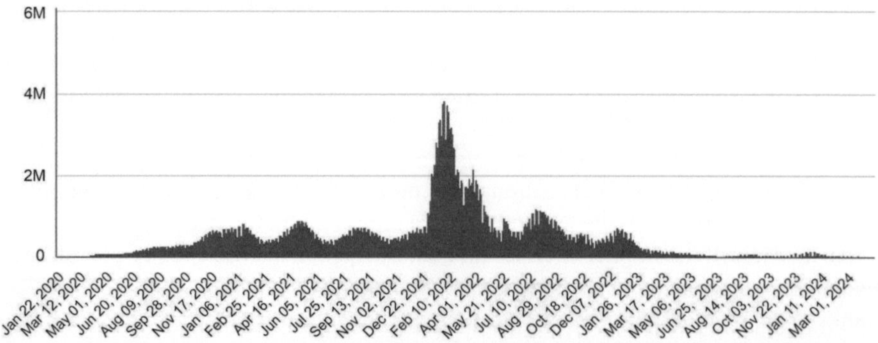

Fig. 3.1 SARS-CoV-2 Daily new cases. *Source* worldometers.info 2024, Access on 24.09.2024

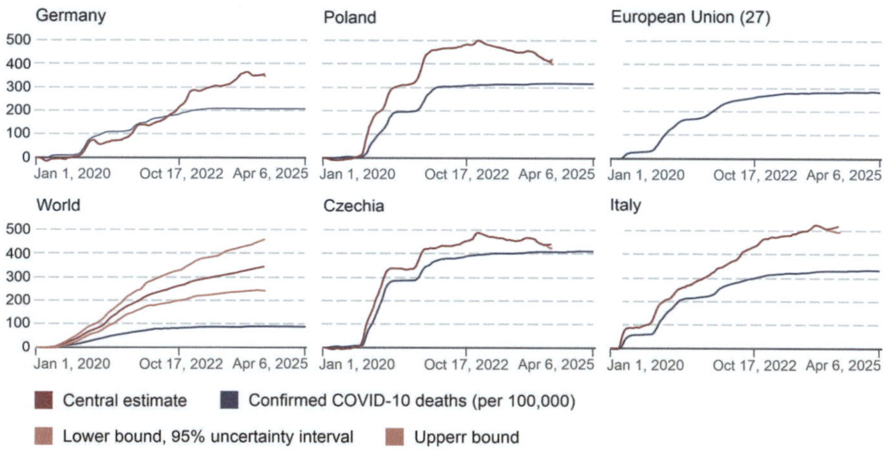

Fig. 3.2 Estimated cumulative excess deaths per 100,000 people during SARS-CoV-2 in selected countries. *Source* The Economist (2024),—processed by Our World in Data Tak sugerują tutaj: https://ourworldindata.org/grapher/excess-deaths-cumulative-economist?overlay=sources

The above data indicate significant differences in virus transmission, which varied greatly between core and peripheral areas due to a range of factors. Considering the set of factors analyzed in this book, certain patterns can be identified.

Transportation accessibility undoubtedly influenced virus transmission. Core areas are typically characterized by higher population density and mobility. Urbanization and better access to public transport networks contributed to higher virus transmission rates (Wang et al. 2021). Public transportation vehicles, with high passenger turnover, crowding, and poor ventilation, led to more frequent infections, especially when social distancing rules were not followed, hygiene guidelines were lacking, or preventive hygiene measures were not adhered to. In peripheral areas, which are less densely populated, the risk of infection was lower. However, the underdeveloped public transportation network and generally weaker infrastructure made access to healthcare and medical assistance more difficult. This lack of accessibility led to delays in starting treatment and increased vulnerability to severe cases of infection. Perceptions of infection risks in public transport shaped new mobility patterns, with many people opting for individual means of transportation or alternative communication methods. E-accessibility played a crucial role in limiting virus transmission in core areas by enabling the rapid dissemination of information, the development of telemedicine, and more effective coordination of public health measures. In cities, high digital connectivity facilitated the swift implementation of protective measures and access to medical advice without requiring in-person visits, thereby reducing infection risks. Conversely, in peripheral areas, limited e-accessibility hindered access to reliable information and healthcare services, leading to higher virus transmission and exacerbating existing health inequalities (Alghamdi et al. 2021). The lack of stable internet infrastructure in these regions not only delayed public health responses but also contributed to misinformation, which could result in lower adherence to sanitary guidelines.

The transmission of SARS-CoV-2 was closely linked to various aspects of human capital, including demographics, health and safety, education, and the labor market. These factors varied significantly between central urban areas and peripheral, particularly rural, regions, influencing how communities responded to the pandemic and how effectively they could limit the spread of the virus. In core areas, where population density and mobility were high, the virus spread more quickly. However, better healthcare infrastructure enabled more effective pandemic management despite the risk of medical facilities becoming overwhelmed (Nicolelis et al. 2021). Conversely, rural areas, while experiencing lower transmission rates due to fewer social interactions, had limited access to healthcare, leading to delayed diagnoses and higher mortality rates (Dawood et al. 2022). The perception of health security also differed depending on the type of area. In core areas, access to better medical resources meant that residents perceived the risk of infection and disease progression differently than those in peripheral areas, where access to healthcare was more limited (Boon et al. 2023). Education levels played a significant role in compliance with protective measures. Urban residents, who were generally more educated and aware of the dangers, were more likely to follow restrictions, whereas communities in peripheral areas were more vulnerable to misinformation. The labor market was another key factor. In cities, a high number of essential workers, such as those in retail, transportation, and healthcare, increased the risk of transmission (Gale 2020). Meanwhile, peripheral regions often had a different labor market structure, frequently based on agriculture and local businesses, which could reduce infection rates but also limit economic resilience during lockdowns (Kuehl et al. 2023).

Social capital also strongly influenced virus transmission. Studies have shown that communities with strong social networks and greater trust in institutions were more effective in implementing sanitary recommendations, which led to lower mortality rates (Elgar et al. 2020). In core areas, where social interactions were more frequent, the rapid dissemination of information could help curb virus transmission. In contrast, weaker social capital in peripheral areas contributed to misinformation and lower adherence to health guidelines (Kristanti et al. 2021). Family-based social capital played a particularly important role in peripheral regions, where knowledge transfer and support systems within families were crucial. Meanwhile, in core areas, residents had access to more diverse sources of information and assistance (Supriyati et al. 2021).

Institutional capital also played a crucial role in the effectiveness of measures aimed at limiting virus transmission, as it influenced the quality of governance, coordination of actions, and the implementation of appropriate preventive measures. Regions with weaker institutional capacities experienced higher transmission rates and worse health outcomes (Aminah and Susilo 2021; Mao 2021; Rahman et al. 2024). The effectiveness of pandemic control mechanisms varied between countries and regions, largely depending on the speed of response, the ability to enforce restrictions, the efficiency of public services, and the preparedness of procedures for an overburdened healthcare infrastructure (Brazier et al. 2023; Jee 2020; Yu et al. 2020b).

One of the key elements of the institutional system is public trust in government authorities, which in this case was reflected in the high acceptance of health policies, vaccine uptake, adherence to sanitary regulations, and compliance with social distancing measures. In countries where trust levels were high—typically wealthier nations with stable economies and established democracies—SARS-CoV-2 vaccination rates were higher, leading to lower virus transmission and, more importantly, milder disease progression. Conversely, in more authoritarian and poorer countries, vaccination rates were lower, and the disease spread more rapidly (Figs. 3.3 and 3.4).

However, some countries, such as Cuba, Brazil, Bangladesh, and Indonesia, stood out as exceptions, achieving high vaccination rates despite lower disposable income levels and weaker institutional capacity compared to Western European nations (https://ourworldindata.org/covid-vaccinations).

Trust is also the foundation of effective communication, facilitating the delivery of accurate messaging to the public. Another aspect of strong institutional foundations is the efficiency of government authorities and public health services. Countries with the necessary institutional capacity were able to conduct vaccination campaigns more quickly and effectively. This explains, for example, why vaccination rates were higher in Italy and Germany compared to Poland (Fig. 3.5).

Logistics and the monitoring of vaccine distribution, even to remote and less-accessible regions, also played a crucial role. In countries with lower institutional capacity (e.g., in the African region), vaccine availability was low, and equitable distribution was significantly hindered. Social inequalities posed an additional challenge, making access to vaccines much more difficult for marginalized groups while

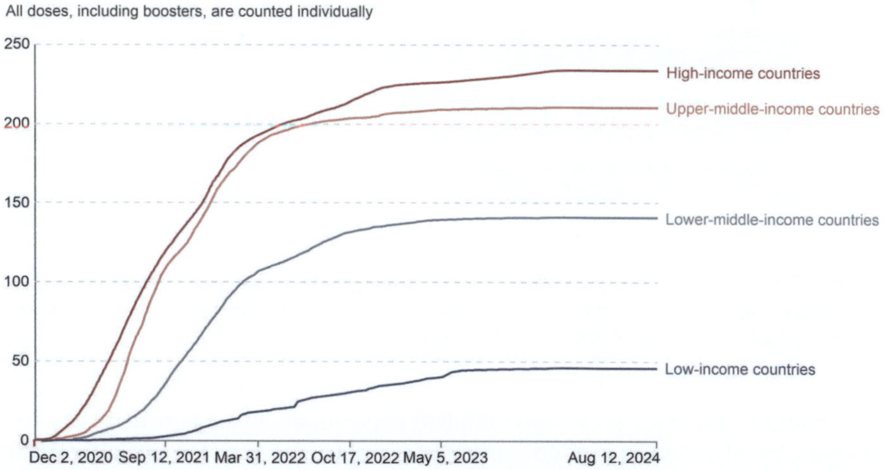

Fig. 3.3 Z SARS-CoV-2 vaccine doses administrated per 100 people, by income group. 2024 *Source* The long-run data on population is based on various sources, described on this page: https://ourworldindata.org/population-source. Access on 23.09.2024

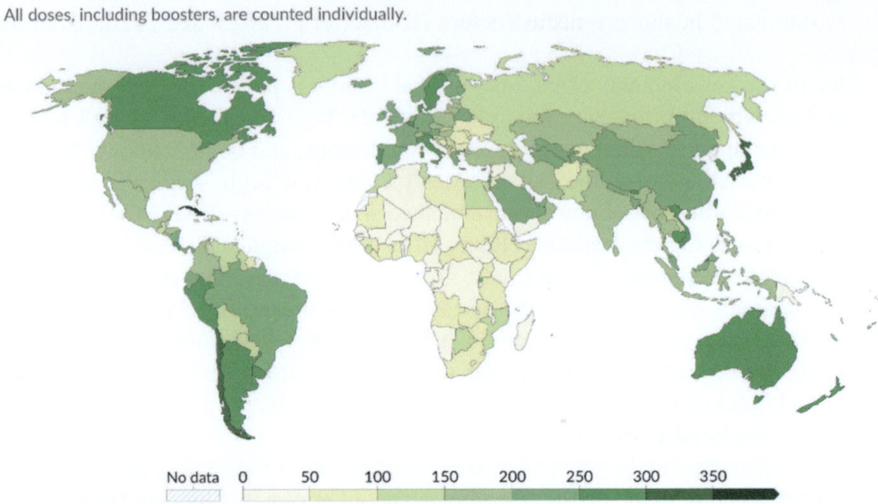

Fig. 3.4 Total SARS-CoV-2 vaccine doses administrated per 100 people (world). 2024 *Source* The long-run data on population is based on various sources, described on this page: https://ourworldindata.org/population-source. Access on 23.09.2024

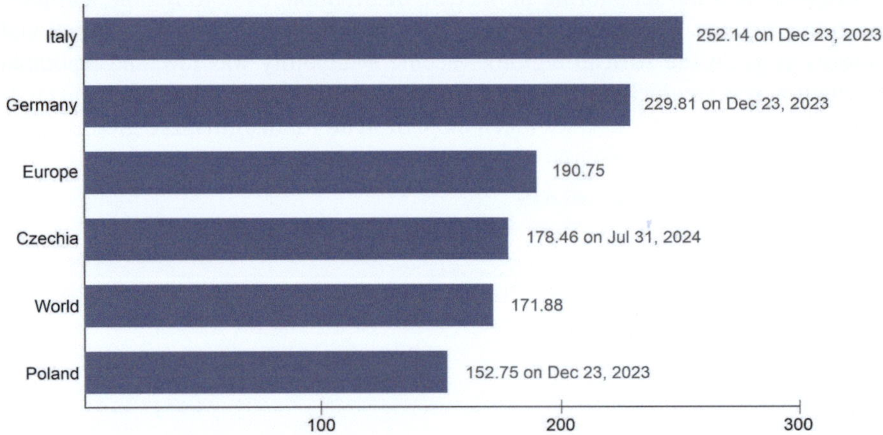

Fig. 3.5 Total SARS-CoV-2 vaccine doses administrated per 100 people (Aug. 2024). *Source* The long-run data on population is based on various sources, described on this page: https://ourworldindata.org/population-source. Access on 23.09.2024

providing privileged access to more affluent populations. In areas with weak institutional capacity, the vaccination process was further complicated by cultural attitudes, the activity of anti-vaccine movements, susceptibility to conspiracy theories, and logistical challenges in vaccine delivery (Yu et al. 2020b). Meanwhile, vaccination campaigns proved effective and led to a gradual easing of restrictions in many regions (Silverman et al. 2022).

Financial capital and the overall economic situation also had a significant impact on virus transmission, influencing the ability to respond to the pandemic in core and peripheral areas. In core areas, local governments with typically larger financial resources could more quickly implement public health measures such as testing, contact tracing, and vaccination campaigns during the pandemic, whereas in peripheral regions, limited budgets hindered effective preventive actions. Areas with lower financial capital, often characterized by poorer housing conditions and higher population density, faced greater challenges in maintaining social distancing, leading to higher virus transmission (Islam et al. 2021).

In core areas, businesses with greater financial resources were able to invest in health and safety measures, such as enhanced sanitation protocols, employee training, and remote work capabilities. Companies that prioritized health and safety were more likely to maintain operations while minimizing the risk of virus transmission among employees and customers. In contrast, businesses with fewer financial resources and lower levels of digitalization, often located in peripheral areas, were less prepared to implement health guidelines, which could have contributed to higher virus transmission (Hidayat and Muniroh 2021).

Social groups with low financial capital, particularly marginalized populations, were disproportionately affected by the consequences of the pandemic (Bach-Mortensen and Esposeti 2021; Sharma et al. 2020a). Numerous lockdowns and other mitigation measures introduced in various countries and regions caused severe disruptions in local and regional economies (Suzuki 2020). Once again, areas with lower-income status were in a relatively worse situation, further exacerbating existing social inequalities. Lower financial capital within a social group was associated with poorer access to healthcare, testing, and vaccines, less-hygienic living conditions, and a higher prevalence of comorbidities, all of which naturally contributed to increased transmission and mortality rates (Ahmed 2023; Hawkins 2020; Roszkowska et al. 2023).

Individuals with lower socio-economic status often work in the informal sector, with limited ability to take sick leave or temporarily suspend work due to illness. Additionally, unexpected expenses related to purchasing protective equipment such as masks and vaccines were difficult to manage. Lower levels of education and social awareness also made such individuals more vulnerable to misinformation, leading to poor decisions regarding safety measures, hygiene practices, and vaccination. Wealthier societies had objectively better opportunities to immediately transition to remote work and education, resulting in better outcomes in terms of maintaining social distancing and thereby limiting virus transmission.

The level of virus transmission was strongly determined by the level of material capital, particularly through the degree of space utilization, the housing market, the state of infrastructure (especially healthcare), climatic factors, and ecosystem goods. High building density in core areas, the concentration of people in enclosed spaces such as public transport, offices, and commercial areas, and cramped housing conditions facilitated virus transmission. In peripheral areas, lower population density was usually accompanied by lower adherence to health norms, which, in cases such as markets, could increase the risk of infection (Yang and Wang 2021). The effectiveness of space utilization in controlling the spread of the virus was largely a function of local governance and compliance with health recommendations.

A key factor in virus transmission and the fight against the pandemic was the level of infrastructure, particularly healthcare facilities and the availability of medical personnel (Bhopalwala et al. 2022; Mouratidis 2021; Muntaner et al. 2023; Sharifi and Khavarian-Garmsir 2020). Healthcare systems faced unprecedented challenges during the pandemic. The demand for healthcare resources surged significantly, leading to rapid shortages of hospital beds, personal protective equipment (PPE), and essential medical supplies (Zhou et al. 2023). The excessive burden on medical personnel and infrastructure shortages resulted in significant disruptions in healthcare services in many countries, including disruptions in primary care, planned procedures, and medical services, as patients often avoided treatment due to fear of infection (Downar et al. 2022). The pandemic also negatively impacted the mental well-being of healthcare workers, who, overwhelmed by workload and risk exposure, frequently reported increased levels of stress and burnout. This further strained the entire healthcare system, leading to dysfunctionality, particularly in less-affluent areas (Bhopalwala et al. 2022; Claponea and Iorga 2023). Shortages in infrastructure, lack of personal protective equipment, ventilators, and intensive care unit (ICU) beds led to higher virus transmission, more severe cases of SARS-CoV-2, and consequently higher mortality rates (Chen et al. 2020; Noorimotlagh et al. 2021; Tagashira et al. 2020). This situation was significantly more challenging for peripheral areas, particularly rural ones, which, due to relatively lower-quality infrastructure, limited availability of medical personnel, and greater distances to specialized healthcare facilities, testing sites, and contact-tracing capabilities, experienced higher virus transmission rates (Bhopalwala et al. 2022; Cao et al. 2023; Sharma et al. 2020b). The unequal availability of healthcare resources and medical personnel exacerbated existing socio-economic disparities in this context (Kim and Bostwick 2020; López-Gay et al. 2021; Martines et al. 2021).

Climatic conditions also played an objective role. Numerous studies have confirmed the relationship between virus transmission and reproduction and climatic factors, particularly temperature, humidity, and wind speed (Khan et al. 2021; Li et al. 2021; Liu et al. 2022a,b; Prata et al. 2020; Rasul 2021). This explains the varying course of the pandemic in different climate zones; higher temperatures and higher humidity were generally associated with reduced virus transmission, while colder climates (such as Poland) experienced a higher number of infections (Barría-Sandoval et al. 2022).

Similar patterns observed on a global scale were also present in Poland, where differences between core and peripheral areas shaped the dynamics of virus transmission. Population density, residents' mobility, the level of e-accessibility, and healthcare availability had a significant impact on the course of the pandemic in various regions of the country. Although the initial dynamics of infections in Poland differed from those observed in Western Europe, the mechanisms of virus spread and the factors facilitating transmission were similar.

The first case of the virus in Poland was recorded in March in the western part of the country, in Zielona Góra (ECDC 2024; Medicover 2024). The onset of the pandemic in Poland occurred slightly later due to the country's more peripheral location relative to major intercontinental air traffic hubs. The initial spatial and quantitative dynamics of infections also differed from those in Western and Southern European countries. The first infections in Poland primarily affected young individuals, including labor migrants working in Western European countries and visiting their families in Poland. Waves of infections among the elderly mainly occurred within organized elderly care systems (Afelt and Kursa 2021). A detailed spatial analysis of virus transmission in Poland was provided by Śleszyński (2022). Over the two years of the pandemic, Poland experienced five consecutive waves of infections, occurring during autumn–winter and winter–spring periods.

The first wave (spring 2020) was characterized by lower infection and mortality rates compared to other European countries. The second wave (autumn 2020) erupted with high intensity and high mortality rates relative to other countries. The third and fourth waves continued to show high mortality, whereas the fifth wave had lower mortality despite a high infection rate, resulting from the virus's evolution and weakening. The situation in each region was similar: the highest concentration of infections occurred in regional capitals, while peripheral areas experienced lower infection rates.

According to preliminary conclusions presented in the study (Śleszyński 2022), factors that facilitated virus transmission included the concentration of the population in areas with the highest population density and dense commuting networks, a high proportion of elderly residents in the centers of large- and medium-sized cities, a significant share of multifamily block housing in medium and large cities with urban planning unfavorable for maintaining spatial social distancing, and the existence of socially disadvantaged areas with low human capital and low epidemiological awareness.

The sixth wave in September 2022 posed a significantly lower threat (Fig. 3.6).

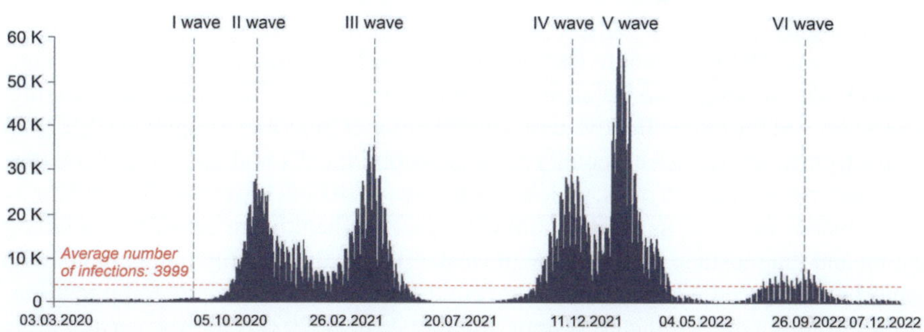

Fig. 3.6 Daily number of SARS-CoV-2 infections in Poland. *Source* https://koronawirusunas.pl. Access 08.11.2024

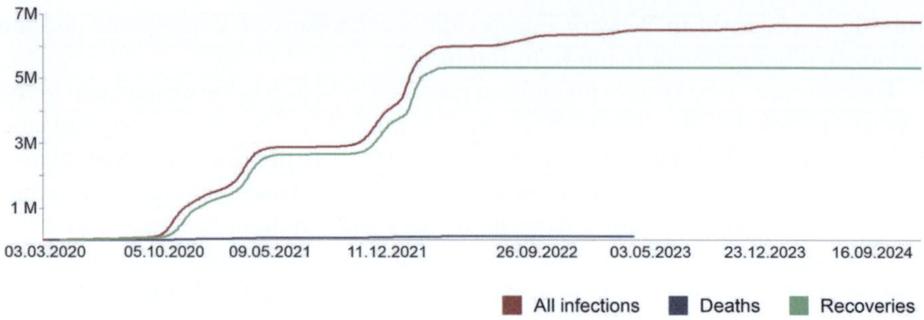

Fig. 3.7 Increase in confirmed SARS-CoV-2 infections in Poland. *Source* https://koronawirusunas.pl. Access 08.11.2024

According to data from the Ministry of Health (2024), from the beginning of the pandemic (March 4, 2020) to October 22, 2024, the number of infections in Poland reached 6,755,937, with 120,772 deaths and 5,335,695 recoveries (Fig. 3.7).

The transmission of the virus in Poland also exhibited spatial variation (Bochenek et al. 2021; Mirskaet al. 2024; Myck et al. 2023; Semczuk-Kaczmarek et al. 2021). Figure 3.8 illustrates the spatial distribution of detected infections in the period from November 24, 2020, to November 1, 2024.

The highest infection rates were recorded in the largest cities and their metropolitan areas, which serve as centers of the economy, education, and administration. These regions are characterized by high population mobility, intense social interactions, and a well-developed transportation infrastructure, all of which contributed to the spread of the virus. The dynamics of virus transmission were also influenced by the polycentric structure of Poland's settlement network, which includes 16 (or 18) regional centers (two regions have two regional centers)—the largest cities—without a dominant monocentric concentration in the capital, as is typical in countries such as Hungary, the Czech Republic, or Slovakia. This structure resulted in virus transmission being dispersed among multiple urban centers rather than concentrated in a single dominant hub, as seen in countries with a more centralized settlement system.

Population density was a key factor influencing this dynamic. Studies have shown that, in highly urbanized regions such as the Silesian Voivodeship, SARS-CoV-2 contributed to 24% of all deaths (Barański et al., 2021). The high number of infections in these regions may also be attributed to better access to healthcare and testing, which allowed for more effective case identification (Ciupa and Suligowski 2021). In contrast, peripheral areas, especially rural and economically underdeveloped regions, exhibited lower infection rates, which may be a result of both lower population density and reduced mobility, as well as limited access to healthcare services, including testing and diagnostics. In regions with weaker medical infrastructure, patients with comorbidities faced a higher risk of severe disease progression and complications, further straining the healthcare system (Raciborski et al. 2021; Orlewska et al. 2022). Moreover, differences in health policies and adherence to restrictions could influence local transmission rates. In regions that implemented stricter measures, infection rates were lower, whereas, in areas with a more relaxed approach to restrictions, the virus spread more rapidly (Gruszczyński and Zatoński 2021). Social and cultural

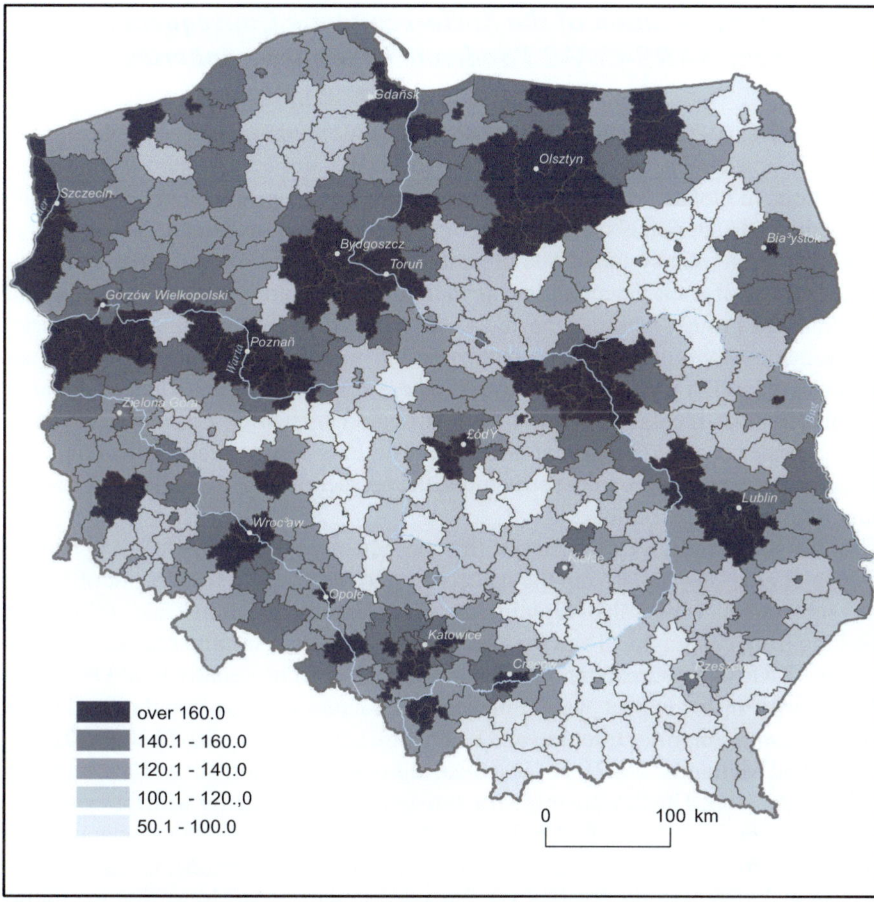

Fig. 3.8 Number of detected infections per 1,000 inhabitants (in the period from November 24, 2020, to November 1, 2024). *Source* Own elaboration

factors also played a role. In regions with low institutional trust and strong resistance to vaccinations, higher infection rates were observed, as confirmed by studies on public attitudes toward the pandemic in Poland (Raciborski et al. 2021).

Environmental factors also warrant attention. Research suggests that higher humidity and lower temperatures may have facilitated virus transmission, explaining the seasonal spikes in infections during colder months (Bochenek et al. 2021). Conversely, during the summer, increased mobility and social activities may have contributed to subsequent waves of the pandemic. The spatial variation in SARS-CoV-2 transmission in Poland thus reflected the combined effects of demographic, socio-economic, and environmental factors. Areas with higher social and institutional capital and better access to public resources were able to manage the crisis more effectively, whereas less-developed regions with limited healthcare and social infrastructure were more vulnerable to the long-term impacts of the pandemic.

3.1.3 *Systematization of the Socio-economic Consequences of the SARS-CoV-2 Pandemic in Inner Peripheries*

The SARS-CoV-2 pandemic has generated wide-ranging and complex socio-economic consequences, which continue to be the subject of intensive research. Disruptions in global value chains, mass business closures, and unprecedented state interventions to support economies have revealed both the weaknesses and resilience of modern economic systems. The impact of the pandemic was particularly evident in the context of social inequalities, as individuals with lower economic status and marginalized groups suffered the most, losing both health and financial security. At the macroeconomic level, the pandemic led to an increase in public debt and the phenomenon of a K-shaped recovery, where some sectors experienced dynamic growth after the recession, while others struggled with deep crises (Nicola et al. 2020; Sumner et al. 2020; Suk et al. 2022). The long-term effects of the pandemic include changes in employment structures, the acceleration of digitalization, and the redefinition of public policies.

In this subsection, we characterize the consequences of SARS-CoV-2 in internal peripheral areas, systematizing them according to the adopted framework of factors and subfactors, while also highlighting their interconnections and differences resulting from geographical and socio-economic contexts. As numerous studies indicate, the SARS-CoV-2 pandemic did not affect all regions equally (ESPON 2022). The severity of its impact was influenced by factors such as population density, urbanization, economic structure, and healthcare capacity (Amdaoud et al. 2021a,b; Florida and Mellander 2022). For instance, regions with a more diversified economy and well-developed healthcare infrastructure tended to cope better with the crisis than those that were less diversified or had weaker healthcare systems (Amdaoud et al. 2021a,b; Qamar et al. 2022). Research by Martin and Gardiner (2021) suggests that the regional effects of the SARS-CoV-2 pandemic will persist long after the immediate health crisis subsides. Areas severely affected by the pandemic may experience prolonged economic slowdowns, increased inequalities, and shifts in urbanization and migration patterns. Therefore, recovery strategies for these areas must adopt a long-term perspective and address all socio-economic repercussions (Bourdin and Levratto 2024; Martin and Gardiner 2021).

The consequences of the SARS-CoV-2 pandemic can be classified in various ways, depending on different criteria. Becla and Czaja (2021) systematize them as presented in Table 3.3.

The consequences of the SARS-CoV-2 pandemic, as presented in the table, clearly correspond to the key development factors of internal peripheries: accessibility, human, social, institutional, financial, and material capital. Each of these resources was directly or indirectly affected by the crisis, highlighting their significance as the foundations for the functioning of various areas. The pandemic not only deepened existing challenges but also revealed new interdependencies among these factors, which influence the ability of internal peripheries to cope with crisis situations. In this context, systematizing the consequences of the pandemic in relation to these

Table 3.3 Consequences of the SARS-CoV-2 pandemic

Criterion of distinction	Consequences	Characteristics
Duration of impact	Short term	Consequences occurring during the pandemic, such as mobility restrictions, declines in revenue, rising unemployment, disruptions in supply chains, and the rapid necessity for economic adaptation
	Long term	Consequences felt after the pandemic ended, including permanent changes in economic structure, acceleration of digitalization processes, reorganization of global supply chains, and an increase in economic protectionism
Level of impact on economy	Microeconomic	Consequences affecting individual economic entities and households, including: • Decreased revenues and profits due to temporary closures or restrictions in operations caused by workforce shortages, supply chain disruptions, or lack of B2B payment liquidity • Giffen behavior, where demand for essential goods increases despite rising prices (hoarding) • Changes in the nature and scope of business activities • Development of economic support programs such as crisis relief packages • Rapid growth of e-commerce and courier services
	Meso-economic	Consequences affecting regions, industries, economic sectors, and entire social groups. This includes the destructive impact of the pandemic on sectors such as transportation, tourism, gastronomy, and hospitality, as well as changes in value chains and production limitations (e.g., in the automotive industry)
	Macroeconomic	Consequences affecting national economies or economies integrated within international organizations (e.g., the EU). These consequences include national-level effects such as a decline in production dynamics due to movement restrictions imposed during the pandemic
	Megaeconomic	Consequences of a global nature affecting groups of nations or entire global socio-economic systems, such as global corporations limiting their operational reach or changes in the processes of global economic integration

(continued)

Table 3.3 (continued)

Criterion of distinction	Consequences	Characteristics
Level of impact on social life	Socio-psychological	Consequences related to deteriorating mental health, stress, uncertainty about the future, and social isolation, which led to a decrease in quality of life and a diminished sense of security among residents
	Socio-demographic	Demographic consequences, such as increased mortality, shortened life expectancy, and declining fertility rates, which affected population structure and growth dynamics
	Impact on social relationships	Weakening of interpersonal relationships due to restrictions in education, social, and healthcare systems, negatively affecting social capital and local community cohesion
Political and awareness-related consequences	Weakening of democratic institutions and rise of populism	Increased popularity of populist political movements and conspiracy theories regarding the origins of the virus and the dangers of vaccines, leading to a decline in trust in democratic and scientific institutions
	Rising interpersonal aggression	Increased aggression, both in public and digital spaces, caused by stress, social isolation, and misinformation, resulting in social polarization and tensions within local communities
	Expansion of ignorance	The spread of attitudes rejecting scientific knowledge, including the disregard for sanitary recommendations and the promotion of unverified information, prolonging the duration of the pandemic and hindering preventive and remedial actions

Source Own elaboration based onBecla and Czaja (2021)

factors allows for a better understanding of how each of them impacted the social and economic situation in peripheral areas and what measures may support their recovery and development. Examining these elements in further analysis will enable a multi-dimensional perspective on the mechanisms of internal peripheries and identify areas that require special attention in future development efforts.

Accessibility

Internal peripheries are characterized by a relatively large time distance from central areas, which we previously referred to as an accessibility deficit in the external dimension. The SARS-CoV-2 pandemic spread more rapidly in metropolitan areas, which are defined by high population density and frequent social interactions, utilizing a dense transport network that is crucial for economic development and tourism (Bourdin and Levratto 2024). The external accessibility deficit in internal peripheries during the SARS-CoV-2 pandemic had both positive and negative effects, which are worth analyzing in a broader context.

On the one hand, the restriction of population and goods flows may have been beneficial for the population living in internal peripheries. Research shows that territories with better transport accessibility, stronger integration into global networks, and more developed infrastructure, which are usually attractive to residents and investors, became particularly vulnerable to the rapid transmission of the virus during the pandemic (Cutrini and Salvati 2021). Lower external accessibility meant a reduced risk of contact with external sources of infection, which may have slowed the spread of the virus in internal peripheries. In response to the threat posed by virus transmission, many governments in Europe and worldwide introduced measures aimed at limiting mobility to slow down the spread of the disease (Hale et al. 2020). Among these measures were restrictions on the movement of people across national borders, although their effectiveness was limited in countries experiencing large local outbreaks of infections (Roelofs et al. 2022; Russell et al. 2021). In some countries, such as Australia and New Zealand, the implementation of more decisive measures, including limiting intraregional mobility between areas with high infection rates, had clear effects in terms of more effective virus control (Bandyopadhyay and Sridhar 2023; Saul et al. 2020). On the other hand, the external accessibility deficit was also associated with significant costs, particularly for border regions, which often meet the criteria of internal peripherality. In Europe, the closure of national borders and the introduction of movement restrictions led to disruptions in the flow of workers and goods, causing these regions to lose approximately 10% of their potential GDP in 2020 (Capello et al. 2023). Residents faced difficulties accessing labor markets, leading to increased unemployment and deteriorating living conditions (Brezdeń 2023; Kolasińska 2022). These economic losses highlight that mobility restrictions, although intended to curb virus transmission, in many cases had a negative impact on the socio-economic development of affected areas. These restrictions also made it more difficult for residents to access essential services such as healthcare, education, and shopping. In particular, rural and less-urbanized areas, which had already struggled with transport issues, found themselves in an even more challenging situation (Dominiak et al. 2022).

Considering the long-term consequences of the pandemic crisis, it appears that internal peripheries have faced increasing risks due to the deepening internal accessibility deficit. During the pandemic, there was a natural preference for individualized car transport over public transportation (Giubergia et al. 2023; Kołsut and Stryjakiewicz 2023; Tubis 2022; Zajfert 2024). The increased number of private vehicle users contributed to air pollution, traffic congestion, and parking issues, especially in urban areas. Globally, pandemic-related lockdowns had a negative impact on public transportation, with many routes being suspended (Capello et al. 2023). The already limited connections between core cities and peripheral areas were discontinued and, after the pandemic subsided, were not reinstated due to financial constraints and decreased public demand. Studies indicate that public transportation to medical facilities suffered the least disruption, whereas all other routes lost significance (Cottreau et al. 2023). On the other hand, the SARS-CoV-2 pandemic also contributed to the development of policies promoting active mobility, such as walking and cycling. Many cities expanded and, more importantly, integrated their cycling and pedestrian infrastructure, marking a step toward more sustainable transportation. These changes have the potential to improve quality of life, reduce greenhouse gas emissions, and encourage long-term shifts in spatial behaviors, leading to greater adoption of active mobility among residents of Functional Urban Areas (Giubergia et al. 2023; Islind et al. 2020; Kellermann et al. 2022; Santana et al. 2023).

E-accessibility became a key factor in overcoming spatial barriers. The pandemic accelerated the digitalization of numerous services across public sectors (e-administration, telemedicine, online education), commercial activities (remote work), and individual use. Digitalization tools, including remote work, digital access to business documents and correspondence, and online meetings, proved to be an extremely effective and viable alternative, enabling many businesses and institutions to continue operations (Dyba and Di Maria 2023; Hynes 2020). They also played a crucial role in facilitating public, social, and individual life organization (Haefner and Sternberg 2020). However, the effectiveness of this alternative relied heavily on high-quality and well-organized digital infrastructure.

The SARS-CoV-2 pandemic highlighted the issue of digital exclusion, which restricted access to remote work opportunities, online education, and telehealth services in various regions. Peripheral regions, where digital infrastructure was typically of lower quality, were particularly disadvantaged (De Backer et al. 2021). An additional challenge was the low level of digital literacy, which limited local communities' ability to effectively use solutions such as e-administration, telemedicine, and online education (Haefner and Sternberg 2020). Many residents in these areas lacked access to high-speed internet or suitable devices, making it difficult for them to benefit from essential services such as telemedicine, e-learning, and e-administration. Research indicates that residents of areas with low technological accessibility struggled to obtain reliable information about SARS-CoV-2, which worsened their health conditions and limited their access to psychological support.

The SARS-CoV-2 pandemic has highlighted the complexity of accessibility issues in peripheral areas, both externally and internally. Mobility restrictions and inadequately developed digital infrastructure significantly hampered the functioning of these areas, exposing digital exclusion and limited access to healthcare, education, and public administration services. At the same time, the pandemic accelerated the digitalization of many services, underscoring the urgent need for investments in broadband internet infrastructure, the development of e-government, and support for digital skills. These actions can reduce inequalities between peripheral and urbanized areas, preparing society for future crises while ensuring sustainable development and improving the quality of life.

Human Capital

In terms of human capital in internal peripheral areas, the SARS-CoV-2 pandemic posed a significant threat. Regions already affected by depopulation and an increasing number of elderly residents were particularly vulnerable to health and life risks. Throughout the pandemic, higher mortality rates were recorded among older age groups (Boterman 2020). The high number of deaths in this demographic further weakened population resources, limiting both the economic and social potential of these regions. The lack of young, working-age individuals to replace the aging population reduced the development capacity of these areas. A shrinking tax base and workforce made these regions increasingly dependent on external support. The SARS-CoV-2 pandemic contributed to declining fertility rates across Europe and altered reproductive behavior patterns (Pomar et al. 2022).

At the same time, the pandemic triggered some short-term migration to less-urbanized areas. During the SARS-CoV-2 crisis, workplaces and schools were partially or entirely closed, prompting some urban residents to move to suburban and rural areas, either to family homes or rented houses and apartments. Faced with the threat of infection, people began to appreciate living in areas with lower population density. The pandemic also led to return migration to peripheral regions; however, these individuals often struggled with a lack of suitable job opportunities (Cross et al. 2022). The possibility of remote work allowed some people to permanently relocate outside core urban areas (Antonietti et al. 2020; Cottreau et al. 2023). However, this potential was not fully realized in peripheral areas, mainly due to infrastructural limitations. These temporary migration trends indicate the potential of peripheral areas in shaping new lifestyle patterns and work arrangements, but they require comprehensive support measures to ensure their long-term, sustainable development.

A major challenge for residents of areas classified as internal peripheries was ensuring their health security. At the onset of the pandemic, the most significant issue was access to healthcare facilities, their capacity, and available equipment. Non-SARS-CoV-2-related healthcare services were almost entirely suspended, while the health conditions of individuals suffering from other illnesses deteriorated. Preventive healthcare was also halted, and in many countries, including Poland, the healthcare system proved to be entirely inefficient (Płonka 2022). The most-affected groups were the elderly and patients with chronic diseases, for whom interruptions in treatment sometimes had irreversible consequences. Since traditional forms of healthcare were hindered by mobility restrictions, telemedicine began to develop intensively. However, in peripheral areas with low e-accessibility, patients encountered significant challenges, mainly due to the lack of stable internet connections and the necessary digital literacy to use telemedical tools. This led to delays in disease diagnosis and treatment, as well as difficulties in obtaining information about SARS-CoV-2 (Bartlett et al. 2022). In regions where healthcare infrastructure was already weak, the pandemic exacerbated public health problems. The lack of access to telemedicine further increased health exclusion in many cases.

The challenges were not limited to physical health alone. Social isolation and pandemic-related stress also affected mental health and residents' sense of security. In peripheral regions, where access to psychological support was limited, increased levels of anxiety and depression further weakened human capital, making it more difficult for individuals to function in both professional and educational spheres (Radzińska 2023). On the other hand, preventive measures such as the mandatory use of protective masks and frequent disinfection of hands and surfaces in public places may have contributed to an increased sense of health security. Studies indicate that these virus transmission prevention measures had an impact on reducing the incidence of other viral and bacterial infections (Radenović et al. 2022). The "state of emergency" declared in many countries, including Poland, paradoxically contributed to a higher sense of public security. Increased law enforcement activity was observed, and the population adhered to movement restrictions, which may have influenced the subjective perception of safety among residents (Cichecka et al. 2024). Research findings indicate that the SARS-CoV-2 pandemic led to changes in crime patterns, resulting in a decline in traditional forms of crime while increasing cybercrime and economic offenses (Macit 2021).

The pandemic had a significant impact on education, slowing down the progress of learning (Betthauser et al. 2023). Faced with the threat to the health and lives of students and teachers, educational establishments, including universities, were closed and replaced with online education, which prevented direct interaction between teachers and students. Across the world, many secondary schools adopted remote education; however, at the primary school level, it was not recommended everywhere. Despite the benefits of this innovative approach to education, it posed several challenges, including the lack of computer equipment for every student in a household, limited access to the internet, and isolation from classmates (Bourdin et al. 2023; Major et al. 2021).

In peripheral areas, which generally have weaker telecommunication infrastructure (e-accessibility), the transition to e-learning revealed additional challenges. Students and teachers often lacked access to appropriate tools or a stable internet connection, making it difficult to effectively use educational platforms. Technical issues, a lack of support from teachers, and the limited ability of parents to assist their children significantly lowered the quality of education (Al-Barakat et al. 2023). Research findings also indicate that the pandemic widened the educational opportunity gap between children from different socio-economic backgrounds (Lalin et al. 2024) and between children living in internal peripheral areas and those in development cores in Poland.

The lower level of satisfaction with remote learning may have long-term consequences for the future educational and career opportunities of residents in peripheral areas. The pandemic highlighted the urgent need for investment in educational infrastructure, including digital tools and training programs for teachers.

The labor market crisis triggered by the pandemic redefined the state and structure of human capital utilization. It was characterized by limited employer activity and a highly uneven impact on workers. On the one hand, it posed a threat to employees in specific industries, while on the other, it provided an opportunity for those who could work remotely. Sectors such as tourism, gastronomy, hospitality, and transport, which play a crucial role in the economies of peripheral areas, were severely affected by pandemic-related restrictions (Sun et al. 2020). In many regions, the freezing of tourism activities and restrictions on population mobility led to a drastic decline in local business revenues and, consequently, to job losses (OECD 2021). A significant issue was the decline in seasonal employment, which is a vital component of the economy in peripheral agricultural and tourism regions. The drop in tourism resulted in reduced demand for seasonal labor and sharp downturns in local economies. The restriction of labor migration due to lockdowns caused local labor markets to lose a substantial portion of their workforce, leading to economic stagnation and increased poverty in low-income communities. Many individuals who had migrated abroad for work before the pandemic lost their primary source of income, exacerbating financial difficulties. During the pandemic, many employers attempted to maintain employment despite the economic downturn by implementing strategies such as reducing working hours or temporarily suspending operations. This so-called labor hoarding was intended to prevent a permanent depletion of human capital, which could hinder post-crisis recovery. However, in peripheral regions, where employers had fewer financial reserves, the effectiveness of these measures was limited. Unlike core regions, where the development of digital tools and the spread of remote work increased efficiency and flexibility in work organization, peripheral areas faced significant challenges due to limited telecommunications infrastructure, which made adaptation difficult (Bołdyrew 2023; Sadecki 2022). Remote work also introduced several risks, particularly in peripheral areas, such as difficulties in balancing professional and household responsibilities, which disproportionately affected women (Sobczak 2021), increased stress and burnout (Dolot 2020) and changes in how individuals view the boundary between work and personal life, ultimately leading to lasting shifts in workplace culture (Bielińska-Dusza and Hamerska 2023; Stankowska 2024). The economic stagnation caused by the crisis also had a profound impact on slowing the entry of graduates into the labor market, leading to even higher unemployment rates among young people (Gruss et al. 2022). The pandemic underscored social inequalities in job accessibility. Workers in roles requiring physical presence, often in low-paying jobs, had limited remote work options and were generally more exposed to infection risks. The crisis revealed that certain groups of workers were not covered by social safety nets, including those employed in various forms of precarious and non-standard work arrangements. In some cases, attempts to access support funds led to improved working conditions (OECD 2021).

The rise in unemployment was mitigated by the implementation of short-term work programs and job-retention initiatives. Widespread use of job-retention schemes helped cushion the impact of the pandemic on labor markets and facilitated the reopening of economies once restrictions were lifted (Ando et al. 2022). However, these programs did not effectively protect vulnerable socio-economic groups, including workers with low levels of education, young people, foreign-born individuals, ethnic minorities, women, and those in precarious or non-standard employment (OECD 2021). Moreover, the SARS-CoV-2 pandemic led to an increase in early retirement decisions across Europe. The decline in workforce participation among older individuals in the Eurozone was particularly evident between the end of 2019 and the beginning of 2021 (Botelho and Weissler 2022).

The pandemic highlighted pre-existing challenges in peripheral areas, exacerbating their demographic, healthcare, educational, and economic difficulties. Population aging, limited access to healthcare services, and inequalities in education and labor market access further weakened the human capital of these regions. At the same time, the pandemic acted as a catalyst for certain positive changes, such as the acceleration of digitalization, which, if accompanied by appropriate infrastructure investments and effectively utilized by local communities, could enhance the development potential of peripheral areas in the future. Some of the consequences of the SARS-CoV-2 pandemic for human capital development can even be seen as opportunities for these regions. The crisis accelerated digital transformation and workplace transformation, bringing new opportunities for workers, entrepreneurs, and regional economies (OECD/European Commission 2021; World Bank 2022). In many cases, remote work gained importance, which in the long term may reduce the demand for office space and reshape employment structures (Glaeser and Cutler 2021; Micek et al. 2023). However, it is important to emphasize that these trends displayed significant spatial disparities.

Social and Institutional Capital

The SARS-CoV-2 pandemic triggered an omnipresent sense of threat to the health, and even the lives, of people, families, and entire local communities. During the crisis, a sense of solidarity within local communities often emerged, leading to mutual assistance in the face of danger. The bonding social capital in internal peripheral areas was activated from the very beginning of the pandemic. Support networks based on family and neighborly relationships played a significant role in providing help with everyday challenges, such as childcare during school closures. Intergenerational solidarity increased as grandparents frequently supported parents forced to work under difficult pandemic conditions (Alonso Ruiz et al. 2022). Community efforts also turned toward health-supporting actions, such as organizing the delivery of medications and food for the elderly and conducting local information campaigns on health protection measures.

While the pandemic strengthened bonding social capital, it also imposed limitations on the development of bridging social capital. Fear and a lack of trust among people, as well as toward institutions, undermined cooperation even among neighboring local communities (Alonso Ruiz et al. 2022; Lucchini et al. 2021). During the pandemic, governments were able to use the "state of emergency" to consolidate their power and introduce new methods of social control; however, some researchers see it as an opportunity for new civil society initiatives and social innovations (Cichecka et al. 2024; Musiał-Karg and Kapsa 2021). Nonetheless, it appears that, since the pandemic, institutions have weakened due to the collapse of interpersonal relationships, and there has also been an increase in the popularity of conspiracy theories that devalue the scientific approach to understanding the world (Becla and Czaja 2021).

The period of the SARS-CoV-2 pandemic marked a time of significant changes in the way leisure time was spent, even in peripheral areas. There were changes related to a reduction in the physical activity of the population, particularly the kind previously done outdoors (Mucha and Mucha 2021). For young people, the lack of access to youth centers, sports fields, or other facilities such as skate parks meant a limitation in leisure opportunities, which affected their mental well-being and level of social integration. On one hand, the pandemic led to an increased interest in home-based activities such as cooking, handicrafts, and gardening. In peripheral areas, residents more frequently returned to traditional activities such as baking bread, making preserves, or engaging in artisanal work, which increased their sense of self-sufficiency, allowed for additional savings (Kandeel et al. 2023), and helped reduce anxiety and improve mental well-being (Pagaoa et al. 2021; Szymańska 2022). This also promoted the building of social bonds, the creation of community, and mutual support (Sinclair et al. 2021). On the other hand, people became accustomed to spending their leisure time at home, using digital tools for entertainment, learning, education, and socializing. The amount of time spent in front of screens increased, used both for work and recreation (Pandya and Lodha 2021). Some patterns of recreational behavior persisted even after the end of the pandemic. Tourist preferences also changed; some tourists permanently shifted their travel choices, forgoing international trips to popular destinations in favor of excursions in their immediate surroundings, with some even purchasing recreational plots (Kugiejko and Kociszewski 2021). The observed changes in leisure time patterns may have contributed to a decrease in the quality of life due to a lack of satisfying social contacts.

The pandemic highlighted the differences in the functioning of institutional capital between core and peripheral areas, both in terms of crisis management and in building the resilience of individual regions. In core areas, thanks to strong public institutions, it was possible to rapidly implement measures to curb the spread of the virus and mitigate the effects of the pandemic. Efficient institutions allowed for quick decision-making, rational resource allocation, and coordination of actions among various stakeholders. In peripheral areas, institutional deficits manifested in many aspects. A significant indicator of institutional weakness was the ability of healthcare institutions to cope with the rising number of infections. The lack of cooperation between different levels of public authority, particularly evident in peripheral areas, often led to fragmented, inconsistent, and uncoordinated responses that hindered effective management of healthcare even at a basic level (Aleluia et al. 2023). The capacity for effective crisis management was also hampered in peripheral areas by a frequent lack of legitimacy and public trust in government/municipal institutions, which significantly affected adherence to health guidelines and, consequently, the effectiveness of preventive measures (Christensen and Ma 2021). The absence of broader consultation practices, or even the neglect of consultations with local communities, led to conflicts or to actions that were ill suited to local contexts (Regmi et al. 2023). Weaknesses in governance structures were also evident in information management and communication. In peripheral areas, where access to reliable information was limited for various reasons, failures in communication had a detrimental impact on public health outcomes (Moussallem et al. 2022).

Another important issue was the ineffective management of aid resources and support for businesses. In many peripheral areas, entities faced difficulties accessing financial support due to the insufficient number of business support institutions and a lack of administrative capacity to rapidly implement crisis measures (Adda 2023). Unlike in core areas, public administration in peripheral regions was often unable to effectively manage relief packages and stimulus programs, leaving businesses with less access to financial resources. Institutional inefficiency led to delays, waste of funds, and insufficient support for entities affected by the pandemic (Zahara et al. 2022). The lack of coordinated policies resulted in many local businesses struggling to survive the pandemic. Shortcomings in the availability of digital technologies and appropriate infrastructure, as well as a shortage of qualified business-supporting companies, meant that officials could not implement e-services as required, healthcare facilities could not offer telemedicine to the necessary extent, and businesses found it more difficult to transition to remote work or digital customer service, which, under lockdown conditions, led to marginalization and a weakening of their competitive position. Deficits also emerged in businesses' ability to innovate, which often determined survival during the pandemic. While in core areas both public administration and businesses implemented new digital solutions and innovative business models essential for adapting to the new reality, peripheral areas lacked the institutional capacity to support digital transformation (Owusu-Oware 2022). This issue was particularly critical for SMEs and family businesses, which, without adequate support for business consulting from regional development agencies, chambers of commerce, or investment funds in peripheral areas, were unable to adapt quickly and effectively to the new conditions (Anggadwita et al. 2022).

Institutional capital played a key role in shaping the economic structure of regions and their ability to respond to crises. In peripheral areas, micro and small enterprises, often family-run, dominate, operating mainly in sectors such as agriculture, small services, tourism, and retail. In contrast, core areas are characterized by a higher share of medium and large enterprises as well as developed sectors of advanced technologies. Companies in peripheral regions tend to have lower capital intensity, weaker digitalization, and less resilience to sudden economic crises. This economic structure is the result of historical and social conditions that shape local institutional capital, thereby influencing the regions' capacity for adaptation and innovation (Proniewski 2012). These differences became evident during the pandemic in the survival strategies of companies. In core areas, large- and medium-sized companies more frequently and effectively took advantage of state support and flexible forms of employment. In contrast, in peripheral regions, where self-employment and sole proprietorships, as well as agriculture, prevail, many entrepreneurs were exposed to temporary suspension or even complete closure of their businesses.

The SARS-CoV-2 pandemic demonstrated how closely social and institutional capital are interconnected. In the face of the crisis, local support networks and intergenerational solidarity played a key role in mitigating the effects of the pandemic, but limited trust in institutions and inconsistent public communication weakened the effectiveness of preventive measures. Where institutions were strong, communities adapted more easily to changes and implemented remedial strategies. In peripheral areas, however, institutional weaknesses, such as a lack of coordination, insufficient economic support mechanisms, and administrative inefficiency, exacerbated the crisis despite the existing social capital. Inadequate digitalization, limited business adaptability, and regulatory uncertainty undermined the community's ability to self-organize and recover. Pandemic experiences indicate that the sustainable development of peripheral areas requires a simultaneous strengthening of institutions and building of social trust. Investments in public administration, digitalization, and business-supporting institutions must go hand in hand with initiatives aimed at promoting social cohesion and civic engagement to better respond to similar crises in the future.

Financial Capital

Financial resources and flows in internal peripheral areas are on average lower than in the development cores. The crisis triggered by the pandemic further deepened these disparities, also because the state's political and financial situation was uncertain (Mahler et al. 2022). The pandemic revealed both structural weaknesses and the need to build more resilient financial mechanisms, especially in lagging areas, which is crucial for their future. The SARS-CoV-2 pandemic had a significantly negative impact on the financial situation of local governments in peripheral areas. These areas, which had already struggled with a limited tax base and low levels of investment, were forced to confront a sharp decline in revenues, particularly from personal income tax (PIT) and corporate income tax (CIT). The economies of these regions, which rely on micro and small enterprises as well as seasonal activities, were particularly vulnerable to the effects of lockdowns. A lack of clear budget information and unforeseen changes in fiscal law exacerbated the difficulties faced by local administrations, forcing them to reduce infrastructural investments that could improve living conditions in peripheral regions (Klimek 2022; Tabis 2023). At the same time, local governments had to increase spending to combat the pandemic. This was evident in the purchase of personal protective equipment, the organization of remote learning, the adaptation of local healthcare facilities to crisis requirements, and support for the most vulnerable social groups (ESPON 2022; Tandon et al. 2020). Consequently, the lack of funds for local development increased the risk of economic stagnation, especially in areas where public investments are a key element of development, such as internal peripheries.

The mobilization of public funds (both local and state) often took the form of compensation or mitigating measures through targeted financial transfers. The initial cash transfers introduced by member states to support workers were emergency responses aimed at mitigating the effects of measures to restrict the spread of SARS-CoV-2. They included wage supplements, job-retention schemes, support for income from self-employment, as well as sickness benefits (Ando et al. 2022). Although these measures had the advantage of limiting the negative effects of the pandemic, they were not intended to chart an original and innovative path to recovery, to create a new balance, or to follow a transformation process that would enable better management than before the crisis. They were more ad-hoc actions aimed at achieving quick, short-term effects rather than long-term measures that would lay the foundation for sustainable development under new conditions. While this policy generally helped to temporarily alleviate the crisis's effects, it did not necessarily create the conditions for building a genuine resilience policy (Bourdin et al. 2023).

Undoubtedly, establishing a resilience policy in the internal peripheral regions requires strengthening the potential of these territories. The economic crisis triggered by the pandemic directly affected the financial situation of local governments, to which the national administration frequently assigned additional tasks without increasing the corresponding subsidies. The situation is well illustrated by the example of Poland, where, partly due to the state of emergency and the rapid, unconsulted introduction of new tax laws, local governments lacked clear information about their budgets. At the same time, the quickly expanding scope of assigned tasks exceeded their financial capacities, resulting in a deepening budget deficit (Klimek 2022; Tabis 2023). In the internal peripheral areas, most economic entities are micro and small enterprises, which are considered more sensitive to economic shocks because they have less own capital and achieve lower profit margins compared to larger, more stable enterprises (Bourdin et al. 2023). Under these conditions, despite the ad-hoc protective measures implemented by the government administration, entrepreneurs had to make difficult decisions regarding workforce reductions to achieve tangible cost savings while continuously striving to generate revenue, carefully planning their budgets, maintaining a customer focus, and seeking a market niche (Chaturvedi and Karri 2022). Companies offering certain services shifted their sales to online platforms, capitalizing on the growing market for home delivery services. Consequently, consumers were able to benefit from the delivery of ready-made food products from restaurants as well as, for example, books from local bookstores. Some entrepreneurs were able to transition to remote work arrangements, a solution that was particularly accessible to skilled employees, which has become the new norm in many cities and regions worldwide, influencing company performance. In some cases, employees became more productive; however, this trend was not observed everywhere, especially in economically weaker areas (Ardi et al. 2024). The pandemic also exposed the difficulties enterprises faced in adapting to new conditions, such as the development of e-commerce and remote work. In peripheral regions, the lack of access to high-speed internet and technology, coupled with limited competencies, prevented many companies from taking advantage of these solutions, a factor that proved crucial for survival and minimizing financial losses in other regions. Although support programs, such as "anti-crisis shields," were available, in practice many entrepreneurs did not take advantage of them due to complicated application procedures and limited access to information (Czechowska et al. 2023). This phenomenon was again more visible in lagging areas than in core regions. Data indicate that the most common pandemic-related disruptions associated that shaped the financial situation of entrepreneurs were caused by a decline in sales, an increase in payment delays compared to the same period in the previous year, as well as disruptions in the supply chain that led to shortages of materials, goods, and services. Other reported disruptions include rising raw material prices, difficulties in importing, as well as challenges in acquiring alternative suppliers and leveraging exports to stimulate demand for products and services (Canton et al. 2021).

According to Mahler et al. (2022), the pandemic had a disproportionate impact on the financial situation of communities in peripheral areas that were already grappling with issues such as transport exclusion, low levels of education, and limited access to financial services. Job losses in low-paid and seasonal sectors, upon which local economies depend, led to a rapid deterioration in the material well-being of many families. Concerns about future financial stability forced residents to reduce expenditures and accumulate savings, which further weakened local markets (Szczerba 2023). Residents in peripheral areas experienced heightened financial stress due to limited access to both financial services and support systems (Cui et al. 2022). Limited access to formal sources of financing resulted in a greater reliance on informal sources, which tend to be more expensive and increase beneficiaries' vulnerability to poverty. An increase in unemployment, coupled with worries about financial futures, could also lead to rising indebtedness, thereby imposing an additional burden on households. Moreover, the poorer quality of transportation and internet infrastructure limited the possibility of engaging in remote work, further negatively impacting residents' financial conditions. Financial resilience is directly linked to the level of education and financial literacy, both of which were often lacking in peripheral areas (Dluhopolskyi et al. 2023). Individuals with limited financial knowledge found it more challenging to cope with the obstacles presented by the pandemic, deepening cycles of indebtedness and financial instability (Yao and Zhang 2023).

The SARS-CoV-2 pandemic not only revealed but also exacerbated the pre-existing financial problems of internal peripheral areas, where a limited tax base and the predominance of microenterprises rendered these regions particularly vulnerable to the crisis's effects. Simultaneously, increased expenditures on combating the pandemic and a shortage of funds for local development heightened the risk of economic stagnation. Although government assistance programs and the digitalization of financial services alleviated some of the negative impacts, limited access to technology and cumbersome support procedures deepened financial exclusion and inequalities. The long-term recovery of these regions' resilience requires strengthening the financial autonomy of local governments, supporting local entrepreneurs in pursuing diversified business activities adapted to changing market conditions, and investing in education, including financial education, and digitalization to reduce the exclusion of local communities.

Material Capital

The SARS-CoV-2 pandemic had a significant impact on various aspects of material capital in internal peripheral areas, leading to changes in the utilization of space, the quality of housing conditions, and the accessibility of ecosystem goods and services. In these regions, a key challenge regarding material capital is the inadequate development and low quality of its infrastructural components (Bănică et al. 2024; Pike et al. 2023). However, these areas are characterized by a relatively better state of the natural environment, the quality of which is becoming an increasingly important factor in location decisions (Budner 2004; Chudak 2009). The sanitary restrictions imposed during the pandemic, mobility limitations, and changes in the structure of work and daily life in local communities not only gave rise to new problems but also opened new opportunities for the development of internal peripheries.

In internal peripheral areas, especially outside the main city, a deconcentration of development was observed, which significantly increased the costs of providing technical infrastructure, including broadband internet. Furthermore, the dispersed settlement pattern hindered the attraction of investors, partly due to the incomplete development of local legal regulations. On the other hand, during the pandemic, these areas were less susceptible to the spread of the virus (Cutrini and Salvati 2021), rendering them attractive to individuals seeking to "escape" from core areas and the principal cities of internal peripheries. The SARS-CoV-2 pandemic contributed to the intensification of suburbanization processes and extensive expansion of development, as a portion of the remote-working population opted to migrate to more affordable locations while retaining their employment in cities (Nagarjun and Sridhar 2023). This dynamic suburbanization and migration to peripheral regions created significant challenges for local communities, necessitating adjustments to infrastructure and public service delivery to accommodate the growing population. The insufficient development of transportation and technical infrastructure significantly hampered daily functioning, thereby contributing to the deepening of social exclusion. Additionally, population growth and changes in consumption patterns led to increased costs related to waste management, particularly in the disposal of medical waste and single-use plastics, which further strained local waste management systems (Sarkoidie and Owusu 2021). The sprawl of development generated new challenges in spatial planning, forcing the adaptation of legal regulations and infrastructure to meet rising demands. The pandemic also revealed difficulties related to the maintenance and modernization of public utility facilities, such as schools, cultural centers, and healthcare institutions, which had already been facing financial constraints. Limited access to healthcare services became a key challenge, as local medical facilities struggled with resource shortages that impaired their ability to respond effectively to the health crisis. Deficiencies in healthcare infrastructure not only heightened risks for residents but also deepened inequalities in access to medical care. Furthermore, the budget crisis resulting from a decline in local tax revenues exacerbated these issues. Consequently, local government institutions, unable to carry out the necessary investments, were forced to curtail their operations, which negatively affected the availability and quality of public services. The closure of public institutions, such as schools and libraries, further limited educational and cultural opportunities. Students from families with lower socio-economic status, often deprived of adequate computer equipment and stable internet access, faced difficulties in participating in remote learning, which could have adversely affected their educational development (Długosz 2022).

During the pandemic, the problem of housing overcrowding became particularly acute, revealing structural deficiencies in housing resources. In countries characterized by unfavorable housing conditions, such as Poland, these effects were especially pronounced, particularly in large cities where high population density and limited apartment sizes significantly hindered families' daily functioning. In peripheral areas, although residents often had access to more spacious housing, issues related to poorer access to digital infrastructure and longer distances to educational and service facilities also negatively affected quality of life (Miszczuk 2011; Myck et al. 2020). During lockdown, entire families were forced to remain within confined living spaces, which impeded the performance of everyday tasks. In urban settings, the lack of designated areas for work and study led to reduced efficiency in remote work and online education, further deteriorating residents' psychological well-being. Parents worked remotely while children participated in online education, generating additional challenges not only related to access to adequate computer equipment but also due to the overloading of internet infrastructure in certain regions, particularly rural ones (Cesarski 2022; Ptaszek et al. 2020). At the same time, the pandemic influenced the dynamics of the real-estate market, both the secondary and primary markets experienced rising housing prices. In cities, this was associated with limited supply and increasing demand for larger dwellings adapted to remote work and learning (Narodowy Bank Polski 2022). In peripheral areas, there was heightened interest in real estate, which in some regions led to price increases and a revitalization of local property markets (Petersen 2024). Moreover, the lack of rental housing and the slowdown in development investments further exacerbated housing availability issues, a situation that was particularly acute in large cities (Narodowy Bank Polski 2022).

A critical component of material capital is the quality of the natural environment, which directly affects societal quality of life. The decline in economic activity during the pandemic contributed to a temporary improvement in air quality and a reduction in greenhouse gas emissions. However, once restrictions were lifted, emissions returned to previous levels or even increased, indicating the transient nature of these changes and the limited effectiveness of the pandemic as a factor in reducing the impact of human activity on the environment (Muhammad et al. 2020). Reduced anthropogenic pressure in certain areas during the pandemic fostered processes of plant succession and the restoration of some natural habitats. This phenomenon was observed across Europe, highlighting the positive impact of reduced human activity on ecosystems (Bouzouina et al. 2022; Lausi 2022). Such changes may offer opportunities for internal peripheral areas, which are characterized by a greater richness of natural resources compared to highly urbanized core areas. The growing interest in settling in these regions, driven by a desire to live close to nature, appears to be a likely direction for socio-spatial changes (Antonetti et al. 2020). Moreover, increased environmental awareness may have spurred greater interest in green spaces as a vital element of quality of life in both cities and suburban areas.

Despite numerous challenges, the pandemic also created new opportunities in the context of sustainable development. Local and regional authorities can leverage the experiences gained to intensify investments in ecosystem goods and services, sustainable construction, and the development of environmentally friendly technologies. Such measures can not only improve residents' quality of life but also enhance the attractiveness of peripheral areas for potential new inhabitants and investors. The SARS-CoV-2 pandemic contributed to an increase in environmental awareness among the public, promoting the adoption of sustainable urban policies such as the development of cycling infrastructure and the creation of new green areas (Sharifi and Khavarian-Garmsir 2020). In light of these transformations, governments and local administrations now have the opportunity to accelerate efforts toward energy transformation and the reduction of greenhouse gas emissions in urban areas.

The SARS-CoV-2 pandemic acted as a catalyst for profound changes in internal peripheral areas, significantly impacting their social, economic, and infrastructural functioning. These regions, which were already struggling with limited access to services, infrastructure, and development resources, experienced even greater marginalization. On one hand, the pandemic crisis accentuated the deficits of these areas; on the other hand, it revealed certain developmental opportunities linked to accelerated digitalization, growing environmental awareness, and changing societal preferences in housing and mobility. Table 3.4 provides a synthetic summary of the socio-economic consequences of the pandemic structured around the five development factors adopted in this study as the basis for characterizing the territorial capital of internal peripheries: accessibility, human capital, social and institutional capital, financial capital, and material capital. The information presented highlights both the short-term effects and the long-term challenges and development opportunities for internal peripheral regions.

3.2 European Cohesion Policy in the Face of the SARS-CoV-2 Pandemic

The European Union's Cohesion Policy plays a crucial role in pursuing economic, social, and territorial cohesion, which is a treaty-based objective of the Community's functioning. It represents one of the most significant EU policies aimed at improving living standards in less developed, including peripheral areas. Interventions under this policy should both reduce developmental disparities and support competitiveness and economic growth, while maintaining alignment with sustainable development principles. However, the wide range of goals within these actions, often perceived as contradictory, complicates their implementation. Consequently, their effectiveness has increasingly come under public criticism, especially amid economic turbulences and crises. Economic slowdowns or recessions exacerbate developmental disparities, leading to growing negative sentiments toward the EU, particularly in economically disadvantaged and peripheral regions, which consequently become lagging areas (Capello and Perucca 2019; Dąbrowski et al. 2019). Erosion of trust is especially noticeable in areas unable to benefit from structural and cohesion funds according to their needs, or in which utilization of these funds has failed to yield expected outcomes. This underscores the need for a more equitable resource distribution and targeted interventions (Crescenzi and Giua 2019; Faragó and Varró 2016) that ultimately better reflect the diverse needs of individual territories.

Table 3.4 Socio-economic consequences of the SARS-CoV-2 pandemic

Factor	Consequences	Description
Accessibility	Social	Restrictions in access to education and public services due to inadequate digital infrastructure have deepened educational and social inequalities (long term)
		Insufficient development of e-services and the digitalization of administration has hindered access to basic public services (short and long term)
		Transition of social and economic activities to the digital sphere has highlighted digital exclusion among residents of peripheral areas (long term)
		Restrictions in access to education and public services due to inadequate digital infrastructure have deepened educational and social inequalities (long term)
		Insufficient development of e-services and digitalization of administration has hindered access to basic public services (short and long term)
		Transition of social and economic activities to the digital sphere has highlighted digital exclusion among residents of peripheral areas (long term)
	Economic	Disruptions in transportation and employee mobility weakened local supply chains, reducing economic efficiency (short term)
		Mobility restrictions within regions increased unemployment levels (short term)
		Termination of public transportation reduced social and economic mobility, exacerbating transport exclusion (long term)
		Increased costs of private transportation raised living costs and air pollution levels (long term)
		Border closures caused income losses in border regions dependent on cross-border trade (short term)
		Lack of effective transport connections limited access to neighboring labor markets and services, impacting quality of life (long term)

(continued)

Table 3.4 (continued)

Factor	Consequences	Description
		Increased logistic costs and delivery delays raised product prices in peripheral regions (short term)
Human capital	Social	High mortality rates among elderly weakened intergenerational bonds and local support networks (long term)
		Limited access to healthcare and preventive measures worsened residents' physical and mental health (long term)
		Lack of psychological support increased the risk of depression and health issues (long term)
		Educational inequalities deepened due to limitations in remote education (long term)
		Graduates faced difficulties entering the labor market, thereby increasing feelings of exclusion (short term)
		Temporary migrations to less-urbanized areas increased interest in these regions but required measures supporting integration (short and long term)
	Economic	High mortality among the elderly reduced labor force resources, weakening regions' economic potential (long term)
		Declining birth rates reduced future labor supply, hindering economic development (long term)
		Temporary return migrations to peripheral areas increased pressure on local labor markets, causing unemployment growth (short term)
		Closure of schools and suspension of vocational training reduced graduates' skill levels and labor market competitiveness (long term)
		Digitalization of work practices (remote work) created new economic opportunities in peripheral regions (long term)
Social and institutional capitals	Social	Activation of local support networks strengthened social solidarity and neighborly assistance (short term)
		Reduction in social activities, declining trust, and anxiety weakened cooperation between communities (long term)

(continued)

Table 3.4 (continued)

Factor	Consequences	Description
		Popularity of conspiracy theories reduced trust in science and public institutions (long term)
		Lack of resources for public institutions weakened their effectiveness and image (long term)
	Economic	Increased costs of organizing local support and crisis activities burdened local government budgets (short term)
		Lack of investment in digitalization of public institutions limited access to e-services (long term)
		Weakening of interregional cooperation reduced economic potential of peripheral regions (long term)
		Underinvestment in local institutions limited their ability to respond to residents' needs (short and long term)
Financial capital	Social	Deepening financial exclusion in communities with limited access to information about support programs (long term)
		Increased stress and uncertainty due to loss of financial stability (short and long term)
	Economic	Declining local government revenues from taxes reduced infrastructure investment capabilities (long term)
		Expenditure on crisis activities, such as purchasing protective equipment and organizing remote learning, increased (short term)
		Micro and small enterprises forced to reduce employment increased unemployment levels (short term)
		Lack of access to high-speed internet limited business adaptation to e-commerce and remote work (long term)
Material capital	Social	Overcrowding of dwellings during lockdowns affected families' quality of life (short term)
		Insufficient healthcare infrastructure restricted access to medical care (long term)

(continued)

Table 3.4 (continued)

Factor	Consequences	Description
		The pandemic emphasized the importance of green technologies and spatial planning (long term)
	Economic	Increased costs of maintenance and modernization of public infrastructure (long term)
		Slowdown in real-estate investments and rising property prices limited housing accessibility (long term)
		The pandemic temporarily reduced greenhouse gas emissions, but the effect was transient (short term)
		Increased costs of medical and single-use plastic waste disposal burdened waste management systems (long term)

Source Own elaboration

The SARS-CoV-2 pandemic has further complicated the landscape of regional development and associated Cohesion Policy intervention tools. This crisis has prompted a reassessment of Cohesion Policy, emphasizing the need to build resilience and adaptive capacity in regional economies (Artelaris and Mavrommatis 2022; Tudorache and Nicolescu 2023), which has become one of the central themes in debates regarding the policy's future. The pandemic highlighted the importance of Cohesion Policy as a recovery instrument, leading to calls for increased financial support to enhance commercial resilience and stimulate local economies (Artelaris and Mavrommatis 2022; Fedajev et al. 2023; Tudorache and Nicolescu 2023). It also underscored the necessity of a more integrated and flexible approach to regional development, enabling faster responses to crises and providing targeted support to regions most affected by economic disruptions. Lessons drawn from the pandemic will likely shape future policy frameworks, emphasizing solidarity, resilience, and adaptability in the face of current and future challenges. Time will tell whether this leads to a strengthened European Union Cohesion Policy or signals the beginning of its end, resulting in the integration of its activities into national envelopes, an outcome already signaled in discussions regarding the Multiannual Financial Framework beyond 2027.

3.2.1 *New Challenges and Redefinition of the Goals of European Cohesion Policy*

Economic and social cohesion were established as objectives of the Community in the Treaty of Rome, becoming development priorities for member states from 1958 onward. European regional policy was designed to strengthen these goals; however, it was effectively initiated only after the establishment of its institutional framework through the creation of the Directorate-General for Regional Policy in the 1960s. Initially, its interventions were limited solely to top-down, ex-post interregional actions. This approach changed dramatically following the financial crisis of 1973, whose consequences prevented continuation of intervention activities as previously practiced. Due to increasing spatial disparities in development, systemic interventions urgently replaced ad-hoc measures. On the one hand, given limited resources, concentrating interventions became essential, both spatially (problematic areas) and substantively (priority intervention goals). On the other hand, new financial instruments had to be launched, leading to the establishment in 1975 of the European Regional Development Fund, which remains the cornerstone of funding intervention measures designed to reduce socio-economic spatial disparities in Europe. The limited effectiveness of top-down programming and regional policy implementation drew attention to the necessity of concentrating activities on bottom-up intraregional actions supported by coordinated interregional initiatives.

Over subsequent decades, the design of European regional policy underwent significant transformations. Initially perceived as a financially and organizationally limited instrument to assist restructuring and development in problematic areas, it has become, since the late 1990s, the primary EU policy tool compensating for free-market actions, targeted primarily at the poorest regions. In later decades, the scope of intervention of European regional policy fundamentally shifted. It evolved into a principal instrument supporting competitiveness across all European regions, with the aim of achieving increasingly ambitious Community development priorities as outlined in successive strategies: the Lisbon Strategy, the Europe 2020 Strategy, and the European Green Deal. However, focusing actions on these objectives contributed to increased spatial disparities in development. Achieving ambitious Community development priorities proved difficult, sometimes impossible, in the weakest areas in social and economic terms. Thus, by directing European regional policy toward supporting the EU's priority development objectives, its core impact, cohesion, including integration of new member states joining after 2004, was significantly weakened, contributing to its current crisis (Churski and Żuber 2022).

A particularly significant trend in the evolution of European regional policy since the mid-2000s has been the strengthening of its territorial dimension, understood as programming integrated public interventions within designated spatial units. This approach involves including not only entire countries or regions, but also cities and rural, mountainous, and island areas in the programming and implementation of development interventions (Bachtler et al. 2019). Territorial cohesion was formally added as an additional dimension of cohesion in the Lisbon Treaty, which entered into force in 2009, providing a formal reference point for establishing mechanisms to support this dimension. Expanding the priorities of European regional policy to include territorial cohesion alongside economic and social cohesion provided the basis for renaming it "Cohesion Policy."

In 2007, in Leipzig, member states adopted the Territorial Agenda of the EU and the Leipzig Charter on Sustainable European Cities, both of which represented a political declaration expressing member states'[1] readiness to implement spatial considerations in development activities at the EU-wide and national levels. Concurrently, the European Commissioner for Regional Policy, Danuta Hübner, commissioned a report concerning Cohesion Policy reform with the aim of fully territorializing its interventions. This led to the Fabrizio Barca Report (Barca 2009), outlining the principles for reorienting Cohesion Policy actions according to a place-based approach. Consequently, a new paradigm for Cohesion Policy emerged, based on comprehensive and tailored interventions adjusted to the specific territorial capital of individual areas, which determines their varying responses to dynamic socio-economic transformations (Churski et al. 2021).

This new paradigm rests on two fundamental assumptions. Firstly, each location possesses its own unique economic, social, cultural, and institutional characteristics, which spontaneously or deliberately shape its territorial capital through targeted developmental interventions (Camagni 2008). Secondly, the effective utilization of this territorial capital requires knowledge and skills in applying developmental policies. Only those who understand which developmental factors require creation or strengthening, and precisely where and when, can enhance the effectiveness of development interventions (Barca et al. 2012). Therefore, the effective implementation of this new Cohesion Policy paradigm hinges upon objectively recognizing territorial characteristics—particularly by identifying spatial variations in socio-economic development. Development interventions must address objectively identified needs rather than applying a unified, "one size fits all" approach (Morgan 2016). Considering the heterogeneous territorial capital of different regions and their diverse needs during the programming and implementation of Cohesion Policy is thus essential for achieving effective and efficient developmental interventions (Rodríguez-Pose 2013).

Unfortunately, operational implementation of these assumptions has faced significant obstacles from the outset, calling into question, despite numerous efforts, the feasibility of their practical application within regional policy interventions (Böhme et al. 2011). Consensus regarding the effectiveness of place-based policies continues to evolve. In this process, the validity of the place-based approach itself is generally undisputed, with greater attention being placed on the necessity of refining its operational aspects (McCann 2023), emphasizing that Cohesion Policy actions must

[1] Another "Territorial Agenda of the European Union 2020: Towards an Inclusive, Smart, and Sustainable Europe of Diverse Regions" was adopted in 2011 in Gödöllő, Hungary. In 2016, the document "Urban Agenda of the EU: Pact of Amsterdam" was adopted, outlining objectives and principles for actions by the EU and member states aimed at the sustainable development of cities. On December 1, 2020, the "Territorial Agenda 2030: A Future for All Places" was adopted in Germany.

remain sensitive to places (Iammarino et al. 2019) and the inhabitants shaping their territorial capital (Churski et al. 2021).

Optimal functioning of Cohesion Policy is hindered by the fact that, beyond its primary scope (that of reducing spatial disparities in socio-economic development), it is simultaneously engaged in additional tasks. As previously noted, it is expected to measurably contribute to achieving broader European development priorities, support reforms in member states, enhance governance through ex-ante conditionalities and linkages with the European Semester, and provide rapid responses to increasingly frequent and severe global crises. Expanding Cohesion Policy responsibilities without corresponding financial and institutional reinforcement has, in practice, led to its weakening. Excessive focus on priority areas from a European perspective, rather than addressing specific territorial needs, has placed the weakest regions in a particularly challenging position, given their distinct potentials and requirements compared to core areas. As a result, while developmental convergence remains observable at the national level among member states, disparities at the regional and especially subregional levels have begun to increase alarmingly.

The limited effectiveness and low efficiency of current Cohesion Policy measures have been starkly exposed by recent global crises, such as the financial crisis, the SARS-CoV-2 pandemic, and Russia's aggression against Ukraine. On the one hand, these events highlighted significant deficits in existing approaches, negatively verifying investment strategies and particularly their inability to build long-term regional resilience (Martin and Sunley 2020). On the other hand, these crises intensified the hazardous trend of strengthening central authorities at the expense of the competency and financial autonomy of local governments, a phenomenon typical of the "mission-oriented" approach. Such centralization destructively affects the possibility of implementing multilevel governance, a fundamental principle underpinning the operation of Cohesion Policy (Hadjimichalis 2021).

3.2.2 *Cohesion Policy Reaction to SARS-CoV-2 Impact*

The SARS-CoV-2 pandemic created an unprecedented situation with global socio-economic impacts, exacerbating spatial, economic, and social inequalities (Adjaottor et al. 2022; Doroudgar et al. 2021), thus providing the rationale for interventions within the framework of Cohesion Policy. The pandemic necessitated coordinated measures aimed at mitigating both immediate and systemic socio-economic consequences (Daqar et al. 2021; Puspa 2021; Setyawati and Ningrum 2021; Yan et al. 2022).

The European Commission promptly took measures to support employees and employers and to sustain the economy during the crisis. At the beginning of 2020, the Commission launched two packages utilizing approximately €37 billion of previously unspent Cohesion Policy funds: the Coronavirus Response Investment Initiative (CRII) in March and the Coronavirus Response Investment Initiative Plus

(CRII +) in April, both quickly approved by the European Parliament and the European Council. These initiatives introduced significant flexibility and simplifications to enhance financial liquidity and facilitate the rapid reallocation of funds toward those areas most in need of assistance, tailored to local conditions. Priority areas included strengthening healthcare systems, labor markets, and support for small- and medium-sized enterprises (SMEs). Specifically, from April 2020 onwards, member states were permitted to apply a 100% EU co-financing rate for projects implemented under Cohesion Policy funds for one year, effectively eliminating the need for national contributions. Additionally, greater flexibility was introduced, allowing transfers between different Cohesion Policy funds, thus enabling more effective responses in key areas such as healthcare, economic support, and employment protection. Although these ad-hoc funding mechanisms were primarily designed for crisis management, they can be regarded as the foundation of and opportunity for economic recovery during a period of profound crisis. Given that economic support through public funding, under conditions of collapsing revenue streams, represents a significant challenge for all countries, the European Commission also introduced several temporary fiscal policy measures to address these challenges:

- On March 19, 2020, the European Commission adopted the State Aid Temporary Framework, giving member states greater flexibility in supporting enterprises that were viable under normal circumstances but faced difficulties due to the pandemic,[2]
- On March 20, 2020, the Commission activated, for the first time in history, the general escape clause under the Stability and Growth Pact. This measure permitted member states to undertake necessary fiscal actions to support their economies without adhering to the standard rules and limitations of the Pact.[3] Consequently, national governments could support income stabilization and finance job-retention schemes, provide assistance to businesses, or invest in healthcare (Daqar et al. 2021; Szczurek 2021),
- On April 9, 2020, the Eurogroup proposed three safety nets totaling EUR 540 billion aimed at supporting workers, businesses, and member states. Among these measures, the temporary Support to mitigate Unemployment Risks in an Emergency (SURE) instrument was established to reduce unemployment risks arising from this extraordinary situation. SURE provided advantageous loans to finance the creation or expansion of national short-time work schemes. Additionally noteworthy are the pan-European guarantee fund established by the European Investment Bank, the reallocation of resources to credit guarantee instruments under the COSME program, pandemic-modified precautionary credit lines within the European Stability Mechanism (Szczurek 2021), as well as a variety of ad-hoc interventions within sectoral policies.

[2] In May 2022, it was announced that the State Aid Temporary Framework would expire after June 30, 2022, with certain exceptions remaining in force until the end of 2023.

[3] The general escape clause was deactivated at the end of 2023 in response to the anticipated recovery of the European economy to pre-pandemic levels, despite persistent threats associated with the war initiated by Russia.

Table 3.5 NextGenerationEU and Multiannual Financial Framework 2021–2027 total allocation per heading (in EUR)

Heading	NextGenerationEU	MFF (billion)
Single market, innovation, and digital	€11.5 billion	€149.5
Cohesion, resilience, and values	€776.5 billion	€426.7
Natural resources and environment	€18.9 billion	€401.0
Migration and border management	–	€25.7
Security and defense	–	€14.9
Neighborhood and the world	–	€110.6
European public administration	–	€82.5
Total	€806.9 billion	€1,210.9

Source European Commission 2024 (Current prices). https://commission.europa.eu/strategy-and-policy/recovery-plan-europe_en. Access on 24.09.2024

Alongside immediate responses, systemic measures were also prepared and implemented. The European Union's key response to the longer-term socio-economic challenges posed by SARS-CoV-2, explicitly aligned with cohesion objectives, was the NextGenerationEU (NGEU) package launched in July 2020, valued at EUR 750 billion (in 2018 prices). Financial support under NGEU is provided to member states through grants and loans within the framework of the Recovery and Resilience Facility (RRF) and grants allocated through the Multiannual Financial Framework (MFF) (Tables 3.3 and 3.5).

Table 3.3 presents the value of the NGEU package relative to the financial framework for the 2021–2027 perspective. The amounts allocated for post-pandemic recovery within the Community are substantial, constituting approximately two thirds of the total Multiannual Financial Framework (MFF) allocation. For the first time in history, the European Union has decided to issue common debt, for which its member states assume collective responsibility. This substantial allocation for pandemic recovery, primarily funded through loans guaranteed by the EU, represents a significant step toward greater fiscal integration. This development has frequently been described as a "Hamiltonian moment"[4] (Groenendijk 2023; Porte and Jensen 2021), as it signals a fundamental shift in fiscal policy and a departure from previous reluctance toward collective borrowing.

The design of the entire NGEU initiative draws from lessons learned during previous economic crises, particularly those in the Eurozone, which exposed significant gaps in the EU's economic governance system. The initiative aims not only to meet the immediate needs of member states and regions in distress but also to support long-term structural reforms aligned with the EU's horizontal objectives regarding sustainable development and digital transformation (Alcidi and Gros 2020; Groenendijk 2023; Porte and Jensen 2021). The emphasis on reforms and investments also

[4] Alexander Hamilton, one of the US's Founding Fathers and the first Secretary of the Treasury during the costly war for independence from Great Britain, advocated for the mutualization of state debts and their assumption by the federal government.

indicates a fundamental shift in the EU's approach to economic crises, moving away from the fiscal austerity that characterized previous crisis responses and recovery programs (Capati 2023; Pritoni et al. 2023; Saraceno 2022).

Importantly, this shift in approach is significant in terms of perceptions of the EU, particularly in Southern European countries, where support had substantially declined following multiple crises and decades of austerity measures. The rapid adoption of NGEU reflected consensus among member states and represented solidarity in the face of the pandemic threat, as well as demonstrating agency and swift crisis-response capability (Lippi and Terlizzi 2023; Woźniakowski 2023). Implementing NGEU involves adherence to rigorous rules, requiring member states to present National Recovery Plans, which the European Commission assesses according to standardized, concrete criteria (Kolliopoulos 2022). The conditionality attached to plan approval and fund expenditure ensures allocation to genuinely significant projects that substantially enhance economic resilience and social and territorial cohesion (Caramani and Cicchi 2024; Ferrera et al. 2021; Nahtigal 2022).

It is worth noting that the solutions adopted within NGEU have emerged as an alternative to previous Cohesion Policy practices, potentially signaling a shift in its operational model for the upcoming financial perspective.

The structure of NGEU encompasses several components. A central element is the Recovery and Resilience Facility (RRF), emphasizing investments in green and digital transformations (Bokhorst and Corti 2023) designed to make the Community stronger and more resilient in the aftermath of the crisis. Its primary objectives include achieving climate neutrality by 2050, advancing the EU's digital transition, and supporting economic growth. A second critical element is the Recovery Assistance for Cohesion and the Territories of Europe (REACT-EU), which provided an additional €50.4 billion to member states within the 2014–2020 perspective and an extra €0.2 billion for Technical Assistance (ECA 2024). These funds represented an additional Cohesion Policy allocation for 2021–2027, serving as a bridge to a long-term recovery plan. The third component involves the expansion of existing programs and the establishment of new ones. All countries can utilize Horizon Europe, which finances research on new medical technologies and vaccines, digitalization and artificial intelligence development, and climate and sustainability projects, serving as a continuation of the Horizon 2020 program. Moreover, member states have access to InvestEU, a continuation of a previous instrument designed to stimulate strategic innovative investments in key EU sectors (digital and green transformations, transport and energy infrastructure, and SME support).

Another component is a newly established instrument, the Just Transition Fund (JTF), targeting regions dependent on coal and fossil fuels to facilitate their transition toward a low-carbon economy. The catalogue of reinforced programs from the previous financial perspective also includes the RescEU component, substantially expanded to enhance the EU's crisis-response capabilities. Finally, another continuing component, reinforced under the third element of NGEU, is the European Agricultural Fund for Rural Development (EAFRD), aimed at increasing agricultural resilience, facilitating recovery after the crisis, and creating conditions for widespread adoption and realization of the green transition.

Each of these components significantly impacts the reduction of disparities between core regions and inner peripheries, contributing to economic recovery and enhancing the resilience of less-developed areas. These instruments have provided not only immediate financial aid during the pandemic but also long-term support for economic transformation, as well as strengthening the developmental potential of regions distant from Europe's major economic centers.

The Recovery and Resilience Facility (RRF), as the central financial tool of NGEU, holds particular importance for peripheral regions, supporting investments in infrastructure, digitalization, and modernization of public services. Its primary aim is to strengthen local economies through investments in advanced technologies, green transition initiatives, and human capital development. For inner peripheries, this translates into financing projects that enhance residents' quality of life, promote local entrepreneurship, and modernize transport and education. With funding from RRF, peripheral regions have been able, and continue, to reduce their dependency on traditional economic sectors and attract new investments, thereby increasing their competitiveness at the European level.

Equally important for supporting development in these areas is the REACT-EU component, explicitly designed to assist the EU's most remote regions. It guarantees additional funding of €30 per inhabitant for these areas, supplementing their standard allocations (Mazur-Kumrić 2022). This aligns closely with the EU's broader objectives regarding cohesion, sustainable development, and solidarity, ensuring that no region or member state is left behind during the recovery process and that spatial disparities do not deepen (Celi et al. 2020). In this context, REACT-EU serves not merely as a financial instrument but as a direct response to the strategic objectives of Cohesion Policy concerning convergence and long-term integrity of the EU.

Horizon Europe plays an important, albeit indirect, role for peripheral regions by supporting the development of innovative technologies and scientific research that have strategic significance for their long-term development. For regions distant from major scientific and technological hubs, Horizon Europe facilitates knowledge transfer and fosters new economic models based on innovation. Consequently, peripheral regions can participate in international research projects and implement cutting-edge technologies, enhancing their competitiveness. Additionally, the program supports the development of regional smart specializations, enabling better alignment of local resources with the demands of a modern economy, thereby reducing dependence on traditional economic sectors.

InvestEU constitutes a crucial instrument supporting private and public investments throughout the EU, but its significance for peripheral regions is particularly notable. This program provides financial tools to enable SME sector development, improved access to capital, and the implementation of infrastructure projects of strategic importance for economic recovery in less-developed regions. Thanks to InvestEU, peripheral areas can attract new investments, expand modern industrial sectors, and foster technological development, which are essential for overcoming their structural backwardness relative to the more developed parts of the EU. The support also includes funding for projects related to energy transition and sustainable development, vital within the EU's climate policy framework.

The Just Transition Fund (JTF) plays a substantial role in peripheral areas historically dependent on single-sector economies, such as coal or heavy industry. The primary goal of the JTF is to support economic diversification and the creation of alternative employment opportunities, reducing the adverse socio-economic impacts of climate transitions. With JTF funding, peripheral regions can finance modernization projects, develop new economic sectors, and improve the skills of local populations, promoting the growth of more resilient and adaptable economies.

RescEU significantly improves resilience in peripheral areas regarding health security and crisis management. For regions far from major medical and logistical centers, investments in strategic reserves of medicines, medical equipment, and personal protective equipment are essential. This program has also enhanced transport and logistical capacities for patients requiring specialized care. Such a mechanism significantly increased health security levels during the pandemic and ensures an effective response to potential future health crises, especially in peripheral regions where access to advanced medical care is limited.

The European Agricultural Fund for Rural Development (EAFRD) plays a key role in reducing developmental gaps between core and peripheral regions by supporting agriculture and rural area development. The program enables modernization of the agricultural sector, development of sustainable food production technologies, and improved living conditions in rural areas. Through EAFRD support, peripheral regions can invest in rural infrastructure, enhance soil quality, and increase the competitiveness of local producers. Additionally, the fund supports rural tourism and local entrepreneurship, making peripheral areas more attractive for living and economic activity.

Due to various initiatives, peripheral regions gained opportunities to strengthen healthcare system resilience, develop telemedicine, enhance digital and transport infrastructure, and accelerate ecological transformation (Bourdin et al. 2024; Madanipour et al. 2021; Weck et al. 2021). Peripheral regions, characterized by structural limitations such as lower transport accessibility, restricted public services, low digitalization levels, limited resources, weaker infrastructure, and low economic diversification, were particularly affected by the negative effects of the crisis. The insufficient availability of territorial capital resources in these areas (Sect. 3.1.1) hindered their adaptation and recovery capacity after the pandemic. In this context, Cohesion Policy measures taken after the SARS-CoV-2 pandemic outbreak were particularly important for peripheral regions. Structural and infrastructure support in peripheral areas represented a fundamental aspect of these initiatives. Investments in digital technologies, healthcare systems, and sustainable technologies created conditions for their long-term development, reinforcing foundations of economic resilience. Equally significant was the shift in fiscal policy. The launch of the NGEU package and associated investments aimed not only at short-term crisis management but also at supporting structural reforms and digital and ecological transformations. Programs and newly established funds, besides providing immediate financial liquidity, aimed at thorough economic reconstruction and adapting economies to future challenges. Immediate responses and systemic adjustments laid the foundations for economic recovery and improved resilience in regions severely affected by the crisis, including peripheral areas. These measures represent a crucial step toward reducing spatial and social inequalities, although their effectiveness largely depends on continued and expanded initiatives in the future. Cohesion Policy must continually evolve to better respond to diverse regional needs and adapt to emerging challenges. The pandemic highlighted both the EU's capacity for rapid response and areas needing further reforms and investments. To build genuinely sustainable foundations for long-term development, it is essential to maintain and deepen actions targeting economic, social, and territorial cohesion within an integrated and solidarity-based Community.

3.2.3 Post-pandemic Cohesion: Future Prospects and Challenges for European Cohesion Policy

The European Cohesion Policy plays a critical role in promoting economic, social, and territorial cohesion throughout the European Union (EU). The significance of this policy was especially highlighted during the SARS-CoV-2 pandemic, as it aimed to counteract deepening inequalities and address new challenges emerging from this global crisis. The pandemic not only exacerbated existing disparities but also forced a reassessment of strategies implemented within the Cohesion Policy framework, ensuring their effectiveness in supporting recovery and enhancing the resilience of diverse regions. The experiences of the pandemic prompted reconsideration of the Cohesion Policy's objectives regarding territorial cohesion. A substantially more integrated approach is necessary, recognizing linkages among regions, the importance of cross-border cooperation, as well as relations and distances between rural and urban areas, and more broadly between core and peripheral regions. Furthermore, the pandemic underscored social cohesion as a fundamental element of Cohesion Policy. The crisis deepened social inequalities, especially among the most vulnerable social groups, highlighting the need for policies promoting inclusiveness and social support (Özçürümez et al. 2021. Studies indicate that the Cohesion Policy cannot focus solely on economic recovery; it must also prioritize social issues to foster a more cohesive European community (Cunico et al. 2020). Such an approach requires considering economic, social, and territorial cohesion as mutually complementary elements, collectively building resilience against future crises. Economic cohesion strengthens the foundations for development and social cohesion integrates diverse groups and reduces inequalities, while territorial cohesion minimizes spatial disparities in access to resources and developmental opportunities. A fully integrated approach that harmoniously combines these three aspects is essential for ensuring that all EU regions can develop sustainably, inclusively, and resiliently in the post-pandemic landscape.

Experiences from yet another crisis revealed that the method of managing and implementing developmental intervention mechanisms remains a significant challenge. Multilevel governance is crucial for ensuring the effective deployment of funds at all implementation levels. Previous crises revealed numerous governance gaps, hindering effective implementation of crisis response instruments. A separate issue concerns the sustainability of funding mechanisms. The economic consequences of the pandemic significantly strained public finances, raising concerns about the EU's capacity for sustained developmental funding (Bachtler and Méndez 2020a, 2020b). Therefore, it is essential to ensure that funds are directed toward projects with a high potential for achieving Cohesion Policy objectives, particularly regarding sustainable development and digital transformation.

Given that the pandemic strongly influenced social attitudes, implementing appropriate communication strategies has become an important element in the recovery process. Such strategies can assist in social communication, identity-building, and fostering a positive image of the Community, as well as increasing support for European integration (Borz et al. 2022). Social participation and further development of civil society are essential for building trust and ensuring that Cohesion Policy remains responsive to the needs of diverse communities. The need for a more flexible and tailored approach to Cohesion Policy has never been more urgent, as regions grapple with the multifaceted impacts of the crisis.

When identifying recommended directions for change and necessary reforms that could effectively address the current challenges of Cohesion Policy, several key recommendations should be considered. These are crucial for efficient and effective programming and implementation aimed at reducing spatial disparities in socio-economic development among EU states and regions:

- First, the Cohesion Policy must acknowledge that prevailing trends in climate, demographic, geopolitical, and technological changes have been significantly altered by recent events such as the financial crisis, the SARS-CoV-2 pandemic, and Russia's invasion of Ukraine, triggering renewed inflation and concerns related to the security of supply chains for food, energy, and raw materials (OECD 2023, Ninth Report 2024). In these new circumstances, development interventions targeting peripheral regions must build long-term resilience, exploiting opportunities for gaining new competitive advantages based on their resources (Lange et al. 2021) and reorienting both global and local linkages (MacKinnon et al. 2022). This significantly alters previous methods of programming and implementing pro-development activities, making them more complex and requiring greater integration (Pike et al. 2023).
- Second, the Cohesion Policy should promote inclusive development opportunities for all, combating economic exclusion and poverty (European Commission: Directorate-General for Regional and Urban Policy, 2024). Every resident should have guaranteed opportunities and access to public services and basic infrastructure, which is becoming increasingly costly and less efficient, especially in sparsely populated or economically declining regions, posing significant challenges to spatial justice (Fritsch et al. 2023). This requires the implementation of non-standard solutions, leveraging not only product innovations, such as demand-responsive transport, but also organizational innovations, such as integrated public transport services and fare systems, as well as social innovations, like

broader public participation in finding socially acceptable solutions for lagging and peripheral areas (OECD 2023).

- Third, it is essential to strengthen local capacities to adequately respond to developmental challenges, a need apparent not only in peripheral areas characterized by deficits in territorial capital and lacking "critical mass" for local development but also in economically robust areas that risk falling into development traps (OECD 2023). Effectively addressing these challenges requires a thorough understanding of their underlying causes, alongside efforts to overcome them by mobilizing local resources and developing new functional relationships, ensuring broad stakeholder involvement (Farole et al. 2018; Rodriguez-Pose et al. 2023).
- Fourth, barriers limiting the impact of growth poles on surrounding areas must be overcome (Floerkemeier et al. 2021), simultaneously mitigating negative consequences of agglomeration effects, such as congestion in core areas and stagnation or decline in surrounding areas (OECD 2023). This calls for greater utilization of functional areas as entities responsible for programming and implementing supra-local interventions that foster functional relationships and improve intervention effectiveness.
- Fifth, it is crucial to intensify the use of the local level as the architect and implementer of development interventions (Farole et al. 2018) and to grant the Cohesion Policy a coordinating role with respect to other national policies (Bachtler et al. 2019). On one hand, this necessitates technical assistance to strengthen local institutional capacities (European Commission: Directorate-General for Regional and Urban Policy 2024), whose weaknesses currently hinder full participation, multi-level accountability, and integrated planning processes across diverse territories (Pinheiro et al. 2022). On the other hand, this highlights the need for political consensus to move away from siloed sectoral development interventions toward fostering thematic and spatial synergies, coordinated through the Cohesion Policy (Churski and Żuber 2022).
- Sixth, the Cohesion Policy must reinforce supra-local cooperation, which is crucial for overcoming spatial fragmentation constraints. Interregional partnerships and cross-border cooperation can enhance resource utilization, knowledge exchange, and collaborative problem-solving, thereby increasing intervention effectiveness and contributing to greater territorial cohesion.

A more integrated approach that considers the specific characteristics of peripheral regions, unlocks their potential, and supports their ability to overcome structural limitations will undoubtedly invigorate their development and enhance the overall integrity of the EU Community.

References

Abdullah W, Zainudin WNRAB, Ismail SB, Zia-ul-haq HM (2022) The impact of microfinance services on Malaysian b40 households' socioeconomic performance: a moderated mediation analysis. Int J Sustain Dev Plan 17(6):1983–1996. https://doi.org/10.18280/ijsdp.170634

Adda HW (2023) Business opportunities during the covid-19 pandemic in Palu city. Int J Manag Stud Soc Sci Res 4(6):231–234. https://doi.org/10.56293/ijmsssr.2022.4543

Adekola J, Clelland D (2019) Two sides of the same coin: business resilience and community resilience. J Contingencies Cris Manag 28(1):50–60. https://doi.org/10.1111/1468-5973.12275

Adjaottor ES, Ahorsu FA, Chen H, Ahorsu DK (2022) Predictors of Covid-19 stress and Covid-19 vaccination acceptance among adolescents in Ghana. Int J Environ Res Public Health 19(13):7871. https://doi.org/10.3390/ijerph19137871

Afelt A, Kursa M (2021) SARS-CoV-2 w Polsce–dynamika przestrzenna i ilościowa w 2020 r. In: Bezpieczeństwo osób starszych w przestrzeni miejskiej. Analiza doświadczeń, wnioski i rekomendacje z uwzględnieniem okresu pandemii SARS-CoV-2, pp 95–116

Ahmed S (2023) Ageing health and socioeconomic determinants: an introspect. Ann Adv Biomed Sci 6(1). https://doi.org/10.23880/aabsc-16000210

Akgün AI, Türkoğlu SP (2023) Intellectual capital and performance of listed firms during the global financial crisis: the effects of legal origin. Int J Organ Anal 32(4):759–785. https://doi.org/10.1108/ijoa-01-2023-3587

Al-Barakat A, Alali R, Karasneh M, Alakashee B, Sami S, Al-Qatawneh El-mneizel A, Alsalhi N, Alhawamdeh M (2023) Learning through digital technology: role of national and international education teachers 2166–3750

Alcidi C, Gros D (2020) Next Generation EU: a large common response to the Covid-19 crisis. Intereconomics 55(4):202–203. https://doi.org/10.1007/s10272-020-0900-6

Aleluia ÍRS, Vilasbôas ALQ, Pereira GE, Nunes FGDS, Aquino R, Nunes CA, Prado NMDBL (2023) State management of primary health care in response to COVID-19 in Bahia, Brazil. Ciência & Saúde Coletiva 28(5):1341–1353. https://doi.org/10.1590/1413-81232023285.12732022en

Alghamdi SM, Alsulayyim AS, Alqahtani JS, Aldhahir AM (2021) Optimising digital health platforms in Saudi Arabia: determinants from the COVID-19 pandemic experience. https://doi.org/10.20944/preprints202109.0393.v1

Allanjawi KM, Yazdanparast A, Alhenawi Y (2023) Consumer financial planning in the post-covid-19 era: the role of emotional and economic vulnerability. J Consum Behav 23(2):976–986. https://doi.org/10.1002/cb.2257

Alonso FG, Carrasco JB, Rúbies IP (2023) Is Spanish depopulation irreversible? Recent demographic and spatial changes in small municipalities. Vienna Yearbook of Population Research 21:277–309. https://doi.org/10.1553/p-9fd9-h7g5

Alonso Ruiz RA, De Jubera Ocón MS, Valdemoros San Emeterio MÁ (2022) Digital Leisure: an opportunity for intergenerational well-being in times of pandemic? J New Approaches Educ Re-Search 11:31–48. https://doi.org/10.7821/naer.2022.1.806

Amate-Fortes I, Guarnido-Rueda A, Molina-Morales A (2017) Crisis and inequality in the European Union. European Review 25(3):438–452. https://doi.org/10.1017/s106279871700014x

Amdaoud M, Arcuri G, Levratto N (2021a) Are regions equal in adversity? A spatial analysis of spread and dynamics of COVID-19 in Europe. Eur J Health Econ 22(4):629–642. https://doi.org/10.1007/s10198-021-01280-6

Amdaoud M, Arcuri G, Levratto N (2021b) Healthcare system and social trust in the fight against COVID-19: the case of France. Eur J Pub Health 31(4):895–900. https://doi.org/10.1093/eurpub/ckab112

Aminah S, Susilo SRT (2021) State capacity in implementing the covid-19 vaccination program in Indonesia. Adv Soc Sci, Educ Hu-Manities Res. https://doi.org/10.2991/assehr.k.210615.014

Ando MS, Balakrishnan MR, Gruss B, Hallaert MJJ, Jirasavetakul LBF, Kirabaeva K, Solovyeva A (2022) European labor markets and the COVID-19 pandemic: fallout and the path ahead. International Monetary Fund

Anggadwita G, Permatasari A, Alamanda DT, Profityo WB (2022) Exploring women's initiatives for family business resilience during the covid-19 pandemic. J Fam Bus Manag 13(3):714–736. https://doi.org/10.1108/jfbm-02-2022-0014

Annoni P, Dominicis LD Khabirpour N (2019) Location matters: a spatial econometric analysis of regional resilience in the European Union. Growth Chang 50(3):824–855. https://doi.org/10.1111/grow.12311

Antonietti R, Falbo P, Fontini F (2020) The relationship between air quality, wealth, and COVID-19 diffusion and mortality across countries (SEEDS Working Papers N.8, 2020)

Antonová B, Zapletalová Š (2014) The economic crisis and company management: influences and consequences. E+M Ekonomie a Management 17(1). https://doi.org/10.15240/tul/001/2014-1-001

Anundsen AK, Nymoen R (2015) Did US consumers 'save for a rainy day' before the great recession? SSRN Electron J. https://doi.org/10.2139/ssrn.2607953

Araújo RJRD, Queiroz DBD, Paulo E, Nobre CJF (2020) The relationship of governmental transfers nature and the fiscal responsibility of Brazilian municipal governments. Contabilidade. Gestão E Governança 23(1):124–140. https://doi.org/10.21714/1984-3925_2020v23n1a8

Arbolino R, Di P, Marani U (2019) Did the governance of EU funds help Italian regional labour markets during the great recession? JCMS: J Common Mark Stud 58(2):235–255. https://doi.org/10.1111/jcms.12900

Ardi A, Cahyadi H, Meilani YFCP, Pramono R (2024) Talent attraction through flexible work anytime from anywhere. J Infrastruct, Policy Dev 8(3):2998. https://doi.org/10.24294/jipd.v8i3.2998

Arpaia A, Kiss A, Palvolgyi B, Turrini A (2016) Labour mobility and labour market adjustment in the EU. IZA J Migration 5:21. https://doi.org/10.1186/s40176-016-0069-8

Artelaris P, Mavrommatis G (2022) Territorial cohesion, the Covid-19 crisis and the urban paradox. Region 9(1):135–146. https://doi.org/10.18335/region.v9i1.403

Arvanitis KA, Holenstein M, Schriber L (2023) Bottom-up resilience: how civic bottom-up initiatives contribute to community resilience. Swiss Yearb Adm Sci 14(1):58–72. https://doi.org/10.5334/ssas.191

Aslan S, Çininulak M (2022) The effect of the economic crises after 2000 on private hospitals in Turkey. J Int Health Sci Manag 8(16):51–58. https://doi.org/10.48121/jihsam.1151987

Ba Z, Du H, Hasi E, Lu X (2023) Changes in land use and ecosystem service value in desert areas of China after reform and opening up. Front Earth Sci 11. https://doi.org/10.3389/feart.2023.1251605

Bach-Mortensen AM, Esposti MD (2021) Is area deprivation associated with greater impacts of Covid-19 in care homes across England? A preliminary analysis of Covid-19 outbreaks and deaths. J Epidemiol Community Health 75(7):624–627. https://doi.org/10.1136/jech-2020-215039

Bachtler J, Méndez C, Wishlade F (2019) Reforming the MFF and Cohesion Policy 2021–27: pragmatic drift or paradigmatic Shift? Eur Struct Invest Funds J 7(3):179–193

Bachtler J, Mendez C (2020b) Cohesion and the EU budget: is conditionality undermining solidarity. In: Governance and politics in the post-crisis European Union, pp 121–139

Bachtler J, Méndez C (2020a) Cohesion Policy: doing more with less. In Wallace H, Pollack MA, Young AR, Roederer-Rynning C (eds) Policy-Making in the European Union (8th ed). Oxford University Press

Badenhorst W (2017) Long-run accounting conservatism and subsequent equity returns. J Econ Financ Sci 9(1):60–75. https://doi.org/10.4102/jef.v9i1.29

Badiru A, Ademola AS, Dambo H (2023) How does economic growth respond to public infrastructure expenditure shocks? Evidence from Svar in Nigeria. J Econ Sci (2.1):53–68. https://doi.org/10.55603/jes.v2i1.a5

Bakchan A, Hacker ME, Faust KM (2021) Resilient water and wastewater infrastructure systems through integrated humanitarian-development processes: the case of Lebanon's protracted refugee crisis. Environ Sci Technol 55(9):6407–6420. https://doi.org/10.1021/acs.est.0c05630

Bandyopadhyay S, Sridhar KS (2023) COVID-19, cities and regions: Is it merely short term or has it changed our cities and regions forever? Reg Sci Policy Pract 15(8):1669–1671. https://doi.org/10.1111/rsp3.12735

Bănică A, Pascariu GC, Kourtit K, Nijkamp P (2024) Unveiling core-periphery disparities through multidimensional spatial resilience maps. The Annals of Regional Science, pp 1–29

Barański K, Brożek G, Kowalska M, Kaleta-Pilarska A, Zejda J (2021) Impact of covid-19 pandemic on total mortality in Poland. Int J Environ Res Public Health 18(8):4388. https://doi.org/10.3390/ijerph18084388

Barca F (2009) An Agenda fora reformed Cohesion Policy. A place based approach to meeting European challenges and expectations. European Comission, Brusells

Barca F, McCann P, Rodríguez-Pose A (2012) The case for regional development intervention: place-based versus place-neutral approaches. J Reg Sci. Wiley Blackwell 52(1):134–152. j.1467-9787.2011.00756.x

Barrett N, Fowles J, Jones PA, Reitano V (2018) Forecast bias and fiscal slack accumulation in school districts. Am Rev Public Adm 49(5):601–613. https://doi.org/10.1177/0275074018804671

Barría-Sandoval C, Ferreira G, Méndez AC, Toffoletto MC (2022) Impact of Covid-19 on deaths from respiratory diseases: Panel data evidence from Chile. Infect Ecol and Epidemiol 12(1). https://doi.org/10.1080/20008686.2021.2023939

Bartlett AH, Makhni S, Ruokis S, Selling MK, Hall L, Umscheid CA, Kao C (2022) Use of clinical pathways integrated into the electronic health record to address the coronavirus disease 2019 (covid-19) pandemic. Infect Control Hosp Epidemiol 44(2):260–267. https://doi.org/10.1017/ice.2022.64

Becla A, Czaja S (2021) Ekonomiczne i społeczne konsekwencje pandemii koronawirusa. Eur J Manag Soc Sci 2(Article 2). https://doi.org/10.56652/ejmss2021.2.6

Benassi F, Salvati L (2019) Economic downturns and compositional effects in regional population structures by age: a multi-temporal analysis in Greek regions, 1981–2017. Qual Quant 53(5):2611–2633. https://doi.org/10.1007/s11135-019-00875-9

Benink H (2020) Global bank capital and liquidity after 30 years of basel accords. J Risk Financ Manag 13(4):73. https://doi.org/10.3390/jrfm13040073

Bernanke BS (1986) Alternative explanations of the money-income correlation. Carn.Roch. Conf Ser Public Policy 25:49–99

Betthauser BA, Bach-Mortensen AM, Engzell P (2023) A systematic review and meta-analysis of the evidence on learning during the COVID-19 pandemic. Nature Human Behaviour.

Bhopalwala H, Akbar A, Dewaswala N, Wisnieski L, Minhas AMK, Hussain A, Bhopalwala A (2022) Outcomes of heart failure in Covid-19 patients: an appalachian experience. Cardiol Res 13(3):162–171. https://doi.org/10.14740/cr1389

Bielińska-Dusza E, Hamerska M (2023) Praca zdalna na tle różnic międzypokoleniowych. Prace Naukowe Uniwersytetu Ekonomicznego We Wrocławiu 67(5):22–37. https://doi.org/10.15611/pn.2023.5.03

Binnashir-Alketbi A, Jimber-del-Río J, Ibañez-Fernández A, Vergara-Romero A (2023) Human resource development and crisis management: the role of communication and learning in the United Arab Emirates. TEM J 1130–1141. https://doi.org/10.18421/tem122-59

Bisello M, Maccarrone V, Fernández-Macías E (2020) Occupational mobility, employment transitions and job quality in Europe: the impact of the great recession. Econ Ind Democr 43(2):585–611. https://doi.org/10.1177/0143831x20931936

Blažek J, Hejnová T, Rada H (2018) The impacts of the global economic crisis and its aftermath on the banking centres of Europe. Eur Urban Reg Stud 27(1):35–49. https://doi.org/10.1177/0969776418807240

Bochenek B, Jankowski M, Gruszczyńska M, Nykiel G, Gruszczynski M, Jaczewski A, Pinkas J (2021) Impact of meteorological conditions on the dynamics of the Covid-19 pandemic in Poland. Int J Environ Res Public Health 18(8):3951. https://doi.org/10.3390/ijerph18083951

Böhme K, Doucet P, Komornicki T, Zaucha J, Świątek D (2011) How to strengthen the territorial dimension of 'Europe 2020' and the EU Cohesion Policy. Report based on the Territorial Agenda 2020. European Commission

Bokhorst D, Corti F (2023) Governing Europe's recovery and resilience facility: Between discipline and discretion. Gov Oppos 1–17. https://doi.org/10.1017/gov.2023.14

Bołdyrew J (2023) Praca zdalna w dobie pandemii i związane z nią wyzwania dla pracowników Remote working in pandemic and its challenges for employees. Zjawiska Niepożądane W Zatrudnieniu Z Perspektywy Prawa Pracy–Wybrane Zagadnienia. https://doi.org/10.18778/8331-307-8.07

Boomgaarden HG, Spanje JV, Vliegenthart R, Vreese CHD (2011) Covering the crisis: Media coverage of the economic crisis and citizens' economic expectations. Acta Politica 46(4):353–379. https://doi.org/10.1057/ap.2011.18

Boon ACM, Harastani HH, Joshi A, Bricker TL, Seehra K, Hassan AO, Diamond MS (2023) Intranasal vaccination of Syrian hamsters with chad-sars-cov-2-s disrupts airborne transmission cycles of Sars-CoV-2. https://doi.org/10.21203/rs.3.rs-3185983/v1

Borca B, Putz L-M, Hofbauer F (2021) Crises and their effects on freight transport modes: a literature review and research framework. Sustainability 13(10):5740. https://doi.org/10.3390/su13105740

Borkowski P, Jazdzewska-Gutta M, Szmelter-Jarosz A (2021) Lockdowned: Everyday mobility changes in response to COVID-19. J Transp Geogr. 90:102906. https://doi.org/10.1016/j.jtrangeo.2020.102906

Borz G, Brandenburg H, Mendez C (2022) The impact of EU Cohesion Policy XE Cohesion Policy on European identity: a comparative analysis of EU regions. European Union Politics 23(2):259–281. https://doi.org/10.1177/14651165221076444

Botelho V, Weissler M (2022) COVID-19 and retirement decisions of order workers in the euro area. ECB Economic Bulletin (6) European Central Bank

Boterman WR (2020) Urban-rural polarisation in times of the corona outbreak? The early demographic and geographic patterns of the SARS-CoV-2 epidemic in the Netherlands. Tijdschr Econ Soc Geogr 111(3):513–529

Boterman W, Gent WV (2014) Housing liberalisation and gentrification: the social effects of tenure conversions in Amsterdam. Tijdschr Econ Soc Geogr 105(2):140–160. https://doi.org/10.1111/tesg.12050

Bourdin S, Levratto N (2024) The spatial footprint of COVID-19 and local policy answers. Reg Stud 58(2):255–262. https://doi.org/10.1080/00343404.2023.2242411

Bourdin S, Cottereau V, Hermet F, Jean-Pierre P, Medeiros E (2023) How have Europe's outermost regions dealt with the economic and social consequences of the COVID-19 crisis? Effects, policies and recommendations. Reg Sci Policy Pract 15(8):1820–1841

Bourdin S, Moodie J, Gassen N S, Evers D, Adobati F, Amdaoud M, Arcuri G, Casti E, Cottereau V, Eva M, Lőcsei H, Iațu C, Ibănescu BC, Jean-Pierre P, Hermet F, Levratto N, Löfving Coll-Martinez E, Psycharis Y, Salenius V, Remete S (2024) COVID-19 as a systemic shock: curb or catalyst for proactive policies towards territorial cohesion? Reg Stud 58(8):1543–1556. https://doi.org/10.1080/00343404.2023.2242387

Bouzouina L, Kourtit K, Nijkamp P (2022) Impact of immobility and mobility activities on the spread of COVID-19: evidence from European countries. Reg Sci Policy Pract 14(S1):6–20. https://doi.org/10.1111/rsp3.12565

Brazier JF, Meehan A, Shield RR, White EM, Grabowski DC, Geng F, Gadbois EA (2023) All hands-on deck: administrator perspectives on managing COVID-19 outbreaks in U.S. nursing homes. J Long-Term Care 297–310. https://journal.ilpnetwork.org/articles/https://doi.org/10.31389/jltc.237

Brezdeń P (2023) Działalność przemysłowa i jej struktura przestrzenna w czasie trwania pandemii covid-19 w Polsce. Stud Ind Geogr Comm Pol Geogr Soc 37(2). https://doi.org/10.24917/20801653.372.1

Brill F, Raco M (2021) Putting the crisis to work: the real estate sector and London's housing crisis. Polit Geogr 89:102433. https://doi.org/10.1016/j.polgeo.2021.102433

Brown R, Cowling M (2021) The geographical impact of the Covid-19 crisis on precautionary savings, firm survival and jobs: evidence from the United Kingdom's 100 largest towns and cities. Int Small Bus J: Res Entrep 39(4):319–329. https://doi.org/10.1177/0266242621989326
Budner W (2004) Lokalizacja przedsiębiorstw. Aspekty ekonomiczno-przestrzenne i środowiskowe. Wydawnictwo Akademii Ekonomicznej w Poznaniu
Buerger C, Lofton ML (2023) The interplay between fiscal institutions and the great recession: evidence from U.S. school districts. AERA Open 9. https://doi.org/10.1177/23328584231189176
Camagni R (2008) Regional competitiveness: towards a concept of territorial Capital. In: Capello R, Camagni R, Chizzolini B, Fratesi U (eds) Modelling regional scenarios for the enlarged Europe: European competitiveness and global strategies. Springer Verlag. Berlin, pp 33–48
Campos EG, Werner-Masters K, Cordova-Buiza F, Paucar-Caceres A (2021) Community ecotourism in rural Peru: resilience and adaptive capacities to the covid-19 pandemic and climate change. J Hosp Tour Manag 48:416–427. https://doi.org/10.1016/j.jhtm.2021.07.016
Canton E, Colasanti F, Durán J, Garrone M, Hobza A, Simons W, Vandeplas A (2021) The sectoral impact of the COVID-19 crisis. An Unprecedented and Atypical Crisis (No. 069). Directorate general economic and financial affairs (DG ECFIN), European Commission
Cao M, Yao Q, Chen B, Ling Y, Hu Y, Xu G (2023) Development of a composite regional vulnerability index and its relationship with the impacts of the Covid-19 pandemic. Comput Urban Sci 3(1). https://doi.org/10.1007/s43762-023-00078-x
Capati A (2023) Covid-19 pandemic and the establishment of the recovery and resilience facility. Polit: IAPSS J Polit Sci 55:19–43. https://doi.org/10.22151/politikon.55.2
Capello R, Perucca G (2019) Cohesion Policy and European identity building: trust as a mediating element. Reg Sci Policy Pract 11(4):637–654. https://doi.org/10.1111/rsp3.12229
Capello R, Caragliu A, Panzera E (2023) Economic costs of COVID-19 for cross-border regions. Reg Sci Policy Pract 15(8):1688–1702
Caramani D, Cicchi L (2024) The relief–reform divide: an analysis of national responses to Covid-19 in seven EU member states' recovery plans. https://doi.org/10.31235/osf.io/guabh
Carlucci M, Chelli F, Salvati L (2018) Toward a new cycle: short-term population dynamics, gentrification, and re-urbanization of Milan (Italy). Sustainability 10(9):3014. https://doi.org/10.3390/su10093014
Celi G, Guarascio D, Simonazzi A (2020) A fragile and divided European Union meets Covid-19: further disintegration or 'Hamiltonian moment'? J Ind Bus Econ 47(3):411–424. https://doi.org/10.1007/s40812-020-00165-8
Cesarski M (2022) Polityka mieszkaniowa i rynek mieszkaniowy w Polsce-gospodarstwa domowe w czasie pandemii COVID-19. Studia z Polityki Publicznej 9(2):101–116
Chaturvedi R, Karri A (2022) Entrepreneurship in the times of pandemic: barriers and strategies. FIIB Business Review 11(1):52–66
Chen R, Huang C (2021) Landscape evolution and it's impact of ecosystem service value of the Wuhan city, China. Int J Environ Res Public Health 18(24):13015. https://doi.org/10.3390/ijerph182413015
Chen R, Xu G, Yang L, Deng Z, Hu Q, Hu H, Wang Z (2020) A new rapid screening program based on risk scores for Covid-19 patients. Intern Emerg Med 16(4):925–932. https://doi.org/10.1007/s11739-020-02534-6
Chen Y (2023) Further analysis in global financial crisis: the panic of 1907 and 2008 subprime mortgage crisis. Adv Econ, Manag Polit Sci 11(1):158–163. https://doi.org/10.54254/2754-1169/11/20230534
Chernick H, Langley A, Reschovsky A (2011) Revenue diversification and the financing of large American central cities. Public Financ Manag 11(2):138–159. https://doi.org/10.1177/152397211101100204
Chita E, Drimili E, Gareiou Z, Milioti C, Vranna A, Poulopoulos SG, Zervas PE (2020) Impact of economic crisis on passenger transportation–case of travelling to the Greek mainland from Crete. Promet–Traffic Transp 32(3):347–360. https://doi.org/10.7307/ptt.v32i3.3255

Chowdhury A, Żuk P (2018) From crisis to crisis: capitalism, chaos and constant unpredictability. Econ Labour Relat Rev 29(4):375–393. https://doi.org/10.1177/1035304618811263
Christensen T, Ma L (2021) Comparing Sars and Covid-19: challenges of governance capacity and legitimacy. Public Organ Rev 21(4):629–645. https://doi.org/10.1007/s11115-021-00510-y
Chudak M (2009) Czynniki lokalizacyjne stymulujące przedsiębiorczość a wielkość i ranga miasta. Prace Naukowe Uniwersytetu Ekonomicznego We Wrocławiu 46:213–221
Churski P, Perdał R, Konecka-Szydłowska B, Herodowicz T (2021) European Regional Development. Springer, Contemporary regional and local perspectives of socio-economic and socio-political changes
Churski P, Żuber P (2022) EU regional policy and its implications for the development of Polish regions, pp 487–521. https://doi.org/10.1007/978-3-031-06108-0_20
Cichecka A, Karolak M, Ufel W (2024) A romantic public tragedy? COVID pandemic and the changes of governance in Poland. Critical Sociology 08969205241258408
Ciupa T, Suligowski R (2021) Green-blue spaces and population density versus covid-19 cases and deaths in Poland. Int J Environ Res Public Health 18(12):6636. https://doi.org/10.3390/ijerph18126636
Claponea RM, Iorga M (2023) Burnout, organizational justice, workload, and emotional regulation among medical and non-medical personnel working in Romanian healthcare units. Behav Sci 13(3):225. https://doi.org/10.3390/bs13030225
Clark-Ginsberg A, Rueda IA, Monken J, Liu J, Hong C (2020) Maintaining critical infrastructure resilience to natural hazards during the COVID-19 pandemic: hurricane preparations by US energy companies. J Infrastruct Preserv Resili-Ence 1(1). https://doi.org/10.1186/s43065-020-00010-1
Climate-resilient infrastructure (2018) OECD Environment Policy Papers. https://doi.org/10.1787/4fdf9eaf-en
Colantoni A, Hălbac-Cotoară-Zamfir R, Halbac-Cotoara-Zamfir C, Cudlín P, Salvati L, Morera AG (2020) Demographic resilience in local systems: an empirical approach with census data. Systems 8(3):34. https://doi.org/10.3390/systems8030034
Collaborative Research (2013) Case studies for ethics in academic research in the social sciences, pp 39–46. https://doi.org/10.4135/9781452269986.n6
Collard F, Fève P (2008) VAR or DSGE Models: which should we choose? Economie & prévision 183–184(2):153–174
Copestake A, Estefania-Flores J, Furceri D (2024) Digitalization and resilience. Res Policy 53(3):104948
Coşkun Y (2022) Housing affordability: an econometric framing for policy discussions. Int J Hous Mark Anal-Ysis 16(2):374–407. https://doi.org/10.1108/ijhma-01-2022-0015
Cottreau B, Adraoui A, Manout O, Bouzouina L (2023) Spatio-temporal patterns of the impact of COVID-19 on public transit: an exploratory analysis from Lyon. France. RegNal Sci Policy and Pract 15(8):1702–1721
Crescenzi R, Giua M (2019) One or many cohesion policies of the European Union? on the differential economic impacts of Cohesion Policy across member states. Reg Stud 54(1):10–20. https://doi.org/10.1080/00343404.2019.1665174
Crespo P, Faytong-Haro M (2022) Impact of the 2007–2008 United States economic crisis on pet ownership. Animals 12(21):3010. https://doi.org/10.3390/ani12213010
Cross K, Evans J, MacLeavy J, Manley D (2022) Analysing the socio-economic impacts of COVID-19: a new regional geography or pandemic enhanced inequalities? Reg Stud Reg Sci 9(1):461–485. https://doi.org/10.1080/21681376.2022.2084447
Cucinotta D, Vanelli M (2020) WHO declares COVID-19 a Pandemic. Acta Biomed 91(1):157–160. https://doi.org/10.23750/abm.v91i1.9397
Cui P, Dong Z, Yao X, Cao Y, Sun Y, Lan F (2022) What makes urban communities more resilient to covid-19? A systematic review of current evidence. Int J Environ Res Public Health 19(17):10532. https://doi.org/10.3390/ijerph191710532

Cunico G, Aivazidou E, Mollona E (2020) European Cohesion Policy XE European Cohesion Policy performance and citizens' awareness: a holistic system dynamics framework. Investig RegNales, J RegNal Res 46:131–162
Curran L, Zignago S (2010) The financial crisis: impact on key clothing markets and suppliers. J Fash Mark Manag: Int J 14(4):530–545. https://doi.org/10.1108/13612021011081724
Cutrini E, Salvati L (2021) Unraveling spatial patterns of COVID-19 in Italy: global forces and local economic drivers. Reg Sci Policy Pract 13:73–108
Czechowska ID, Hajdys D, Ślebocka M (2023) Nieprawidłowości w umowach o kredyt konsumencki w Polsce, kontekst ochrony konsumenta na rynku usług finansowych. Finanse I Prawo Finansowe (2):215–231. https://doi.org/10.18778/2391-6478.s2.2023.10
Dąbrowski M, Stead D, Mashhoodi B (2019) EU Cohesion Policy can't buy me love? Exploring the regional determinants of EU image. Reg Sci Policy Pract 11(4):695–712. https://doi.org/10.1111/rsp3.12237
Daqar MAMA, Constantinovits M, Arqawi S, Daragmeh A (2021) The role of fintech in predicting the spread of Covid-19. Banks Bank Syst 16(1):1–16. https://doi.org/10.21511/bbs.16(1).2021.01
Dawood FS, Mkhencele T, Sun K, Viboud C, Bhiman JN, Buys A, Wolter N (2022) Sars-cov-2 incidence, transmission, and reinfection in a rural and an urban setting: results of the phirst-c cohort study, South Africa, 2020–21. Lancet Infect Dis 22(6):821–834. https://doi.org/10.1016/s1473-3099(22)00069-x
De Backer T, Van Nieuwenhuyse B, De Bacquer D (2021) Antihypertensive treatment in a general uncontrolled hypertensive population in Belgium and Luxembourg in primary care: Therapeutic inertia and treatment simplification. The SIMPLIFY Study. Plos ONE 16(4):e0248471. https://doi.org/10.1371/journal.pone.0248471
De S, Mohapatra S, Ratha D (2021) Sovereign credit ratings, relative risk ratings and private capital flows: evidence from emerging and frontier markets. Stud Econ Financ 38(4):873–898. https://doi.org/10.1108/sef-10-2020-0437
de la Porte C, Jensen MD (2021) The next generation EU: an analysis of the dimensions of conflict behind the deal. Soc Policy and Adm 55(2):388–402. https://doi.org/10.1111/spol.12709
Dellepiane-Avellaneda S (2014) The political power of economic ideas: the case of 'expansionary fiscal contractions. Br J Polit Int Relat 17(3):391–418. https://doi.org/10.1111/1467-856x.12038
Demirci Ş, Konca M, Yetim B, İlgün G (2019) Effect of economic crisis on suicide cases: an ARDL bounds testing approach. Int J Soc Psychiatry 66(1):34–40. https://doi.org/10.1177/0020764019879946
Dimelli D (2023) Recycling urban wastelands in Greek cities. Challenges and opportunities. Contesti. Città Territori Progetti (1):51–73. https://doi.org/10.36253/contest-14315
Długosz P (2022) Zdalne nauczanie wśród uczniów szkoły podstawowej na obszarach peryferyjnych w Europie Środkowo-Wschodniej. Edukacja 3(163):123–140
Dluhopolskyi O, Пахненко O, Lyeonov S, Semenog A, Artyukhova N, Cholewa-Wiktor M, Jastrzębski W (2023) Digital financial inclusion: covid-19 impacts and opportunities. Sustainability 15(3):2383. https://doi.org/10.3390/su15032383
Dolls M, Fuest C, Peichl A, Wittneben C (2022) Fiscal consolidation and automatic stabilization: new results. IMF Economic Review 70(3):420–450. https://doi.org/10.1057/s41308-022-00162-y
Dolot A (2020) Wpływ pandemii covid-19 na pracę zdalną–perspektywa pracownika. E-Mentor 83(1):35–43. https://doi.org/10.15219/em83.1456
Dominiak J (2022) Wpływ pandemii covid-19 na zmiany w sektorze usług w Polsce. Stud Ind Geogr Comm Pol Geogr Soc 36(2). https://doi.org/10.24917/20801653.362.8
Dong J (2023) The tension between Chinese peasants' family capital reproduction strategies and massification in higher education. Int J Parents Educ 9(1). https://doi.org/10.54195/ijpe.18236
Doroudgar S, Hernandez FE, Kali-Rai R, Yoshizuka K (2021) Multi-cultural project initiative: an inclusive representation of coronavirus guidelines for America. Res Social Adm Pharm 17(4):816–818. https://doi.org/10.1016/j.sapharm.2020.04.006

Downar J, Boese K, Lalumiere G, Bercier G, Leduc S, Rice J, Amit A, Charbonneau V (2022) A clinical response team providing support to long-term care homes with Covid-19 outbreaks in Eastern Ontario–a cohort study. Canadian Geriatrics J 25(2):171–174. https://doi.org/10.5770/cgj.25.561

Drakaki M, Tzionas P (2017) Community-based social partnerships in crisis resilience: a case example in Greece. Disaster Prev Manag: Internat J 26(2):203–216. https://doi.org/10.1108/dpm-09-2016-0190

Drop J (2001) Rodzaje szoków gospodarczych a unia walutowa types of economic shocks and the Monetary Union. Zeszyty Naukowe, Akademia Ekonomiczna w Krakowie 589:21–35

Dyba W, Di Maria E (2023) Assessment and support of the digitalization of businesses in Europe during and after the COVID-19 pandemic. RegNal Sci Policy and Pract 1–16. https://doi.org/10.1111/rsp3.12717

ECA (2024) Adapting Cohesion Policy rules to respond to COVID-19. Special Report. https://www.eca.europa.eu/Lists/ECADocuments/SR23_02/SR_Covid_II-ReACT_EU_EN.pdf

ECDC (2024) Timeline of ECDC's response to Covid-19. https://www.ecdc.europa.eu/en/covid-19/all-reports-covid-19

The Economist (2024) Processed by Our World in Data. "Central estimate" [Coronaviurs]. The Economist, "Excess Deaths Model" [original data]

Efthymiou D, Antoniou C (2017) Understanding the effects of economic crisis on public transport users' satisfaction and demand. Transp Policy 53:89–97. https://doi.org/10.1016/j.tranpol.2016.09.007

Eleuteri F, Arduino LS (2014) Individual, community and societal effects of the global financial crisis: a bioecological model. J Risk Anal Cris Response 4(4):203–213. https://doi.org/10.2991/jrarc.2014.4.4.2

Elgar FJ, Stefaniak A, Wohl MJA (2020) The trouble with trust: time-series analysis of social capital, income inequality, and Covid-19 deaths in 84 countries. Soc Sci Med 263:113365. https://doi.org/10.1016/j.socscimed.2020.113365

ESPON (2022) Territorial impacts of COVID-19 and policy responses in European regions and cities. Final Report. https://archive.espon.eu/sites/default/files/attachments/TERRCOV_Final%20Report_Main.pdf

European Commission: Directorate-General for Regional and Urban Policy. (2024). Forging a sustainable future together–Cohesion for a competitive and inclusive Europe–report of the high-level group on the future of Cohesion Policy. Publications Office of the European Union. https://data.europa.eu/doi/https://doi.org/10.2776/974536

Fadhil GA, Burhan AM (2021) Investigating the effects of economic crisis on construction projects in Iraq. E3S Web of Conferences 318:02005. https://doi.org/10.1051/e3sconf/202131802005

Faragó L, Varró K (2016) Shifts in EU Cohesion Policy and processes of peripheralization: a view from central eastern Europe. Eur Spat Res Policy 23(1):5–19. https://doi.org/10.1515/esrp-2016-0001

Farole T, Goga S, Ionescu-Heroiu M (2018). Rethinking lagging regions. Using Cohesion Policy to deliver on the potential of Europe's regions. World Bank

Fathi S, Sajadzadeh H, Sheshkal FM, Aram F, Pintér G, Felde I, Mosavi A (2020) The role of urban morphology design on enhancing physical activity and public health. Int J Environ Res Public Health 17(7):2359. https://doi.org/10.3390/ijerph17072359

Fedajev A, Kostić M, Milošević I, Vesić T, Stevanovic A, Rădulescu M, Stefan M (2023) Evaluating the outcomes of monetary and fiscal policies in the EU in times of crisis: a PLS-SEM approach. Sustainability 15(11):8466. https://doi.org/10.3390/su15118466

Feldkircher M (2014) The determinants of vulnerability to the global financial crisis 2008–2009: credit growth and other sources of risk. J Int Money Financ 43:19–49. https://doi.org/10.1016/j.jimonfin.2013.12.003

Ferguson SH, Yurkowski DJ, Young BG, Fisk AT, Muir DCG, Zhu X, Thiemann GW (2020) Comparing temporal patterns in body condition of ringed seals living within their core

geographic range with those living at the edge. Ecography 43(10):1521–1535. https://doi.org/10.1111/ecog.04988
Fernández-Albertos J, Kuo A, Balcells L (2013) Economic crisis, globalization, and partisan bias: evidence from Spain. Int Stud Q 57(4):804–816. https://doi.org/10.1111/isqu.12081
Ferrera M, Miró J, Ronchi S (2021) Walking the road together? EU polity maintenance during the Covid-19 crisis. West Eur Polit 44(5–6):1329–1352. https://doi.org/10.1080/01402382.2021.1905328
Fisk JM, Harris PA, Kuks SMM, Morris JC, Vinke-De Kruijf J (2023) Framing water infrastructure for climate resilience: governance dimensions and challenges. Public Works Manag Policy 29(2):133–145. https://doi.org/10.1177/1087724x231212556
Floerkemeier H, Spatafora N, Venables A (2021) Regional disparities, growth, and inclusiveness. IMF Working Paper, WP/21/38
Florida R, Mellander C (2022) The geography of COVID-19 in Sweden. Ann Reg Sci 68(1):125–150. https://doi.org/10.1007/s00168-021-01071-0
Fochezatto A, Sanguinet ER, Batistela P, Valdes R (2023) Income leakage regional effects: supply and demand shocks during the pandemic in Brazil and Chile. Geographies 3(4):654–672. https://doi.org/10.3390/geographies3040034
Ford MT, Ihrke DM (2016) The impact of Wisconsin's act 10 on municipal management in smaller municipalities: views from local elected officials. Public Policy Adm 33(2):170–189. https://doi.org/10.1177/0952076716683763
Frankenberg E, Thomas D (2017) Human capital and shocks: evidence on education, health and nutrition. https://doi.org/10.3386/w23347
Fritsch M, Kahila P, Nemeth S, Scott JW (eds) (2023) Spatial justice and cohesion. The role of place-based action in community development (1st ed). Routledge. https://doi.org/10.4324/9781003229681
Furberg D, Ban Y, Nascetti A (2019) Monitoring of urbanization and analysis of environmental impact in Stockholm with sentinel-2a and spot-5 multispectral data. Remote Sens 11(20):2408. https://doi.org/10.3390/rs11202408
Gaag NVD, Beer JD (2014) From demographic dividend to demographic burden: the impact of population ageing on economic growth in Europe. Tijdschrift Voor Economische en Sociale Geografie 106(1):94–109. https://doi.org/10.1111/tesg.12104
Gailhard İU, Kataria K (2014) Economic crisis and labour force transition to inactivity: a comparative study in German rural and urban areas. Stud Agric Econ 116(1):25–32. https://doi.org/10.7896/j.1320
Gale P (2020) Thermodynamic equilibrium dose-response models for MERS-CoV infection reveal a potential protective role of human lung mucus but not for SARS-CoV-2 XE SARS-CoV-2. Microbial Risk Analysis 16:100140. https://doi.org/10.1016/j.mran.2020.100140
Garbelli M (2018) Financial crisis, environmental turbulence and international trade. The case of Italy. Europ Scient J ESJ 14(7):1–13. https://doi.org/10.19044/esj.2018.v14n7p1
Giacomini E, Marinelli N, Riccetti L (2023) The impact at stake: risk and return in publicly listed social impact firms. Bus Ethics, Environ and Responsib 32(2):713–741. https://doi.org/10.1111/beer.12511
Giacomini R (2013) The relationship between DSGE and VAR models, cemmap working paper, No. CWP21/13. Centre for Microdata Methods and Practice (cemmap). https://doi.org/10.1920/wp.cem.2013.2113
Giannakis E, Mamuneas TP (2022) Labour productivity and regional labour markets resilience in Europe. Ann Reg Sci 68(3):691–712. https://doi.org/10.1007/s00168-021-01100-y
Giannakis E, Efstratoglou S, Antoniades A (2018) Off-farm employment and economic crisis: evidence from Cyprus. Agriculture 8(3):41. https://doi.org/10.3390/agriculture8030041
Giubergia D, Bin E, Diana M (2023) Changes in mobility choices during the first wave of the COVID-19 pandemic: a comparison between Italy and Sweden. arXiv:2303.07803
Giuliani M (2020) Economy or austerity? Drivers of retrospective voting before and during the great recession. Int Polit Sci Rev 43(2):173–190. https://doi.org/10.1177/0192512120919138

Glaeser EL, Cutler DM (2021) Survival of the city: living and thriving in an age of isolation. Penguin Press
Gloger A, Morgan GT, Lach N, Jackson SF, Urban R, Rolston IA (2021) A connected community approach: citizens and formal institutions working together to build community-centred resilience. Int J Environ Res Public Health 18(19):10175. https://doi.org/10.3390/ijerph181910175
Glückler J, Panitz R (2023) Live music in the time of corona: on the resilience and impact of a philharmonic orchestra on the urban economy. Sustainability 15(4):3611. https://doi.org/10.3390/su15043611
Gorina E, Maher C, Joffe MD (2017) Local XE Local fiscal distress: measurement and prediction. Public Budg Financ 38(1):72–94. https://doi.org/10.1111/pbaf.12165
Grazioli M (2022). batti il 5!: grassroots strategies against the administrative invisibilization of Rome's housing squatters before and during the pandemic. Radic Hous J 4(1):31–50. https://doi.org/10.54825/wqah3246
Groenendijk N (2023) Revenue capacity of the EU: taxes, tax sharing, and resource pooling. Polit Gov 11(4):6–16. https://doi.org/10.17645/pag.v11i4.7240
Gruss JJH, Jirasavetakul LBF, Kirabaeva K, Klein N, Lariau A, Liu LQ, Solovyeva A (2022) European labor markets and the COVID-19 pandemic. International Monetary Fund
Gruszczyński Ł, Zatoński M (2021) Do regulations matter in fighting the covid-19 pandemic? Lessons from Poland. Eur J Risk Regul 12(4):739–757. https://doi.org/10.1017/err.2021.53
Guzal-Dec D, Zwolińska-Ligaj M (2023) How to deal with crisis? Place attachment as a factor of resilience of urban–rural communes in Poland during the Covid-19 pandemic. Sustainability 15(7):6222. https://doi.org/10.3390/su15076222
Hadjimichalis C (2021) An uncertain future for the post-Brexit, post-COVID-19 European Union. Eur Urban Reg Stud 28(1):8–13. https://doi.org/10.1177/0969776420968961
Haefner L, Sternberg R (2020) Spatial implications of digitization: state of the field and research agenda. Geogr Compass 14(12):12544. https://doi.org/10.1111/gec3.12544
Hälbac-Cotoară-Zamfir R, Egidi G, Salvia R, Salvati L, Sateriano A, Giménez-Morera A (2021) Recession, local fertility, and urban sustainability: results of a quasi-experiment in Greece, 1991–2018. Sustainability 13(3):1052. https://doi.org/10.3390/su13031052
Hale T, Petherick A, Phillips T, Webster S (2020) Variation in government responses to COVID-19, version 4.0. Blavatnik School of Government Working Paper
Halikiopoulou D, Vasilopoulou S (2016) Breaching the social contract: crises of democratic representation and patterns of extreme right party support. Gov Oppos 53(1):26–50. https://doi.org/10.1017/gov.2015.43
Hannan SA (2017) The drivers of capital flows in emerging markets post global financial crisis. J Int Commer, Econ Policy 8(02):1750009. https://doi.org/10.1142/s1793993317500090
Hardy J, Imani Y, Zhuang B (2017) Regional resilience and global production networks in China: an open political economy perspective. Compet Chang 22(1):63–80. https://doi.org/10.1177/1024529417744177
Harnani S, Rusminingsih D, Damayanti L (2022) The role of human capital in education, environment, and economic. Asia Pac J Manag Educ 5(2). https://doi.org/10.32535/apjme.v5i2.1620
Hawkins D (2020) Social determinants of Covid-19 in Massachusetts, United States: an ecological study. J Prev Med Public Health 53(4):220–227. https://doi.org/10.3961/jpmph.20.256
He D, Miao P, Qureshi NA (2022) Can industrial diversification help strengthen regional economic resilience? Front Environ Sci 10. https://doi.org/10.3389/fenvs.2022.987396
Hendrick R, Wang S (2018) Use of special assessments by municipal governments in the Chicago metropolitan area: is leviathan tamed? Public Budg Financ 38(3):32–57. https://doi.org/10.1111/pbaf.12190
Hermann C (2014) Structural adjustment and neoliberal convergence in labour markets and welfare: the impact of the crisis and austerity measures on European economic and social models. Compet Chang 18(2):111–130. https://doi.org/10.1179/1024529414z.00000000051

Hidayat R, Muniroh H (2021) Gender role in mediating capital relations and information technology investment with SME financial performance during the Covid-19 pandemic. In: proceedings of the 2nd Borobudur international symposium on humanities and social sciences, BIS-HSS 2020. https://doi.org/10.4108/eai.18-11-2020.2311743

Hinkley S, Weber R (2020) Incentives and austerity: how did the great recession affect municipal economic development policy? Urban Affairs Review 57(3):820–855. https://doi.org/10.1177/1078087420964254

Holgersen S (2013) Urban responses to the economic crisis: confirmation of urban policies as crisis management in Malmö. Int J Urban Reg Re-Search 38(1):285–301. https://doi.org/10.1111/1468-2427.12029

Hörisch F, Obert P (2020) Social capital and the impact of the recent economic crisis: comparing the effects of economic and fiscal policy developments. Social Policy and Administration 54(7):1141–1159. https://doi.org/10.1111/spol.12590

Hu X, Yang C (2019) Institutional change and divergent economic resilience: path development of two resource-depleted cities in China. Urban Studies 56(16):3466–3485. https://doi.org/10.1177/0042098018817223

Hu X, Li L, Dong K (2022) What matters for regional economic resilience amid Covid-19? evidence from cities in northeast China. Cities 120:103440. https://doi.org/10.1016/j.cities.2021.103440

Hu MG, Zhang H, Tang J, Yan S (2023) Zoning and optimization strategies of land spatial ecological restoration in Liangjiang new area of Chongqing based on the supply–demand relationship of ecosystem services. Land 12(6):1255. https://doi.org/10.3390/land12061255

Huikari S, Miettunen J, Korhonen M (2019) Economic crises and suicides between 1970 and 2011: time trend study in 21 developed countries. J Epidemiol Community Health 73(4):311–316. https://doi.org/10.1136/jech-2018-210781

Hynes W (2020) Bouncing forward: a resilience approach to dealing with COVID-19 and future systemic shocks. Environ Syst Decis 40(2):174–184. https://doi.org/10.1007/s10669-020-09776-x

Iammarino S, Rodriguez-Pose A, Storper M (2019) Regional inequality in Europe: Evidence, theory and policy implications. J Econ Geogr 19(2):273–298. https://doi.org/10.1093/jeg/lby021

Ijeoma N (2017). The effect of global financial crisis on the performance of Nigerian stock exchange. Int J Account Econ Stud 5(1):46–50. https://doi.org/10.14419/ijaes.v5i1.7344

Imperiale AJ, Vanclay F (2016) Experiencing local community resilience in action: learning from post-disaster communities. J Rural Stud 47:204–219. https://doi.org/10.1016/j.jrurstud.2016.08.002

Islam A, Sayeed MA, Rahman K, Zamil S, Abedin J, Saha O, Hassan MM (2021) Assessment of basic reproduction number (r0), spatial and temporal epidemiological determinants, and genetic characterization of Sars-CoV-2 in Bangladesh. Infect Genet Evol 92:104884. https://doi.org/10.1016/j.meegid.2021.104884

Islind AS, Óskarsdóttir M, Steingrímsdóttir H (2020) Changes in mobility patterns in Europe during the COVID-19 pandemic: Novel insights using open source data. arXiv:2008.10505

Jee Y (2020) Interim evaluation of South Korea's response to Covid-19 and preparation for a post-Covid-19 world in public health. Korean J Policy Stud 35(3):169–190. https://doi.org/10.52372/kjps35308

Johansson AC (2011) Financial XE financial markets in East Asia and Europe during the global financial crisis. The World Economy 34(7):1088–1105. https://doi.org/10.1111/j.1467-9701.2011.01366.x

Jun K, Bryer TA (2016) Facilitating public participation in local governments in hard times. Am Rev Public Adm 47(7):840–856. https://doi.org/10.1177/0275074016643587

Kaicker N, Gupta A, Gaiha R (2023) Recreation and social obligations during Covid in India: does social distancing act as a barrier? Rev Dev Econ 28(2):527–543. https://doi.org/10.1111/rode.13064

Kampe L, Goehlich V, Wüst K (2021) The impact of culture on the perception of young Japanese and Germans in the management of the corona crisis by their government. Public Organ Rev 21(4):759–778. https://doi.org/10.1007/s11115-021-00592-8

Kandeel A, Fahim M, Deghedy O, Roshdy WH, Khalifa MK, Shesheny RE, Abdel-Ghaffar K (2023) Resurgence of influenza and respiratory syncytial virus in Egypt following two years of decline during the Covid-19 pandemic: Outpatient clinic survey of infants and children, October 2022. Public Organ Rev 23(1). https://doi.org/10.1186/s12889-023-15880-9

Kaneko S, Hara K, Sato S, Nakashima T, Kawazoe Y, Taguchi M, Hara T (2021) Association between preoperative toe perfusion index and maternal core temperature decrease during cesarean delivery under spinal anesthesia: a prospective cohort study. https://doi.org/10.21203/rs.3.rs-547619/v1

Kang Y, Na K (2018) Determinants of labor productivity in emerging markets: Evidence from pre– and post-financial crisis Mexico. Gadjah Mada International J Business 20(3):259. https://doi.org/10.22146/gamaijb.28891

Kapelko M, Stefanou SE (2017) The impact of the 2008 financial crisis on dynamic productivity growth of the Spanish food manufacturing industry. An Impuls Response Analysis Agric Eco-Nomics 48(5):561–571. https://doi.org/10.1111/agec.12357

Kar S, Bhattacharya B (2011) Shocks, economic growth and the Indian economy. Working Papers id:4319, eSocial Sciences

Kass A, Crosby AH, Parker B (2022) A vicious cycle: fiscal intervention, pension underfunding, and instability in (re)making racialized geographies. Environ Plan A: Econ Space 55(7):1780–1798. https://doi.org/10.1177/0308518x221084995

Kazana-McCarthy J (2021) Beyond the 'stop gap': Young (adult) women's experiences of living with parents in the aftermath of the Greek austerity crisis. Fam, RelatShips Soc 10(2):343–358. https://doi.org/10.1332/204674319x15759631828766

Kellermann R, Conde DS, Rößler D, Kliewer N, Dienel HL (2022) Mobility in pandemic times: exploring changes and long-term effects of COVID-19 on urban mobility behavior. Transp Res Interdiscip Perspec-tives 15:100668

Kestens K, Van Cauwenberge P, Bauwhede HV (2011) Trade credit and company performance during the 2008 financial crisis. Accounting and Finance 52(4):1125–1151. https://doi.org/10.1111/j.1467-629x.2011.00452.x

Khan IM, Haque U, Zhang W, Zafar S, Wang Y, He J, Sun H, Lubinda J, Rahman MS (2021) Covid-19 in China: risk factors and r0 revisited. Acta Trop 213:105731. https://doi.org/10.1016/j.actatropica.2020.105731

Kholyavin A (2017) 'Nothing special… just a crisis': framing social policies in Latvia 2008–2010. J Soc Policy Stud 15(2):297–308. https://doi.org/10.17323/727-0634-2017-15-2-297-308

Kienzler D, Schmid KD (2014) Hysteresis in potential output and monetary policy. Scott J Polit Econ 61(4):371–396. https://doi.org/10.1111/sjpe.12050

Kim SJ, Bostwick WB (2020) Social vulnerability and racial inequality in Covid-19 deaths in Chicago. Health Educ Behav 47(4):509–513. https://doi.org/10.1177/1090198120929677

Kim Y, Warner ME (2017) Geographies of local government stress after the great recession. Soc Policy Adm 52(1):365–386. https://doi.org/10.1111/spol.12307

Klein M, Linnemann L (2018) Macroeconomic effects of government spending: the great recession was (really) different. SSRN Electron J. https://doi.org/10.2139/ssrn.3239358

Klimek M (2022) Funkcjonowanie jednostek samorządu terytorialnego podczas pandemii COVID-19 The functioning of local government units during the COVID-19 pandemic. Studia Politologiczne, Studia i Analizy 64. https://doi.org/10.33896/SPolit.2022.64.6

Knodt M, Fraune C, Engel A (2021) Local XE "Local" governance of critical infrastructure resilience: types of coordination in German cities. J Contingencies Cris Manag 30(3):307–316. https://doi.org/10.1111/1468-5973.12386

Knuepling L, Sternberg R, Otto A (2024) Rural areas as winners of COVID-19, digitalization and remote working? Empirical evidence from recent internal migration in Germany. Camb J Reg, Econ So-ciety

Kolasińska E (2022) Niepewność miejsca pracy w czasie pandemii covid-19 na podstawie dzienników pandemicznych. Ruch Prawniczy, Ekonomiczny i Socjologiczny 84(4):227–240. https://doi.org/10.14746/rpeis.2022.84.4.14

Kolliopoulos A (2022) Drafting national recovery plans and stakeholder involvement: An opportunity for strengthening EU legitimacy? Region and Periphery 13(13):91–108. https://doi.org/10.12681/rp.30755

Kołsut B, Stryjakiewicz T (2023) The economic geography of the car market. The automobile revolution in an emerging economy (2nd ed). Routlege, Taylor & Francis Group.

Komornicki T (2011) Przemiany mobilności codziennej polaków na tle rozwoju motoryzacji Changes in the daily mobility of Poles against the background of the development of motorization. Prace Geograficzne nr 227, PAN, Instytut Geografii i Przestrzennego Zagospodarowania Kraju, im. Stanisława Leszczyńskiego, Warszawa

Kristanti D, Edison E, Yunanto MK, Alfiandri A, Utari DS, Samnuzulsari T, Yudiatmaja WE (2021) Strengthening social capital of urban community during Covid-19 disaster. E3S Web Conf 331:01013. https://doi.org/10.1051/e3sconf/202133101013

Kuehl PJ, Dearing J, Werts A, Cox J, Irshad H, Barrett EG, Tucker S, Langel SN (2023) Design and validation of an exposure system for efficient inter-animal sars-cov-2 airborne transmission in Syrian hamsters. Microbiology Spectrum 11(6). https://doi.org/10.1128/spectrum.04717-22

Kugiejko M, Kociszewski P (2021) Organizacja czasu wolnego w obliczu pandemii COVID–19–doświadczenie różnych pokoleń (osób młodych i seniorów) Organization of free time in the face of the COVID-19 pandemic–the experience of different generations (young people and seniors). Turystyka Kulturowa 4(121):157–182

Kurmann A, Otrok C (2013) News shocks and the slope of the term structure of interest rates. American Econ Review 103(6):2612–2632. https://doi.org/10.1257/aer.103.6.2612

Lalin SAA, Ahmed MNQ, Haq SMA (2024) The effects of the COVID-19 pandemic on students' academic performance and mental health: An overview. Reg Sci Policy Pract 16(7):100046

Lange T (2014) Social capital and job satisfaction: The case of Europe in times of economic crisis. Eur J Ind Relat 21(3):275–290. https://doi.org/10.1177/0959680114542907

Lange B, Hülz M, Schmid B, Schulz C (eds) (2021) Postgrowth geographies. Columbia University Press, Spatial relations of diverse and alternative economies. Social and Cultural Geography

Lausi L, Amodio M, Sebastiani A, Fusaro L, Manes F (2022) Assessing cultural ecosystem services during the Covid-19 pandemic at the Garden of Ninfa (Italy). Annali di Botanica 12:63–75

Lavesson N (2016) When and how does commuting to cities influence rural employment growth? J Reg Sci 57(4):631–654. https://doi.org/10.1111/jors.12324

Lee JY, Yahiaoui D, Lee K, Cooke FL (2022) Global talent management and multinational subsidiaries' resilience in the Covid-19 crisis: Moderating roles of regional headquarters' support and headquarters–subsidiary friction. Hum Resour Manage 61(3):355–372. https://doi.org/10.1002/hrm.22100

Leite I (2019) Drugs and society: a public health issue and the new approaches in Portugal and Uruguay. American Internat J Contemp Res 9(1). https://doi.org/10.30845/aijcr.v9n1p3

Lettau M, Ludvigson S (2011) Shocks and crashes. NBER Working Paper 16996. www.nber.org/papers/16996

Li M, Zhang Z, Cao W, Liu Y, Du B, Chen C, Wang X (2021) Identifying novel factors associated with Covid-19 transmission and fatality using the machine learning approach. Sci Total Environ 764:142810. https://doi.org/10.1016/j.scitotenv.2020.142810

Li M, Wang X (2022) How regions react to economic crisis: regional economic resilience in a Chinese perspective. Sage Ope 12(4). https://doi.org/10.1177/21582440221142507

Li Y, Zhang P, Lo K, Tan J, Yang Q (2022) Borderland economic resilience under Covid-19: Evidence from China–Russia border regions. Int J Environ Res Public Health 19(20):13042. https://doi.org/10.3390/ijerph192013042

Lippi A, Terlizzi A (2023) The marble cake of social services in Italy and Spain: Policy capacity, social investment, and the national recovery and resilience plans. Eur J Soc Secur 25(2):178–195. https://doi.org/10.1177/13882627231169266

Liu D, Tai Q, Wang Y, Miao P, Zhang L, Su B (2022a) Impact of air temperature and containment measures on mitigating the intrahousehold transmission of Sars-CoV-2: A data-based modelling analysis. BMJ Open 12(4):e049383. https://doi.org/10.1136/bmjopen-2021-049383

Liu Q, Su Z, Huang W (2022b) Analysis of the influencing factors of the high-quality utilization of territorial space based on the perspective of spatial equilibrium: a case study of Hunan province. China Sustainability 14(19):12818. https://doi.org/10.3390/su141912818

Lombe M, Newransky C, Safadi NS (2013) Assessing the effects of an asset-building program in resource-constrained households: the case of Sub-Saharan Africa. J Soc Serv Res 40(1):15–28. https://doi.org/10.1080/01488376.2013.842952

López-Gay A, Spijker J, Cole H, Marqués AG, Triguero-Mas M, Anguelovski I, Marí-Dell'Olmo M, Álamo-Junquera D, López-Gallego F, Borrell C (2021) Sociodemographic determinants of intraurban variations in Covid-19 incidence: the case of Barcelona. J Epidemiol Community Health 76(1):1–7. https://doi.org/10.1136/jech-2020-216325

Lucchini L, Centellegher S, Pappalardo L et al (2021) Living in a pandemic: Changes in mobility routines, social activity and adherence to COVID-19 protective measures. Sci Rep 11:24452. https://doi.org/10.1038/s41598-021-04139-1

Lusardi A, Schneider D, Tufano P (2011) Financially fragile households: evidence and implications. Brook Pap Econ Act 2011(1):83–134. https://doi.org/10.1353/eca.2011.0002

Luz JAA, Reis JGMD, Leite FDA, Araújo KWFD, Moritz G (2016) Effects of transport infrastructure in the economic development. IFIP Adv Inf Commun Technol 633–640. https://doi.org/10.1007/978-3-319-51133-7_75

Ma Y (2024) Impact of digital economy development on labor force employment. Highlights in Business. Econ Manag 33:672–677. https://doi.org/10.54097/65xnfm06

Macartney H (2011) Finance unravelled: a historical materialist analysis of EU public policy. Compet Chang 15(1):48–70. https://doi.org/10.1179/102452911x12905309382012

Macit R (2021) The impact of COVID-19 on Crime. In Bozkurt V, Dawes G, Gülerce H, & Westenbroek P (eds) The societal impacts of covid-19: a transnational perspective. Istanbul University Press

MacKenzie C, Garavan TN, Carbery R (2014) The global financial and economic crisis. Adv Dev Hum Resour 16(1):34–53. https://doi.org/10.1177/1523422313508925

MacKinnon KR, Kia H, Salway T, Ashley F, Lacombe-Duncan A, Abramovich A, Enxuga G, Ross LE (2022) Health care experiences of patients discontinuing or reversing prior gender-affirming treatments. JAMA Netw Open 5(7):e2224717. https://doi.org/10.1001/jamanetworkopen.2022.24717.PMID:35877120;PMCID:PMC9315415

Madanipour A, Shucksmith M, Brooks E (2021) The concept of spatial justice and the European union's territorial cohesion. Eur Plan Stud 30(5):807–824. https://doi.org/10.1080/09654313.2021.1928040

Mahler DG, Yonzan N, Lakner C (2022) The impact of COVID-19 on global inequality and poverty. World Bank. https://doi.org/10.1596/1813-9450-10198

Major LE, Eyles A, Machin S et al (2021) Learning loss since lockdown: variation across the home nations. Centre for economic performance, London school of economics and political science

Mandelkern R (2015) Explaining the striking similarity in macroeconomic policy responses to the great recession. Comp Polit Stud 49(2):219–252. https://doi.org/10.1177/0010414015606734

Mansor M, Sabri MF, Mansur M, Ithnin M, Magli AS, Rahim HA, Janor H (2022) Analysing the predictors of financial stress and financial well-being among the bottom 40 percent (b40) households in Malaysia. Int J Environ Res Public Health 19(19):12490. https://doi.org/10.3390/ijerph191912490

Mao Y (2021) Political institutions, state capacity, and crisis management: a comparison of China and South Korea. Int Polit Sci Rev 42(3):316–332. https://doi.org/10.1177/0192512121994026

Marangos J, Triarchi E (2022) Students' comprehension of the Greek economic crisis through an introductory macroeconomics course. Panoeconomicus 69(1):121–142. https://doi.org/10.2298/pan190526010m

Marmet S, Wicki M, Gmel G, Gachoud C, Daeppen J, Bertholet N, Studer J (2020) The psychological impact of the COVID-19 crisis on young Swiss men participating in a cohort study. https://doi.org/10.31234/osf.io/kwxhd

Martin R (2011) Regional economic resilience, hysteresis and recessionary shocks. J Econ Geogr 12(1):1–32. https://doi.org/10.1093/jeg/lbr019

Martin R, Sunley P (2014) On the notion of regional economic resilience: conceptualization and explanation. J Econ Geogr 15(1):1–42. https://doi.org/10.1093/jeg/lbu015

Martin R, Sunley P (2020) Regional economic resilience: evolution and evaluation," Chapters. In: Bristow G, Healy A (eds) Handbook on Regional Economic Resilience, chapter 2. Edward Elgar Publishing, pp 10–35

Martin R, Sunley P, Gardiner B, Tyler P (2016) How regions react to recessions: resilience and the role of economic structure. Reg Stud 50(4):561–585. https://doi.org/10.1080/00343404.2015.1136410

Martin R, Gardiner B (2021) The resilience of Britain's core cities to the great recession (with implications for the COVID recessionary shock). In: Economic resilience in regions and organizations, pp 57–89. https://doi.org/10.1007/978-3-658-33079-8_3

Martines MR, Ferreira RV, Toppa RH, Assunção LMD, Desjardins MR, Delmelle E (2021) Detecting space–time clusters of COVID-19 in Brazil: mortality, inequality, socioeconomic vulnerability, and the relative risk of the disease in Brazilian municipalities. J Geogr Syst 23(1):7–36. https://doi.org/10.1007/s10109-020-00344-0

Martínez MÁ, Wissink B (2022) The outcomes of residential squatting activism in the context of municipalism and capitalism in Madrid and Barcelona (2015–2019). J Urban Aff 45(1):65–83. https://doi.org/10.1080/07352166.2022.2115921

Matharu SK, Juneja D (2021) Factors impacting resilience of women entrepreneurs in India in the face of COVID-19. Vis: J Bus Perspect 28(4):448–458. https://doi.org/10.1177/09722629211043299

Mazur-Kumrić N (2022) Post-COVID-19 recovery and resilience building in the outermost regions of the European Union: towards a new European strategy. EU Comp Law Issues ChallS Ser. https://doi.org/10.25234/eclic/22443

McCann P (2023) How have place-based policies evolved to date and what are they for now? Background paper for the OECD-EC High-level expert workshop series on "Place-based policies for the future. https://www.oecd.org/cfe/regionaldevelopment/place-based-policies-for-the-future.htm

McDonald BD (2018) Local XE "Local" governance and the issue of fiscal health. State Local GovMent Rev 50(1):46–55. https://doi.org/10.1177/0160323X18765919

McQuade P, Schmitz M (2016) The great moderation in international capital flows: a global phenomenon? SSRN Electron J. https://doi.org/10.2139/ssrn.2839797

Medicover (2024) Pandemia koronawirusa na świecie i w Polsce, kalendarium Coronavirus pandemic in the world and in Poland, A timeline. https://www.medicover.pl/o-zdrowiu/pandemia-koronawirusa-na-swiecie-i-w-polsce-kalendarium,7252,n,192

Medina-García C, de la Fuente R, Broeck PVD (2021) Exploring the emergence of innovative multi-actor collaborations toward a progressive urban regime in Madrid (2015–2019). Sustainability 13(1):415. https://doi.org/10.3390/su13010415

Mia A, Sangwan S, Hussain AHMB, Malim NAK (2021) Rural–urban financial inclusion: Implications on the cost sustainability of microfinance lenders. Manag Decis Econ 43(6):1899–1911. https://doi.org/10.1002/mde.3498

Micek G, Małochleb K, Wojnar K, Smętkowski M (2023) The effects of the COVID-19 pandemic on the coworking spaces in Poland. In: European narratives on remote working and coworking during the COVID-19 pandemic: a multidisciplinary perspective. Springer, pp 78–82. https://doi.org/10.1007/978-3-031-26018-6_8

Miković N (2022) The impact of the COVID-19 pandemic on trends in the European banking sector. Trendovi U Poslovanju 10(1):59–68. https://doi.org/10.5937/trendpos2201059m

Minixhofer P, Stangl R (2021) Green infrastructures and the consideration of their soil-related ecosystem services in urban areas. A Systematic Literature Review. Sustainability 13(6):3322. https://doi.org/10.3390/su13063322
Mirska B, Zenczak M, Nowis K, Stolarek I, Podkowiński J, Rakoczy M, Handschuh L (2024) The landscape of the COVID-19 pandemic in Poland emerging from epidemiological and genomic data. Scientific Reports 14(1). https://doi.org/10.1038/s41598-024-65468-5
Miszczuk A (2011) Demograficzne uwarunkowania peryferyjności regionu (na przykładzie Polski wschodniej) Demographic determinants of regional peripherality (on the example of Eastern Poland). Prace Naukowe Uniwersytetu Ekonomicznego We Wrocławiu 168:472–484
Mitchell D, Davis W, Hendrick R (2020) Learning from the Joneses: The professional learning effect of regional councils of government on municipal fiscal slack in suburban Chicago. Public Budg Financ 41(2):3–21. https://doi.org/10.1111/pbaf.12275
Mobility Patterns and Mode Choice Preferences during the COVID-19 Situation (PDF). Accessed 17 Apr 2025. https://www.researchgate.net/publication/357751695_Mobility_Patterns_and_Mode_Choice_Preferences_during_the_COVID-19_Situation
Mohammad KU, Khan MR (2021) Bank capital structure dynamics and COVID-19: evidence from South Asia. iRASD J Econ 3(3). https://doi.org/10.52131/joe.2021.0302.0045
Mohan S, Hutson AD, Macdonald IAW, Lin CC (2019) Impact of macroeconomic indicators on housing prices. Int J Hous Mark Anal-Ysis 12(6):1055–1071. https://doi.org/10.1108/IJHMA-09-2018-0070
Morgan K (2016) Nurturing novelty: Regional innovation policy in the age of smart specialisation. Environ Plan C: Polit Space 35(4):569–583. https://doi.org/10.1177/0263774X16645106
Morrison C, Ramsey E, Bond D (2017) The role of social entrepreneurs in developing community resilience in remote areas. J Enterprising Communities: People Places Glob Econ 11(1):95–112. https://doi.org/10.1108/JEC-02-2015-0020
Mostafavi A (2017) A system-of-systems approach for integrated resilience assessment in highway transportation infrastructure investment. Infrastructures 2(4):22. https://doi.org/10.3390/infrastructures2040022
Motoc A (2022) Family businesses SMEs through COVID-19 crisis management. Int J Bus Manag 17(3):121. https://doi.org/10.5539/ijbm.v17n3p121
Moudud-Ul-Huq S, Ahmed K, Chowdhury MAF, Sohail HM, Biswas T, Abbas F (2021) How do banks' capital regulation and risk-taking respond to COVID-19? Empirical insights of ownership structure. Int J Islam Middle East Financ Manag 15(2):406–424. https://doi.org/10.1108/IMEFM-07-2020-0372
Mouratidis K (2021) How COVID-19 reshaped quality of life in cities: a synthesis and implications for urban planning. Land Use Policy 111:105772. https://doi.org/10.1016/j.landusepol.2021.105772
Moussallem M, Zein-El-Din A, Hamra R, Rady A, Asmar MK, Bou-Orm IR (2022) Evaluating the governance and preparedness of the Lebanese health system for the COVID-19 pandemic: a qualitative study. BMJ Open 12(6):e058622. https://doi.org/10.1136/bmjopen-2021-058622
Mucha B, Mucha M (2021) Aktywność fizyczna w dobie pandemii COVID-19 Physical activity during the COVID-19 pandemic. In: Nowak K, Szalonka K (eds) Zdrowie i styl życia: Ekonomiczne i społeczne i zdrowotne skutki pandemii [Health and lifestyle: economic, social, and health effects of the pandemic. Wrocław University Press, pp 45–62
Muhammad S, Long X, Salman M (2020) COVID-19 pandemic and environmental pollution: a blessing in disguise? Sci Total Environ 728:138820. https://doi.org/10.1016/j.scitotenv.2020.138820
Muntaner C, Gunn V, Prins S (2023) An emerging storm? Increased health inequities in the context of racialized patriarchal capitalism, deaths of despair, and COVID-19. Med Res Arch 11(10). https://doi.org/10.18103/mra.v11i10.4414
Murphy E, Mercille J (2019) (Re)making labour markets and economic crises: The case of Ireland. Econ Labour Relat Rev 30(1):22–38. https://doi.org/10.1177/1035304619829015

Musiał-Karg M, Kapsa I (2021) Debate: voting challenges in a pandemic. Poland. Public Money and Management 41(1):6–8. https://doi.org/10.1080/09540962.2021.1834549

Musil R (2014) European global cities in the recent economic crisis. Tijdschr Econ Soc Geogr 105(4):492–503. https://doi.org/10.1111/tesg.12106

Myck, M., Oczkowska, M., Garten, C., Król, A., & Brandt, M. (2023). Deaths during the first year of the COVID-19 pandemic: insights from regional patterns in Germany and Poland. BMC Public Health, 23(1), 177.

Naffa H, Dudás F (2023) Does ESG improve crisis resilience? Empirical evidence of global emerging equity markets during the COVID-19 crisis. Period Polytech Soc Manag Sci 32(1):17–27. https://doi.org/10.3311/PPso.19147

Nagarjun KB, Sridhar KS (2023) COVID-19 and the urban housing affordability, evidence from select Indian cities. Reg Sci Policy Pract 15(8):1864–1881

Nahtigal M (2022) EU recovery plans, inclusive knowledge economy and overcoming regional disparities. Lex Localis, J Local Self-Government 20(4):1171–1189. https://doi.org/10.4335/20.4.1171-1189(2022)

Narodowy Bank Polski (2022) Raport o sytuacji na rynku nieruchomości mieszkaniowych i komercyjnych w Polsce w 2020 r Report on the situation in the residential and commercial real estate market in Poland in 2020. https://nbp.pl/wp-content/uploads/2022/09/raport_2020.pdf

Nicola M, Alsafi Z, Sohrabi C, Kerwan A, Al-Jabir A, Iosifidis C, Agha M, Agha R (2020) The socio-economic implications of the coronavirus pandemic (COVID-19): a review. Int J Surg 78:185–193. https://doi.org/10.1016/j.ijsu.2020.04.018

Nicolelis MAL, Raimundo RLG, Peixoto PS, Andreazzi CS (2021) The impact of super-spreader cities, highways, and intensive care availability in the early stages of the COVID-19 epidemic in Brazil. Scientific Reports 11(1). https://doi.org/10.1038/s41598-021-92263-3

Ninth Report on Economic, Social and Territorial Cohesion (2024) Directorate-General for Regional and Urban Policy, European Commission. https://ec.europa.eu/regional_policy/information-sources/cohesion-report_en

Noorimotlagh Z, Jaafarzadeh N, Martínez SS, Mirzaee SA (2021) A systematic review of possible airborne transmission of the COVID-19 virus (SARS-CoV-2) in the indoor air environment. Environ Res 193:110612. https://doi.org/10.1016/j.envres.2020.110612

Nübler L, Austrian K, Maluccio JA, Pinchoff J (2020) Rainfall shocks, cognitive development and educational attainment among adolescents in a drought-prone region in Kenya. Environ Dev Econ 26(5–6):466–487. https://doi.org/10.1017/S1355770X20000406

OECD (2021) Building inclusive labour markets: active labour market policies for the most vulnerable groups. OECD Publishing

OECD/European Commission (2021) The missing entrepreneurs 2021: Policies for inclusive entrepreneurship and self-employment. OECD Publishing. https://doi.org/10.1787/71b7a9bb-en

OECD (2023) The longstanding geography of inequalities. OECD Publishing. https://www.oecd.org/en/publications/oecd-regional-outlook-2023_92cd40a0-en.html

Okkola S, Brunelle C (2018) The changing determinants of housing affordability in oil-booming agglomerations: a quantile regression investigation from Canada, 1991–2011. Hous Stud 33(6):902–937. https://doi.org/10.1080/02673037.2017.1408784

Oravský R, Tóth P, Bánociová A (2020) The ability of selected European countries to face the impending economic crisis caused by COVID-19 in the context of the global economic crisis of 2008. J Risk Financ Manag 13(8):179. https://doi.org/10.3390/jrfm13080179

Orlewska K, Kozieł D, Klusek J, Orlewska E (2022) Burden of COVID-19 mortality and morbidity in Poland in 2020. Int J Environ Res Public Health 19(9):5432. https://doi.org/10.3390/ijerph19095432

Our World in data (2024). Accesed 23 Sep 2024. https://ourworldindata.org/covid-vaccinations

Overton M, Bland RL (2014) The great recession's impact on credible commitment. State Local Gov Rev 46(4):282–297. https://doi.org/10.1177/0160323X14559713

Owusu-Oware E (2022) Home-based e-working under COVID-19 pandemic conditions: Insights from corporate organizations in Ghana. Electron J Inf Syst Dev Ctries 89(3). https://doi.org/10.1002/isd2.12253

Özçürümez S, Hoxha J, İçduygu A (2021) Does social cohesion solve forced migration riddles? Troubled concepts and constrained practices in Turkey. Migration Stud 9(3):597–616. https://doi.org/10.1093/migration/mnaa011

Pagaoa M, Grey JA, Torrone E, Kreisel K, Stenger M, Weinstock H (2021) Trends in nationally notifiable sexually transmitted disease case reports during the US COVID-19 pandemic, January to December 2020. Sex Transm Dis 48(10):798–804. https://doi.org/10.1097/OLQ.0000000000001506

Pandya A, Lodha P (2021) Social connectedness, excessive screen time during COVID-19 and mental health: a review of current evidence. Front Hum Dyn 3:684137. https://doi.org/10.3389/fhumd.2021.684137

Pani SS (2022) House price dynamics under lower leverage: the case of metropolitan cities in India. Int J Hous Mark Anal 17(3):814–836. https://doi.org/10.1108/IJHMA-10-2022-0161

Park J, Park S, Maher CS (2018) The effects of tax and expenditure limitations (TELs) on municipal fiscal outcomes during a period of fiscal distress. Public Financ Manag 18(1):84–110. https://doi.org/10.1177/152397211801800105

Park JH (2023) How did local governments recover their financial condition? Lessons from citizen participation. Transylv Rev Adm Sci 68(E):96–113. https://doi.org/10.24193/tras.68E.6

Partridge MD, Ali K, Olfert MR (2010) Rural-to-urban commuting: three degrees of integration. Growth Chang 41(2):303–335. https://doi.org/10.1111/j.1468-2257.2010.00528.x

Pavel A, Moldovan B, Kourtit K, Nijkamp P (2020) Urban or rural: Does it make a difference for economic resilience? A modelling study on economic and cultural geography in Romania. Sustainability 12(9):3776. https://doi.org/10.3390/su12093776

Petersen AM (2024) Shift in house price estimates during COVID-19 reveals effect of crisis on collective speculation. EPJ Data Sci 13(1):47. https://doi.org/10.1140/epjds/s13688-024-00471-4

Picchio M, Santolini R (2020) Fiscal rules and budget forecast errors of Italian municipalities. Eur J Polit Econ 64:101921. https://doi.org/10.1016/j.ejpoleco.2020.101921

Pike A, Béal V, Cauchi-Duval N, Franklin R, Kinossian N, Lang T, Velthuis S (2023) Left behind places': a geographical etymology. Reg Stud. https://doi.org/10.1080/00343404.2023.2167972

Pinheiro F, Hartmann D, Boschma R, Hidalgo CA (2022) The time and frequency of unrelated diversification, research policy, 51(8). ISSN 104323:0048–7333. https://doi.org/10.1016/j.respol.2021.104323

Pizzuto P (2020) The role of regional competitiveness in shaping the heterogeneous impact of the great recession. Reg Sci Policy Pract 12(2):267–290. https://doi.org/10.1111/rsp3.12196

Plichta J, Sagan A, Plichta G (2022) Szoki egzogeniczne a kształtowanie relacji na rynku B2C z perspektywy struktury wartości konsumentów Exogenous shocks and shaping B2C market relations from the perspective of consumer value structure. Uniwersytet Ekonomiczny w Krakowie

Płonka M (2022) The impact of the coronavirus pandemic on the healthcare system in Poland. In: Post-COVID economic revival. Springer, vol 2, pp 311–323

Pomar L, Favre G, De Labrusse C, Contier A, Boulvain M, Baud D (2022) Impact of the first wave of the COVID-19 pandemic on birth rates in Europe: a time series analysis in 24 countries. Hum Reprod 37(12):2921–2931. https://doi.org/10.1093/humrep/deac215

Poměnková J, Klejmova E, Kučerová Z (2019) Cyclicality in lending activity of Euro Area in pre– and post–2008 crisis: a local-adaptive-based testing of wavelets. Balt J Econ 19(1):155–175. https://doi.org/10.1080/1406099X.2019.1596466

Pongsakornrungsilp S, Pongsakornrungsilp P, Kumar V, Maswongssa B (2021) The art of survival: Tourism businesses in Thailand recovering from COVID-19 through brand management. Sustainability 13(12):6690. https://doi.org/10.3390/su13126690

Prata DN, Rodrigues W, Bermejo PH (2020) Temperature significantly changes COVID-19 transmission in (sub)tropical cities of Brazil. Sci Total Environ 729:138862. https://doi.org/10.1016/j.scitotenv.2020.138862

Prescott P, Gjerde KP (2022) The impact of state fiscal policy on States' resilience exiting the great recession. Public Finance Review 51(1):3–43. https://doi.org/10.1177/10911421221126054

Price L, Rae D, Cini V (2013) SME perceptions of and responses to the recession. J Small Bus Enterp Dev 20(3):484–502. https://doi.org/10.1108/jsbed-04-2013-0057

Pritoni A, Bitonti A, Montalbano G (2023) Change of government and interest groups' preference attainment on the formulation of the Italian national recovery and resilience plan (NRRP): From Conte to Draghi. Italian Political Science Review/Rivista Italiana Di Scienza Politica 53(3):333–350. https://doi.org/10.1017/ipo.2023.7

Proniewski M (2012) Rozwój regionów perfyeryjnych w Unii Europejskiej: Wybrane aspekty. Wydawnictwo Uniwersytetu w Białymstoku

Ptaszek G, Stunża GD, Pyżalski J, Dębski M, Bigaj M (2020) Edukacja zdalna: Co stało się z uczniami, ich rodzicami i nauczycielami? Gdańskie Wydawnictwo Psychologiczne

Puspa SM (2021) Behavior increase body immunity. Muhammadiyah International Public Health and Medicine Proceeding 1(1):827–830. https://doi.org/10.53947/miphmp.v1i1.140

Qamar A, Collinson S, Green A (2022) COVID-19 disruption, resilience and industrial policy: the automotive sector in the West Midlands. Regional Studies 1–15

Raciborski F, Samel-Kowalik P, Gujski M, Pinkas J, Arcimowicz M, Jankowski M (2021) Factors associated with a lack of willingness to vaccinate against Covid-19 in Poland: a 2021 nationwide cross-sectional survey. Vaccines 9(9):1000. https://doi.org/10.3390/vaccines9091000

Radenović T, Radivojević V, Krstić B, Stanišić T, Živković S (2022) The efficiency of health systems in response to the COVID-19 pandemic: evidence from the EU countries. Problemy Ekorozwoju 17(1):7–15

Radzińska J (2023) Zmiana postrzegania i doświadczania solidarności w pandemii covid-19. Studia Socjologiczne. https://doi.org/10.24425/sts.2022.142636

Rahman M, Alam M, Luies SK, Ferdous S, Mamun Z, Rahman MJ, Bidwas D, Ananya T, Kamal A, Chowdhury R, Russel Khan E, Hohnston D, Ahmed T (2024) Institutional capacity assessment in the lens of implementation research: capacity of the local institutions in delivering WASH services at Cox's Bazar district. Bangladesh. Plos One 19(2):e0297000. https://doi.org/10.1371/journal.pone.0297000

Ramey VA (2016) Macroeconomic shocks and their propagation. In Taylor JB, Uhlig H (eds) Handbook of Macroeconomics. Elsevier, pp 71–162. https://doi.org/10.1016/bs.hesmac.2016.03.003

Ramírez RR, Merodio JAM, Estriégana R (2022) What do urban and rural hotel managers say about the future of hotels after Covid-19? The new meaning of safety experiences. Cities 120:103492. https://doi.org/10.1016/j.cities.2021.103492

Ramos-Vielba I, Woolley R (2018) Local XE "Local" factors shaping public sector industrial relations in a context of crisis: a comparison of municipal administrations in Spain. Industrielle Beziehungen Zeitschrift Für Arbeit Organisation und Management 25(3–2018):343–362. https://doi.org/10.3224/indbez.v25i3.04

Rasul A (2021) Relationship between climatic variables and reproduction number (r0) of confirmed Covid-19 cases. https://doi.org/10.21203/rs.3.rs-682756/v1

Raudsepp-Hearne C, Peterson GD, Tengö M, Bennett EM, Holland TG, Benessaiah K, Pfeifer L (2010) Untangling the environmentalist's paradox: why is human well-being increasing as ecosystem services degrade? Bioscience 60(8):576–589. https://doi.org/10.1525/bio.2010.60.8.4

Ray DM, MacLachlan I, Lamarche RH, Srinath K (2016) Economic shock and regional resilience: continuity and change in canada's regional employment structure, 1987–2012. Environ Plan A: Econ Space 49(4):952–973. https://doi.org/10.1177/0308518x16681788

Reed E (2020) Economic shocks definition and examples. https://finance.yahoo.com/news/economic-shocks-definition-examples-221454349.html

Regmi S, Bertone MP, Shrestha P, Sapkota S, Arjyal A, Martineau T, Raven J, Witter S, Baral S (2023) Understanding health system resilience in responding to a pandemic: experience and lessons from an evolving context of federalization in Nepal. https://doi.org/10.21203/rs.3.rs-2918720/v1

Reinhart H, Putra RD, Rafida MR (2023) Karst ecosystem services and land-covers dynamic: case study from Karst of Tuban and Tuban regency, Jawa Timur, Indonesia. IOP Conf Ser: Earth Environ Sci 1190(1):012043. https://doi.org/10.1088/1755-1315/1190/1/012043

Rhouiri M, Meyabe M, Yousfi F, Saidi H, Marghich A, Aiboud-Benchekroun B, Madhat F (2023) Stakeholders' involvement, organizational learning and social innovation: factors for strengthening the resilience of Moroccan cooperatives in the post- COVID-19 era. Sustainability 15(11):8846. https://doi.org/10.3390/su15118846

Rodríguez-Pose A (2013) Do institutions matter for regional development? Reg Stud 47(7):1034–1047

Rodríguez-Pose A, Dijkstra L, Poleman H (2023) The geography of EU discontent and the regional development trap. Working Papers. A series of short papers on regional research and indicators produced by the directorate-general for regional and urban policy. WP 03/2023. European Commission. European Union. Luxembourg

Roelofs B, Ballas D, Haisma H, Edzes A (2022) Spatial mobility patterns and COVID-19 incidence: A regional analysis of the second wave in the Netherlands. Reg Sci Policy Pract 14:21–40

Rost K, Dong F, Dickinson LM (2010) Stakeholder benefit from depression disease management: differences by rurality? J Behav Health Serv Res 38(1):114–121. https://doi.org/10.1007/s11414-009-9204-0

Roszkowska S, Kula B, Pawelec N, Święczkowski M, Tomaszuk-Kazberuk A, Bachórzewska–Gajewska H, Kuźma Ł (2023) Environmental and socioeconomic determinants of heart failure. Kardiologia Polska 81(2):160–167. https://doi.org/10.33963/kp.a2022.0251

Rubio M, Carrasco-Gallego JA (2016) Liquidity, interest rates and house prices in the euro area: a DSGE analysis. J Eur R Estate Res 9(1):4–25. https://doi.org/10.1108/jerer-03-2015-0014

Rückert A, Labonté R (2014) The global financial crisis and health equity: early experiences from Canada. Glob Health 10(1):2. https://doi.org/10.1186/1744-8603-10-2

Russell TW, Wu JT, Clifford S, Edmunds WJ, Kucharski AJ, Jit M (2021) Effect of internationally imported cases on internal spread of COVID-19: a mathematical modelling study. The Lancet Public Health 6(1):e12–e20. https://doi.org/10.1016/S2468-2667(20)30263-2

Sadecki W (2022) Wpływ pandemii Covid-19 na rozwój pracy zdalnej w branży małych i średnich przedsiębiorstw. Przedsiębiorczość-Edukacja 18(2). https://doi.org/10.24917/20833296.182.10

Salvia R, Egidi G, Salvati L, Rodrigo-Comino J, Quaranta G (2020) In-between 'smart' urban growth and 'sluggish' rural development? Reframing population dynamics in Greece, 1940–2019. Sustainability 12(15):6165. https://doi.org/10.3390/su12156165

Sanne JM, Ekholm HM, Rahmberg M (2021) Contextualizing resilience indicators–comparable across organizations yet specific to context. J Risk Res 24(12):1652–1667. https://doi.org/10.1080/13669877.2021.1913632

Santana C, Botta F, Barbosa H, Privitera F, Menezes R, Di Clemente R (2023) COVID-19 is linked to changes in the time–space dimension of human mobility. Nat Hum Behav 7(10):1729–1739

Saraceno F (2022) The return of fiscal policy. https://doi.org/10.54394/whwg2947

Sarkodie SA, Owusu PA (2021) Impact of COVID-19 pandemic on waste management. Environ Dev Sustain 23(5):7951–7960

Sathurshan M, Saja AMA, Thamboo J, Haraguchi M, Navaratnam S (2022) Resilience of critical infrastructure systems: a systematic literature review of measurement frameworks. Infrastructures 7(5):67. https://doi.org/10.3390/infrastructures7050067

Saul A, Scott N, Crabb BS, Majumdar SS, Coghlan B, Hellard ME (2020) Impact of Victoria's stage 3 lockdown on COVID-19 case numbers. Med J Aust 213(11):494-496.e1. https://doi.org/10.5694/mja2.50872

Semczuk-Kaczmarek K, Ryś-Czaporowska A, Sierdziński J, Kaczmarek Ł, Szymański FM, Płatek AE (2021) Association between air pollution and Covid-19 mortality and morbidity. Intern Emerg Med 17(2):467–473. https://doi.org/10.1007/s11739-021-02834-5

Setyawati D, Ningrum MY (2021) Community knowledge and compliance in doing prevention of COVID-19. South East Asia Nursing Research 3(1):16. https://doi.org/10.26714/seanr.3.1.2021.16-22

Sharifi A, Khavarian-Garmsir AR (2020) The COVID-19 pandemic: impacts on cities and major lessons for urban planning, design, and management. Sci Total Environ 749:142391. https://doi.org/10.1016/j.scitotenv.2020.142391

Sharma S, Cain J, Sakhuja A, Schaefer G, Krupica T, Sarwari AR (2020a) Guidance for healthcare providers managing Covid-19 in rural and underserved areas. J Racial Ethn Health Disparities 7(5):817–821. https://doi.org/10.1007/s40615-020-00820-9

Sharma V, Scott J, Kelly J, VanRooyen MJ (2020b) Prioritizing vulnerable populations and women on the frontlines: COVID-19 in humanitarian contexts. Int J Equity Health 19:66. https://doi.org/10.1186/s12939-020-01186-4

Silverman R, Ceci A, Cohen AP, Helmick M, Short EE, Bordwine P, Friedlander MJ, Finkielstein CV (2022) Vaccine effectiveness during outbreak of Covid-19 alpha (b.1.1.7) variant in men's correctional facility, United States. Emerg Infect Dis 28(7):1313–1320. https://doi.org/10.3201/eid2807.220091

Simona-Moussa J, Ravazzini L (2018) From one recession to another: longitudinal impacts on the quality of life of vulnerable groups. Soc Indic Res 142(3):1129–1152. https://doi.org/10.1007/s11205-018-1957-5

Sinclair AH, Hakimi S, Stanley ML, Adcock RA, Samanez-Larkin GR (2021) Pairing facts with imagined consequences improves pandemic-related risk perception. Proc Natl Acad Sci 118(32). https://doi.org/10.1073/pnas.2100970118

Śleszyński P (2022) Geografia pandemii. Zróżnicowanie i dynamika rozprzestrzeniania się COVID-19 w Polsce (marzec 2020-marzec 2022). Studia Politologiczne 65. https://doi.org/10.33896/SPolit.2022.65.1

Smith BG, Krishna A, Smith SB (2020) Relational immunity? Examining relationship as crisis shield. J Contingencies Cris Manag 29(1):89–98. https://doi.org/10.1111/1468-5973.12324

Sobczak A (2021) Remote work in the Covid-19 pandemic issues of efficiency and social inequality. Family Upbringing 24(1):147–159. https://doi.org/10.61905/wwr/170404

Song C, Huang H (2021) Rural resilience in China and key restriction factor detection. Sustainability 13(3):1080. https://doi.org/10.3390/su13031080

Srinivasan P, Mariappan K, Devakumar C (2014) On the temporal causal relationship between macroeconomic variables. Sage Open 4(1). https://doi.org/10.1177/2158244014525419

Stan M (2022) The impact of the pandemic crisis on employment in the context of urbanization. Tech Soc Sci J 33:492–505. https://doi.org/10.47577/tssj.v33i1.6977

Stankowska M (2024) Praca zdalna a macierzyństwo. Analiza korzyści i trudności wynikających z pracy zdalnej w opinii matek małych dzieci. Roczniki Nauk Społecznych 52(2):81–95. https://doi.org/10.18290/rns2024.0019

Steiner A, Atterton J (2014) The contribution of rural businesses to community resilience. Local Econ: J Local Econ Policy Unit 29(3):228–244. https://doi.org/10.1177/0269094214528853

Steiner A, Atterton J (2015) Exploring the contribution of rural enterprises to local resilience. J Rural Stud 40:30–45. https://doi.org/10.1016/j.jrurstud.2015.05.004

Stengel S, Roth C, Breckner A, Cordes L, Weber S, Ullrich C, Peters-Klimm F, Wensing M (2022) Resilience of the primary health care system–German primary care practitioners' perspectives during the early Covid-19 pandemic. BMC Prim Care 23(1). https://doi.org/10.1186/s12875-022-01786-9

Suder M, Kusa R, Duda J, Dunska M (2022) How small printing firms alleviate impact of pandemic crisis? Identifying configurations of successful strategies with fuzzy-set qualitative comparative analysis. Entrep Bus Econ Rev 10(2):61–80. https://doi.org/10.15678/eber.2022.100204

Suk KIM, Min XU, David J (2022) Highly unusual features of the 2020 US Recession. J Public Adm, Financ and Law (23)

Sumner A, Hoy C, Ortiz-Juarez E (2020) Estimates of the impact of COVID-19 on global poverty. UNU-WIDER Working Paper 2020/43

Sun X, Wandelt S, Zhang A (2020) How did COVID-19 impact air transportation? A first peek through the lens of complex networks. J Air Transp Manag 89:101928

Supriyati S, Habibah K, Wahyuni AS, Wahab RAA, Halim KS, Nugroho EA, Soddiq MS (2021) Family social capital on public respond to COVID-19 in Indonesia. J Community Empower Health 4(3):196. https://doi.org/10.22146/jcoemph.64567

Sutton J, Arku G (2022) The importance of local characteristics: an examination of Canadian cities' resilience during the 2020 economic crisis. Canadian Geographies/Géographies Canadiennes 66(4):712–727. https://doi.org/10.1111/cag.12780

Suzuki Y (2020) Faculty opinions recommendation of cell entry mechanisms of Sars-Cov-2. Fac Opin–Post-Publ Peer Rev Biomed Lit. https://doi.org/10.3410/f.737912526.793577430

Szczerba WM (2023) CSR activities of cooperative banks in Poland during the covid-19 pandemic and state security. Nat Secur Stud 28(2):71–80. https://doi.org/10.37055/sbn/169715

Szczurek M (2021) Reakcja Unii Europejskiej na COVID-19 jako wypełnienie braków w konstrukcji Europejskiej Unii Gospodarczej i Walutowej. Gospodarka Narodowa 4(308):83–102. https://doi.org/10.33119/GN/143262

Szymańska AI (2022) Wpływ pandemii COVID-19 na zachowania rynkowe młodych dorosłych w obszarze sharing economy. Studies of the Industrial Geography Commission of the Polish Geographical Society 36(2). https://doi.org/10.24917/20801653.362.16

Tabis D (2023) Wpływ pandemii COVID-19 na sytuację finansową polskich samorządów [Online]. E-Wydawnictwo. Prawnicza i Ekonomiczna Biblioteka Cyfrowa. Wydział Prawa, Administracji i Ekonomii Uniwersytetu Wrocławskiego. http://repozytorium.uni.wroc.pl/Content/140205/PDF/08_D_Tabis_Wplyw_pandemii_wirusa_Covid-19_na_sytuacje_finansowa_polskich_samorzadow.pdf

Tagashira Y, Takamatsu A, Hasegawa S, Uenoyama Y, Honda H (2020) A survey of preparedness against coronavirus disease 2019 (Covid-19) in hospitals in Tokyo, Japan, with healthcare personnel with Covid-19 and in-facility transmission. Infect Control Hosp Epidemiol 42(6):746–750. https://doi.org/10.1017/ice.2020.1304

Tandon A, Roubal T, McDonald L, Cowley P, Palu T, de Oliveira Cruz V, Eozenou P, Cain J, Teo HS, Schmidt M, Pambudi E (2020) Economic impact of COVID-19. World Bank. https://doi.org/10.1596/34572

Taylor SA, Stephens H, Grossman D (2021) The opioid crisis and economic distress: consequences for population change. J Reg Sci 62(2):541–577. https://doi.org/10.1111/jors.12576

Tilmann H, Richter N, Hunecke M (2021) Heavy crisis, new perspectives? Investigating the role of consumption, time wealth and meaning construction during countrywide COVID-19 lockdown in Germany. https://doi.org/10.31234/osf.io/e5684

Tomao A, Mattioli W, Fanfani D, Ferrara C, Quaranta G, Salvia R, Salvati L (2021) Economic downturns and land-use change: a spatial analysis of urban transformations in Rome (Italy) using a geographically weighted principal component analysis. Sustainability 13(20):11293. https://doi.org/10.3390/su132011293

Tong K, Li L, Breivik K, Wania F (2022) Ecological unequal exchange: quantifying emissions of toxic chemicals embodied in the global trade of chemicals, products, and waste. Environ Res Lett 17(4):044054. https://doi.org/10.1088/1748-9326/ac5f95

Totouom A (2023) Oil dependency, political institutions, and urban–rural disparities in access to electricity in Africa. Nat Res Forum 47(1):114–133. https://doi.org/10.1111/1477-8947.12272

Treviño JOM, Cuellar CY (2021) Informality, gender employment gap, and Covid-19 in Mexico: identifying persistence and dynamic structural effects. Revista Mexicana De Economía Y Finanzas 16(3):1–25. https://doi.org/10.21919/remef.v16i3.636

Tubis AA (2022) The impact of COVID-19 on public transport in Polish cities on the example of Wroclaw. In: Intelligent solutions for cities and mobility of the future. Springer International Publishing, pp 143–155

Tudorache AT, Nicolescu L (2023) Insights about the effects of Covid-19 on international trade during the main pandemic years in Romania and Poland. Sustainability 15(11):8726. https://doi.org/10.3390/su15118726

Tušl M, Brauchli R, Kerksieck P, Bauer GF (2021) Impact of the Covid-19 crisis on work and private life, mental well-being and self-rated health in German and Swiss employees: a cross-sectional study. https://doi.org/10.21203/rs.3.rs-141755/v1

Uyarra E, Valdaliso JM (2021) Agency, institutions and regional resilience: an approach from the Basque region. In: Towards resilient organizations and societies, pp 277–304. https://doi.org/10.1007/978-3-030-82072-5_11

Vindigni G, Mosca A, Bartoloni T, Spina D (2021) Shedding light on peri-urban ecosystem services using automated content analysis. Sustainability 13(16):9182. https://doi.org/10.3390/su13169182

Volkos P, Symvoulakis EK (2021) Impact of financial crisis on mental health: a literature review 'puzzling' findings from several countries. Int J Soc Psychiatry 67(7):907–919. https://doi.org/10.1177/00207640211011205

Waldron R (2018) Capitalizing on the state: the political economy of real estate investment trusts and the 'resolution' of the crisis. Geoforum 90:206–218. https://doi.org/10.1016/j.geoforum.2018.02.014

Waldron R, Redmond D (2015) Stress in suburbia: counting the costs of Ireland's property crash and mortgage arrears crisis. Tijdschr Econ Soc Geogr 107(4):484–501. https://doi.org/10.1111/tesg.12170

Waldron R, Redmond D (2016) "We're just existing, not living!" mortgage stress and the concealed costs of coping with crisis. Hous Stud 32(5):584–612. https://doi.org/10.1080/02673037.2016.1224323

Wang X, Chen Y (2024) More rational, more attractive: industrial structure rationalization and migrant workers' employment choices in China. Sustainability 16(7):2746. https://doi.org/10.3390/su16072746

Wang L, Xu C, Wang J, Jia-jun Q, Yan M, Zhu Q (2021) Spatiotemporal heterogeneity and its determinants of COVID-19 transmission in typical labor export provinces of China. BMC Infect Dis 21(1). https://doi.org/10.1186/s12879-021-05926-x

Warren G, Grigaliūnaitė S (2023) Leveraging social capital to catalyse post-pandemic anti-fragility: the "Vilnius model." J Place Manag Dev 16(4):577–601. https://doi.org/10.1108/jpmd-11-2022-0104

Weck S, Madanipour A, Schmitt P (2021) Place-based development and spatial justice. Eur Plan Stud 30(5):791–806. https://doi.org/10.1080/09654313.2021.1928038

World Bank (2022) World development report 2022: finance for an equitable recovery

Woźniakowski TP (2023) No borrowing without taxing? Fiscal solidarity of Next Generation EU in light of the American experience. Polit Gov 11(4):73–81. https://doi.org/10.17645/pag.v11i4.7233

Wu R, Possingham HP, Yu G, Jin T, Wang J, Yang F, Zhao H (2019) Strengthening China's national biodiversity strategy to attain an ecological civilization. Conserv Lett 12(5). https://doi.org/10.1111/conl.12660

Xie Y, Wu H, Yao R (2023) The impact of climate change on the urban–rural income gap in China. Agriculture 13(9):1703. https://doi.org/10.3390/agriculture13091703

Yan A, Howden K, Mahar A, Glidden C, Garland SN, Oberoi S (2022) Gender differences in adherence to Covid-19 preventative measures and preferred sources of Covid-19 information among adolescents and young adults with cancer. Cancer Epidemiol 77:102098. https://doi.org/10.1016/j.canep.2022.102098

Yang C, Wang J (2021) Transmission rates and environmental reservoirs for COVID-19–a modeling study. J Biol Dyn 15(1):86–108. https://doi.org/10.1080/17513758.2020.1869844

Yao R, Zhang J (2023) Employment status and financial resilience during the Covid-19 pandemic. Int J Bank Mark 41(5):992–1009. https://doi.org/10.1108/ijbm-08-2022-0371

Yeşin P (2015) Capital flow waves to and from Switzerland before and after the financial crisis. Swiss J Econ Stat 151(1):27–75. https://doi.org/10.1007/bf03399413

Yin H, Peng X, Dixon P, Rimmer MT (2014) The economic effects of facilitating the flow of rural workers to urban employment in China. Pap Reg Sci 93(3):619–643. https://doi.org/10.1111/pirs.12004

Yu S, Kim D (2021) Changes in regional economic resilience after the 2008 global economic crisis: the case of Korea. Sustainability 13(20):11392. https://doi.org/10.3390/su132011392

Yu Z, Razzaq A, Rehman A, Shah A, Jameel K, Mor RS (2020a) Disruption in global supply chain and socio-economic shocks: a lesson from Covid-19 for sustainable production and consumption. Oper Manag Res 15:233–248. https://doi.org/10.1007/s12063-021-00179-y

Yu JS, Jeong HJ, Kim SJ, Lee JY, Choe YJ, Choi EH, Cho EH (2020b) Sustained vaccination coverage during the coronavirus disease 2019 epidemic in the Republic of Korea. Vaccines 9(1):2. https://doi.org/10.3390/vaccines9010002

Zahara Z, Muslimin M, Buntuang PCD (2022) The impact of marketing innovations and business plans on business sustainability during the Covid-19 pandemic. Innov Mark 18(3):121–135. https://doi.org/10.21511/im.18(3).2022.11

Zajfert M (2024) Urban transport during the COVID-19 pandemic: a case study of Poland. Public Transport 1–24

Zarębski P, Czerwińska-Jaśkiewicz M, Klonowska-Matynia M (2022) Innovation in peripheral regions from a multidimensional perspective: evidence from the middle Pomerania region in Poland. Sustainability 14(14):8529. https://doi.org/10.3390/su14148529

Zavras D (2020) Studying healthcare affordability during an economic recession: the case of Greece. Int J Environ Res Public Health 17(21):7790. https://doi.org/10.3390/ijerph17217790

Ženka J, Pavlík A, Slach O (2017) Resilience of metropolitan, urban and rural regions: a central European perspective. GeoScape 11(1):25–40. https://doi.org/10.1515/geosc-2017-0003

Zhang D, Nan Y, Zhu W, He C, Liu Z, Yang Y (2023) Indirect impacts of future urban expansion on ecosystem services in the transnational area of China, North Korea, and Russia. Land Degrad Dev 34(8):2334–2348. https://doi.org/10.1002/ldr.4611

Zhao L, Rasoulinezhad E, Sarker T, Taghizadeh-Hesary F (2022) Effects of Covid-19 on global financial markets: evidence from qualitative research for developed and developing economies. Eur J Dev Res 35(1):148–166. https://doi.org/10.1057/s41287-021-00494-x

Zhou Q, Zhang S, Deng W, Wang J (2023) Has rural public services weakened population migration in the Sichuan-Chongqing region? spatiotemporal association patterns and their influencing factors. Agriculture 13(7):1300. https://doi.org/10.3390/agriculture13071300

Zhu TXC, Zhang W, Sharp B (2023) Analysis of the spatial–temporal evolution of the digital economy and its impact on the employment structure in China from 2001 to 2020. Sustainability 15(12):9619. https://doi.org/10.3390/su15129619

Chapter 4
A Theoretical-Conceptual Model and Its Application in Researching Inner Peripheries

Abstract In this chapter, we present an original synthesis of the theoretical and conceptual findings outlined in previous chapters. This forms the foundation for our operational approach to identifying and characterizing the functioning of inner peripheries as applied in the empirical analysis conducted in this study. The starting point for these findings is the set of assumptions underpinning our triangulated theoretical-methodological-narrative framework. This framework integrates the concept of geographical peripherality, the theory of the Modern World System, the concept of organized proximity, the foundational economy approach, and the theory of spatial justice in a territorial dimension. These perspectives are utilized to define and explain the functioning dynamics of inner peripheries. Based on this theoretical foundation and our interpretation of inner peripheries, we outline the operational concretization of these assumptions applied in our research methodology. This includes the spatial framework for delimiting inner peripheries, the analytical framework characterizing their condition and transformations, and the role we assign to inner peripheries in addressing the challenges of development policy programming in response to the consequences of development shocks. Our findings enable us to present a theoretical-conceptual model and the assumptions for its empirical operationalization, which serve as the basis for the research methodology adopted in this study. This methodology aims to identify, classify, and characterize the functioning of inner peripheries while considering the impact of development shocks, particularly those arising from the SARS-CoV-2 pandemic.

Keywords Theoretical-methodological-narrative triangulation · Theoretical-conceptual model of inner peripheries · Stages and scope of empirical analysis

P. Churski et al., *Inner Peripheries in the New Conditions of Cohesion Post-SARS-CoV-2*, Economic Geography,
https://doi.org/10.1007/978-3-032-10936-1_4

4.1 Theoretical-Conceptual Model of Inner Peripheries: An Attempt at Original Synthesis

The identification and analysis of the functioning of inner peripheries conducted in this study require the application of selected theoretical assumptions rooted in the theoretical and methodological contributions of economics and socio-economic geography. Additionally, these assumptions are grounded in the evolving paradigm of Cohesion Policy under the influence of successive development shocks, including the consequences of the SARS-CoV-2 pandemic.

In our view, inner peripheries represent one category of lagging areas whose emergence is a consequence of the spatial differentiation patterns in socio-economic development processes (Fig. 4.1). The distinctiveness of inner peripheries within this category of areas lies in the fact that the loss or stagnation of their developmental potential is rooted in social factors. Developmental peripherality, whether inner or external, stems not only from limited transport accessibility and considerable time distances from economic centers but, more critically, from a restricted range of functional linkages and in their sustainable establishment being impossible or highly challenging.

Consequently, these peripheries result from both deficits in organized proximity and the influence of geographical peripherality. The characteristics of these areas include lower efficiency in the functioning of the territorial socio-economic system, reduced access to public goods and services, and, as a result, an overall lower quality of life (well-being). These conditions lead to the declining competitive potential of inner peripheries, which, in the long term, results in their permanent marginalization. However, it is important to note that, unlike external peripheries, whose socio-economic situation is determined by objective factors related to their location, the developmental situation of inner peripheries is easier to improve. Interventions

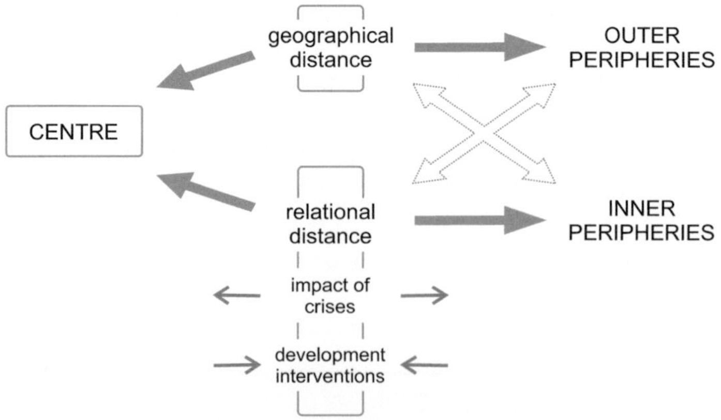

Fig. 4.1 Regularities in the formation of internal and external developmental peripheries. *Source* Own elaboration

in this area should focus on improving the connections between these areas and development cores, as well as on integrating them into value chains that are evolving under the influence of global development shocks. These efforts must also align with contemporary Cohesion Policy objectives, which aim to reduce developmental disparities in accordance with the principle of promoting spatial justice. Additionally, they should protect elements of economically weaker areas' foundational economy. This is particularly crucial during periods of intense global development shocks, such as the consequences of the SARS-CoV-2 pandemic, which exacerbate spatial development disparities and create socially unacceptable outcomes. Under these conditions, development interventions targeting inner peripheries should primarily aim to enhance their long-term adaptability and build resilience to the consequences of future global crises.

We propose studying inner peripheries based on a theoretical-conceptual model that emphasizes key theoretical assumptions as well as the spatial and analytical frameworks used in identifying and characterizing the functioning of inner peripheries (Fig. 4.2).

Theoretical Assumptions

The proposed approach to defining inner peripheries necessitates the use of selected theoretical assumptions from various conceptual frameworks to explain the patterns

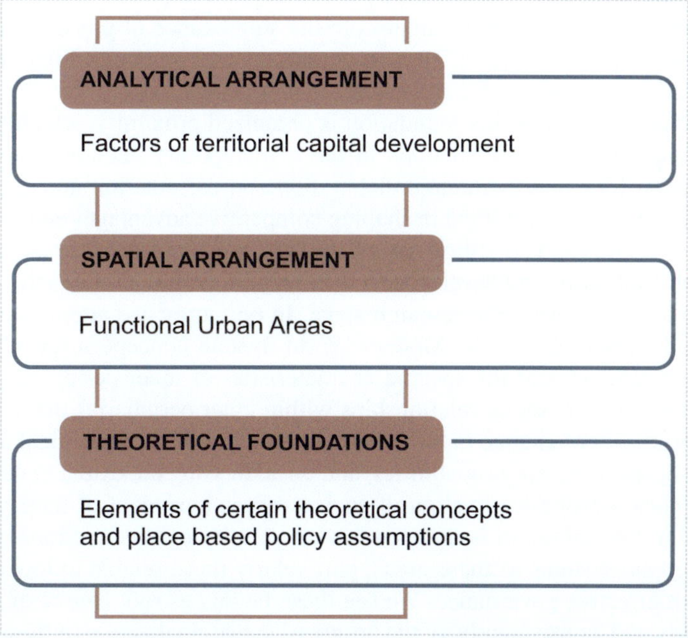

Fig. 4.2 General structure of the theoretical-conceptual model of inner peripheries. *Source* Own elaboration

of their emergence and transformations. Our approach aligns with the research pluralism observed in contemporary science, including in the field of geographical studies. This pluralism is characterized by an eclectic adoption of concepts from diverse scientific theories to comprehensively explain the phenomena under investigation. By adopting this approach, we establish a theoretical foundation underpinning our proposed theoretical-methodological-narrative triangulation. We believe this framework offers an opportunity for a more comprehensive explanation and understanding of the role of inner peripheries within the contemporary spatial disparities of socio-economic development.

The delimitation of inner peripheries and the explanation of the conditions influencing their functioning cannot be achieved without considering the interwoven assumptions of contemporary concepts, including geographical peripherality and organized proximity, which help in understanding the impact that physical and relational accessibility has on the formation of spatial disparities in socio-economic development. We argue that accessibility, understood as a consequence of both physical proximity and organized proximity, is crucial for the identification and characterization of inner peripheries. Despite the often-emphasized specificity of inner peripheries, we maintain that their delimitation and functioning remain significantly influenced by their geographical distance from core areas. Our explanation of these patterns draws upon the rich body of work in location theory, growth pole theory, and core-periphery theory, highlighting the importance of agglomeration effects and scale effects determined by accessibility in shaping developmental processes within spatial contexts. These theories underscore the significance of physical distance in these relationships (Sect. 2.2.1). Thus, geographical proximity constitutes one of the two foundational elements of the concept of inner peripheries.

The second element of this foundation is organized proximity, which highlights the consequences of transformations in the contemporary economy and society, where relationships that facilitate collaboration and the creation and exchange of knowledge are becoming critical in shaping competitive advantages and influencing the living standards and conditions of inhabitants. We consider these factors fundamental to the formation and development of inner peripheries, although they are challenging to identify during the research stage. In our study, we attempt to leverage and adapt the principles of the Modern World System concept at the subregional level, taking into account the specific characteristics of inner peripheries. Specifically, we focus on the social relationships within inner peripheries that result from complex processes of change in the economy, demographics, or political conditions that shape decision-making capabilities, and on addressing the extent to which these affect the socio-economic situation of such areas, independent of their geographical location. Our attention is directed toward identifying dysfunctions within the socio-economic systems of these areas, particularly those related to low efficiency or a lack of effective governance. We see these factors as root causes of relational peripherality and, consequently, of the progressive marginalization of these areas. In this context, the theoretical framework of organized proximity is particularly valuable, as it helps us better understand the role of relational proximity in shaping spatial disparities in socio-economic development. In our view, organized proximity emphasizes the importance of the ability to create and maintain interactions, both within the communities inhabiting a given area and between these communities and their external environment. These abilities are crucial for fostering the everyday collaboration needed to address ongoing challenges and for establishing external relationships that integrate these areas into evolving supra-local networks of economic and social interdependencies. Such capacities are shaped by the logic of belonging and the logic of similarity (Sect. 2.2.1). By considering these perspectives, we aim to incorporate the specific characteristics of inner peripheries into our research, including a broad interpretation of the peripheralization process, which is not solely dependent on geographical location. However, this presents a significant challenge to the empirical operationalization of these assumptions.

The analysis of the functioning of inner peripheries and the identification of their needs regarding the programming and implementation of developmental interventions aimed at achieving a socially acceptable level of spatial development disparities requires the incorporation of additional elements into the theoretical foundations of the proposed theoretical-conceptual model. Our choices are driven by the necessity to include theoretical components that can explain the patterns associated with contemporary transformations in socio-economic development processes. These transformations involve broadening the conceptual scope of development and redefining its goals toward improving the quality and conditions of life for residents, rather than solely maximizing the level and growth rate of economic output. This shift creates the basis for formulating alternative concepts, which, in our view, should be employed for the identification and analysis of the functioning of inner peripheries. We argue that, in this context, the foundational economy concept and the spatial justice concept are particularly useful.

The functioning of inner peripheries is, in our view, largely based on the utilization of foundational economy infrastructure, which enables the provision of goods and services consumed universally. The importance of this segment of resources and activities to the economy of inner peripheries is underscored by the principles of the foundational economy concept. We believe these principles should be integrated into the theoretical framework that facilitates the explanation of the condition and transformations of this category of lagging areas. This integration is crucial for understanding the role of the foundational economy within the structure of public services that shape the living standards and conditions of residents in inner peripheries. Additionally, it helps explain the potential for shaping their developmental trajectories. It is also essential for grasping the significance of this segment of their economy in building resilience to crises, including the consequences of the SARS-CoV-2 pandemic (Sect. 2.2.1). In our opinion, this justifies the need to incorporate the principles of this theory into the structure of the proposed theoretical-conceptual model of inner peripheries and to utilize them for understanding and explaining the functioning of these areas.

Furthermore, based on the assumption that developmental interventions aimed at reducing spatial disparities in socio-economic development to a socially acceptable level must consider contemporary shifts in development policy objectives shaped by successive global shocks disrupting the functioning of territorial socio-economic systems at all spatial scales (Chap. 3), we emphasize the need to equalize opportunities and developmental prospects rather than outcomes. To this end, we integrate the spatial justice theory into our theoretical-conceptual model of inner peripheries. This framework provides insights into the position of inner peripheries in contemporary discussions on spatial inequalities in development and their social consequences. We regard this as particularly significant given the exacerbation of these disparities under the influence of successive crises, including the effects of the SARS-CoV-2 pandemic, which have led to heightened social dissatisfaction. Accordingly, in our theoretical assumptions, we aim to address the need to ensure equal developmental opportunities for all residents in every area. We believe it is necessary to apply the principles of distributive justice, procedural justice, and intergenerational justice in evaluating the functioning of inner peripheries. In our view, this approach facilitates a better understanding of the focus of developmental interventions dedicated to these areas, which should be determined by the condition and structure of territorial capital and never implemented independently of identified contextual conditions. To meet these needs and leverage the principles of spatial justice, we propose the application of a spatial minimum standard as a solution that ensures, on the one hand, developmental opportunities for every resident and, on the other hand, limits economically inefficient public spending on pro-development initiatives in locations devoid of development potential (Sect. 2.2.1).

Finally, against the backdrop of the theoretical foundation of inner peripheries, defined both in terms of structure and scope, we aim to present the key principles of the contemporary paradigm of the EU Cohesion Policy, which is undergoing fundamental transformations under the influence of development shocks such as the SARS-CoV-2 pandemic (Chap. 3). In our view, this allows us to highlight the growing importance of inner peripheries in the programming and implementation of developmental interventions, driven by the alignment of their specific needs with the evolving focus of pro-development activities. In this context, we emphasize the significance of a territorially oriented approach in effectively and efficiently programming and implementing interventions aimed at building the long-term adaptability and resilience of specific territories to the impacts of successive global development crises. We argue that place-based policy is the most appropriate approach for addressing developmental interventions targeted at inner peripheries. By adopting the concept of territorial capital, which forms the analytical framework used to characterize the functioning of inner peripheries, we find no justification for the emerging proposition in the literature to shift from place-based to people-based policy. In our view, territories inhabited by residents who utilize resources and create relational developmental environments become, through the agency of these residents, the subjects of intervention. In our analysis of the functioning of inner peripheries, we focus on the manifestations of an integrated approach to development policy for these areas and the use of ad-hoc intervention instruments directed at them during crises. However, we also emphasize the critical importance of these actions in shaping these areas' long-term adaptability and resilience to the consequences of future global crises. The focus on peripheral areas in terms of creating conditions for their long-term development must, in our view, align with the strategic objectives of Cohesion Policy regarding convergence, as well as the long-term cohesion and integrity of the EU.

Based on this structured framework, we build the theoretical foundation for identifying and explaining the functioning of inner peripheries and for determining the direction of developmental interventions targeted at them (Fig. 4.3).

Spatial Framework

Based on the theoretical foundations constructed to provide a better understanding of the functioning of inner peripheries, we propose adopting a new approach to the

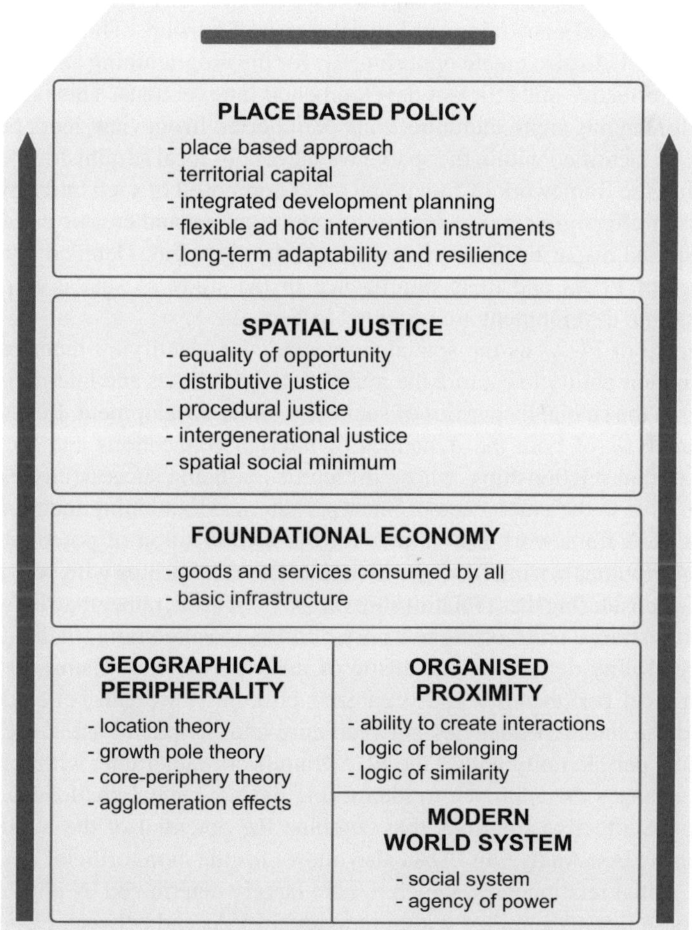

Fig. 4.3 Theoretical-conceptual model of inner peripheries: theoretical foundations. *Source* Own elaboration

spatial framework for their delimitation. The spatial framework is treated as the second fundamental element of our theoretical-conceptual model of inner peripheries. We believe that the primary spatial framework that should be used for analyzing spatial disparities in socio-economic development is the supra-local FUA framework.

A supra-local functional approach that transcends administrative boundaries allows for the identification of the scope of spatial relationships and socio-economic activities, which often extend beyond the formal borders of local administrative units. This approach forms the basis for fostering potential opportunities for cooperation among units within functional areas and for enabling their effective interaction both vertically and horizontally. This leads to the conclusion that, from empirical and applied perspectives alike, FUAs represent potential tools for intervention. They enable the aggregation of the capacities of individual administrative units and the optimal management of territorial capital resources, thereby increasing the chances of overcoming local constraints and developmental barriers. Through supra-local collaboration, FUAs also create opportunities for the programming and implementation of more effective and efficient developmental interventions. This is particularly important for lagging areas, including inner peripheries. In our view, inner peripheries should not be identified within the spatial framework of local administrative units but rather within the framework of functional areas composed of such units. We believe this approach offers greater developmental opportunities and creates conditions for reducing spatial disparities in socio-economic development. Detailed insights into the concept of FUAs and their significance in the study of spatial disparities in socio-economic development are provided in Sect. 2.2.1.

The choice of FUAs as the spatial framework for identifying inner peripheries stems from their ability to capture the multifaceted processes and interdependencies that underpin the spatial disparities of socio-economic development. FUAs facilitate effective analysis of both the dynamics of internal connections and the development of external relationships, whose difficulties in being successfully established often contribute to the emergence of inner peripheries. Examining inner peripheries within the FUA framework also allows for the identification of potential developmental opportunities arising from the dynamics of relationships with core areas. We believe that considering these relationships at the FUA level, rather than between individual administrative units, offers new possibilities for overcoming challenges related to the accessibility deficits characteristic of inner peripheries. Using the concepts of geographical peripherality and organized proximity, we can better define and understand the interdependencies between core and peripheral areas. Considering geographical peripherality within the FUA framework rather than within individual territories changes the approach to identifying developmental challenges, enabling us to propose effective solutions that combine the potential of the administrative units within FUAs. Analyzing FUAs also allows for the monitoring of key resource flows and spatial relationships, which remain largely determined by physical accessibility. These flows, examined within individual FUAs and between FUAs of inner peripheries and those of core areas, are evaluated based on growth pole theory and core-periphery theory, emphasizing agglomeration effects. Both infrastructural and organizational conditions, crucial for mitigating the effects of geographical peripherality, are considered. Addressing these factors within the FUA framework, rather than within individual administrative units, highlights new opportunities for overcoming existing limitations in this regard. A key advantage of using FUAs as the spatial framework for delimiting inner peripheries is their greater ability to identify social and economic interactions through the concept of organized proximity. FUAs offer a better understanding of how the logic of belonging and the logic of similarity shape the developmental dynamics of inner peripheries, particularly in the context of building supra-local relationships and utilizing governance capabilities. Analyzing spatial dynamics within FUAs both within inner peripheries and between them and more developed FUAs helps identify the deficits that limit their ability to fully participate in the socio-economic processes shaping their developmental trajectories. The lack of functional connections with more developed cores contributes to the marginalization of these areas. We fully agree with the recommendations in recent Cohesion Reports, which emphasize the need to target developmental programming and interventions at the FUA level. In our view, FUAs are particularly significant for increasing the effectiveness and efficiency of territorially oriented development policy (place-based policy). They offer a spatial perspective that incorporates territorial capital and the specific local resources and socio-economic conditions, creating new opportunities for programming integrated developmental actions. These actions are more effective and efficient at the supra-local FUA level than at the local level of individual administrative units. Identifying developmental barriers and uncovering dormant or lost potential within FUAs enables the use of flexible intervention tools that can strengthen the long-term adaptability and resilience of inner peripheries. Taking into account the territorial capital of FUAs based not only on local resources but also on the relationships between cores and surrounding areas opens new possibilities for reducing spatial disparities in socio-economic development. This approach facilitates better use of local resources and a more targeted implementation of intervention measures, creating conditions for effectively addressing developmental disparities, which are increasingly pronounced and consequential at the subregional level. In our opinion, this represents a more realistic path toward achieving territorial justice, which aims to ensure equal developmental opportunities for residents of peripheral areas. FUAs provide a more appropriate framework than the local level for assessing to what extent residents of these areas have access to foundational economy resources and services, such as education, transportation, and healthcare, which are critical for shaping developmental opportunities for every resident. Our recommendations for implementing a spatial social minimum, outlined in the concluding recommendations of this book, are also based on the FUA framework. In our view, this approach allows for the precise identification of the basic needs of peripheral residents while also enabling the formulation of strategies and actions to shape a spatially just development policy.

Operationally, for the purposes of our analysis of inner peripheries, we defined FUAs as territorially contiguous areas encompassing a central city and at least one subordinate municipality connected to it through population flows. In defining population flows, we considered commuting for work and bidirectional migration. We applied a travel-time isochrone criterion of 1 h to the central city, extending this to 1.5 h for FUAs with capitals in the largest cities. The inclusion of a municipality in an FUA was based on an analysis of the strongest connections, derived from intermunicipal matrices of migration and commuting patterns. Using this methodology, we identified 413 FUAs in Poland, which together fully covered the country's territory. An analysis of the socio-economic characteristics of these areas, focusing on traits that describe inner peripheries, allowed us to delimit this category of lagging regions and develop their typology.

Analytical Framework

The third fundamental element of our theoretical-conceptual model is the analytical framework, which serves as the basis for characterizing the functioning of inner peripheries. We believe that the most appropriate and comprehensive approach to analyzing the condition and transformations of socio-economic development processes in inner peripheries is provided by the concept of territorial capital, which we adapt in our study using a modified systematization of development factors. We argue that territorial capital, understood as a unique, immobile, or difficult-to-replicate resource socially created based on the development factors of a given area through functional relationships (both interpersonal and interinstitutional), constitutes the most effective analytical framework for inner peripheries. Importantly, territorial capital should not be viewed as a simple sum of resources that serve as development factors for a given area. Instead, it is created by residents who, through their knowledge, skills, and competencies, shape a relational development environment by inhabiting, utilizing, and governing the territory. In this way, the area transcends being merely a physical substrate for developmental processes and becomes a relational environment for those processes. The FUA spatial framework we have adopted is particularly suited to the proposed system for analyzing the condition and transformations of development processes. Focusing on the importance of relationships, which underpin the functional approach, allows us to move beyond a simple inventory of development factors at the local level. This approach offers several advantages:

1. **Beyond individual local administrative units** We do not limit our analysis to the resources of individual local administrative units but consider the potential for their complementary utilization.
2. **Relational dynamics** We emphasize the importance of relationships within the internal structure of an FUA (core–surrounding areas within the FUA categorized as inner peripheries) and between FUAs (those categorized as inner peripheries and other FUAs) in shaping developmental conditions and creating a relational development environment.
3. **Addressing developmental deficits** We seek to identify the causes of developmental deficits and explore opportunities for leveraging developmental potential at the supra-local level, overcoming local development constraints.

Our inquiry is organized around development factors that, in our view, best characterize the developmental situation of inner peripheries. This structure enables us to analyze and understand the state and dynamics of inner peripheries effectively while identifying actionable strategies for their development.

The selection of the development factors for inner peripheries that form the structure of our analytical framework was purposeful. It directly relates to the territorial characteristics of these areas, emphasizing both their developmental determinants and barriers that shape their current state and changes in their developmental situation, organized into subfactors describing dimensions of their developmental condition (Table 4.1). We consider accessibility a key factor for the development of inner peripheries, particularly emphasizing the distinction between internal transport accessibility and external transport accessibility within the FUA framework. Additionally, we highlight the growing importance of e-accessibility, which is especially critical for inner peripheries, as it mitigates the effects of geographical peripherality and strengthens the conditions for organized proximity. The most critical development factor is, in our view, human capital. Its significance is fundamental in our conceptualization of territorial capital as the space within FUAs created by residents and serving as a relational development environment. In specifying this factor, we focus on demographics, health, safety, and education, as these directly influence the quality of human capital and its ability to shape the relational space that determines developmental processes in the inner peripheries we study. We also consider the labor market situation in these areas, which reflects the degree to which human capital within FUAs is engaged in economic processes. Complementing human capital is the third development factor we adopt: social and institutional capital. Its role in the development of inner peripheries is unique due to its importance in shaping relational connections, the foundation of organized proximity. A deficit or dysfunction in these connections is a significant cause of the formation of inner peripheries. Given the challenging measurability of this factor, we examine social activity and leisure time usage to gauge the state of social capital. For institutional capital, we analyze business infrastructure, economic structure, and activity to assess its condition. These considerations refine our framework for identifying the subfactors of social and institutional capital in inner peripheries. The remaining two development factors in our analytical framework are financial capital and material capital. Financial capital is considered a classic development factor. Its condition determines the ability of inner periphery residents to meet their needs and thus influences their developmental opportunities. We analyze this factor through the financial status of local government units, businesses, and residents, enabling us to evaluate the financial condition and developmental potential of the studied inner peripheries. Material capital is also classified as a traditional development factor and is considered the most classic for inner peripheries. Its condition, particularly the extent of deficits, is a fundamental element in identifying their developmental constraints. In detailing this factor, we examine land-use quality to assess spatial development conditions that create a relational development environment. We also consider housing conditions to determine the extent to which basic housing needs are met. Additionally, we analyze ecosystem goods and services, treating them as a significant subfactor of material capital whose importance is growing due to global transformations in developmental processes and their conditions. The application of this analytical framework to describe the developmental situation of inner peripheries, focusing on their deficits, is summarized in Table 4.1. Detailed considerations regarding the concept of territorial capital and its analysis using selected development factors for characterizing the socio-economic situation of inner peripheries are provided in Sect. 2.2.3.

Operationally, for the purposes of our analysis of inner peripheries, the analytical framework of territorial capital is applied at two stages of the empirical study: the national level and the case study level. At the national level, this framework is used to organize the assumptions of the FUA typology, based on quantitative data from public statistics describing the socio-economic characteristics of these areas. At the case study level, it is applied to structure the identification and characterization of the developmental trajectories of inner peripheries, also conducted using data from public sources. The proposed analytical framework is further utilized as the foundation for in-depth case study research, which relies on original and extensive source material collected during fieldwork conducted in four selected FUAs representing first- and second-degree inner peripheries in Poland. At every stage of the empirical study, we assume that the structure of the relational territorial capital of inner peripheries, generated by the residents of these areas, should be analyzed within the framework of five development factors: accessibility, human capital, social and institutional capital, financial capital, and material capital. Wherever possible, depending on data availability, these factors should be detailed with subfactors that highlight the distinctive characteristics of these areas.

4.1.1 The Empirical Application of the Theoretical-Conceptual Model of Inner Peripheries

The theoretical-conceptual model of inner peripheries presented by us serves as the foundation for the empirical research the results of which are showcased in this book. The application of the adopted assumptions regarding the theoretical foundations, spatial framework, and analytical framework at every stage and step of the research process enabled the achievement of the empirical, methodological, and application-oriented objectives outlined in this study (Table 4.2).

The process of empirical analysis aimed at achieving empirical, methodological, and applicational objectives allowed for the verification of the research hypotheses formulated in the study (Table 4.3).

We divided the empirical analysis into two stages and a sequence of six research steps.

In the first stage of empirical research conducted at the national level, we implement the first two research steps, which involve the delimitation and typology of inner peripheries at the national level in Poland. These steps are based on an original approach to identifying and typologizing inner peripheries. This approach relies on

Table 4.1 Developmental dimensions of inner peripheries within the framework of selected development factors (components of territorial capital) and subfactors of development

Factors/Subfactors	Examples
Accessibility • Internal transport accessibility • External transport accessibility • E-accessibility	• Insufficient and low-quality transport infrastructure, lack of spatial mobility • Absence of public transport connections at the supra-local (FUA) and regional levels • Long travel times to growth and development centers • Inadequate access to broadband Internet • Limited access to public services
Human capital • Demographic situation • Health and safety • Education • Labor market	• The depopulation process combined with societal aging • Insufficient access to healthcare services • Limited access to quality education • Labor market mismatches • Low workforce participation • High unemployment rates, particularly long-term unemployment
Social and institutional capital • Social activity • Leisure time utilization • Business infrastructure • Economic structure and activity	• Relatively high levels of bonding social capital combined with a deficit in bridging social capital • Low levels of social activity • Dominance of micro and small enterprises • Deficient access to business support services • Insufficient supra-local collaboration among informal and formal institutions
Financial capital • Financial situation of local government • Financial situation of enterprises • Financial situation of residents	• Limited financial resources and flows across all groups of entities (local government units, entrepreneurs, residents) • Poor financial situation of local government units, resulting in low creditworthiness • Poor financial condition of enterprises, leading to insufficient investment resources among local entrepreneurs • Low-income levels and low consumption among residents
Material capital • Quality of land use • Housing conditions • Ecosystem goods and services	• Lack of economic foundations for initiating developmental processes (resource deficits and poor provision of technical and social infrastructure) • Low quality of land use and lack of spatial planning • Insufficient housing stock and low growth dynamics of this stock • Access to green areas, including forests and protected zones • Environmental challenges, such as high levels of air pollution, particularly during the winter season, caused by low emissions

Source Own elaboration

Table 4.2 Empirical, methodological, and application-oriented objectives

Objectives	Symbol
Empirical goals	
Delimitation and typology of spatial development disparities within the framework of Functional Urban Areas, leading to the identification of inner peripheries	EG1
Determination of the developmental trajectories of inner periphery areas in relation to more developed areas, with particular emphasis on their vulnerability and resilience to the SARS-CoV-2 crisis and its socio-economic consequences	EG2
Determination of the intensity of the diffusion of developmental processes in spatial terms between inner periphery areas and urban centers serving as subregional hubs of functional areas, as well as the factors facilitating and hindering these processes	EG3
Determination of the susceptibility of individual industries and economic sectors to the shocks of the SARS-CoV-2 crisis, along with an assessment of the observed and anticipated economic and social consequences	EG4
Identification of developmental factors and barriers shaping socio-economic transformations that influence the living standards and conditions of residents in selected inner peripheries, considering the consequences of the SARS-CoV-2 crisis shock	EG5
Methodological goals	
Preparation of a model indicator system enabling the identification of inner periphery areas and monitoring the developmental processes occurring in these areas	MG1
Development of principles and criteria for the delimitation and typology of inner periphery areas	MG2
Development of principles and criteria for identifying the developmental trajectories of inner periphery areas	MG3
Preparation of assumptions and testing of a monitoring system for socio-economic changes in inner periphery areas	MG4
Application-oriented goals	
Formulation of recommendations regarding the orientation and implementation methods (including the selection of instruments) of Cohesion Policy developmental interventions in inner periphery areas, with particular emphasis on the role of their urban centers in creating conditions for networking and the diffusion of developmental processes	AG1

Source Own elaboration

the spatial structure of FUAs delineated based on population flows related to migration and commuting. The spatial structure of FUAs serves as the foundation for identifying inner peripheries at a supra-local level, showing the structure of subregional socio-economic development disparities. The typology and characterization of the spatial units identified as inner peripheries are based on an analytical framework utilizing selected development factors. These factors are treated as components of territorial capital, which represents the relational development environment created by the inhabitants through the engagement of their resources in development processes. The selection of factors is deliberate and linked to their potential use in

Table 4.3 Research hypotheses

Hypotheses	Symbol
The inner peripheries identified within the system of functional areas are characterized by varying sizes of both core cities and their surrounding areas	H1
The development trajectories of inner peripheries differ from those of more developed areas and exhibit greater sensitivity to the socio-economic consequences of the SARS-CoV-2 crisis	H2
The spatial spread and diffusion of development processes between inner peripheral areas and urban centers serving as subregional hubs of functional areas are constrained by factors unfavorable to these processes, which have gained prominence as a result of the SARS-CoV-2 crisis	H3
Inner peripheries demonstrate a specific vulnerability of individual industries and economic sectors to the shocks of the SARS-CoV-2 crisis, resulting in distinct economic and social consequences compared to more developed areas	H4
Social and institutional capital is the most critical factor shaping the socio-economic transformations that determine the living standards and conditions of residents in inner peripheral areas. Its importance increases under the impact of the consequences of the SARS-CoV-2 crisis shock	H5
Ensuring the effectiveness and efficiency of Cohesion Policy interventions targeted at inner peripheral areas requires a greater focus on changing implementation instruments rather than the principles of programming this policy	H6

Source Own elaboration

describing the specific development conditions of inner peripheries, as defined by the adopted definition of these areas. Determining the state and changes of development factors and the relationships between them is based on analyzing the status and dynamics of changes in indicator values that identify their characteristics. The scope of findings in this regard varies depending on data availability (Fig. 4.4). The interpretation of the results obtained from these first two research steps, facilitated by the application of the adopted theoretical-conceptual model, of inner peripheries enables the formulation of unique and highly valuable empirical findings:

The spatial structure of FUAs covers the entire territory of Poland, enabling the identification of the extents of functional relationships occurring at a supra-local level across the country.

By explaining the patterns of the obtained spatial distribution and adopting it as a basis for identifying spatial development disparities, within which we search for inner peripheries, we aim to fully utilize the findings of growth pole theory and the core-periphery model. These are based on agglomeration effects, which form the foundation of the concept of geographical peripherality. Treating FUAs as areas linked to central cities, we aim to capture functional connections while emphasizing the importance of a relational approach, including organized proximity. We identify these connections based on residents' daily interactions, providing a picture of actual spatial patterns rather than formal administrative divisions. In the delimitation procedure, the criterion of dominant population flows to the nearest city within the functional network assigns each municipality to the closest central hub based on

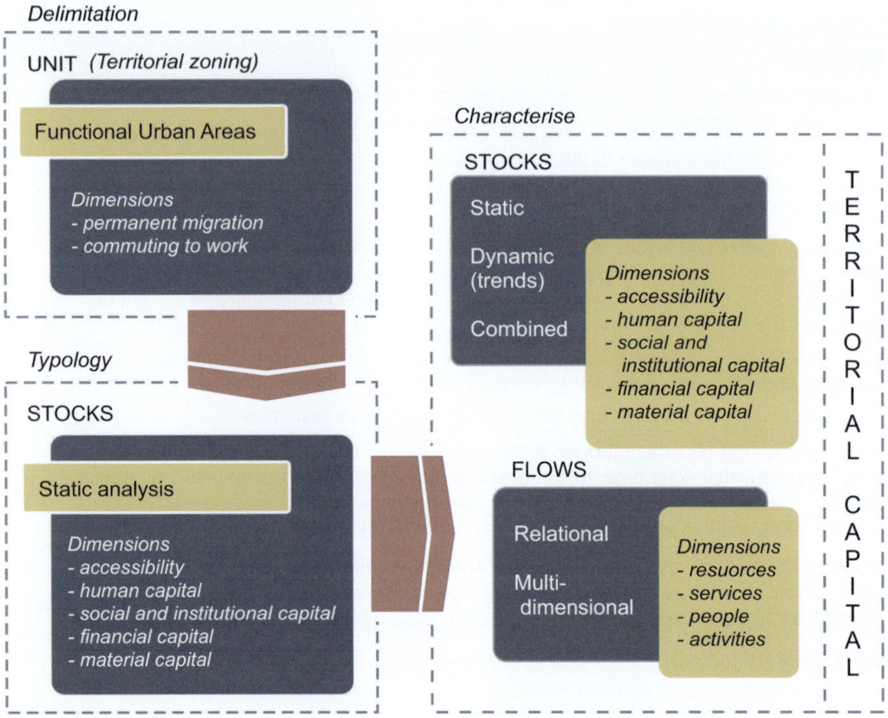

Fig. 4.4 Assumptions of the original identification and typology of inner peripheries. *Source* Own elaboration

real population flow data. This minimizes artificial extensions and unjustified spatial expansion of FUAs around large cities. Such a division enables a more balanced spatial analysis, allowing smaller urban centers to also be identified as functional hubs influencing their surrounding municipalities. The conducted delimitation, exhaustive and cohesive, allows every municipality to be included in a contiguous spatial FUA, creating conditions for improved efficiency and effectiveness in programming and implementing place-based development policies. It also shapes opportunities for strengthening territorial equity across all areas. Adopting the assumption of travel-time isochrones to the central city enables a more precise definition of areas with intense functional ties, reflecting practical opportunities for integration and socio-economic interactions among residents. We assume that areas closer to central hubs are better positioned to effectively utilize resources, enhancing their adaptability to crisis impacts and fostering long-term resilience. This approach also aids in identifying deficits in the foundational economy, which influences residents' living standards and conditions while playing a crucial role in shaping territorial resilience. Our methodology allows for the analysis of all basic local units, regardless of their geographic location or degree of functional connectivity. Including all municipalities provides a more comprehensive view of spatial inequalities, facilitating better explanations of the causes and dynamics of peripheralization and marginalization processes. In analyzing the resulting distribution of FUAs, we pay particular attention to the diversity in the sizes of the delimited areas and their central cities, attempting to identify the dependencies stemming from their development conditions, which are shaped by both infrastructure levels and the potential for creating and strengthening organized proximity. In this regard, we utilize core-periphery theory assumptions and a notion of proximity that is expanded to incorporate its relational dimension to explain the principles governing the obtained spatial distribution. We also consider the alignment of the spatial distribution with Poland's administrative divisions, drawing conclusions based on the principles of place-based development policy. This leads us to propose adjustments to the supra-local administrative division and the scope of competencies for supra-local administrative units, which form part of our final recommendations summarizing the study's findings. Our adopted delimitation procedure has resulted in the creation of an analytical tool that provides a fuller understanding of the spatial conditions influencing development processes. This tool can be used to identify inner peripheries effectively.

Our typology of FUAs provides a new perspective on the spatial development disparities in Poland, within which we identify inner peripheries. By adopting the spatial structure of FUAs, we emphasize the patterns of development processes at a supra-local level, considering relational connections. We aim to explain these patterns using the concepts of geographical peripherality, which focus on physical distance and core-periphery relations, and organized proximity, which highlights new possibilities arising from an expanded interpretation of proximity. To identify developmental types of FUAs and pinpoint inner peripheries, we apply our custom definitions, which also form the basis of the analytical framework used. Developmental types are derived using a framework based on five development factors, viewed as components of relational territorial capital, with indicators reflecting the distinctive characteristics of inner peripheries. The resulting distribution of developmental types is interpreted using our theoretical-conceptual model, with two types classified as inner peripheries. Within this category, we distinguish two subtypes: first-degree inner peripheries with relatively worse developmental conditions and second-degree inner peripheries with relatively better conditions. We focus on differences across developmental types of FUAs, paying particular attention to inner peripheries. Variations in access to agglomeration benefits and the ability of residents to meet basic needs, influenced by resource disparities, are interpreted through internal and external transport accessibility and e-accessibility. These factors affect the ability to mitigate geographical peripherality and foster organized proximity. Significant deficits in these areas are crucial in identifying inner peripheries. Leveraging the insights of the Modern World System theory, we highlight the role of social and institutional capital in shaping spatial development processes. When classifying specific FUAs as core, intermediate, or inner periphery areas, we analyze the socio-economic situation, focusing on the level of social activity as an indicator of institutional deficits that limit the effectiveness of undertaken actions. Differences in access to basic services across FUA developmental types are interpreted as indicators of the state of the foundational economy in these territories. Identified deficits in this area, which affect the living standards and conditions of residents, serve as criteria for distinguishing inner peripheries. Financial and material capital is assessed in terms of its ability to ensure developmental opportunities for residents, which, according to the concept of territorial justice, should be guaranteed at a basic level for all. These resources also shape territorial adaptability and resilience to change. Notable deficits in this regard within certain FUA developmental types justify their classification as inner peripheries. The new perspective on spatial development disparities in Poland that emerges from our analysis underscores the importance of inner peripheries at the supra-local level and expands upon existing empirical findings in this area. However, it should be noted that the delimitation and typology of inner peripheries that are based on the first two research steps rely on quantitative data, which significantly limits the granularity of the findings and necessitates further refinement through case studies.

In the second stage of empirical research conducted at the case study level, we carry out the remaining four research steps, which involve: selecting case studies, analyzing development trajectories, conducting an in-depth analysis of case studies using qualitative data, and formulating recommendations and guidelines for development policies dedicated to inner peripheries.

The selection of case studies was based on the identification of FUAs classified as first- and second-degree inner peripheries that meet the criterion of representativeness for their developmental type. We selected two case studies for each type of inner periphery, identifying their socio-economic situation using the adopted analytical framework based on five development factors treated as components of the relational territorial capital of these areas. Our focus was on the characteristics included in the adopted definition of inner peripheries. Particular importance was given to the broader concept of accessibility, which influences the state of geographical peripherality and organized proximity and their impact on developmental processes. For the latter, we also considered the state of human capital, as well as social and institutional capital. We also took into account the diversity of financial and material capital, which conditions the fulfillment of basic needs for residents. This consideration stems from its role in determining the state of the foundational economy and in creating developmental opportunities for the population, shaping conditions for territorial equity. Additionally, we considered the size diversity of FUAs and their central cities, as well as their locations in various historical and geographical regions of Poland, which directly affect developmental conditions. This approach allowed us to select diverse case study areas. The in-depth analysis of these areas not only provided a better understanding of the dynamics of developmental processes occurring in inner peripheries but also served as a foundation for formulating recommendations for territorial development policies targeted at these lagging areas, accounting for their internal diversity.

In the fourth research step, we focused on identifying the development trajectories of inner peripheries, structured according to the adopted analytical framework. The aim was twofold: on the one hand, to determine the course of developmental transformations in the two identified types of FUAs classified as first- and second-degree inner peripheries relative to the trends observed in FUAs belonging to developmental types with better socio-economic conditions. On the other hand, we directed our attention to selected case study areas, aiming to identify their unique developmental characteristics. By analyzing changes in the values of indicators representing the five components of territorial capital over a short time period 2017–2023 (which included the period of the pandemic, the years immediately preceding it, and those following it), we sought to verify the assumption of greater susceptibility of inner peripheries to the developmental shock caused by the pandemic compared to central areas. Confirmation of this would be a divergence in the levels of indicators describing the quality of individual components of territorial capital between peripheral, core, and intermediate areas. At the same time, we sought to interpret the impact of local development factors specific to the individual FUAs selected as case studies on the distinctive transformations occurring in them, as compared to the broader group of peripheral FUAs.

The fifth research step pertains to a comprehensive field study program conducted across four case studies of inner peripheries. This included an analysis of existing data, such as those contained in: strategic, programmatic, and planning documents that detail the challenges and development priorities of the studied areas; surveys of residents; in-depth interviews with local government leaders; and focus group discussions with entrepreneurs, local officials, and community leaders. All tools used for this research were structured within the adopted analytical framework, considering five development factors as components of relational territorial capital. The qualitative field research provided an opportunity to determine the significance of the studied factors in shaping the relational development environment of the analyzed territories, which could not be fully captured during the quantitative research phase. Particular attention was paid to changes in accessibility that influence the diffusion of development processes, the adaptability and resilience of the local economy, and the transformations in development factors that affect the living standards and conditions in the studied inner peripheries as a result of the SARS-CoV-2 pandemic shock. In assessing the situation of the studied inner peripheries, we examined the impact of geographical peripherality on their development, determined by barriers related to distance from economic centers and limitations in generating agglomeration effects. Special attention was given to changes in these aspects related to the pandemic. Interviews with local government representatives and entrepreneurs offered insights into how functional isolation limits access to markets and public services, posing significant challenges for inner peripheries in terms of crisis response and adaptation. This analysis facilitated the identification of key actions that could improve the inner cohesion of FUAs and foster stronger connections with FUAs in more developed regions, thereby enhancing their adaptive response to future developmental shocks. The findings from direct research were also aimed at identifying the capacity of the studied areas to create conditions for organized proximity. We sought to explore the potential for building supra-local networks of collaboration and strengthening interactions among diverse social and economic groups. This is critical for reinforcing organized proximity, which mitigates the impact of geographical peripherality on development processes and is equally important for fostering adaptability to developmental shocks. We attempted to identify barriers to the formation of the logic of belonging and the logic of similarity, highlighting how they affect difficulties in building support networks and strengthening social relationships. This, in turn, hinders the developmental potential of inner peripheries. We also examined how these barriers have evolved due to the SARS-CoV-2 pandemic. This approach enabled us to identify potential mechanisms for supporting the relational inclusion of inner peripheries in connections with more developed regional structures, which is vital for strengthening their autonomous development and building long-term resilience to shocks. By adapting the principles of the Modern World System theory, we analyzed peripherality from the perspective of agency and resources available to inner peripheries. By assessing the decision-making autonomy of these areas relative to core regions, we identified potential political strategies to enhance local and supra-local agency and their ability to actively shape their future within broader socio-economic structures, quickly mitigate shocks, and build long-term resilience. Through interviews and surveys, we gathered opinions on access to and the quality of basic services such as education, healthcare, social services, transportation, and housing. Identified deficiencies in foundational economic infrastructure were interpreted using the foundational economy concept, emphasizing their impact on residents' living standards and conditions, as well as their role in reducing the vulnerability of these areas to crises. We also used the field research to gain insights into residents' perceptions of developmental opportunities and changes in this regard under the effects of the SARS-CoV-2 pandemic. We focused on how residents perceive distributive justice (equal access to resources and public services) and procedural justice (opportunities to participate in decision-making processes), which enhances their agency in programming and implementing development actions. Deepening this aspect, we aimed to identify unique developmental challenges and barriers in the studied territories. Our exploration aimed to establish the foundations for place-based development actions tailored to the specificities of these areas. During interviews, we emphasized identifying current capabilities and future needs for flexibly adapting development interventions to local realities, considering the diverse socio-economic contexts of the studied inner peripheries. We assessed the extent to which current policies and intervention tools effectively address the specific needs of these areas and identified changes that could support their long-term development and integration. This approach also helped us recognize the varied consequences of the pandemic across the identified types of peripheries and assess how current intervention tools meet local needs under the impact of developmental shocks. To identify optimal approaches, we inquired about supra-local collaboration and its importance in combining potentials and overcoming barriers, both in daily conditions and during the unique pressures of the SARS-CoV-2 pandemic. We also gathered conclusions and opinions on modifying public policies, particularly regional policies, to ensure their increased effectiveness and impact in supporting the long-term stability and adaptive capacity of inner peripheries in the face of future crises.

The final, sixth research step is of a synthesizing and validating nature and focuses on formulating recommendations and guidelines for development policy. Its findings, based on the results of our research conducted on the inner peripheries' FUAs in Poland, are addressed to all levels of Cohesion Policy programming and implementation, from the European level to regional and local levels. The results of our research provide a strong foundation for formulating policy recommendations aimed at improving the effectiveness and efficiency of Cohesion Policy interventions targeted at inner peripheries. These recommendations are intended to strengthen sustainable development, enhance resilience, and improve the quality of life for residents. In detailing these recommendations, we fully utilize the assumptions of our adopted theoretical-conceptual model. Particular emphasis is placed on the need to improve the effectiveness and efficiency of Cohesion Policy interventions by highlighting the unique opportunities for programming and implementation using our spatial FUA framework. Our recommendations consider the necessity of reducing the impact of geographical peripherality on contemporary spatial development disparities, including ongoing processes of peripheralization and marginalization, as evidenced by our research findings. Using insights into the current state and the need for changes in conditions that foster organized proximity, we propose specific Cohesion Policy interventions to strengthen the creation and functioning of supra-local networks of collaboration and to foster interactions between diverse social and economic groups. We highlight the importance of these actions in not only improving the current developmental situation of inner peripheries but also building their long-term adaptability and resilience to crises. Significant attention is given to supporting the foundational economy within the framework of Cohesion Policy, emphasizing its role in determining living standards and conditions while also noting its impact on building resilience and adaptability of inner peripheries in the face of global crises. Recognizing the need to pursue territorial justice, our recommendations advocate for shifting Cohesion Policy interventions from equalizing outcomes to equalizing developmental opportunities. In this context, we propose implementing a "spatial minimum standard," which, at the FUA level, could fundamentally change the developmental situation of inner peripheries. Our recommendations aim to strengthen the place-based policy approach, grounded in territorial specificity, emphasizing the need to tailor policies and interventions to the unique geographic, social, and economic contexts of these areas. We stress the importance of considering territorial capital, which is created by residents who leverage their resources to form a relational developmental environment. Furthermore, we underline the need to increase the share of flexible intervention tools within Cohesion Policy. Such tools allow for rapid adaptation of developmental interventions to changing socio-economic conditions, enabling more effective crisis response and supporting the long-term stability and resilience of peripheral areas. By focusing on these aspects, our recommendations aim to enhance the ability of Cohesion Policy to address the specific needs of inner peripheries effectively, fostering their integration and long-term development while strengthening their capacity to adapt to and recover from future shocks.

The structure of the procedure underlying the empirical application of the theoretical-conceptual model of inner peripheries, organized into the stages and steps of the completed study, along with the associated objectives, hypotheses, and methods used, is presented in Table 4.4. This table also indicates where the results of each stage and step of the empirical analysis are described in subsequent sections of this book, which we assume will assist readers in quickly locating the relevant content.

Table 4.4 Stages and steps of empirical analysis

Stages and steps of empirical analysis	Research tasks/Chapter containing results	Goals	Hypotheses	Methods
Stage I—National level				
Step 1—*Delimitation of FUAs*	1. A multistage delimitation of municipalities leading to the identification of inner peripheral areas in Poland	EG1 MG1 MG2		– Indicator methods – Modified method of highest flow linkages
	5. Inner peripheries in Poland ***5.2 New delimitation of urban functional areas***			
Step 2—*Typology of FUAs*	1. Operationalization of socio-economic development factors shaping developmental processes in inner peripheral areas 2. Multicriterion typology of inner peripheral areas in Poland, leading to their classification with consideration of types of peripherality	EG1 MG1 MG2	H1	– Indicator methods – Hierarchical clustering methods – Multidimensional classification methods (Gaussian mixture model clustering analysis)
	5. Inner peripheries in Poland ***5.3 Inner peripheries in the typology of Functional Urban Areas***			
Stage II—Case study level				
Step 3—*Selection of case studies*	1. Purposeful selection of inner peripheral areas representing the identified types of peripherality	MG1		– Indicator methods – Distribution analysis
	5. Inner peripheries in Poland ***5.4 Case studies of inner peripheries***			
Step 4—*Identification of development trajectories of inner peripheries*	1. Analysis of dynamics of selected territorial capital indicators in various types of FUAs between 2017 and 2023	EG2 MG1 MG3	H2	– Indicator methods – Dynamic analysis

(continued)

Table 4.4 (continued)

Stages and steps of empirical analysis	Research tasks/Chapter containing results	Goals	Hypotheses	Methods
	6. Development processes of the inner periphery in the period of development shocks ***6.1 Developmental trajectories***			
Step 5—*In-depth case studies using qualitative data*	1. In-depth case studies using qualitative data 2. Operationalization of socio-economic development factors shaping developmental processes in inner peripheral areas 3. Analysis of the resilience and vulnerability of inner peripheral areas and their core cities to global socio-economic change trends 4. Assessment of the role of the SARS-CoV-2 crisis in shaping socio-economic transformations in designated inner peripheral areas in Poland 5. Identification and monitoring of factors and barriers affecting the development trajectories of underdeveloped areas	EG3 EG4 EG5 MG1 MG4	H3 H4 H5	– Analysis of existing data (desk research) – Survey technique (CATI) – In-depth interview (IDI) – Focus group research (FGI)
	6. Development processes of the inner periphery in the period of development shocks ***6.2 Accessibility and its changes*** ***6.3 Economy and its transformations*** ***6.4 Inhabitants, living conditions and their changes***			

(continued)

Table 4.4 (continued)

Stages and steps of empirical analysis	Research tasks/Chapter containing results	Goals	Hypotheses	Methods
Step 6—*Formulation of recommendations and guidelines for development policy*	1. Formulation of recommendations regarding the directions and methods for implementing development interventions in inner peripheral areas in Poland, considering the social and economic consequences of the SARS-CoV-2 crisis 2. Formulation of guidelines for shaping urban policies for subregional functional centers	AG1	H6	– Validation and synthesis
	7. Challenges of European Cohesion Policy post-SARS-CoV-2 ***7.1 The role of inner peripheries against economic, social and political challenges post-2027 period*** ***7.2 Recommendation for operational change***			

Source Own elaboration

Chapter 5
Inner Peripheries in Poland

Abstract This chapter presents the first empirical stage of a project that consists of two steps: the delimitation of Functional Urban Areas and their typology in order to identify the inner peripheries in Poland. We first briefly present the state of knowledge on the regional development disparities in the country. After that, we describe how we divided the territory of Poland into 413 Functional Urban Areas (FUAs) based on the inter-municipal movement of population: commuting and mirgration. Following a discussion of the resulting division of Poland into FUAs and confronting it with the administrative division of the country, we use the FUAs as the building blocks for the typology of territories based on their multidimensional core-peripherality characteristics, which, in turn, were used to identify territories that we label as inner peripheries. The classification of FUAs and the identification of two types of inner peripheries based on a set of 47 indicators is described in the third section of the chapter and then confronted with the previous studies on the regional development disparities in Poland. The final part of the chapter outlines how we selected four FUAs classified as inner peripheral ones for further investigation as case study areas and provides a description of these areas.

Keywords Functional Urban Areas (FUA) · Multidimensional peripherality · Case study areas · Poland

5.1 Spatial Differences in Development in Poland

Socio-economic development is not uniform across space. An internal spatial differentiation of development characterizes each country, and regional inequalities vary significantly across countries (Wei 2015; Floerkemeier et al. 2022). They are primarily shaped by a set of factors specific to a given country, which exhibit spatial and temporal variability. These factors can reinforce or counteract each other, driving regional convergence or divergence (Wei 2015).

P. Churski et al., *Inner Peripheries in the New Conditions of Cohesion Post-SARS-CoV-2*, Economic Geography,
https://doi.org/10.1007/978-3-032-10936-1_5

In the case of large countries, such as Poland, regional differentiation is usually significant and contributes to limiting the efficiency of the functioning of the socio-economic systems of these countries. Large or growing regional development disparities, manifested at various levels of spatial aggregation (regions, subregions, local systems), lead to the creation of areas that are caught in a "regional development trap" (see Diemer et al. 2022, Rodriguez-Pose et al. 2024). Such areas include inner peripheries, characterized by numerous development deficits, which suffer from widening disparities in the level and quality of life of residents.

Poland, a country inhabited by nearly 38 million people, is one of the largest (6th position) and fastest-growing economies in the European Union. The country's location in Central Europe and numerous historical events, including 123 years under the power of neighboring empires, 44 years of communist regime, and the turmoil of the systemic transformation after 1989, led to the formation of visible development disparities. These disparities are still visible, as proven in numerous analyses, despite its recent history having been peaceful and stable and the implementation of numerous regional policy projects.

Socio-economic differences between regions had been formed since the Middle Ages. Across centuries, western regions incorporated Western European novelties such as urbanization and capitalism earlier than the country's east (Churski et al. 2021). However, visible socio-cultural borders within the Polish territory were formed between 1795 and 1918, when Poland did not exist as an independent state but belonged to three empires. The western part of Poland belonged to Prussia (German Empire as of 1871), the eastern part to Russia, and the southern part to Austria. The period of partitions coincided with important socio-economic transformations, including the start of industrialization, which accelerated urbanization, the transition of society from feudal to capitalist, and the establishment of universal education. The territories administered by the various invaders belonged to three completely different political and, thus also, economic systems; they differed in terms of the level of urbanization and the diversity of the economic structure (Fig. 5.1, Churski and Perdał 2022). In 1918, after the end of World War I and the fall of the three empires, Poland gained independence. The newly established Second Polish Republic survived until only September 1939, when Nazi Germany invaded. After World War II, the political borders of Poland were changed; Poland lost a significant part of its territories in the east to the USSR and gained smaller but better-developed areas in the west (Fig. 5.1). In addition, Poland was incorporated into the Soviet sphere of influence, and a socialist regime was established in the politics and economy. Those spatial changes led to a massive forced migration: the expulsion of Germans from now-Polish territory and a parallel repatriation of Poles from territories joined to the Soviet Union. Moreover, these movements overlapped with accelerated migration to rapidly industrializing districts under the centrally planned communist economy.

The political transition from authoritarian communism to a democratic system in Poland started as a result of the Round Table Agreements in 1989, which started a series of non-violent revolutions in the socialist bloc. At that time, parliamentary democracy was restored in Poland, and the central planning system was replaced

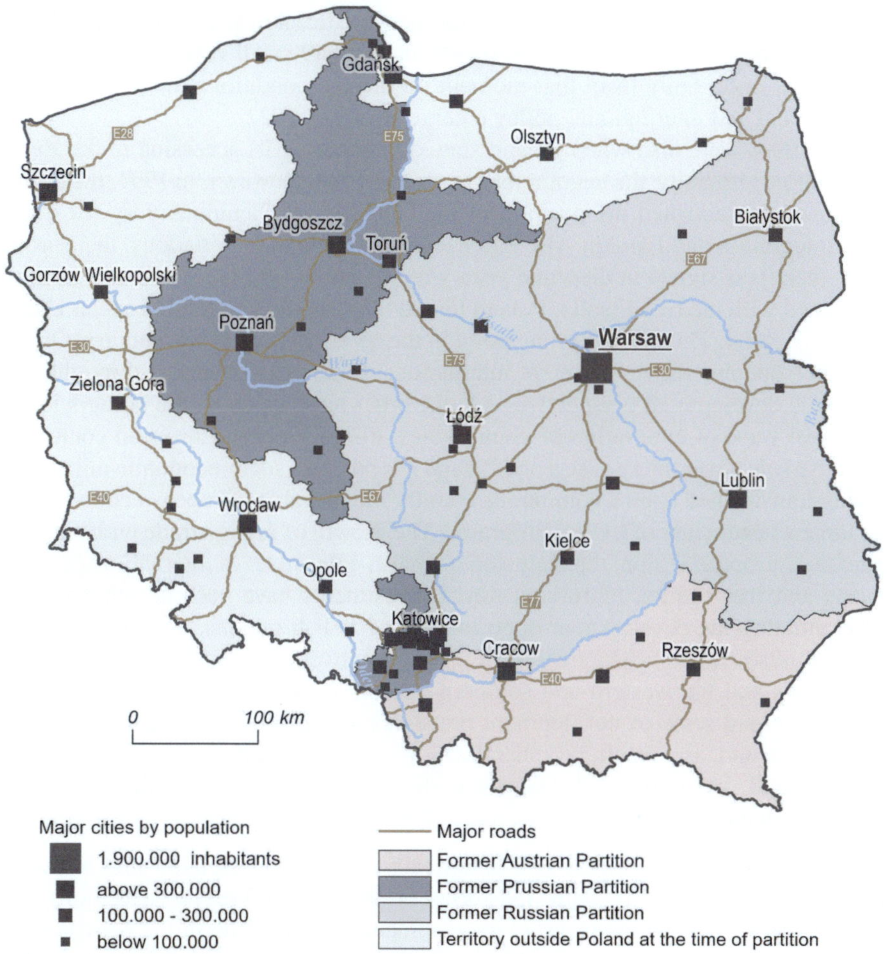

Fig. 5.1 Historical regions in Poland. *Source* Own elaboration

with a market economy, which sought to integrate market mechanisms with intervention intended to minimize the negative social impact of the market. The country moved away from a socialist economy, favoring a capitalist, free-market model. As a result of the implementation by Tadeusz Mazowiecki's government of the so-called Balcerowicz Plan (designed following ideas of Jeffrey Sachs), the Polish economy, striving to achieve macroeconomic stabilization, underwent shock therapy (Churski et al. 2021). During the economic transition period, especially the first half of the 1990s, spatial indifference dominated economic policy: state authorities attempted to keep neutrality toward the modernization processes in the spatial system (Korenik 2022). As a result, development disparities increased entirely uncontrolled in the face of the lack of regional policy intervention.

The pursuit of regional policy in Poland became feasible after the public administration reform carried out in 1999, which replaced 49 small voivodeships with 16 new, larger ones. Only from that moment on did the conditions for implementing intraregional and interregional policy emerge (Korenik 2022).

The 1990s were also when Poland started negotiating its accession to the European Union. Officially, the negotiations started in 1998. However, in 1991, the Polish government established the position of the Government Plenipotentiary for European Integration and Foreign Aid. After several rounds of negotiations, the Europe Agreement was signed in the same year, which gave Poland the status of a country associated with the EU. Finally, Poland joined the structures of the European Union on May 1, 2004. This was a turning point in the recent history of the country. On the one hand, opening the economy for international trade and investments brought new economic development perspectives. On the other hand, opening the borders led to a massive wave of emigration of young Poles, filling Western European countries' labor markets. Over the last 30 years, Poland has passed a socio-economic milestone, having transformed from a communist country into a stable European economy and the largest beneficiary of EU aid programs. The growth of foreign trade with partners from the European Union, the inclusion of Polish enterprises in international supply chains and the location of foreign direct investments have become other factors differentiating the development opportunities of Polish regions, as these development impulses first reached areas with more attractive locations: well-connected to large cities and the western and southern borders of the country. The differences in the pace and scale of development processes still persist, leading to increasing polarization and, as a result, peripheralization of economically weak areas.

In Poland, we observe development differences that may be described and explained in terms of:

1. **Historical factors**: There is a historically established division into a better-developed "Poland A" (the territories of the former Prussian partition) and "Poland B" with a lower level of development (the territories of the Austrian and Russian partitions). On the other hand, the consequences of the post-socialist transformation negatively affected the areas of north-western Poland (Domański 2021; Churski and Perdał 2022).
2. **Functional factors**: The spread of development processes from central cities toward rural areas and smaller towns, following the core–periphery model (Churski and Perdał 2022).
3. **Urbanization factors**: The degree of urbanization is an aggregate property of the area that reflects the level of socio-economic development (Czyż 2012; Churski and Perdał 2022).
4. **Cultural and historical factors**: Within this group, we could point to the attitudes related to the level of religiosity, family (e.g., consent to informal relationships or the birth of children outside of marriages), place attachment, family history, and finally, political preferences (Gorzelak 2021).
5. **Efficiency of local governments** and their dedication to improving residents' well-being (Gorzelak 2021).

Development inequalities in Poland are a complex and widely discussed topic. Regardless of the course of development processes and their results, spatial disparities resulting from spontaneous polarization are present, and they should be the subject of continuous in-depth studies and analyses. These studies should aim to identify changing development factors and barriers determined by the influence of external megatrends that shape the state of specific regional and local resources that create individual areas' territorial capital (see Sect. 2.3.3).

Based on the literature review of works addressing spatial differentiation of the level of development in Poland, we identified differences in the methodological approaches applied, which were determined by differences in the definitions and objectives of the research. The analysis of these studies allowed us to distinguish four main aspects of the research, which can be used for their systematization:

1. **Territorial reference system**. The analysis of spatial differentiation requires the definition of a basic territorial unit. It is necessary to decide whether the analysis will be based on the units of the administrative division of the country, functional units, or geometric fields. Researchers in Poland mostly use administrative division units for pragmatic reasons, mainly due to the availability of statistical data. In recent years, the popularity of the functional approach has grown, as mentioned in Sect. 2.3.2. When local administrative units are used, the decision on the level of administrative division in use is also important. The analysis of interregional differences is based on a comparison of voivodeships (approximately corresponding to the NUTS 2 level) or subregions (NUTS 3). The intraregional level analyses are performed at the level of counties (LAU 1) and mostly communes (LAU 2). An important topological choice is between the exhaustive (covering all spatial units) or non-exhaustive character of the analysis. Limited analyses usually refer to units at the same hierarchical level, distinguished either based on settlement, administrative, or size criteria.
2. **Subject of the analysis**. The study's primary objective may be the identification of the level of regional differentiation. Then, the result is the detection of areas with high or low levels of development. In the second approach, regional development factors constitute the basis for explaining the disparities in other issues, such as the standard of living, quality of life, and others. Defining the subject of the study also requires deciding whether the analysis would be based solely on one group of indicators, e.g., relating to the economic dimension, or whether a multicomponent analysis would be performed based on a comparison of development conditions.
3. **Nature of the analysis**. The determination of the state and directions of changes in development processes is based on the analysis of either resources or relations. The resource-based approach takes into account the state of development and the level of advancement of development processes at the moment of time adopted for the analysis ("state at" or change in time). The relational approach, on the other hand, is based on assessing the intensity and quality of flows of goods, services, and human capital between units using a hierarchical or a-hierarchical approach. A "mixed" approach is also possible, in which the analysis is based

simultaneously on the analysis of resources and flows. It provides a broader factual perspective.

4. **Method of systematization**. The result of the conducted analysis is the partition of the set of spatial units, usually using the procedure of logical division. This division must meet three conditions: separation, exhaustiveness, and exclusivity (Bremer 2004). The final effect of grouping is classification or typology. Classification refers to the assignment of units based on the values of the studied indicators. On this basis, spatial units are assigned to appropriate classes of adopted threshold values, which need to be determined by the classification. The number of classes may be predetermined or derived from the analysis results. On the other hand, the application of the typological approach consists in determining the object composition of individual types based on the similarity of certain features of their structures. As a result of typologization, spatial units are assigned to one of the types representing a specific composition of indicator values.

Considering the above, we have prepared a list presenting the characteristics of selected existing empirical studies analyzing the spatial distribution of development processes (Table 5.1). The presented list is not an exhaustive presentation of the research on development differences in Poland. We have only listed studies published within a relatively short time horizon which, therefore, take into account the current conditions stimulating development processes in Poland.

Regional differences in terms of the level of socio-economic development were revealed in each of the studies cited above. On the scale of the entire country, and especially at the subregional level, there is still a visible difference between western and eastern Poland. This is the result of the impact of many factors, among which the most important are the lack of large urban centers and the low level of urbanization of the regions of eastern Poland, as well as the monofunctional structure of the economy with a dominant agricultural sector and the historical factor that "connects" these two issues (the Russian partition). At the level of interregional analyses, the described distance is less noticeable and to a small extent corresponds with the historical divisions according to the borders of the partitions.

A development underperformance is also visible in Western Poland, and especially in the north of the country. Since the beginning of the transformation period, i.e., after 1989, as a result of structural problems, these areas have been characterized by the highest unemployment rates in the country and a specific set of socio-cultural characteristics of their inhabitants, referred to as the learned helplessness, which translates into low quality of human capital, a multitude of social problems and low dynamics of economic growth (Domański 2021). This part of the country was not covered by a special EU aid program designed for Eastern Poland's regions to help equalize development opportunities.

Moreover, spatial analyses conducted for smaller local government units revealed the most evident differences in the core-periphery relationship. This type of spatial pattern is characteristic not only for Poland. Large cities and their suburbs concentrate the greatest development potential and perform numerous functions for the benefit

Table 5.1 Review of selected empirical studies on regional development disparities in Poland

Authors	Territorial reference system	Subject of the analysis	Nature of the analysis	Method of systematization	Results
Czapiewski and Janc (2009)	– Administrative units—poviats and communes (LAU 1 and 2) – Intraregional analysis – Non-exhaustive analysis	– Identification of inner peripheries based on the analysis of spatial differentiation of the level of socio-economic development, the level of infrastructure development and population flows – Multi-indicator analysis	– Mixed approach: analysis of resources supple-mented by the analysis of flows of human capital	– Classification approach	– Inner peripheries were described as zones where the influence of large cities is weakened; these areas are located primarily on the borders of voivodeships, due to low spatial accessibility to the growth centers of the regions
Czyż (2012)	– Administrative units—subregions (NUTS-3) – Intraregional analysis – Exhaustive analysis	– Analysis of the level of socio-economic development based on the level of economic development and standard of living – Multi-indicator analysis	– Resources approach	– Classification approach	– The diagnosis of the spatial distribution of the group of indicators showed that most of the country is characterized by a low or very low level of development, especially in eastern, northern and north-western Poland – Subregions distinguished by a high and very high level of development occur around the largest cities (development enclaves)

(continued)

Table 5.1 (continued)

Authors	Territorial reference system	Subject of the analysis	Nature of the analysis	Method of systematization	Results
Bański and Mazur (2016)	– Administrative units—communes (LAU 2) – Intraregional analysis – Non-exhaustive approach (excluding cities and Functional Urban Areas of major cities)	– Assessment of the level of socio-economic development – Multiindicator analysis	– Resources approach with partial analysis of the temporal change of indicator	– Typological approach	– Analysis of the spatial differentiation of the group of indicators allowed for identification of rural areas participating in development and those requiring support for development processes, which account for most of the country by area – Application of the location approach to cities and their functional areas separated accessible and peripheral rural areas – A challenging situation, in the context of the level of socio-economic development, concerns areas in north-western Poland, but also areas primarily on the borders of voivodeships, at a large distance from the growth centers of the regions
ESPON (2017)	– Non-administrative units—subregions (NUTS-3) – Intraregional analysis – Exhaustive analysis	– Identification of inner peripheries based on the analysis of accessibility to the centers of development and services and the indicators of economic development – Multiindicator analysis	– Resource approach	– Classification approach	– Areas in which at least one peripherality factor was diagnosed were considered to be inner peripheries. The accumulation of all three factors constitutes the characteristic of deeply peripheral areas – The distribution of NUTS 3 peripheral areas is uneven. The largest number of inner peripheral units is located in north-western and central-southern Poland. The existence of inner peripheries is clearly visible in border areas of the entire country

(continued)

Table 5.1 (continued)

Authors	Territorial reference system	Subject of the analysis	Nature of the analysis	Method of systematization	Results
Śleszyński et al. (2017)	– Administrative units—communes (LAU 2) – Intraregional analysis – Exhaustive analysis	– Identification of the areas of strategic intervention: growth areas and problem areas – Multiindicator analysis	– Resource approach	– Typological approach	– Growth areas were indicated deductively, based on the analysis of strategic documents and expert knowledge – Problem areas were indicated based on the analysis of 3 groups of indicators: natural, economic, and social, based on which 7 groups of problem areas were identified, including mixed ones – In macrospatial terms, problem areas are concentrated primarily in eastern and northern Poland
Domański (2018)	– Administrative units—voivodeships – Interregional analysis – Exhaustive analysis	– Analysis of the level of economic development – Multiindicator analysis	– Resource approach considering temporal changes	– Classification approach	– An analysis of spatial differentiation of changes in GDP value and their changes over time showed significant differences in economic development in Poland – Economic growth of Poland is accompanied by increasing interregional disparities. The slowest-developing regions of the country turned out to be the voivodeships in northern and western Poland, while growth concentrates in regional centers

(continued)

Table 5.1 (continued)

Authors	Territorial reference system	Subject of the analysis	Nature of the analysis	Method of systematization	Results
Komornicki et al. (2019)	– Administrative units—poviats and communes (LAU 1 and 2) – Intraregional analysis – Non-exhaustive analysis	– Identification of border problem areas – Multiindicator analysis	– Resources approach	– Typological approach	– Identification of 3 types of border areas: borders of opportunities (mainly in the western part of the country), borders of moderate disadvantages, and borders of disadvantages (in northern, southern and eastern parts of the country). For the latter areas, the peripheral location is a development barrier
Bożek et al. (2021)	– Administrative units—voivodeships; – Interregional analysis – Exhaustive analysis	– Analysis of the level of economic development – Multiindicator analysis	– Resources approach considering temporal changes	– Classification approach	– The analysis of spatial variation of indicators, considering their temporal changes, proved the persistence of considerable disproportions in the regions' development level – The economic situation improved significantly in all voivodeships, yet the growth was uneven. Development disparities grew
Churski et al. (2021)	– Administrative units—communes (LAU 2) – Intraregional analysis – Exhaustive analysis	– Analysis of the level of socio-economic development – Multi-indicator analysis	– Resources approach	– Classification approach	– Diagnosis of the disparities in the level of socio-economic development (5 groups in the classification) – The differences in the level of socio-economic development are visible in the relation East–West of Poland (influence of the historic factor) and core-periphery

(continued)

Table 5.1 (continued)

Authors	Territorial reference system	Subject of the analysis	Nature of the analysis	Method of systematization	Results
Stanny et al. (2021)	– Administrative units—communes (LAU 2) – Intraregional analysis – Non-exhaustive analysis, excluding cities	– Assessment of the level and structure of the socio-economic development	– Resources approach	– Typological approach	– The diagnosis of the spatial distribution of the group of indicators confirmed that rural areas are socio-economically diverse: the closer they are to regional centers, the more favorable their development characteristics are – The development of rural areas is uneven; this is reflected in the set of territorial units differing in the occurrence and intensity of development indicators in the relationships between these indicators
Śleszyński (2021)	– Administrative units—communes (LAU 2) – Intraregional analysis – Exhaustive analysis	– Assessment of the quality of life – Multiindicator analysis	– Resource approach	– Classification approach	– The diagnosis of the spatial variation of indicators proved the spatial polarization of the country – The role of the historical factor is diminishing – The distribution of the areas characterized by low quality of life reflects the distribution of problem areas (Śleszyński et al. 2017), which points at the inertia of the development processes in Poland

(continued)

Table 5.1 (continued)

Authors	Territorial reference system	Subject of the analysis	Nature of the analysis	Method of systematization	Results
Churski and Perdał (2022)	– Administrative units—poviats (LAU 1) – Intraregional analysis – Exhaustive analysis	– Quality of life – Multiindicator analysis	– Resources approach considering temporal changes	– Classification approach	– Analysis of the disparities in quality of life (6 classification groups). Differences in the quality of life in Poland are visible along an east–west divide (influence of the historic factor) and a core–periphery divide – Identification of the changes in quality of life over time. No visible changes in the spatial distribution of quality of life
Heffner and Twardzik (2022)	– Administrative units—communes (LAU 2) – Intraregional analysis – Non-exhaustive approach (only rural areas and small towns)	– Assessment of the level of socio-economic development – Multiindicator analysis	– Resources approach	– Typological approach	– Results of the typology based on the results of the typology of other authors, led to the identification of peripheral areas threatened by progressive marginalization. These areas occur primarily in eastern and northern Poland – In southern and western Poland, there are areas that were also considered peripheral, although they respond to Cohesion Policy – In the spatial context, the socio-economic development of rural areas and small towns is mainly associated with the proximity of the metropolitan areas or attractive towns (e.g., tourist or business location)

(continued)

Table 5.1 (continued)

Authors	Territorial reference system	Subject of the analysis	Nature of the analysis	Method of systematization	Results
Spychała and Spychała (2022)	– Non-administrative units—subregions (NUTS-3) – Intraregional analysis – Exhaustive analysis	– Analysis of the level of regional development – Multi-indicator analysis	– Resources approach considering temporal changes	– Classification approach	– Analysis of the spatial distribution of the group of indicators and the inclusion of changes over time showed a large differentiation of regional development in Poland. The subregions associated with the largest cities in the country are the best developed; they also observed the largest increase in the socio-economic level. As a result, development disparities in the country have increased
Piekutowska (2023)	– Administrative units—regions (NUTS 2) – Interregional analysis – Exhaustive analysis	– Analysis of the level of regional development – Multi-indicator analysis	– Resources approach	– Classification approach	– Analysis of the spatial differentiation of the group of indicators showed large disparities in regional development in Poland. These disparities are permanent – The most decisive factor of the regional differentiation of the Polish economy is inequality on the labor market
Pietrzykowski (2024)	– Administrative units—regions (NUTS 2) – Interregional analysis – Exhaustive analysis	– Analysis of the level of socio-economic development – Single-indicator analysis	– Resources approach considering temporal c hanges	– Classification approach	– Analysis of spatial differentiation of selected economic indicators and their changes over time showed that development inequalities gradually decrease in Poland. However, polarization is still clear; the voivodeships of eastern Poland still lag behind economically, even though their level of development has noticeably increased in the analyzed period

Source Own elaboration

of their surroundings. However, smaller cities have less economic potential, which does not suffice to stimulate the development of the surrounding areas.

The validity of the above literature review is confirmed by the fact that although the studies we have cited are based on different assumptions and use different research procedures, all of them lead to an assessment of the differences in socio-economic development in Poland. Each analysis showed more or less profound spatial differences in development, noticeable primarily in subregional systems, with a similar distribution of advantaged and disadvantaged areas.

In our review of studies on identifying spatial differentiation of socio-economic development in Poland, we found a very small number of studies (only 5) relating to peripheral areas, including only two that mentioned inner peripheries. Still, even those studies which mention inner peripheries, do not treat them as complex socio-economic phenomenon as proposed in this paper.

An example of such study is that of Czapiewski and Janc (2009), whose aim was to present the internal boundaries of the peripheries of socio-economic development. The authors based their study on the assumption that intraregional boundaries differentiate a given area into parts characterized by more favorable and less favorable socio-economic conditions. These boundaries largely overlap with the boundaries of voivodeships, but they depend to the greatest extent on the spatial accessibility (measured by distance) of a given area to the central city. Therefore, they designate zones of weakening of the influence of large urban centers. The presented delimitation is therefore an example of identification of external peripheries, according to our understanding, although taking into account the interregional scale of analysis. Similarly, Komornicki et al. (2019) studied what we would call external peripheries. Their analysis aimed to classify border areas in the country and designate problem areas that require potential development support. Bański and Mazur (2016) distinguished urban functional areas of the largest cities in Poland (voivodeship capitals), and then conducted a typology on the remaining territory of the country, distinguishing areas with good accessibility and peripheral areas. Both groups were additionally differentiated based on the development dynamics, location conditions and dominant functional structure. The typology of Heffner and Twardzik (2022), in turn, is not based on the authors' original research. It is a result of a compilation (cartographic imposition) of several typologies of rural areas conducted in previous years by other research teams. As a result, the authors of this meta-analysis distinguished several types, 2 of which had peripheral features: marginalized peripheries and peripheral systems responding to development policy.

The first and so far the only delimitation of the inner peripheries, in the sense of "disconnected" areas, in Poland was carried out by ESPON (PROFECY—Processes, Features and Cycles of Inner Peripheries in Europe 2007). Its assumptions were discussed in detail in Sect. 2.1. The obtained spatial distribution of the three types of internal peripheries in Poland covered almost 35% of the country's area. However, it seems that the adopted units of spatial analysis (NUTS 3) were too large; the use of LAU units would certainly have provided a more accurate presentation of spatial disparities.

Moreover, in the literature, we can find research from other scientific disciplines on social peripheries conceptually linked to our understanding of inner periphery. These are research on social peripheries conducted by sociologists (Wódz et al. 2016) or lost spaces present in urban planning (Racon-Leja 2010). The terms mentioned refer to "disconnected" areas that meet the criteria of the definition of internal peripheries. In each of these cases, the boundaries of the periphery have not been determined, and the territorial units to which the analyses refer were considered peripheral *in antecessum.* It seems justified to determine new, current boundaries of areas recognized as internal peripheries, taking into account the specificity of the country's regional and local resources that make up its territorial capital.

5.2 New Delimitation of Urban Functional Areas

According to the theoretical-conceptual model presented in Chap. 4, we assumed that when identifying inner peripheries in Poland, we use local functional territorial systems as spatial units of analysis. We called them Functional Urban Areas, which include towns and the surrounding municipalities connected to them by functional links. There have been many attempts to delimit such functional areas of cities in Polish geography and spatial planning. However, few of them were exhaustive, meaning that each municipality is assigned to a central city. Such an approach requires abandoning the hierarchical way of assigning to central centers, common in previous studies, in favor of single-level/horizontal regionalization. In our project, we assumed that all cities, regardless of their size and place in the hierarchical settlement structure, may form their functional areas, and each territory can be assigned to a functional area of a city. Our delimitation was based on units of the third level of administrative division (LAU; municipalities) (Fig. 5.2). It is based on commuting matrices (2011 and 2016, according to Statistics Poland), full inter-municipal migration matrices (2011–2020, also according to the Central Statistical Office), and data on travel times between localities determined using the Google Maps Distance Matrix tool (2022, Churski et al. 2023). The diversity of data collection and publication procedures in various sources forces us to use different dates of observation.

Polish geographers and regional planners have a long tradition of research on urban regions (Czyż 2009; Churski et al. 2023). Over time, they used different terms to describe the zones or areas of urban influence, the most popular of which were urban agglomerations, metropolitan areas, and functional urban regions/areas. The term urban agglomeration refers to areas of population concentration (Iwanicka-Lyra 1969, Dziewinski and Kosinski 1964). Urban agglomeration can be defined as a compact settlement complex that includes an urban core surrounded by smaller towns and rural settlements. It is characterized by a high population concentration, a high level of urbanization, and the prevalence of non-agricultural functions. The core city should have strong connections with the surrounding area in terms of the availability of labor markets and service facilities (Dziewoński et al. 1977, Korcelli

Fig. 5.2 Administrative division of Poland. *Source* Own elaboration

1976, Rykiel 1978). Numerous studies on agglomerations addressed the terminological considerations and the delimitation of agglomerations. Although the terminology that defines urban agglomerations clearly emphasizes the functional relations of a city with its surroundings, the delimitation of such was mainly based on the morphological aspect (Czyż 2009). As Korcelli (1973, p. 158) states, the concept of urban agglomeration "contains primarily morphological content," which equates it with the related concept of "urbanized area." The morphological, demographic, and socio-economic aspects were primarily used in the empirical delimitations of urban agglomerations performed mainly in the 1960s and 1970s (Chramiec 1966; Iwanicka-Lyra 1969; Leszczycki et al. 1971).

Functional links are more emphasized than the morphological aspect in the concept of metropolitan areas. It is defined as a system of core-environment relations based on the flows of people, goods, and information and an area of diffusion of social and economic phenomena (Korcelli 1973, 1976). The metropolitan areas should be characterized by the presence and considerable strength of socio-economic links between residential, labor, leisure, etc. centers throughout the internal system and its functional integration by both the endogenous function and the exogenous functions located in different parts of the metropolitan area (Czyż 2009; Markowski and Marszał 2006; Smętkowski et al. 2009; Szmytkie 2013). Empirical studies on the delimitation of metropolitan areas came later than those on urban agglomerations, though resulted in similar numbers of less than ten (Smętkowski et al. 2009) to about 20 metropolitan areas in Poland (Gontarski 1980; Swianiewicz and Klimska 2005).

Korcelli (1976) was the first in Poland to consider the issue of metropolity with reference to the concept of daily urban systems (Daily Urban Systems, Berry 1973). He predicted that due to progressive urbanization and the development of functional links, the entire area of the country would be covered by the extent of functional urban regions, the centers of which would not only be large cities. His concept of functional urban regions was close to the concept of metropolitan areas. What separates them is the different nature of functional links tying the region together. The concept of a metropolitan area puts pressure on the links of economic organizations, the development of higher-order (metropolitan) functions and their specialization, while the concept of a functional urban region refers more to direct contacts and mobility of residents within the framework of the city daily urban systems (Hołowiecka and Szymańska 2008; Korcelli 1976; Śleszyński 2014).

In recent years, research on the relations of cities with their surroundings has focused on the very similar concept of Functional Urban Areas (FUA). The interest in FUAs is largely policy-led and related to the changing model of the EU regional policy, which is increasingly oriented toward territorially integrated management, emphasizing exploiting the potentials of urban centers and transferring them to their functional areas. In contrast to metropolitan areas, FUA is also applied to functional areas of smaller cities (Bartosiewicz 2016; Ilnicki and Janc 2021). Studies on Functional Urban Areas in Poland also refer to the issue of their morphological and functional polycentricity (Bański et al. 2012, Marcińczak and Bartosiewicz 2018).

The presented theoretical achievements, especially the concept of Functional Urban Areas, are increasingly important and widely used in the practice of spatial policy. The concept of Functional Urban Areas was adopted for the Concept for Spatial Development of the Country 2030 (Ministry of Regional Development 2011). This document assumes that "a Functional Urban Area is a spatially continuous settlement system composed of administratively separate units. It includes a compact urban area and a functionally related urbanized zone" (p. 191). These functional relationships are a consequence of population movements related to migration flows and the functioning of the labor market, as well as access to public services (e.g., commuting to schools). The need to delimit FUAs, as elements of spatial planning, is important to form partnerships and develop Integrated Territorial Investments (ITAs) strategies needed to obtain EU subsidies. In 2007, ESPON published a report presenting a delimitation of FUAs within EU member states. The delimitation included 151 FUAs of Poland's largest cities. Still, due to the inaccessibility of, e.g., commuting data at that time, the delimitation was not based on the criterion of functional links (Śleszyński 2013).

Among the many attempts to delimit urban functional areas, few studies in Poland are based on measuring functional links between cities and surrounding areas, exhaustive (which means that each municipality is assigned to a central city) and recent. As these three prerequisites were essential for our work, we decided to perform our own delimitation based on the theoretical concept of an FUA and the methodical experience presented in previous Polish studies. Below, we outline the core methodical assumptions we made when preparing the delimitation procedure: the choice of the basic spatial unit of analysis, choosing between zonal and nodal approaches, between deductive and inductive classification, between exhaustive and non-exhaustive divisions, between hierarchical and a-hierarchical grouping, as well as between typological and regional classification. Later, we will outline the detailed workflow of the analysis and present the resulting division of Poland into functional spatial units.

The first decision when attempting to delimit functional areas is determining the basic spatial unit of analysis, which will be later aggregated into FUAs. Most studies were based on administrative subdivisions. Sometimes, the reference units were artificially delimited basic fields (Dziewoński and Kosiński 1964). Poland has three-level administrative division and splits into 16 voivodeships, 380 poviats and 2477 municipalities (of which about 1000 are fully or partially urban and the rest are rural). To keep as much detail as possible, we took municipalities as units of reference for our analysis. It would be more accurate to delimit some of the smaller FUAs based on smaller territorial units (statistical localities, villages, geodesic precincts), but it would significantly limit the scope of statistical data we could obtain for such defined FUAs. Our delimitation ignores higher-level administrative divisions—poviats and voivodeships—which do not always reflect functional connections but are instead a result of political bargaining (Zaborowski 2016). Instead, following our supra-local functional approach, we decided that each FUA must contain a central city and at least two municipalities. However, not all the cities would have their own FUAs, as they can as well belong to other cities' FUAs.

Functional areas can be delimited as zonal or nodal regions (Chojnicki and Czyż 1992; Rychłowski 1967). Zonal regions are homogeneous because all their member units are characterized by similar features, e.g., socio-economic characteristics, demographics, or land-use structures. Such an approach was common in identifying agglomerations. Nodal regions are closer to the definition of functional urban regions/areas or metropolitan areas: they are delimited based on the strength of interactions between central cities and the surrounding areas. To measure the functional interactions between municipalities and the central city, the strength of flows of, for example, information, goods, or people are used. The daily flows of people related to work, education, or use of services are particularly commonly used, defining the daily urban system (Berry 1973; Śleszyński 2013). Due to data availability, Polish researchers use mostly flows to measure functional linkages (Potrykowska 1989; Smętkowski 2007; Śleszyński 2013). Their intensity and spatial extent account for the strength of the impact of the core city (Komornicki et al. 2013; Sudra 2018). The functional criterion also considers migration, which, although not daily travel, may be related to daily spatial mobility due to the interconnectedness of different forms of mobility (Kaufmann 2002). Migration, especially within the same region, can result from or generate trips to work, school, and services and to visit family or friends. In our delimitation, following our supra-local functional approach to designing the units of analysis, we accepted the nodal definition of the region and used commuting to work and bidirectional migration as a proxy of functional links.

Commuting to work is the only circular flow of people for which Statistics Poland creates a municipal linkage matrix. This dataset is based on tax returns and is published once every five years. It has some flaws: every commuting survey is conducted according to a slightly modified methodology, there is no information on pairs of municipalities with negligible flows (less than ten people), the matrix includes only information on the commuting of wage earners, but not the self-employed persons and entrepreneurs. Finally, the registered addresses of residence and place of work do not necessarily correspond to the actual ones (Statistics Poland 2010, Śleszyński 2012). In particular, these inadequacies apply to remote work or work in different locations. Hence, we can suppose that the commuting matrix over-estimates the number of commuters to large cities, which host the headquarters of large companies whose employees work in different locations. The second source of functional link information was the inter-municipal migration matrix. We looked at round-trip migration between the central city and the municipality due to the theoretical assumption of the reciprocity of migration, which we also observed in the data (return streams match one-way migration streams). Urban regions of different sizes and locations may be characterized either by the prevalence of migration to the center or suburbanization. Although for the purpose of the delimitation of supra-local functional areas we were only interested in short-distance migrations (within a potential FUA), they could not be a priori separated from interregional ones, which would interest us in only after delimiting FUAs, in order to evaluate the relations between individual FUAs.

The next methodical decision was to determine the way to classify the units (municipalities): between deductive classification (by division) and inductive classification (by clustering). The first, more popular approach assumes assigning municipalities to a previously defined set of central cities (defined either based on an administrative criterion, e.g., voivodeship capitals, or size) based on, e.g., a defined threshold of strength of functional links. Deductive classifications are usually also non-exhaustive classifications, meaning that not all spatial units are included in the delimitation. We tried to obtain an exhaustive classification to determine the assignment of all municipalities into FUAs. Moreover, in the supra-local functional approach to defining FUAs we could not assume the number or a list of central cities a priori. For these two reasons, we excluded deductive classification. In the inductive classification we performed, municipalities are freely grouped into functional areas only based on the adopted criteria. This approach, however, is more complex, as it requires a method to translate the functional linkages matrix into a grouping.

The most commonly used methods to assign units to functional regions based on the functional linkages are the variations of the highest flow method (Nystuen and Dacey 1961). In principle, it allows us to assign localities to central cities to form a hierarchical regionalization—each unit can be assigned to an immediate central city, which in turn can be assigned to a higher-order central city. In our case, however, as we seek to delimit inner peripheries based on the lowest-level territorial functional entities, we are concerned with single-level (a-hierarchical) regionalization—dividing the country into FUAs of the same level. Therefore, we assigned municipalities to the lowest-level central cities. As a starting point, we took a set of all municipalities as potential FUA members and all cities as potential FUA centers or members. To assign a municipality to a central city, we compared the strength of its functional links with all cities in Poland. FUA capital cities ("independent" cities) were selected based on the presence of at least one subordinate municipality that is not also the capital of its own FUA. To avoid the cyclicity of the assignment ("mutual" assignments of cities), we considered the sizes of opposite commuting streams in contentious situations (the larger commuting target is dominant over the larger commuter provider).

The nature of the data used, both on commuting (probably an overrepresentation of commuting to large cities) and migration (regional combined with interregional), leads to the fact that the areas of large cities have a large range, certainly, given temporal accessibility, beyond the range of daily systems. Still, population flows are not the only criterion to FUA delimitation, we also consider physical accessibility the necessary prerequisite of supra-local collaboration, which is also essential in delimiting supra-local functional areas. Therefore, we added the maximum travel-time criteria for the municipality and the central city to limit the FUAs. Referring to previous studies (Komornicki et al. 2009, 2018), we considered an isochrone of one hour as the limit of a convenient one-day commute to work, services or school. However, in order to take into account the fact that commuting times in the regions of the largest cities (including Warsaw) often exceed this length, and that setting a fixed limit for all cities would artificially limit the scope of FUAs, we adopted an isochrone of one and a half hours for cities larger than 500,000 residents. To determine the distance, we used the Google Maps Distance Matrix API, which allows for measuring actual commute times, considering traffic conditions (including congestion).

The last essential methodical choice was between the typological classification and regional classification. Typological classification attributes objects to classes based on the similarity of certain features (Parysek 1982). The resulting classes of municipalities would not need to be spatially continuous. Thus, the basic rule distinguishing functional areas, i.e., the principle of spatial compactness, would not be kept. The condition of spatial compactness is met, in turn, if we perform a regional classification. In this method, the grouping of spatial units leads to forming spatial classes consisting of contiguous units (Runge 2007) with similar properties. Like most previous authors, we assumed that our division should be a regional classification. Since, in practice it is not a common border but rather communication cohesion that creates physical accessibility, matters for functional links, and creates the potential for supra-local collaboration (Zaborowski 2016) which are essential in our approach to supra-local functional areas, we established that municipalities in one FUA must border each other and have a direct road connection.

Taking into account our above-discussed methodical decisions, we defined a Functional Urban Area (FUA) as a spatially contiguous area comprising a central city and at least one subordinate municipality linked to it by population flows (commuting and reciprocal migration based on Statistics Poland data) and located within the isochrones of a one-hour commute to the central city (or 1.5 h for the largest cities). Compared to previous delimitations of Functional Urban Areas in Poland, the novelty of such an approach lies in the use of functional bounds indicator and the exhaustive nature of the classification, which will assign all lowest-level administrative divisions into one of the functional units.

Thus, in the procedure for dividing the country into FUAs (Fig. 5.3), we first calculated commuting, migration, and aggregate population flow matrices between all 2477 municipalities and ranked potential central cities for all municipalities. We did this based on the 2011 and 2016 Statistics Poland commuting matrices and full inter-municipal migration matrices for 2011–2020 according to Statistics Poland (we summarized migration over this period). The population flow rates were calculated using the following formulas:

$$CI_{m\to c} = \frac{C_{m\to c}}{\sum C_{m\to}}$$

$$MI_{m\leftrightarrow c} = \frac{M_{m\to c} + M_{c\to m}}{\sum M_{\to m} + \sum M_{m\to}}$$

$$FI_{m\to c} = \frac{CI_{m\to c} + MI_{m\leftrightarrow c}}{2}.$$

The commuting index (CI) from municipality m to central city c is the ratio of the number of commutes (C) in that direction to the sum of commutes to all communes from commune g. The migration index (MI) between municipality m and the central city c is the ratio of the number of migrants (M) between these cities in both directions to the sum of the migration turnover of municipality m. The aggregate population flow index (FI) is the arithmetic mean of the commuting and migration indexes.

Based only on the ranking of the population flow index, we conducted the first assignment of municipalities to potential central cities. Each municipality was assigned to the city connected to it by the highest value of the population flow index. 429 cities had at least one municipality assigned to them.

Cities that were themselves central cities for other municipalities were also assigned to other central cities, so in the second step, we joined these central cities to their own FUAs. This led to the reassignment of 215 municipalities and the creation of 27 temporary FUAs composed of only one municipality, which was against the assumed definition of FUA.

In the third step, we eliminated such single-municipality MOFs by returning the municipalities to FUAs to which they belonged in the first stage. In two specific cases of pairs of mutually assigned municipalities, we determined the central city based on the positive commuting balance within the pair of municipalities.

In the fourth stage, we used the time distance criterion to reassign municipalities that are too distant from the central city of the FUA. We operationalized the time

Fig. 5.3 Procedure of Functional Urban Areas (FUA) delimitation. *Source* Own elaboration

distance condition of 1 h and 1.5 h for central cities larger than 500,000 residents, such that it must take no more than an hour (one and a half hours) by car to get from the municipal office of the subordinate municipality to the city hall of the central city on a weekday when leaving at any full hour, according to Google Maps Distance Matrix API data. This tool enabled us to determine the travel time between two points using the same algorithm as Google Maps navigation, based on information about the actual speed of cars obtained from drivers' cell phones with active location functions and Google apps. To save time and API queries, we checked the travel times for one in the winter of 2021/2022, with travel start hours set to 6, 7, 8, 15, 16, and 17. Among the time calculation algorithms available via API, we choose best_ guess mode, providing the most likely travel time. 109 municipalities turned out to be located too far from the center of the FUA to which they had been assigned so far and were therefore included in the FUA of the city to which the next highest population flow index connected them. It reduced the size of 17 FUAs of large cities (e.g., Warsaw FFUA lost 46 municipalities). It also led to the creation of 15 new FUAs, so the previous steps were re-applied: including capital cities in their own FUAs and eliminating FUAs composed of a single municipality. After that, another iteration distance check with parallel control for other conditions changed the affiliation of 6 municipalities.

The final step of the analysis was verifying the FUA's spatial compactness. FUAs were considered spatially compact if all municipalities bordered each other and were linked by a road. In case of discontinuity, a municipality was to the FUA of the next city in terms of population flow rate index. At this stage, we did not create new FUAs, but only modified existing ones, minimizing the number of transferred municipalities and simultaneously controlling for other definitional conditions. After this final step, 31 municipalities changed their assignment to FUAs. This resulted in eliminating three FUAs, and their number reached the final 413.

The delimitation procedure led to the grouping of 2477 municipalities into 413 Functional Urban Areas (FUAs, Fig. 5.4), each consisting of between 2 and 92 municipalities (FUA Warsaw). The area of FUAs ranges between 14 (FUA Żyrardów) and 8.8 thousand km^2 (FUA Warsaw), and the population (according to 2021 Census) between 7.5 thousand (FUA Kosów Lacki) and 3.3 million (FUA Warsaw). The population density of the FUAs range between 18.5 (FUA Ustrzyki Dolne) and 3537 persons/km^2 (FUA Żyrardow). The largest FUAs are those formed around the largest cities and the capital cities of voivodeships.

The number of FUAs and their average size is similar to the country's second-level administrative division units—poviats. Three hundred and seven of FUA's central cities are also the capitals of poviats. Many FUA borders overlap with poviat borders. However, the main difference between FUAs and poviats is in the variability in the FUA size and its correlation to the size of the central city. Administrative units have roughly the same size regardless of the size of the capital city. FUAs, in turn, differ significantly in size, and their extent is strongly correlated to the size of the central city: larger cities have larger areas of influence than smaller ones. The borders of FUAs resemble administrative divisions in western and northern Poland, with a historically well-developed network of middle-sized cities. In Mazowsze region (around Warsaw), on the other hand, FUA layout differs from the administrative structure the most, due to the dominance of Warsaw in the settlement network and the underdevelopment of medium-sized cities. The size variation among FUAs is a natural result of our definition being based on supra-local functional connections, as larger cities tend to influence larger areas than do smaller cities. We expect smaller

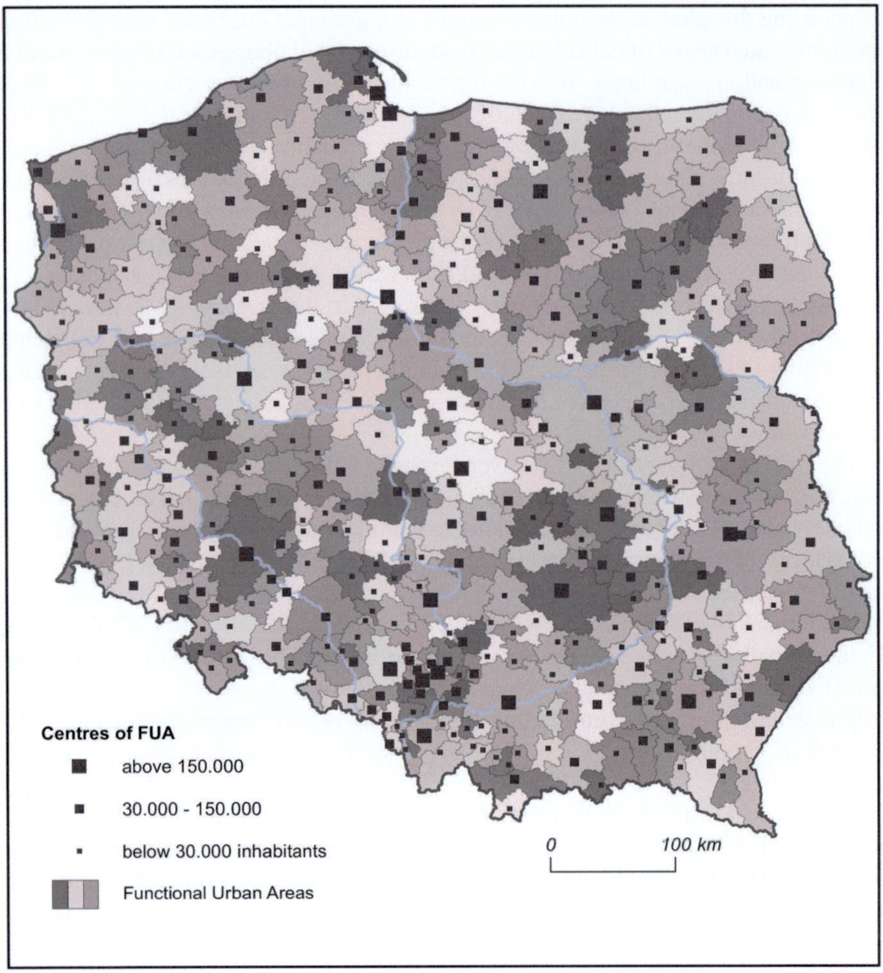

Fig. 5.4 Functional Urban Areas (FUAs) of Poland. *Source* Own elaboration

cities with smaller functional areas to have weaker connections with major cities and therefore fall within the definition of inner peripheries.

Our delimitation of Functional Urban Areas (FUAs) in Poland has several features that distinguish it from the previously published divisions:

- It is independent of the administrative divisions into poviats and voivodeships, which enables us to capture functional connections, which emphasizes the importance of a relational approach, including organized proximity.
- It is an exhaustive classification based on functional relations, enabling the identification of the extent of functional relationships occurring at a supra-local level across the country.

- The FUAs may be operational in the context of an integrated territorial approach to regional policy, because, in the delimitation, we reflected practical criteria such as the territorial and travel-time compactness of the functional areas, in addition to actual interactions among residents.

The delimitation of FUAs aimed to construct spatial units that will be used in the further steps of the analysis to determine spatial variations in socio-economic development and distinguish the inner peripheries. The next section of this chapter presents the procedure of grouping FUAs into types and distinguishing those that meet the criteria of inner peripheries.

5.3 Inner Peripheries in the Typology of Functional Urban Areas

After dividing Poland into FUAs, we attempted to identify areas that fulfill the definition of inner peripheries. The basis of the procedure was the set of development factors elaborated in Sect. 2.3.3 and presented in our theoretical-conceptual described in Chap. 4. We first looked for the indicators that could be used to measure each of the five factors and 17 subfactors listed in the previous chapters. Due to the diversity of the factors in question, we could not limit our search to public statistics published by Statistics Poland. Instead, we also reached for less-popular statistical databases of various public institutions and big-data sources. When choosing the variables, we considered data availability, recency, and spatial resolution, which allowed us to summarize them into FUAs. We drew the values of the variables from 15 different data sources, including the public database of Statistics Poland, unpublished public registers (e.g., of the Ministry of Finance) and previous research publications.

Overall, we took into account 47 variables representing various dimensions of territorial capital, quantifying five major development factors and 17 subfactors. The choice of indicators was designed to reflect the full extent of the subfactors that reflect the five types of territorial capital required to identify the inner peripheries, as described in Sect. 2.3. The complete list of variables assigned to development subfactors and factors is presented in Table 5.2.

Each development subfactor was quantified by one to four variables. The accessibility factor was divided into internal and external transport accessibility and e-accessibility in our conceptual model. We measured internal transport accessibility by the car travel time from all communes to the FUA central city (using Google Maps Distance Matrix API), the average distance to schools and an indicator of car ownership. In external accessibility, we took into account car travel times to major cities outside of FUAs by employing traffic models developed by IGSO PAS (Komornicki et al. 2018), as well as public transport connections (Kołsut and Stryjakiewicz 2023). E-accessibility was quantified by wire and mobile internet connection access and quality.

Table 5.2 Representation of territorial capital by variables used in the classification of Functional Urban Areas (FUAs)

Capital/factor	Subfactor	Variables
Accessibility	Internal transport accessibility	Average car travel time from all communes to FUA central city
		Average distance of pupils' homes to primary schools
		Passenger cars per 1,000 inhabitants
	External transport accessibility	Multimodal accessibility Indicator
		Car travel time from FUA central city to nearest regional capital of voivodeship
		Car travel time from FUA central city to nearest 100k + city
		Number of daily public transport connections from FUA central city to regional city (capital of voivodeship) during a working day
	E-accessibility	Share of population with access to broadband internet
		Indicator of internet access points without access to broadband
		Share of population with access to mobile LTE network
Human capital	Demographic situation	Population in post-working age per 100 inhabitants of working age
		Share of population of pre-working age
		Rate of natural increase per 1000 population
		Net registered migration increases per 1000 inhabitants
	Health and safety	Number of outpatients using healthcare advice per 10,000 inhabitants
		Identified crimes per 10,000 inhabitants
		Car accidents and collisions per 10,000 inhabitants

(continued)

Table 5.2 (continued)

Capital/factor	Subfactor	Variables
	Education	Places in kindergartens per 1,000 children 3–6 years old
		Number of pupils per class in primary schools
		Average result of the matura exam in mathematics
		Average result of the matura exam in English
	Labor market	Share of unemployed in population of productive age
		Share of employed population of productive age
		Share of population of productive age running personal businesses
Social and institutional capital	Social activity	Non-governmental organizations per 10,000 inhabitants
		Graduates of courses run by culture institutions per 10,000 inhabitants
	Leisure	Sport and recreation institutions per 10,000 inhabitants
		Attendees of sport and cultural events per 10,000 inhabitants
	Business infrastructure	Business environment institutions per 10,000 businesses (all NACE sections)
		All-year-round hotel bed-places per 1,000 inhabitants
	Economic structure and activity	Share of new economic entities in creative sector among all new entities (NACE J59, J60, M71, M73, M74, R90)
		Number of economic entities per 1,000 inhabitants of productive age

(continued)

Table 5.2 (continued)

Capital/factor	Subfactor	Variables
		Number of large (>49 employees) economic entities per 10,000 inhabitants
Financial capital	Financial situation of local governments	Own income of local governments per inhabitant
		Investment expenditures of local governments per inhabitant
		Net operational surplus of local governments
	Financial situation of businesses	Businesses income per taxpayer
	Financial situation of inhabitants	Personal income per adult inhabitant
Material capital	Land use	Urbanized area per 1,000 inhabitants (CLC level 2 classes: 1.1, 1.2, and 1.3)
	Housing situation	Dwellings completed per 1000 inhabitants
	Housing quality	Share of population in dwellings with central heating
		Share of housing buildings connected to sewerage system
		Usable floor space per inhabitant
	Ecosystem services	Share of population with large (>25 ha) green areas within 1 km of home
		Share of protected areas in commune's area
		Deviation from the norm of atmospheric pollution (mean of BaP, PM10, PM25)
		Number of applications to "Pure Air" program per 1,000 single-family homes

Source Own elaboration

In the human capital demographic subfactor, we took into account public statistics on three data dimensions: demographic (age) structure, natural increase, and permanent migration. Health and safety were measured by one health indicator (patients per population) and two public safety metrics (crimes and car accidents). In measuring education, we distinguished access (places in kindergartens and crowing of schools) and results of education (results of matura exam after secondary school data provided by the Central Examination Board). Finally, labor market quality was measured by unemployment, contact and self-employment.

Within the social and institutional capital, nine variables represented four subfactors. We measured social activity by the number of non-governmental organizations and graduates of courses run by cultural institutions. Access to leisure was identified based on the number of sports and recreation institutions and attendees of sports and cultural events. The number of business environment institutions and hotels was considered business infrastructure. Finally, economic activity and structure were measured by the number of economic entities (including large ones) and the share of those within the creative sector.

We measured financial capital mostly based on unpublished data from the Ministry of Finance using three indicators of the finance of local governments (own income, investments expenditures, and net operational surplus), business income per taxpayer as an indicator of the financial situation of businesses, and personal income per adult inhabitant to measure the financial situation of inhabitants.

Material capital was the most complex measurement in terms of the diversity of data sources. The efficiency of land use was measured with the use of Corine Land Cover data. Housing situation (dynamics of new construction) and quality (floor area per inhabitant and technical infrastructure in housing) were measured using Statistics Poland data. Finally, we quantified ecosystem services provision using three data dimensions: accessibility to green areas, level of air pollution, and pro-ecological investments.

Individual variables had varied and sometimes very asymmetrical distributions, which would hinder future attempts to synthesize the multidimensional database; therefore, after the examination of histograms and Q-Q plots, we transformed most variables using natural logarithm, square root or more complex transformations to obtain distributions similar to natural distribution (Table 5.3).

We attempted to reduce the number of variables to decrease the complexity of the dataset and enable the identification of peripheral FUAs. Considering two possible approaches to variable reduction, i.e., calculating latent variables (through principal component analysis) on the one hand and reducing the number of variables taken into account based on systematic criteria on the other hand, we choose the latter approach. In doing so, we attempted to satisfy two conditions: removing variables closely correlated with other variables in the set of grouping variables and preserving variables representing all development factors of the data set and preferably all subfactors.

Therefore, we designed a two-stage variable selection procedure (Fig. 5.5). The first step was the analysis of clusters of variables. After data matrix transposition, hierarchical cluster analysis was applied to the variables with the agglomeration grouping method, using Euclidean distance and the complete linkage method. After visual analysis of the cluster dendrogram, we decided to adopt the cut-off level, distinguishing 22 clusters of variables. Then, we selected one variable from each group to eliminate redundant variables, but we still had all development factors and most subfactors. The number of variables per factor ranged from two (social and institutional capital) to eight (human capital). The choice of the variables was,

Table 5.3 Representation of territorial capital by variables used in the classification of Functional Urban Areas (FUAs[a])

Variable	Data source	Date of data	Symbol and transformation in Fig. 5.6	Transformation for classification	Included in grouping (F score)
Average car travel time from all communes to FUA central city	Google Maps Distance Matrix API[b] (data collection as in the delimitation of FUAs)	March 2022	AI_GMAPS	Square root	Yes (15.8)
Average distance of pupils' homes to primary schools	Statistics Poland	31.12.2018	AI_SCHOOL	Square root	No
Passenger cars per 1,000 inhabitants	Adam Mickiewicz University (Kołsut and Stryjakiewicz 2022)	31.12.2020	AI_CARS	None	Yes (33.0)
Multimodal Accessibility Indicator	IGSO PAS[c]; Komornicki et al. (2018)	2020	AE_MMAI	Square root	Yes (59.9)
Car travel time from FUA central city to nearest regional capital of voivodeship	IGSO PAS; Komornicki et al. (2018)	2019	AE_VOIV	Square root	No
Car travel time from FUA central city to nearest 100 k + city	IGSO PAS; Komornicki et al. (2018)	2019	AE_100K	Square root	Yes (67.1)
Number of daily public transport connections from FUA central city to regional city (capital of voivodeship) during a working day	Data obtained from e-podroznik.pl—a major online timetable aggregator and ticket sales agent	01.02.2022	AE_PUBL	Natural logarithm	No
Share of population with access to broadband internet	Office of Electronic Communications	31.12.2020	AD_BROAD	None	No
Indicator of internet access points without access to broadband	Office of Electronic Communications	31.12.2020	AD_NO	Square root	Yes (46.4)
Share of population with access to mobile LTE[d] network	Office of Electronic Communications	31.12.2020	AD_LTE	25th power	No

(continued)

Table 5.3 (continued)

Variable	Data source	Date of data	Symbol and transformation in Fig. 5.6	Transformation for classification	Included in grouping (F score)
Population in post-working age per 100 inhabitants of working age	Statistics Poland	31.12.2020	HD_POST	Natural logarithm	Yes (37.3)
Share of population of pre-working age	Statistics Poland	31.12.2020	HD_PRE	Natural logarithm	No
Rate of natural increase per 1000 population	Statistics Poland	Average 2018–20	HD_NI	Natural logarithm after adding 20	Yes (49.7)
Net registered migration increases per 1000 inhabitants	Statistics Poland	Average 2018–20	HD_MIGR	Natural logarithm after adding 10	No
Number of outpatients using healthcare advice per 10,000 inhabitants	Statistics Poland	Average 2018–20	HH_PAT (thousands)	None	No
Identified crimes per 10,000 inhabitants	Polish police (National Police Headquarters)	2018–20	HH_CRIME	Natural logarithm	Yes (31.3)
Car accidents and collisions per 10,000 inhabitants	Polish police (National Police Headquarters)	2018–20	HH_ACCID	Natural logarithm	Yes (4.0)
Places in kindergartens per 1,000 children 3–6 years old	Statistics Poland	31.12.2018	HE_KIND	None	Yes (55.89)
Number of pupils per class in primary schools	Statistics Poland	31.12.2020	HE_CLASS	None	No
Average result of the matura exam in mathematics	Central Examination Board	31.12.021	HE_MAT	2nd power	Yes (20.6)
Average result of the matura exam in English	Central Examination Board	31.12.2021	HE_ENG	3rd power	No
Share of unemployed in population of productive age	Statistics Poland	31.12.2020	HL_UNEMP	Natural logarithm	Yes (45.3)

(continued)

Table 5.3 (continued)

Variable	Data source	Date of data	Symbol and transformation in Fig. 5.6	Transformation for classification	Included in grouping (F score)
Share of employed population of productive age	Ministry of Finance	31.12.2020	HL_EMPL	None	No
Share of population of productive age running personal businesses	Ministry of Finance	31.122020	HL_BUSI	Natural logarithm	Yes (87.7)
Non-governmental organizations per 10,000 inhabitants	National Court Register	31.12.2021	SA_NGO	Natural logarithm	No
Graduates of courses run by culture institutions per 10,000 inhabitants	Statistics Poland	Average 2018–20	SA_COURS	Natural logarithm	Yes (7.2)
Sport and recreation institutions per 10,000 inhabitants	Statistics Poland	31.12.2020	SL_REKR	Natural logarithm	No
Attendees of sport and cultural events per 10,000 inhabitants	Statistics Poland	Average 2018–20	SL_EVENT (thousands)	Natural logarithm	No
Business environment institutions per 10,000 businesses (all NACE sections)	Statistics Poland	31.12.2020	SB_ENVIR (thousands)	Natural logarithm	No
All-year-round hotel bed-places per 1,000 inhabitants	Statistics Poland	31.12.2020	SB_HOTEL	Natural logarithm	Yes (13.6)
Share of new economic entities in creative sector among all new entities (NACE J59, J60, M71, M73, M74, R90)	Statistics Poland	31.12.2020	SS_CREAT	None	No
Number of economic entities per 1,000 inhabitants of productive age	Statistics Poland	31.12.2020	SS_ENTIT	Natural logarithm	No
Number of large (>49 employees) economic entities per 10,000 inhabitants	Statistics Poland	31.12.2020	SS_LARGE	None	No

(continued)

Table 5.3 (continued)

Variable	Data source	Date of data	Symbol and transformation in Fig. 5.6	Transformation for classification	Included in grouping (F score)
Own income of local governments per inhabitant	Statistics Poland	31.12.2020	FL_OWN (thousands PLN)	Natural logarithm	No
Investment expenditures of local governments per inhabitant	Statistics Poland	31.12.2020	FL_INW (thousands PLN)	Natural logarithm	Yes (24.0)
Net operational surplus of local governments	Statistics Poland	31.12.2020	FL_SURP	8th power	Yes (8.1)
Businesses income per taxpayer	Ministry of Finance	31.12.2020	FB_INC (thousands PLN)	Natural logarithm	No
Personal income per adult inhabitant	Ministry of Finance	31.12. 2020	FP_INC (thousands PLN)	Natural logarithm	Yes (152.6)
Urbanized area per 1,000 inhabitants (CLC level 2 classes: 1.1, 1.2, and 1.3)	CORINE Land Cover	31.12.2018	ML_URB	Natural logarithm	No
Dwellings completed per 1,000 inhabitants	Statistics Poland	Average 2018–20	MS_NEW	Natural logarithm	Yes (76.3)
Share of population in dwellings with central heating	Statistics Poland	31.12.2020	MQ_CH	2nd power	Yes (86.1)
Share of housing buildings connected to sewerage system	Statistics Poland	31.12.2020	MQ_SEWER	None	No
Usable floor space per inhabitant	Statistics Poland	31.12.2020	MQ_SPACE	Natural logarithm	No
Share of population with large (>25 ha) green areas within 1 km of home	IGSO PAS (Śleszyński 2021)	2018	ME_POP	1.3rd power	Yes (1.1)
Share of protected areas in commune's area	Statistics Poland (Śleszyński 2021)	31.12.2020	ME_AREA	Natural logarithm	No

(continued)

Table 5.3 (continued)

Variable	Data source	Date of data	Symbol and transformation in Fig. 5.6	Transformation for classification	Included in grouping (F score)
Deviation from the norm of atmospheric pollution (mean of BaP[e], PM10[f], PM25[f])	Chief Inspectorate for Environmental Protection (Śleszyński 2021)	Average 2019–20	ME_NORM	Natural logarithm	No
Number of applications to "Pure Air" program per 1,000 single-family homes	National Fund for Environmental Protection and Water Management	2021	ME_PURE	Natural logarithm	Yes (18.8)

[a] Functional Urban Areas
[b] Application Programming Interface
[c] Institute of Geography and Spatial Organization, Polish Academy of Science
[d] Long-Term Evolution
[e] Benzopyrene
[f] Particulate matter
Source Own elaboration

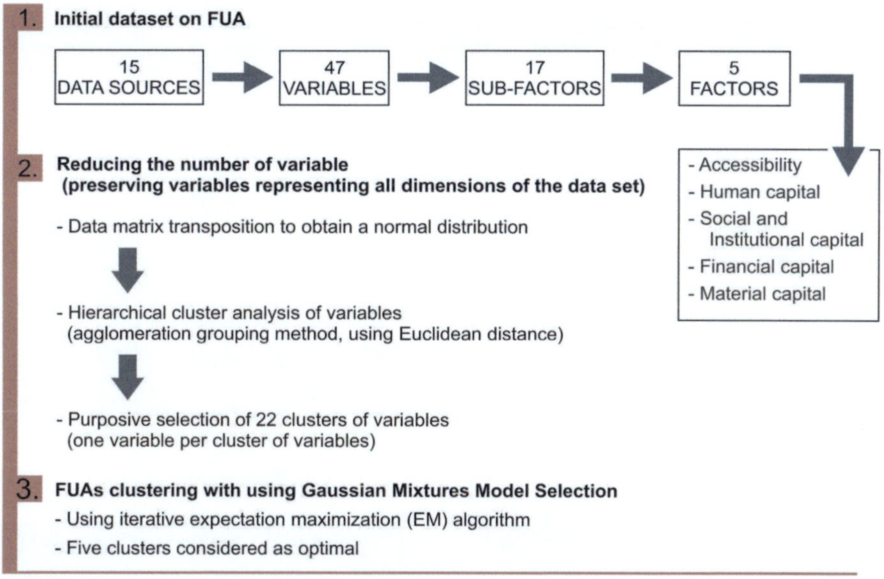

Fig. 5.5 Stages of typology of FUAs. *Source* Own elaboration

therefore, largely subjective yet informed by the results of variable clustering and correlation analysis. The selected 22 are marked in the last column of Table 5.3.

Next, we used the reduced number of variables to group FUAs into types. To do so, we considered various methods of cluster analysis: hierarchical (Ward); centroid-based (k-means); density-based (DBSCAN and OPTICS); and distribution-based (Gaussian mixtures) (Bishop 2006). Considering the possibilities of interpretation and application of the clustering method results for identifying and classifying inner peripheries, we chose the Gaussian mixtures method, implemented in the Python Scikit-learn package. This model uses an iterative expectation maximization (EM) algorithm. Given an assumed number of clusters, the algorithm detects clusters of values of individual variables in multidimensional data space, assuming that they are sampled from the number of normal (Gaussian) distributions corresponding to the number of clusters. Compared with other popular methods (hierarchical and k-means), the method is more effective for detecting clusters of different sizes (with different standard deviations) and shapes. This feature was helpful for our study, as we could a priori assume the similarity in the numbers of units assigned to each group or the same level of internal consistency of each group.

We investigated the results of cluster analyses for the number of clusters from two to seven, using various types of variable matrices (Gaussian Mixture Model Selection). The results were examined regarding the possibility of interpreting the results and comparing heuristic clustering quality measures (BIC, Caliński-Harabasz and Davies-Bouldin indexes, see Appendix C). Finally, we found that using a spherical variance matrix and grouping FUAs into five clusters was optimal. The clustering was

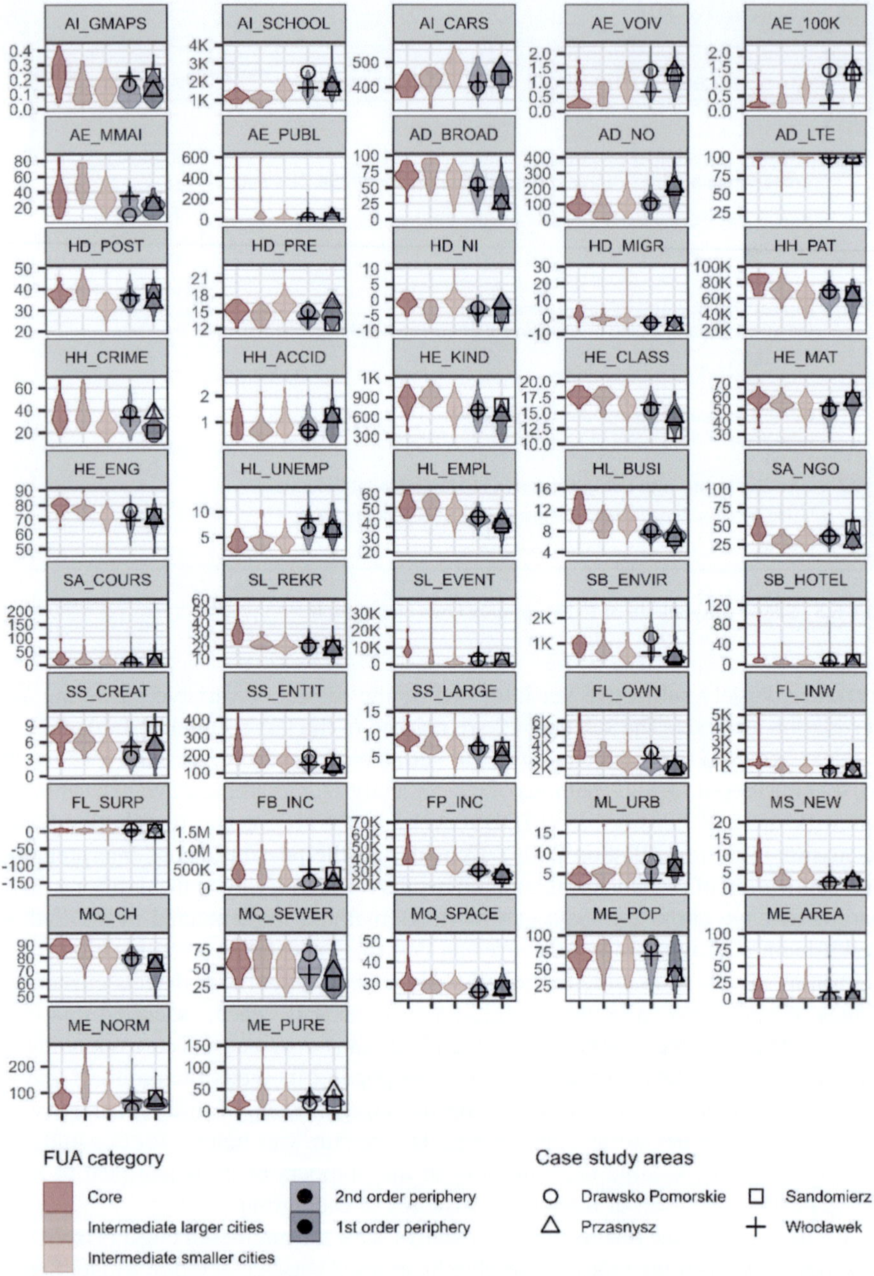

Fig. 5.6 Characteristics of territorial capital in five types of Functional Urban Areas. *Source* Own elaboration

repeated on multiple sets of random input parameters for the clustering procedures, converging to the same results each time.

After applying the clustering procedure, we measured the influence of each of the variables on the cluster assignment using F-statistic (see the last column of Table 5.2). Financial (e.g., personal income), economic (e.g., the share of the population running personal income) and housing variables (percentage of the population with central heating and number of new population) turned out to be the most influential, followed by external transport accessibility characteristics. On the other hand, characteristics remotely related to economic development, such as access to green areas or traffic safety, were the least decisive in assigning FUAs to groups. Despite the assumed complex nature of the definition of inner peripheries, the level of economic development and activity remains the most influential dimension of their distance to the central areas.

After investigating the basic characteristics (Fig. 5.6) and geographic distribution of FUAs (Fig. 5.7) and named them core areas, two types of intermediate areas and two types of peripheral areas, these last being most interesting for us. Each type will be briefly characterized below.

FUAs formed around all 18 capital cities of 16 administrative regions (voivodeships) of Poland (two regions have two capital cities), as well as seven more cities (25 altogether), were included in the core areas group. Apart from regional capitals, subregional centers such as Koszalin and Bielsko-Biała and cities within the zone of influence of major agglomerations (Gdynia) entered the group of central cities of core areas. Core areas consist of a total of 484 communes. Due to their central location and well-developed transport infrastructure, they usually have high external accessibility. Yet, large FUAs of this kind may experience problems in internal accessibility. Their economic activity and financial resources indicators usually overtake all other groups of FUAs. The labor market is flexible, and unemployment is low. The quality of life is also high due to a favorable housing situation and broad access to public services, translating into a relatively high level of human capital. One challenge of such areas is high anthropogenic impact on the environment resulting from high population density, lowering the perceived quality of life due to environmental problems and transport congestion.

The first group of transitional FUAs are intermediate areas of large and middle-sized cities. They are formed around towns, usually of above 30,000 inhabitants. Thirty-nine FUAs of this type include 250 communes. The group includes functional areas of most of the cities of the Upper Silesian conurbation in the south of the country—the industrial district whose economy was formerly based on hard coal mining, steelmaking, and machine industry and has now transformed into a mixed industrial-service economy. Other FUAs within this group are formed around middle-sized industrial cities in other parts of Poland. These areas' urbanization and industrial heritage translate into well-developed technical and social infrastructure. Thus, transport accessibility, access to public services, and housing quality are above average. Yet, economic restructuring left deficits in finance and entrepreneurship: in these regards, intermediate areas of larger cities lag behind core areas. This group

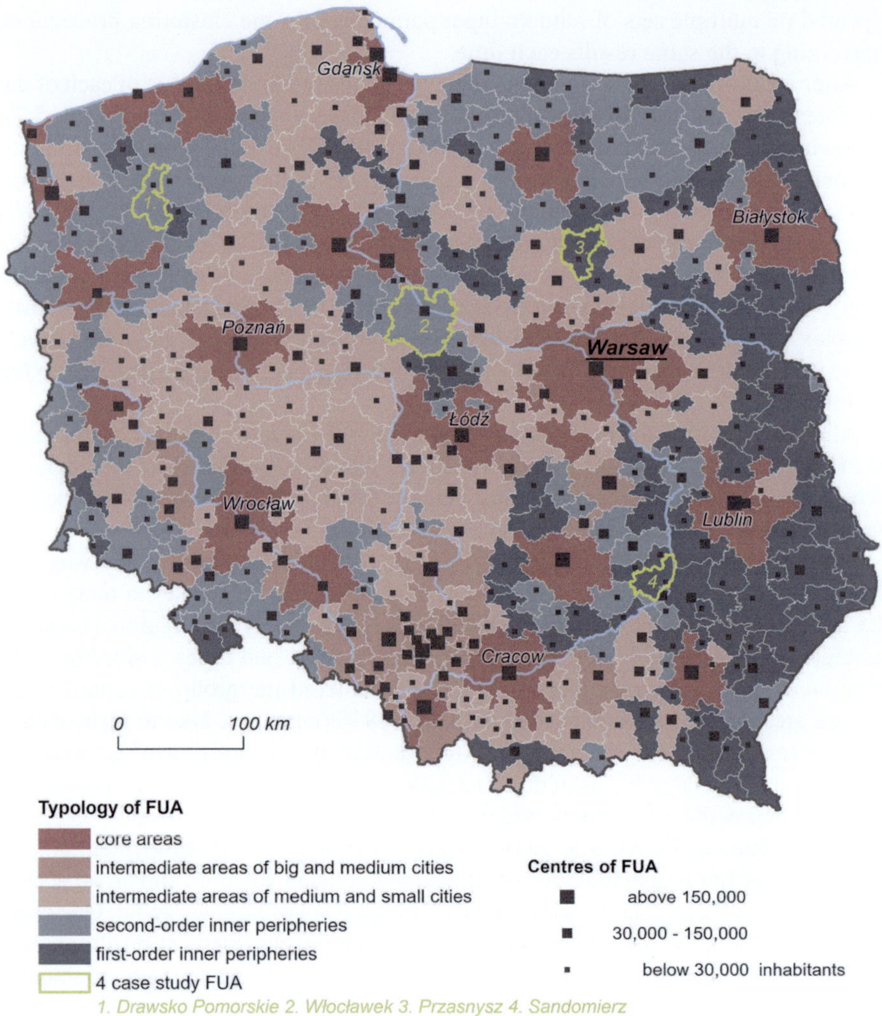

Fig. 5.7 Classification of Functional Urban Areas and the selection of case study areas. *Source* Own elaboration

of FUAs also struggles with demographic problems, such as a negative birth rate despite a relatively advantageous labor market.

Another type of areas located between cores and peripheries on the territorial capital scales are the intermediate areas of middle and small cities. Central cities of FUAs within this group are of fewer than 150,000 inhabitants, predominantly poviat capitals with between 20,000 and 50,000 inhabitants. One hundred forty-one FUAs in this group comprise 781 communes. Historical conditions visibly shape the distribution of these areas: their most extensive concentration is in the central-western part of the country and the north, which roughly corresponds to the territories occupied by Germany from the nineteenth century until 1920. Another factor is geographic proximity to core areas, e.g., Warsaw, Łódź and Kraków. The FUAs in this group are typically characterized by high multimodal accessibility, yet they sometimes face low transport accessibility of the central city. Their labor market is relatively strong, yet average income streams and activity are below expectations. They have average access to public services, but a favorable demographic situation manifested in a relatively low share of people of post-working age (Fig. 5.5). The challenges for such areas is to diversify and stimulate the economy, improve transport accessibility, and improve the scope and quality of public services.

We labeled two groups of FUAs characterized by low levels of territorial capital as inner peripheries. Those with higher levels of indicators among the two groups were called second-order inner peripheries. There are 112 FUAs formed of 467 communes within this group. The central cities of those FUAs are medium-sized and small cities with fewer than 150,000 inhabitants, typically 10,000–50,000. These regions are primarily located in north-eastern and north-western Poland, regions with the lowest population density in the country and which experienced difficulty in undergoing economic transformation in the 1990s, particularly due to the high degree of collectivization of agriculture. Those areas have relatively high internal accessibility, low efficiency or even lack public transport. The labor market is typically inflexible and declining, and incomes are low, as is economic activity. Public finances are highly dependent on central budget revenues, the access to public services is limited, and they have low quality. The housing market is weak, and housing infrastructure is below average. The areas experience demographic stagnation (Fig. 5.5.). Second-order peripheral areas face challenges of creating stable sources of financing for public investments, improving the quality and accessibility of public services, and adapting to negative demographic trends.

First-order inner peripheries are characterized by the most unfavorable values of most indicators of territorial capital. Ninety-six FUAs of this kind comprise 495 communes. The central cities of those FUAs are small cities (fewer than 30,000 inhabitants), located mainly in the eastern part of the country, historically economically less-developed than the west, and which additionally underwent the most challenging and prolonged economic transformation. These areas are the geographical peripheries of the European Union. They are characterized by relatively low external and average internal accessibility, a low level of development of digital communication infrastructure, a stagnant labor market, a lack of large-scale economic activity, low individual purchasing power and own income of local governments, which contributes to the problems in maintaining budget stability even with limited public investments. First-order peripheral areas are also highly dependent on the central budget, technical, and social infrastructure is average, the housing situation is bad, and access to public services is limited. These areas experience significant demographic regression. On the other hand, the level of public safety is exceptionally good (Fig. 5.5). The challenge of these areas is primarily to look for competitive advantages that enable the generation of local income, counteracting the demographic depression.

Notably, these two types of inner peripheries do not represent qualitatively different types of peripherality and do not reflect the distinction between physical distance-based external peripheries and relationally-defined inner peripheries as described in Chap. 2. Instead, they are based on the different intensities of the deficits of territorial capital characterizing inner peripherality. Yet, in the case of the first-order inner peripheries, the territorial capital deficits are even more apparent than those of the second-order inner peripheries.

Inner peripheries within the two categories include FUAs of various sizes. First-order peripheral FUAs have between two and 22 municipalities with an area of between 182 and 2,280 km^2 and a population between 9,600 and 183,200 inhabitants. Second-order peripheral FUAs may be larger, with up to 30 municipalities, 86–3,902 km^2, and 11,100–251,500 inhabitants. Intermediate-area FUAs have similar sizes (up to 30 municipalities, 46–3,225 km^2) and similar or larger populations (15,100–675,700). The core areas are the largest in all measures (up to 146 municipalities, 11,600 km^2 and 3.9 million inhabitants). All major cities are within core areas, while smaller cities (up to 279,000 inhabitants) are capitals of intermediate areas. Second-order peripheral cities are of up to 113,000 inhabitants, and first-order ones are of up to 58,000. However, the centers of peripheral FUAs also include smaller towns with as few as 6,000 inhabitants (second-order) and 4,300 (first-order).

Our delimitation and typology of inner peripheries are distinctive from the previous studies on spatial differentiation of socio-economic development in Poland cited in Sect. 5.1. There, we reviewed various approaches to studying regional development disparities to identify those studies that were concerned, even indirectly, with the areas of inner peripheries in Poland. We distinguished five works that meet this criterion: ESPON (2007), Czapiewski and Janc (2009), Bański and Mazur (2016), Komornicki et al. (2019), Heffner and Twardzik (2022). We will now discuss our delimitation and typology with their findings.

All the studies cited above were based on an administrative reference system, primarily due to access to published statistical data. The ESPON typology was based on the largest spatial units, subregions (NUTS3), while the other authors worked on poviats and municipalities. In our delimitation, we abandoned the administrative reference system in favor of the originally delimited Functional Urban Areas, which, although composed of municipalities, are distinctive from the administrative division into poviats and reflect the actual areas of influence of the local cities.

In all five cited studies, the differentiation of the level of socio-economic development of the areas was examined based on multi-indicator analysis. The selection of indicators was usually conditioned by access to published statistical data. In our study, we also analyzed many deliberately selected indicators, but we integrated many different data sources, including unpublished data. We generated some of the essential data on accessibility ourselves using an application to generate data on travel times on a given route in real-time, and we also obtained tax data from the Ministry of Finance and crime data from the Polish police. Naturally, we also referred to the Statistics Poland public databases, but integrating various data sources is another innovative trait of our research approach.

In order to determine the status and directions of changes in development processes, most of the cited authors performed static or dynamic analyses of resources, taking into account the state of development and the level of advancement of development processes. In our typology, we used two approaches simultaneously: analyzing and assessing the status of territorial capital resources in FUAs and analyzing human capital flows (commuting to work, internal migration).

The result of the five works to which we compare our delimitation and typology of inner peripheries were typological divisions, in which spatial units were assigned to previously defined types representing a specific set of indicator values. One of the earliest of the cited studies (Czapiewski and Janc 2009) was a classification in which the classes of communes and counties were determined based on the values of the studied indicators. Moreover, only one cited study was an exhaustive territorial division, while four were based on non-exhaustive analyses. Our delimitation was an exhaustive and separate typology; each commune was assigned to one central city of an FUA because we believe that each city, regardless of size, can create its own functional area. As a result of the cited typologies and classifications, researchers gave different names to the areas that can be compared with the first- and second-order inner peripheries that we identified (Table 5.4).

Our study was inspired by the results of the ESPON typology (2007), which distinguished three types of inner peripheries: enclaves with low economic potential, areas with poor access to public utility services, and areas lacking relational proximity. However, our typology based on the FUA reference system is more territorially detailed and based on up-to-date data. The internal peripheries distinguished by Czapiewski and Janc (2009) as zones of weakened influence of large cities located at the borders of voivodeships (NUTS 2) partially overlap with our research results due to the presence of areas of transport exclusion. The typology proposed by Bański and Mazur (2016) was based on rural functional areas, from which rural peripheral areas with various functions (production, consumption, mixed) requiring support for development processes were distinguished. In our results, inner peripheries largely cover rural areas where development problems have been diagnosed, mainly in the following areas: demographic, social, and economic. The typology of Komornicki et al. (2019) divides border areas into areas that function at the borders of disadvantage and those that function the borders of moderate disadvantage. In our typology, border areas, especially along the eastern and north-eastern borders of Poland, are identified as inner peripheries. In the work of Heffner and Twardzik (2022), marginalized peripheries were designated as depopulated rural areas with a low level of socio-economic development, a dominant agricultural function, and difficulties in accessing basic and public services. Additionally, they distinguished peripheral systems responding to development policy, defined as rural areas in the initial phase of depopulation, with a significant role of the agricultural sector in the local economy, and characterized by problems with access to public and higher-order services. This approach is consistent with the functional approach to our typology of first- and second-order internal peripheries.

Table 5.4 Representation of territorial capital by variables used in the classification of Functional Urban Areas (FUAs)

	Selected existing studies (ESPON 2007, Czapiewski and Janc 2009, Bański and Mazur 2016, Komornicki et al. 2019, Heffner and Twardzik 2022)	Our study
Territorial reference system	Administrative and statistical units NUTS 3, LAU 1, and LAU 2	Functional Urban Areas
Subject of the analysis	Assessment of the level of socio-economic development; multiindicator analysis	Assessment of the level and structure of territorial capital; multiindicator analysis
Nature of the analysis	Analysis of resources	Analysis of resources and flows of the territorial capital
Method of systematization	Typology and classification	Classification
Results	(1) Inner peripheries as enclaves with low economic potential, areas with poor access to public services, areas lacking relational proximity (2) Inner peripheries as zones of weakened influence of large cities (3) Rural peripheral areas with various functions (production, consumption, mixed) requiring support for development processes (4) Border areas (borders of moderate disadvantages; borders of disadvantages) (5) Marginalized peripheries, a peripheral system responding to development policy	First-level and second-level inner peripheries

Source Own elaboration

Summing up, our measurement of the differentiation of the level of territorial capital, leading to the identification of inner peripheries, is an original study distinguishable from the previous studies from the perspectives of basic and applied research alike. This is due to three main premises:

- The methodology for determining peripheral areas was adapted to the research assumptions, which allowed for the identification of areas based on the analysis of five development factors (territorial capitals). The applied approach which takes into account multicriterion analysis, allows for a broad view of regional inequalities. In our opinion, omitting or neglecting any of the group of factors would reduce the quality of the analysis.
- Studies of regional development differences based on administrative and existing statistical units (LAU 1, LAU 2, and NUTS) do not include "functional wholes" (in the meaning of Functional Urban Areas), in which interconnected socio-economic activity takes place and for this reason have lower utility.

- Studies using predetermined administrative or statistical reference units ignore the variety of the sizes of zone of influence of individual local centers. In our delimitation and classification, we noticed that peripheral Functional Urban Areas are typically smaller than intermediate and far from the core-area FUAs formed around the major cities.

5.4 Case Studies of Inner Peripheries

After distinguishing the types of FUAs according to the territorial capital characteristics, we selected four case study FUAs on which we focused our further interest. The informed selection was based on several criteria. First, we choose two cases of FUAs from each type of inner peripheries: first- and second-order inner peripheries. Second, we attempted to find the cases that lie within the typical values for each group of FUAs of most territorial capital indicators. In the case of the first-order peripheries, these are values within the lowest (least favorable) quartile of most indicators. In the case of the second-order inner peripheries, the values usually lie within the second quartile. Third, we looked for case study areas in different regions of the country that typify large concentrations of peripheral FUAs in Poland, and with different historical characteristics. Fourth, we considered FUAs that differed in size and in the size of their central city.

Applying the criteria led us to the selection of the FUAs of Włocławek and Drawsko Pomorskie as representatives of the second-order inner peripheries and Sandomierz and Przasnysz representing first-order inner peripheries. As Fig. 5.6 presents, the territorial capital indicators are typically within the second quartile for the two former and, for the latter two, within the lowest quartile for the entire population of FUAs.

Sandomierz is located in south-eastern Poland, within the largest concentration of FUAs characterized as peripheral (mostly first-order peripheries, Fig. 5.7), which corresponds to the low level of socio-economic development of the whole region (five voivodeships in Eastern Poland belong to the economically weakest regions in the entire EU). This is associated with the historical heritage: nineteenth-century Russian and Austrian territories outside of the major cities lagged behind German territories in the west and north, resulting in deficits of infrastructure, agricultural development, and urbanization that lasted for decades.

Przasnysz is located in north-eastern Poland, where historical factors linked with distance to major agglomerations and low population density contribute to forming the second cluster of peripheral FUAs (both first- and second-order peripheries). Despite being located in the same voivodeship with the main economic center of Poland—Warsaw—it is a part of the outer belt of areas beyond the immediate influence of the Polish capital.

Drawsko Pomorskie represents another significant concentration of second-order peripheries in historically well-suited yet thinly populated Western and Central

Table 5.5 Basic characteristics of case study FUAs

FUA	FUA category	Number of communes	Area (km^2)	Population (2023)	Population of central city (2023)
Włocławek FUA	Second-order inner periphery	20	2,131	212,632	100,807
Drawsko Pomorskie FUA	Second-order inner periphery	2	890	23,160	11,224
Przasnysz FUA	First-order inner periphery	6	848	39,805	16,357
Sandomierz FUA	First-order inner periphery	9	670	69,471	21,273

Source Own elaboration

Pomerania, which experienced relative economic decay during the economic transformation of the 1990s due to, e.g., a large proportion of state-owned farms that were dissolved after the economic transformation creating an economic void and unemployment crisis.

Włocławek FUA has central location along the north–south axis of the country but stays outside the sphere of influence of the most dynamic economic cores. Włocławek is also the largest city among the centers of the case study FUAs, with 101,000 inhabitants (2023) and has the largest area of influence, reaching 20 communes with 213,000 inhabitants. It also served as an important industrial center and a regional capital for a long time. Between 1975 and 1998, it was the capital city of one of the 49 voivodeships of Poland, thus representing the type of peripherality based on lost administrative position.

The remaining case study FUA centers are smaller, having between 11,000 and 21,000 inhabitants. They form smaller FUAs comprising between two and nine municipalities and totaling between 23,000 and 69,000 inhabitants (Table 5.5). All FUA centers are also capital cities of poviats (additionally, Włocławek constitutes an entire urban poviat in its own right). Even though the FUA borders resemble those of poviats, the actual functional relations appear to bypass the administrative borders, and often, the borders of the FUA do not align with those of the poviat. Below, we will characterize the location, population, territorial structure, natural conditions, history, and economic characteristics of each FUA selected as a case study area.

5.4.1 *Włocławek FUA: A Middle-Sized City that Lost Its Administrative Function*

Włocławek is a city with county rights, classified as a medium-sized city with a population exceeding 100,000 residents. However, the city has been deeply affected by a significant population decline. Over the past 20 years, it has experienced a decrease of more than 20,000 inhabitants. This trend partially stems from migration of residents to the functional area outside the city's administrative boundaries and partially from migration to other urban centers offering better employment opportunities and, consequently, higher living standards. The FUA of Włocławek encompasses the largest core city and represents the largest functional area among the case studies analyzed. It consists of 20 administrative units, including cities, urban–rural municipalities, and rural municipalities (Table 5.6). The total area of the Włocławek FUA is 2,131 km^2, with a population exceeding 212,000 residents. This places it above average in terms of both size and population compared to other FUAs. The spatial extent of the Włocławek FUA does not align with the boundaries of Włocławek Poviat, as it excludes certain municipalities within the county (e.g., Izbica Kujawska) while incorporating municipalities from other counties, such as Aleksandrów Poviat (Bądkowo, Waganiec), Lipno Poviat (Bobrowniki, Dobrzyń nad Wisłą, Wielgie), and Radziejów Poviat (Osięciny, Topólka) (Fig. 5.8).

The Włocławek FUA is located within the Southern Baltic Lake Districts, at the intersection of three physical geographical units: Chełmno-Dobrzyń Lakeland, the Toruń–Eberswalde Ice Marginal Valley, and the Wielkopolskie Lakeland. This area is divided by the course of the Vistula River, Poland's largest river. The northern part includes Dobrzyń Land, which overlaps two geographical regions: the Dobrzyń Lakeland and the Urszulewo Plain. This region is characterized by highly diverse post-glacial landforms and numerous lakes, with agricultural land on relatively poor soils interspersed with forests and a dense network of rivers and streams. In the central and southern parts, the Włocławek FUA encompasses the Kujawy region, mainly Eastern (White) Kujawy and partially Western (Black) Kujawy, covering fragments of four physical geographical regions: the Płock Basin and the Nieszawa Gap of the Vistula River (part of the Toruń–Eberswalde Ice Marginal Valley, situated along the Vistula), and the Inowrocław Plain and the Kuyavian Lakeland (part of the Wielkopolskie Lakeland in the southern section of the FUA). The terrain here is predominantly flat, typical of post-glacial ground moraine landscapes, with gently undulating topography marked by the occasional moraine hills and sandy-gravel ridges characteristic of moraine uplands. Due to the presence of high-quality soils, including chernozems, brown soils, and podzolic soils, this region is intensively utilized for agriculture, making it one of the most important agricultural areas in Poland. The eastern part of the Włocławek FUA is distinguished by the presence of a large forest complex with numerous lakes, protected as part of the Gostynin-Włocławek Landscape Park. This park is a critical element of the ecological corridor connecting the Kampinos National Park with the Bydgoszcz Forest and, further, with Tuchola Forest. Its natural value and landscape diversity make the area particularly attractive for tourism and recreation. The park contains various morphological forms, including subglacial channels, eskers, and terrace levels of the Vistula River. There is one of the largest inland dune complexes in Poland originating from the post-glacial period. Within the Gostynin-Włocławek Landscape Park, there are over 40 lakes, including Gościąż Lake, globally unique for its distinct sediment stratification. Rakutowskie Lake and its surrounding wetlands are internationally important bird

Table 5.6 Communes within Włocławek FUA

No	Name	Area (km^2)	Population (2023)
1	Włocławek (city)	85	100,807
2	Bądkowo (rural commune)	80	4,049
3	Baruchowo (rural commune)	106	3,290
4	Bobrowniki (rural commune)	95	2,977
5	Boniewo (rural commune)	78	3,169
6	Brześć Kujawski (rural part of commune)	144	6,479
	Brześć Kujawski (urban part of commune)	7	4,525
7	Choceń (city)	100	7,605
8	Chodecz (rural part of commune)	120	3,733
	Chodecz (urban part of commune)	2	1,705
9	Dobrzyń nad Wisłą (rural part of commune)	110	5,113
	Dobrzyń nad Wisłą (urban part of commune)	6	1,962
10	Fabianki (rural commune)	76	10,243
11	Kowal (city)	5	3,278
12	Kowal (rural commune)	115	3,858
13	Lubanie (rural commune)	69	4,311
14	Lubień Kujawski (rural part of commune)	149	5,445
	Lubień Kujawski (urban part of commune)	2	1,307
15	Lubraniec (rural part of commune)	146	5,947
	Lubraniec (urban part of commune)	2	2,696
16	Osięciny (rural commune)	123	7,180
17	Topólka (rural commune)	103	4,488
18	Waganiec (rural commune)	55	4,415
19	Wielgie (rural commune)	134	6,657
20	Włocławek (rural commune)	219	7,393
	Total	2,131	212,632

Source Own elaboration

protection areas, particularly for waterfowl and wetland species (in the "Rakutowskie Bogs"), and are part of the European Natura 2000 network.

Ethnographically, the Włocławek FUA encompasses two regions. The northern part includes Dobrzyń Land, inhabited by the Dobrzyniacy, a territorial community in historic Poland, situated north-east of Włocławek and bordered by the Vistula River (to the south and south-west), the Drwęca River (to the north and north-west), and the Skrwa River (to the east). The central and southern parts include Kujawy, inhabited by the Kujawiacy, a region situated between Greater Poland, Pomerania, Dobrzyń Land, and Mazovia that is culturally closest to Greater Poland. Geographically, Kujawy is a historic region located between the upper Noteć River, Lake Gopło, and the left

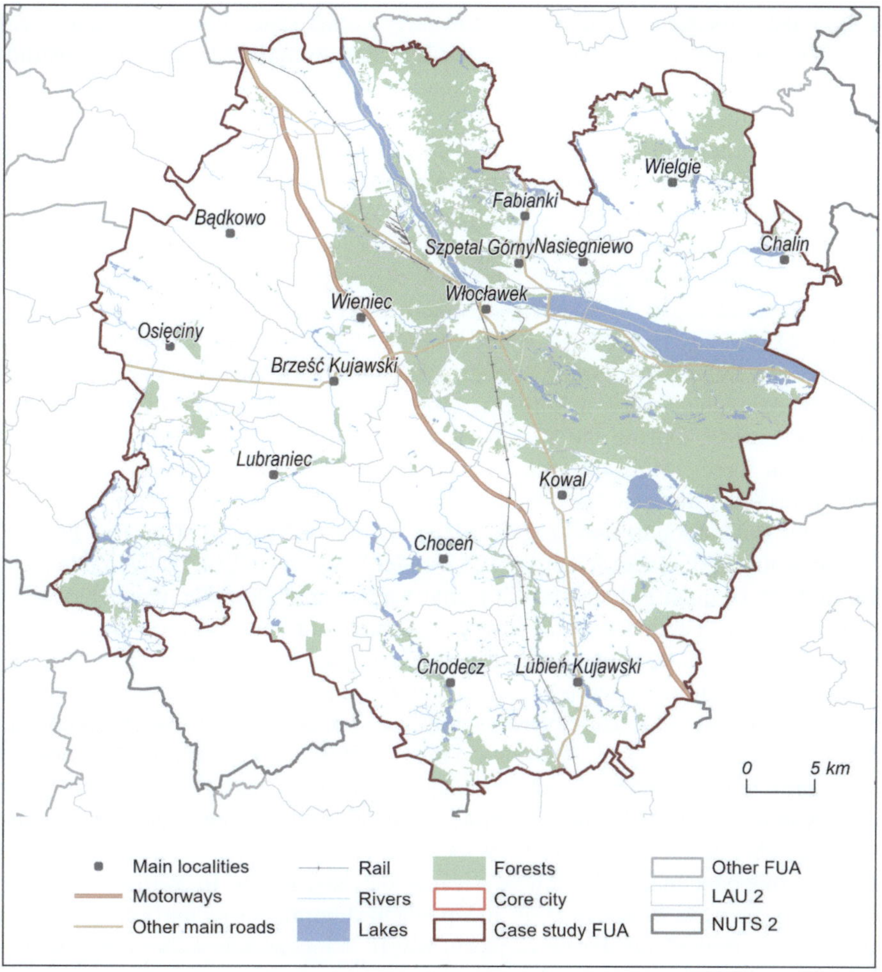

Fig. 5.8 Map of Włocławek FUA. *Source* Own elaboration

bank of the Vistula (from the Skrwa River mouth in the south to the Wda River mouth in the north). Its historical capital was Kruszwica, later moved to Włocławek in the early twelfth century when the seat of bishops was relocated (Radzimski 2017).

Włocławek is one of the oldest cities in Poland. The city's name originates from the masculine name Władysław and its Polish variant Włodzisław. It first appeared as Władysławia in the Polish chronicle written in Latin by Gallus Anonymous between 1112 and 1116 CE. As early as the tenth century BCE, a settlement of the Lusatian culture existed in the area, followed by a settlement of the Pomeranian culture around the fifth century BCE. By the first half of the eleventh century at the latest, a fortified settlement was established, protected by two rivers, a moat, and an earth-and-wood rampart. During the reign of Bolesław Chrobry (992–1025), Włocławek was one of the most important strongholds in the early Polish state, alongside Poznań and Gniezno. The creation of the Włocławek Diocese around 1123 and the construction of a Romanesque cathedral significantly influenced the city's development and political history. In 1255, Duke Casimir I of Kujawy granted Włocławek city rights under Chełmno Law, though historians speculate that this may in fact have happened earlier. The subordination of the city to the affluent and powerful bishops of Włocławek shaped its history for centuries. Włocławek competed economically with the Teutonic-controlled Toruń located further down the Vistula River. During the Polish–Teutonic War, in 1329, Teutonic forces invaded and put Włocławek to the flame, including its Romanesque cathedral. The destruction was so extensive that in 1339 the city had to be re-chartered under Chełmno Law. On the ruins of the Romanesque cathedral, a Gothic cathedral was built, which remains one of the city's symbols. After the Treaty of Toruń in 1466, the city experienced its most intense period of growth. During the Swedish invasion in 1657, Włocławek was torched by Swedish forces. More devastation occurred during wars in the early eighteenth century, bringing decay to the city. By 1789, the city had only 1,325 residents.

During Poland's partitions (1772–1918), Włocławek was first part of Prussia, then, after the Congress of Vienna (1815), under Russian control. Its favorable location midway through Polish lands in fertile Kujawy and along the navigable lower Vistula River with abundant water resources made it attractive for investment during the industrial revolution. Investments included a chicory processing plant, factories producing agricultural machinery, instruments, and physical apparatuses, as well as breweries, a glassworks, a brickyard, sawmills, and other plants. The river port on the Vistula was visited by up to 150 ships at a time. In 1899, the Cassirer brothers opened a pulp mill, which quickly became the major industrial plant of Włocławek. By the early twentieth century, faience, ceramic, chemical, printing, metal, and food industries were also developing. In 1900, Włocławek had 22,971 residents. After Poland regained independence, Włocławek became the largest city in Warsaw Province (excluding Warsaw, which was an independent administrative unit) and the 18th largest city in Poland. On the eve of World War II, Włocławek had about 68,000 residents. The war inflicted severe damage on the city, not only to its infrastructure but also its population. Nearly 40% of its residents were displaced or killed, including most of its Jewish population of at least 12,000 in 1939.

After the war, the communist government seized thousands of homes, hundreds of production facilities and workshops, and numerous social, cultural, and religious properties. Reconstruction efforts peaked between 1947 and 1949, followed by industrial expansion during the six-year plan (1950–55). Factories for ceramics, faience, metal products, and vanillin were established, with major expansions to the "Celuloza" paper mill, then one of the largest in Poland. In 1967, construction of a new chemical fertilizer factory started, and in 1969 the Vistula dam was completed as part of the still-unfinished Lower Vistula Cascade project. During the communist era, Włocławek's concentration of industry earned it the label "red city." The economic transition of the 1990s, involving reorientation of international trade and privatization errors, led to skyrocketing unemployment. The city's decline was exacerbated by its losing provincial capital status in 1998. By 2016, Włocławek was classified as a city losing its socio-economic functions and warranting special intervention (Śleszyński 2016). Efforts to revitalize the city, such as establishing the Włocławek Economic Development Zone and investments by Anwil and by Orlen, failed to achieve the desired outcomes. Depopulation and economic decline continued, making Włocławek's situation unique among Polish cities of its size. This decline was a focus of Piotr Witwicki's book *Znikająca Polska* (Vanishing Poland), where Włocławek was dubbed "Poland's Detroit" (Witwicki 2021).

The Włocławek FUA is located in central Poland, within the Kuyavian-Pomeranian Voivodeship (NUTS 2). Włocławek is situated 63 km from Toruń, 106 km from Bydgoszcz, and 162 km from Warsaw (by road). The city is in close proximity to other large and medium-sized urban centers that serve as FUA capitals, including Toruń (191,000 inhabitants), Płock (112,000), and Konin (71,000). Other main towns within neighboring FUAs, aside from these three larger cities, belong to the medium- and small-sized categories: Kutno (42,700 inhabitants), Koło (19,700), Gostynin (16,900), Lipno (13,400), Aleksandrów Kujawski (11,500), Ciechocinek (10,400), and Radziejów Kujawski (5,100). The Włocławek FUA is well-connected by road, with the A1 motorway serving as a major north–south transportation artery in Poland, linking the Tricity area with Silesia and the southern border with the Czech Republic. Through connections with other national roads and highways, the A1 provides excellent accessibility between Włocławek, regional hubs like Toruń and Bydgoszcz, the national capital Warsaw, and other major cities. Additionally, two major east–west national roads traverse the FUA: National Road 10, connecting Szczecin with Warsaw, and National Road 62, linking Strzelno with Siemiatycze (Fig. 5.9). Włocławek has had access to standard-gauge rail since 1862. Currently, it is served by railway line 18 connecting Kutno and Piła, with direct rail links to major cities such as Gdynia, Bydgoszcz, Toruń, Szczecin, Kołobrzeg, Piła, Warsaw, Lublin, Chełm, Rzeszów, Kraków, Katowice, Bielsko-Biała, and Zakopane. A narrow-gauge railway line connecting the sugar factory in Brześć Kujawski to Włocławek was operational starting in 1908 but was dismantled in 1978 during the communist period, as it was considered outdated. Recent plans for the construction of the Central Communication Port (CPK), intended to become the largest aviation hub in Central and Eastern Europe, included projects for high-speed rail connections to link many Polish cities. For Włocławek, the proposed rail projects included (1) a line connecting Płock—Włocławek—Grudziądz—Tczew—the Tricity (Northern CMK)—Lębork—Słupsk, and (2) a line linking Włocławek—Toruń—Bydgoszcz—Piła—Szczecinek—Białogard—Kołobrzeg as part of "spoke no. 1" integrating Pomerania and Kuyavia. These plans would significantly reduce travel times to regional hubs, Warsaw, and other major cities in Poland. However, following a change in government in 2024, the decision was made to continue the first phase of the Central Communication Port construction without implementing the nationwide rail component.

The Włocławek FUA is located in the southern part of the Kuyavian-Pomeranian Voivodeship, bordering the Mazovian, Łódź, and Greater Poland Voivodeships (NUTS 2). The Kuyavian-Pomeranian Voivodeship was established in 1999 and is one of two regions in Poland with dual capitals, Bydgoszcz and Toruń. The

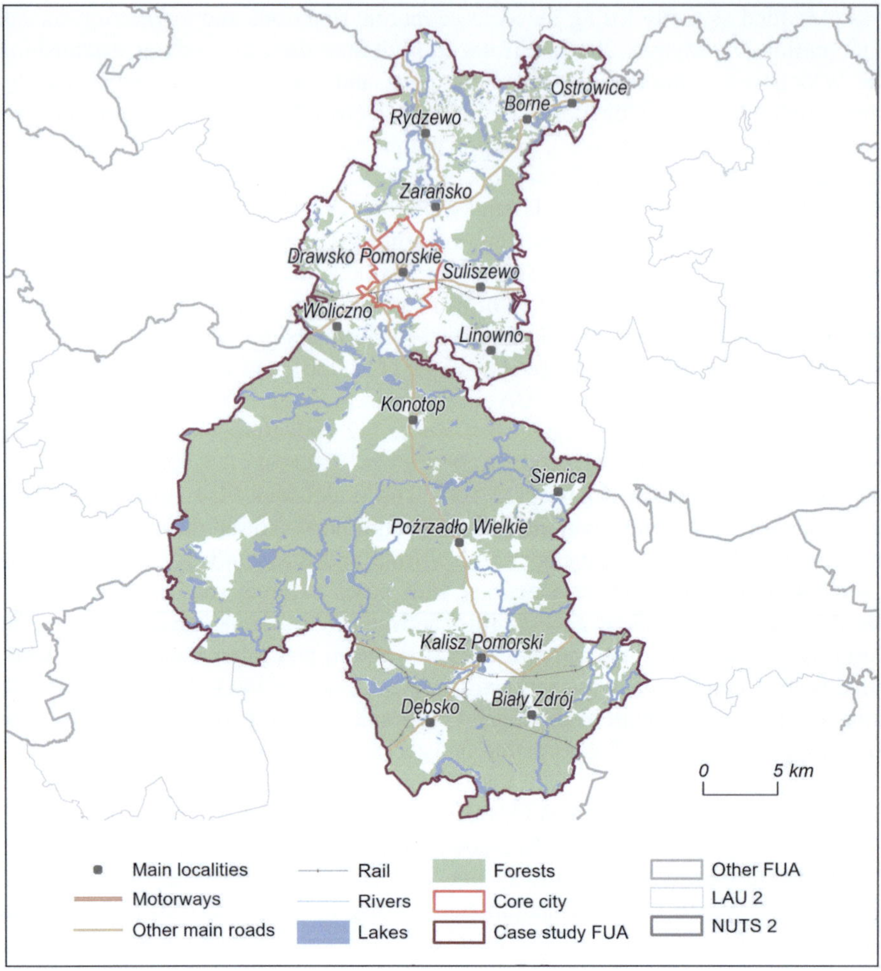

Fig. 5.9 Map of Drawsko Pomorskie FUA. *Source* Own elaboration

intense rivalry, or de facto antagonism, between these two cities hampers their development and negatively affects the growth of the entire region, including the Włocławek FUA. The Kuyavian-Pomeranian Voivodeship covers an area of 17,971 km^2, which constitutes 5.5% of Poland's total area. The region is home to approximately 2.0 million inhabitants, also representing 5.5% of the country's population. Economically, the Kuyavian-Pomeranian Voivodeship is among the less-developed regions in Poland. According to Statistics Poland (GUS), its regional gross domestic product (GDP) in 2022 amounted to PLN 133.9 billion (€31.5 billion), accounting for 4.4% of Poland's GDP (ranking 8th among voivodeships). The GDP *per capita* (PPS) was €23,100 in 2022, which was only 65% of the EU average, qualifying the region as a less-developed area under the European Union's Cohesion Policy criteria. According to the *Strategy for the Development of the Kuyavian-Pomeranian Voivodeship 2030+— Acceleration Strategy 2030 +* (2020), Włocławek, alongside Grudziądz and Inowrocław, has been classified as a regional center complementing voivodeship capitals. These centers are intended to stimulate regional development and act as core hubs for supra-local labor markets while serving as centers of social and economic activity. Their development is expected to directly impact smaller urban centers and rural areas functionally connected to them. These centers are envisioned to provide business, administrative, cultural, educational, knowledge-based, and healthcare services at a regional level and, in some cases, at the national level. In the regional policy of the Kuyavian-Pomeranian Voivodeship, Włocławek and other regional centers are categorized as "medium-sized cities." As part of the territorialization of regional policy actions, the Włocławek FUA has been designated as an Area of Strategic Intervention (ASI). However, its spatial extent in official strategies is significantly smaller than the functionally identified area. Włocłwek ASI, in the strategy, encompasses only ten municipalities—half of those being functionally connected to Włocławek. It is worth emphasizing, however, that the majority of units we have included in the Włocławek Functional Urban Area (FUA) have been identified as areas of negative socio-economic conditions at the regional level (Regional Development Strategy of the Kuyavian-Pomeranian Voivodeship... 2020—Annex No. 2).

The Włocławek FUA is experiencing deeply negative demographic trends, particularly in its peripheral areas. For every 100 working-age individuals, there are 37 post-working-age ones, and the natural growth rate is −4.6‰, a result of an extremely high decline in birth rates. These negative demographic trends are accompanied by stagnation in the construction sector (Fig. 5.6).

Economically, the Włocławek FUA is characterized by poor performance. Personal income stands at 30,400 PLN *per capita*, representing 75% of the value recorded in the regional centers of Kuyavian-Pomeranian Voivodeship (Bydgoszcz and Toruń) and 44% of the value corresponding to the Warsaw FUA. Only 7.4% of residents are engaged in entrepreneurial activities, and the unemployment rate (8.6%) is significantly higher than the average for peripheral areas (Fig. 5.6). Among economic sectors, services dominate over industry within the FUA. In the service sector, trade is predominant. Key industries driving Włocławek's economic development include chemicals, construction, food processing, and metalworking. The largest employers in the Włocławek FUA are among the top 10 enterprises in the Kuyavian-Pomeranian Voivodeship. One notable employer is the Polish retail network Lewiatan, which operates nearly 3,000 stores nationwide, employing 25,000 people. Its headquarters in Włocławek generate revenues exceeding PLN 15.3 billion (€3.6 billion). Another major firm is Anwil, part of the Polish oil giant Orlen. Anwil, the largest employer in Włocławek and one of the largest in the region, produces nitrogen fertilizers and PVC, employing over 1,200 people in Włocławek and contracting work to numerous local companies. Anwil's revenues exceed PLN 5.7 billion (€1.3 billion). Other notable enterprises include: WIKA Polska, a successor to the Manometer and Thermometer Factory founded by Ignacy Ciechurski in 1916, employing over 800 workers and generating revenues of PLN 579 million (€136.2 million); Salamander Window & Door Systems SA, a global manufacturer of windows and doors, employing more than 600 workers in Włocławek and achieving revenues of PLN 837 million (€196.9 million); WIRECO POLAND Sp. z o.o., a successor to the Wire Products Factory founded in 1895 by C. Klauke, known for producing steel wires and cables, employing over 500 people and generating revenues of PLN 417 million (€98.1 million); and Guala Closures DGS Poland SA, a leading global manufacturer of aluminum closures for glass packaging, employing over 500 people and generating revenues of PLN 733 million (€172.5 million).

Outside Włocławek, a significant concentration of smaller enterprises is found in the Brzeska Special Economic Zone located near the A1 motorway junction. This area is experiencing strong economic growth, surpassing the performance of the Włocławek Economic Development Zone—Industrial and Technological Park (WSRG-PPT), which is part of the Pomeranian Special Economic Zone. Companies in the Brzeska Special Economic Zone include: Kongsberg Automotive (automotive components manufacturer), LPP Logistics (fashion logistics), RTE (bicycle manufacturer), Anwis (provider of shading technology solutions), H&S Steel (steel profiles, sheets, and tape manufacturer and distributor), Kleib (construction chemicals producer). Despite these developments, the labor market in Włocławek and its functional area has not fully recovered since the economic transformation of the 1990s and the recession following the loss of administrative functions in 1999. The local labor market continues to experience above-average unemployment levels, despite the presence of many secondary and vocational schools, as well as higher education institutions, including the State Academy of Applied Sciences, Kuyavian University, and a branch of the Higher School of Computer Science in Łódź in Włocławek.

5.4.2 Drawsko Pomorskie FUA: One of the Most Sparsely Populated Areas in Poland

Drawsko Pomorskie is a town in north-western Poland, located in the West Pomeranian Voivodeship. It serves as the seat of Drawsko Poviat and the urban–rural municipality, with a population of approximately 16,000 (Table 5.7). Together with the Kalisz Pomorski municipality, it forms aFunctional Urban Area (Fig. 5.9) a total population of around 23,000 and a territory covering 890 km^2. Both the number of municipalities within the FUA and its population make it one of the smaller Functional Urban Areas in Poland.

The FUA area is located within the Drawsko Lakeland, which is part of the West Pomeranian Lakeland, as well as within the Drawsko Plain, belonging to the physical geographical macroregion of the South Pomeranian Lakeland (Kondracki 1965, Drawsko Pomorskie 2025a). The FUA borders other municipalities of Drawsko County, such as Złocieniec and Wierzchowo, as well as the municipalities of Łobez, Połczyn-Zdrój, and Świdwin. This area is characterized by diverse terrain shaped by

Table 5.7 Communes within Drawsko Pomorskie FUA

No	Name	Area (km^2)	Population (2023)
1	Drawsko Pomorskie (urban part of commune)	22	11,224
	Drawsko Pomorskie (rural part of commune)	387	5,097
2	Kalisz Pomorski (urban part of commune)	12	4,039
	Kalisz Pomorski (rural part of commune)	469	2,800
	Total	890	23,160

Source Own elaboration

the last Baltic glaciation. The most distinctive landscape features are moraine hills exceeding 150 m above sea level and numerous post-glacial lakes. One example is Lake Drawsko, one of the deepest lakes in Poland, reaching nearly 80 m in depth. The most significant river in the area is the Drawa, which, together with lakes such as Lubie, Zarańsko, Dołgie, and Gągnowo, forms an extensive water system that supports both the fishing industry and water-based recreation, including kayaking. Drawa, a right-bank tributary of the Noteć River, creates picturesque gorges within the municipality's borders, resembling mountainous landscapes, making it one of the most renowned kayaking routes in Poland.

The region's soils are diverse. The moraine areas are characterized by fertile soils that are favorable for agriculture, with dominant brown and podzolic soils formed on glacial tills. In contrast, sandy soils dominate the outwash plains, offering lower agricultural value and often supporting pine forests. The area is distinguished by a high degree of forestation, with a significant portion covered by forests and protected areas, including the Drawsko Landscape Park and numerous Natura 2000 sites. As a result, the region has high tourism and recreational potential, particularly in nature-based, water, and active tourism.

The history of Drawsko Pomorskie and Kalisz Pomorski dates back to the early Piast era, when these lands were part of the Polish state (Drawsko Pomorskie 2025b). Following the period of feudal fragmentation, they came under the rule of the Duchy of Pomerania and later Brandenburg, which initiated intensive colonization in the thirteenth century. In 1297, Drawsko was granted town privileges under Magdeburg Law by the Brandenburg Margraves Otto IV and Conrad I, marking the beginning of its development as a trade and craft center. Over the centuries, the region was influenced by Brandenburg, the Kingdom of Poland, the Teutonic Order, and the Knights Hospitaller, giving it strategic significance. The seventeenth century brought waves of disasters—devastating fires, the Thirty Years' War, and plagues—which led to significant population and economic losses. The town rebounded in the eighteenth and nineteenth centuries, gaining administrative importance with the establishment of Drawsko County in 1815. The expansion of road and rail infrastructure, including the Koszalin–Szczecin connection in 1877, contributed to economic growth. After World War I, the town found itself on the periphery of Germany, slowing its development. However, in the 1930s, with the construction of the Pomeranian Wall, it became an important strategic point. In 1945, it was captured by the Polish Army and subsequently incorporated into Poland. The immediate post-war years brought destruction caused by the stationed Soviet troops.

Today, the Functional Urban Area of Drawsko Pomorskie is a region rich in history, reflected in its urban layout, historical landmarks, and local traditions. The heritage of medieval settlement, Brandenburg colonization, nineteenth-century expansion, and the hardships of the twentieth century shape its identity. Drawsko Pomorskie and Kalisz Pomorski draw from this history to create an attractive cultural and tourist offering, appealing to both residents and visitors.

The Functional Urban Area (FUA) of Drawsko Pomorskie benefits from a favorable transportation network. The national road 20 runs through its territory, connecting Stargard with Szczecinek, alongside several regional roads: 148 (Drawsko Pomorskie—Łobez), 162 (Zarańsko—Świdwin), 173 (Drawsko Pomorskie—Połczyn-Zdrój), and 175 (Drawsko Pomorskie—Kalisz Pomorski). Railway connections are provided by line 385 (Runowo Pomorskie—Szczecinek), facilitating access to regional economic and administrative centers. The distances to major cities in the region are: Szczecin (390,000 inhabitants)—100 km, Koszalin (105,000 inhabitants)—100 km, Kołobrzeg (41,000 inhabitants)—80 km, and Poznań (540,000 inhabitants)—200 km. Bus services are operated by PKS Złocieniec, but their availability, especially in the evening hours, is limited. Due to the insufficient public transport service, many residents rely primarily on private vehicles.

The Functional Urban Area (FUA) of Drawsko Pomorskie, as part of the Central Zone Partnership (ZMP 2021), is integrated into a broader territorial cooperation framework within the West Pomeranian Voivodeship. Recognized in the Regional Development Strategy as a key subregional hub, Drawsko Pomorskie plays a crucial role in military, tourism, and economic development, particularly in wood processing, forestry, and agriculture. At the same time, it is situated within a Strategic Intervention Area due to persistent socio-economic challenges, including high unemployment, economic stagnation, and demographic decline, exacerbated by the legacy of liquidated State Agricultural Farms (PGRs) and its peripheral location (Pomorze Zachodnie 2023). Its strategic significance lies in enhancing functional linkages, improving mobility, and attracting infrastructure and economic investments to counter depopulation and foster local entrepreneurship. Through balanced, polycentric regional growth, the city contributes to improving access to public services in peripheral areas, supporting broader efforts in territorialized development policies.

The Functional Urban Area (FUA) of Drawsko Pomorskie is one of the most sparsely populated regions in Poland. Its population density is significantly lower than both the national and regional averages due to vast forested areas, extensive agricultural land, and a relatively low number of inhabitants. The region is experiencing a clear trend of depopulation and an aging population. Several factors contribute to these demographic changes: young people are increasingly migrating to larger cities for education and better job opportunities, birth rates are steadily declining, and the proportion of post-working-age residents is growing. While external migration has a relatively minor impact, some residents do move abroad, though not on the same scale as from other Polish regions. Recently, there has been a modest influx of new residents, primarily individuals seeking tranquility and natural surroundings.

The economy of the FUA is based on several key sectors: the military, forestry, the wood industry, agriculture, trade, and tourism. The most significant employer and economic driver is the Combat Training Center in Drawsko Pomorskie, one of the largest and most important military training grounds in Europe. This site plays a strategic role within NATO and the Polish Armed Forces, attracting military personnel from across the world. The permanent presence of military units and training exercises supports a range of local services, including transportation, hospitality, accommodation, and technical and administrative services. A considerable number of residents in Drawsko Pomorskie and Kalisz Pomorski are directly or indirectly employed in institutions and companies linked to the military. The training ground also generates high demand for rental properties, affecting the local real-estate market.

The FUA is rich in forest resources, which support the development of forestry and the wood industry. A significant part of Kalisz Pomorski is covered by forests managed by the State Forests, ensuring stability in this sector. The wood industry encompasses both raw material extraction and processing, including sawmill operations, construction components, furniture manufacturing, and wood-based products. Several larger companies operate in the area, alongside numerous smaller craft-based businesses.

Agriculture is an important economic sector, particularly in the southern part of the FUA, especially in Kalisz Pomorski, where farms focus on crop production and livestock breeding. Key crops include cereals, rapeseed, and fodder plants, alongside a growing number of organic farms contributing to sustainable agriculture. The region also hosts dairy farms and cattle and pig breeding operations. Expansive meadows and pastures favor animal husbandry and fodder production.

Trade and services play a vital role in the local economy, concentrated mainly in the urban centers of Drawsko Pomorskie and Kalisz Pomorski. Both towns have shopping centers, markets, and service points catering to residents, tourists, and military personnel stationed at the training ground. The region supports a mix of small, family-run businesses and larger retail chains. The service sector includes gastronomy, transportation, logistics, IT, and administrative services, responding to the needs of an evolving local market.

Tourism, while not the main sector of the economy, is playing an increasingly significant role in the region's development. The Drawskie Lakeland, with its numerous lakes, forests, and protected areas such as the Drawski Landscape Park, offers excellent conditions for active tourism.

The Functional Urban Area (FUA) of Drawsko Pomorskie has considerable development potential, driven by the synergy of various economic sectors. The modernization of road and rail infrastructure, increased investment in modern technologies within the wood processing industry and agriculture, as well as the further development of tourism, can contribute to the region's economic growth. However, challenges remain, particularly demographic issues related to the outflow of young people to larger cities, as well as the need for further economic diversification to enhance the region's resilience to potential crises.

5.4.3 *Przasnysz FUA: In the Shadow of Warsaw Agglomeration*

Przasnysz is a poviat capital city with a population of 16,000 inhabitants (Table 5.8), which forms a Functional Urban Area together with five nearby rural communes (Przasnysz rural commune, Czernice Borowe, Jednorożec, Krasne and Krzynowłoga Mała, Fig. 5.10). Altogether, the FUA has an area of 848 km^2 and a population of 40,000 inhabitants, locating it above the average in size but around the average in population for an FUA. Przasnysz FUA encompasses Prasnysz poviat with the exception of the urban–rural municipality of Chorzele.

The area is located within the Northern Masovia Lowland. Central and south-eastern parts of the FUA belong to the geographic region of Ciechanów Heights and Mława Hills and are flat with several moraine hills and cut by the Węgierka River flowing from the north-west toward the south-east. Fertile soils are built on glacial tills, so most of the land is used for farming and plant production, with wheat and sugar beats being major crops. The north-eastern part of the FUA is covered by sandy and poor soils covered by the pine forest of Kurpie Plain (Solon et al. 2018). There are wetlands in the valleys of the river Orzyc and other rivers. North-eastern forested parts are partially included in the European Nature 2000 framework. Apart from forestry, this part of the FUA is the fringe of Poland's largest cattle grazing area, covering the north-eastern part of the country. Forestry and isolated character of the region formed a specific regional culture called "Kurpie" in the north-eastern part of the region and neighboring regions (Kondracki 1965).

As a local center for a farming region, Przasnysz was granted city rights by Mazowsze duke Janusz in 1427 and developed as a royal town and a trading center in the sixteenth century due to its location on the crossroads of major trading routes

Table 5.8 Communes within Prasnysz FUA

No	Name	Area (km^2)	Population (2023)
1	Przasnysz (city)	25	16,357
2	Czernice Borowe (rural commune)	120	3,491
3	Jednorożec (rural commune)	235	6,607
4	Krasne (rural commune)	101	3,270
5	Krzynowłoga Mała (rural commune)	185	3,235
6	Przasnysz (rural commune)	182	6,845
	Total	848	39,805

Source Own elaboration

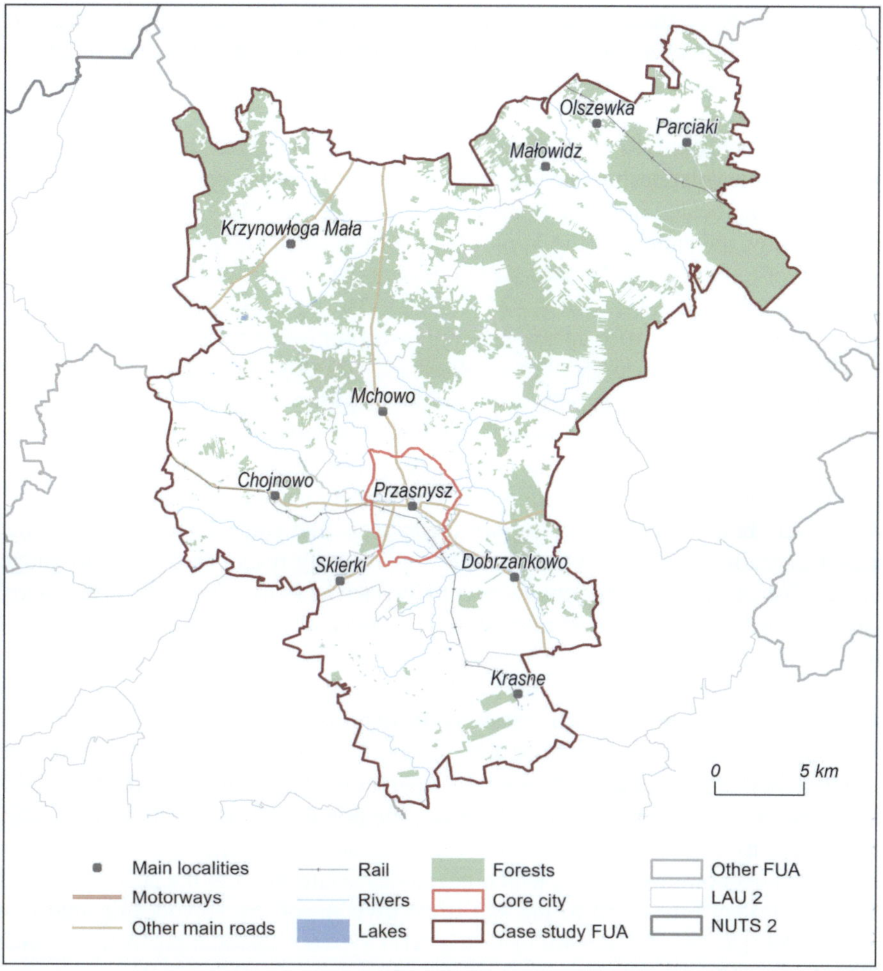

Fig. 5.10 Map of Przasnysz FUA. *Source* Own elaboration

connecting Toruń with Lithuania and Warsaw with Olsztyn and Prussia. The seventeenth century brought decay to the city due to a fire in 1613 followed by the Swedish invasion of 1657. As of 1815, Przasnysz belonged to Russia and slowly gained significance as a local center of farming, trade, and craft. After the failure of the January Uprising, a significant Russian military contingent and local administration were located here, starting the new military function of the town and its position as poviat capital, both of which continue to this day. At the turn of the twentieth century, Przasnysz had already more than 10,000 inhabitants of mixed nationality. During World War I, the city was subject to heavy fights between Russian and German armies. Resulting destructions, together with the subsequent Polish–Soviet War and the deportation of the Jewish population during the Second World War, halted urban development. Nevertheless, some public investments were made in infrastructure and education in the interwar period. The growth returned in 1960s, when the first large industrial plant, a branch of the High Voltage Equipment Manufacturing Plants (ZWAR), was built together with a vocational school. Between 1975 and 1998, the city's importance diminished due to the division of the urban areas between Ostrołęka and Ciechanow voivodeship and losing its poviat status, which was regained in 1999.

The area is located north of Mazowieckie Voivodeship, 103 km (road distance) from Warsaw, the capital of the Voivodeship and Poland. There are no other major cities closer to Przasnysz. Middle-sized ones are the capitals of neighboring FUAs: Ciechanów (42,000 inhabitants)—25 km and Ostrołęka (48,000)—52 km away. Other nearby FUAs are formed around similar sizes or even smaller towns (Mława—30,000, Maków Mazowiecki—9,000, Wielbark—3,000). The region is not accessible by any motorway or expressway, and the major arteries are National Road 57, connecting it to Warsaw, and voivodeship Road 544 and 617, connecting Przasnysz with nearby towns (Fig. 5.9). Przasnysz has never been accessible by standard-gauge railway. A narrow-gauge connection with the nearby manor Krasne and railway hub Mława was built in 1915 to transport sugar beets and for military purposes, but passenger traffic was discontinued in 1985. A recently reactivated standard-gauge railway Ostrołęka-Szczytno crosses part of Przasnysz FUA. There are preliminary plans to build a new Zegrze–Przasnysz railway line, which would reduce the travel time to Warsaw from the current 1.5–2 h by car to a little over 1 h.

The FUA is a characteristic representative of the outer, northern, area within Mazowsze voivodeship. The voivodeship is the largest in Poland (35,600 km^2, 11.3% of the area of Poland), the most populous one (5.5 million inhabitants, 14.6% of the country's population) and boasts the largest economy (regional product of PLN 0.7 trillion is 23.1% of Polish GDP). However, the population, particularly economic products, is mostly concentrated in the region's capital city, Warsaw. The elevation of regional economic statistics, which negatively contributed to the accessibility of Cohesion Policy funds, led to the statistical division of the voivodeship into two regions (NUTS-2 units) according to the NUTS2016 classification in force since 2018. Ten poviats forming Warsaw agglomeration comprise 8.7% of Poland's population and 17.5% of GDP, which *per capita* is twice as high as the Polish average and 162% as high as EU average. The regional unit, where Przasnysz belongs, lies below both EU (73%) and Polish average (94.7%).

In the regional strategy (Strategia Rozwoju Województwa Mazowieckiego 2030+), Przasnysz Poviat is located within the Ostrołęka problem Area of Strategic Intervention together with four other poviats. Low economic productivity, unfavorable demographic dynamics, high share of agriculture and food sector in the economy, and large protected areas characterize this north-eastern peripheral area of Mazowieckie voivodeship. Strategic instruments that are aimed to improve the socio-economic situation of the area are focused on the improvement of transport accessibility (including rail and road infrastructure and public transport), and the protection and utilization of the natural environment (including the development of ecological farming, agritourism and other forms of tourism, and the development of technical infrastructure).

Przasnysz FUA faces demographic problems typical for peripheral areas. There are as many as 33 persons of post-productive age per 100 of productive age and a negative natural increase resulting from a dramatic decrease in birthrates. The negative demographic trends are followed by stagnation in the construction sector and are associated with low social activity, e.g., low participation in courses for adults (Fig. 5.6).

Among the outer part of Mazowieckie voivodeship, Przasnysz is even among those with the least favorable values of economic indicators. Personal income of PLN 27,600 per inhabitant is c. 40% of that in Warsaw. Only 7.1% of the population own businesses, and the unemployment rate is close to the average for peripheral areas (6.6%). The obstacle to economic development may be the poorly developed business infrastructure, e.g., less than one hotel bed per thousand inhabitants—among the lowest values in Poland (Fig. 5.6). The largest employers of the FUA (apart from public entities—the military unit and poviat hospital) are two industrial manufacturers located in Przasnysz. One of them is formerly ZWAR, now owned by ABB, a multinational machinery manufacturer. The factory produces electrical equipment, primarily fuses, exported to multiple countries. It has around 400 employees, including skilled experts. The second major production company is bike manufacturer Kross, a private firm established in 1990 as a bike shop and later as a manufacturer selling pushbikes and electric bikes in Polish and multiple international markets. Around 500 employees work in Przasnysz in the firm's headquarters and a production plant. Larger industrial firms avoid the city despite the existence of a special economic zone in the rural commune, which hosts a branch of the Faculty of Power and Aeronautical Engineering of the Technical University in Warsaw.

Apart from farming, manufacturing and public services, commercial services are the last major sector of the economy in Przasnysz FUA. Around Przasnysz, many small firms have benefitted from the expansion of the Polish road transport sector in European markets. Retail is the last important sector of the local economy—the largest retail center, Part Targowa, opened in Przasnysz in 2023.

5.4.4 Sandomierz FUA: Historic Center Now Located on the Regional Border

Sandomierz is a city with county rights, home to 21,000 residents (Table 5.9). Together with eight neighboring municipalities, including three urban–rural (Klimontów, Koprzywnica, Zawichost) and five rural (Dwikozy, Lipnik, Obrazów, Samborzec, Wilczyce, Fig. 5.11), it forms a Functional Urban Area. The total area of the FUA covers 670 km^2 and has a population of 69,000, placing it above the average population size for FUAs. The spatial extent of the Sandomierz FUA encompasses Sandomierz County, excluding the rural municipality of Łoniów, but including the rural municipality of Lipnik from the Opatów County.

Table 5.9 Communes within Sandomierz FUA

No	Name	Area (km^2)	Population (2023)
1	Sandomierz (city)	29	21,273
2	Dwikozy (rural commune)	85	8,159
3	Klimontów (urban part of the commune)	5	1,758
	Klimontów (rural part of the commune)	95	5,707
4	Koprzywnica (urban part of the commune)	18	2,368
	Koprzywnica (rural part of the commune)	51	3,850
5	Lipnik (rural commune)	81	4,787
6	Obrazów (rural commune)	71	6,086
7	Samborzec (rural commune)	85	8,047
8	Wilczyce (rural commune)	70	3,430
9	Zawichost (urban part of the commune)	20	1,596
	Zawichost (urban part of the commune)	60	2,410
	Total	670	69,471

Source Own elaboration

The Sandomierz FUA is located in south-eastern Poland on the Sandomierz Upland. It occupies the eastern part of the Kielce Upland macroregion and borders the Sandomierz Basin, part of Northern Subcarpathia. The geological foundation of the Sandomierz Upland is an extension of the folded belt of the Świętokrzyskie Mountains, with outcrops present in the form of an escarpment on the left bank of the Vistula River in the north-eastern part of Sandomierz and the southern part of the Dwikozy municipality, known as the Góry Pieprzowe (Pepper Mountains)—the central part of the upland features picturesque loess-covered terrain characterized by undulating topography. Erosional forms are common, cutting through loess surfaces in the form of ravines, including spectacular Queen Jadwiga Ravine and Piszczele Ravine (recognized as historical monuments since 2017). Deeper valley incisions form valleys continuously drained by the left tributaries of the Vistula River, such as the Koprzywianka and Opatówka rivers (Strzyż 2021). The left-bank part of Sandomierz is situated in the Nadwiślańska Lowland, which belongs to the Sandomierz Basin.

The Sandomierz Upland is characterized by specific climatic conditions conducive to plant growth. The most significant climatic factors include the length of the frost-free period, solar radiation, temperature distribution, and total precipitation (Suszyna 2000). Soil and climatic conditions make the Sandomierz Upland a highly utilized agricultural area for cultivating cereals, sugar beets, vegetables, soft fruits, stone fruits, apples, and vineyards. The dominance of this land use has led to a reduction in forest and grassland communities (Ruraż and Panek 2016). The natural potential vegetation primarily consists of Central European deciduous, pine, and lowland riparian forests. The Sandomierz FUA boasts significant natural and cultural tourism values. The region is beautiful due to Sandomierz, often called "Little Rome," as

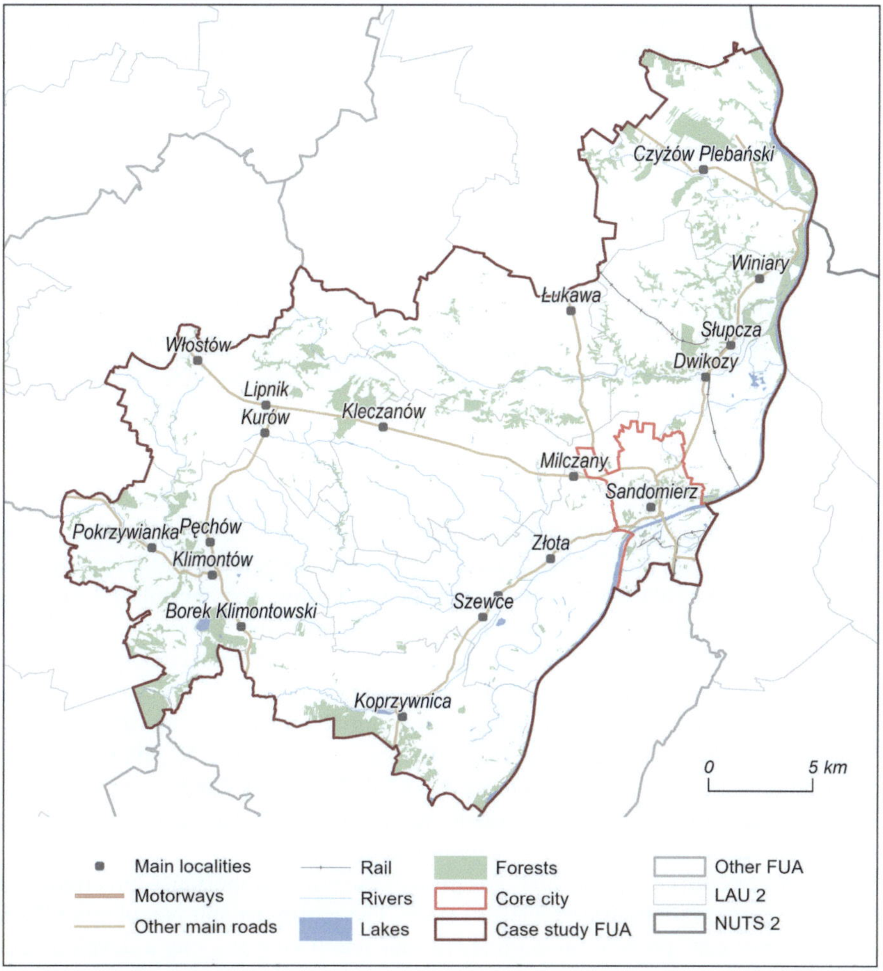

Fig. 5.11 Map of Sandomierz FUA. *Source* Own elaboration

it is situated on seven hills. The area has a 3,000-year settlement history, including nearly 1,000 years of history as a royal city (Strzyż 2021).

Sandomierz is one of the oldest cities in Poland, with its origins linked to its location at the intersection of major trade routes of medieval Europe connecting Ruthenia with Western Europe and Pomerania with Hungary (Kiryk and Chyła 2019). Situated on a high left bank of the Vistula River, Sandomierz was protected from flooding and had a river crossing at the foot of the urban hill. The Vistula was a crucial transport artery leading from the south to the north, reaching Gdańsk, the largest Baltic port. This strategic position facilitated the development of long-distance trade, which contributed to the city's wealth (Okoński 2016). The nearby town of Zawichost previously had a river crossing, but it lost significance in favor of the Sandomierz crossing. Zawichost was formerly a castellany with a market settlement located near the river crossing on a high limestone terrace of the Vistula's left bank. The earliest mention of Sandomierz dates back to the eleventh century, and, by the twelfth century, it had become the capital of one of Poland's provinces during the period of feudal fragmentation. In 1286, Sandomierz was granted city rights under Magdeburg Law, contributing to its economic and urban development. The nearby town of Koprzywnica is known for its Cistercian abbey, founded in the twelfth century, which played a significant role in the region's economic and cultural development. The third and smallest town within the Sandomierz FUA is Klimontów, which dates back to the thirteenth century. From the seventeenth century, it was owned by the Ossoliński family, who contributed to its development by founding, among other structures, the Collegiate Church of St Joseph. The neighboring villages of Dwikozy, Lipnik, Obrazów, Samborzec, and Wilczyce also have medieval origins, with their inhabitants primarily engaged in agriculture, supplying food to nearby towns (Okoński 2016).

Sandomierz reached its peak prosperity in the fifteenth and sixteenth centuries when it functioned as a significant commercial and administrative center of the Crown of the Kingdom of Poland. Due to its fertile loess soils and advantageous location on the Vistula River, Sandomierz became a hub for agricultural exports, including grain, and a trading center for wine, salt, and other goods (Kochanowicz and Sosnowska 2011). The city's decline began in the seventeenth century, primarily due to the Swedish invasions, which devastated the region. In 1655, Sandomierz was occupied and plundered by Swedish forces, significantly weakening its economy. Subsequent conflicts, including the Northern Wars and wars with Russia in the eighteenth century, further contributed to its downfall (Muszyńska and Wijaczka 1996). Following the partitions of Poland, Sandomierz came under Austrian rule and, later, after the Congress of Vienna, under Russian control.

In the nineteenth century, the city lost its administrative importance but continued to develop as an agricultural and craft center. The partitions and Napoleonic Wars led to a significant population decline, further hampering the city's development (Kowalski 1993). The twentieth century brought additional challenges. During World War I, Sandomierz was in the conflict zone, leading to the destruction of infrastructure. After Poland regained independence in 1918, efforts were made to rebuild the city with investments in education and infrastructure. However, World War II resulted in further losses, including the deportation of the Jewish population and the destruction of historical structures (Merta-Staszczak 2020). Post-war reconstruction gradually restored Sandomierz's position. The 1960s and 1970s saw investments in food processing industries and tourism development through historical site restoration and promoting the region's natural attractions. Since the 1999 administrative reform, Sandomierz has regained county status, strengthening its role as a local administrative, economic, and tourist center (Farciak 2017).

Sandomierz FUA is located in south-eastern Poland within the Świętokrzyskie Voivodeship, approximately 90 km (by road) from Kielce (population 183,000), the regional capital. The city is relatively distant from major regional centers: 85 km north of Rzeszów, 120 km south-west of Lublin, 160 km north-east of Kraków, and 200 km south-east of Warsaw. However, it is in proximity to other medium-sized urban centers that form FUAs, such as Tarnobrzeg (43,700 residents) 15 km away, Stalowa Wola (55,000) 30 km away, and Nisko (21,200) 40 km away). All the nearest Functional Urban Areas are located in the Podkarpackie Voivodeship, with which Sandomierz signed a quadrilateral agreement, subsequently establishing the "Four-City Association" Partnership in July 2022. This partnership aims to foster cooperation for joint development beyond the boundaries of the voivodeship. The closest major centers forming an FUA within the Świętokrzyskie Voivodeship include Ostrowiec Świętokrzyski (62,000 inhabitants) 45 km from Sandomierz, Starachowice (44,400) 75 km away, and Skarżysko-Kamienna (41,200) 95 km away. These three cities, situated in the historical Old Polish Industrial Region, have experienced a decline in industrial significance but are increasingly recognized as centers of post-industrial tourism.

The Sandomierz functional area suffers from poor transport accessibility, with no motorways or expressways running through it. However, the planned S74 expressway is expected to traverse the Łódzkie, Świętokrzyskie, and Podkarpackie Voivodeships, connecting Sulejów in Łódzkie Voivodeship to Nisko in Podkarpackie Voivodeship. This road will link two other expressways: S12 (intersecting the A1 Amber One motorway) and S19 (Via Carpatia, connecting Białystok to Barwinek on the Slovak border). Currently, the Sandomierz FUA is served by two national roads: 77 connecting Lipnik with Przemyśl via Sandomierz, and 79, which runs from Warsaw through Kraków to Bytom, crossing the north-western part of the FUA. Sandomierz has a railway station on line 25 linking Łódź Kaliska with Dębica, facilitating connections to cities such as Skarżysko-Kamienna, Ostrowiec Świętokrzyski, and Mielec. The Sandomierz station is located 3 km from the city center on the left bank of the Vistula River. It operates six daily passenger trains: two to Warsaw, two to Przemyśl, one to Stalowa Wola, and one to Rzeszów. The nearest major international airports are over 200 km away in Warsaw and Kraków, while smaller airports in Rzeszów and Lublin are more than 100 km away. The Eastern Green Velo Cycling Route, the longest cycling route in Poland (over 2,000 km in length), runs through the Sandomierz FUA.

The Sandomierz FUA is situated in the south-eastern part of the Świętokrzyskie Voivodeship (NUTS 2), bordering the Podkarpackie and Lublin Voivodeships. The Świętokrzyskie Voivodeship is one of five voivodeships in Eastern Poland, a region characterized by the lowest socio-economic development indicators and thus designated for exceptional support under the EU's Cohesion Policy for 2007–27. The Świętokrzyskie Voivodeship Development Strategy 2030+ identifies strategic intervention areas, including nine medium-sized towns experiencing socio-economic decline (including Sandomierz) and areas threatened by long-term marginalization. Sandomierz is classified as a subregional center with a moderately unfavorable socio-economic situation. Nevertheless, the Sandomierz Urban Functional Area has been designated as a strategic intervention area for the Świętokrzyskie Voivodeship. Its scope differs from the results of our delimitation and includes only four rural municipalities adjacent to Sandomierz: Dwikozy, Obrazów, Samborzec, and Wilczyce. Other municipalities within the Sandomierz FUA have been incorporated into the strategic intervention area "Vistula Valley," which serves as one of Poland's most important ecological corridors. This area is almost entirely protected under the Natura 2000 network but faces environmental constraints, including flood-prone areas and a high concentration of landslides (Świętokrzyskie Voivodeship Strategy 2030 + 2021).The Sandomierz Urban Functional Area exhibits highly unfavorable demographic trends. The demographic dependency ratio for the post-working-age population is 39, and the natural population decline rate stands at −5‰, the highest among the case studies analyzed (Fig. 5.6). Additionally, a housing sector crisis is evident in the studied area. The economic profile of the Sandomierz FUA is reflected in its land-use distribution, with agricultural land constituting 85% of the total area. This region is heavily reliant on agriculture and agricultural processing, with a particular focus on orcharding and horticulture—orchards account for as much as 26% of agricultural land, compared to the national average of less than 2%. Personal income levels in the Sandomierz FUA average 25,800 PLN *per capita*, the lowest among the case studies examined. This represents 70% of the value recorded in the regional capital, Kielce, and only 38% of that in the Warsaw FUA. Additionally, only 6.5% of the working-age population is engaged in entrepreneurial activity, one of the lowest rates in Poland. The entire FUA exhibits a low degree of industrialization, with its largest company being Pilkington Automotive Poland in Sandomierz. Pilkington has been producing glass since 1826 and was acquired by the Japanese NSG Group in 2006. It is now one of Poland's largest automotive glass manufacturers, employing over 1,700 workers in Sandomierz. The second primary industry is aggregate extraction and processing (limestone, dolomite, and sandstone). The PBI Group, a consortium of Polish family-owned companies, specializes in aggregate extraction, road construction, and producing concrete, paving stones, prefabricated elements, and mineral-asphalt mixtures. The company employs over 1,000 people. Other processing companies in the Sandomierz FUA include the Sambor Fruit Processing Plant in Samborzec, a juice producer, and the "Dwikozy" Fruit and Vegetable Industry Plant S.A. in Dwikozy. Sandomierz, as one of Poland's oldest cities, plays a significant role in the regional tourism economy. Its unique urban and architectural complex, which has survived numerous wartime destructions, attracts many tourists interested in history and culture. The cultural heritage potential of Sandomierz and its surroundings has led to an increase in the number of visitors, fostering the development of tourism infrastructure, including accommodation, gastronomy, and related services. Tourism growth has also contributed to increased employment in this sector (Farciak 2017).

The four presented case study areas represent two types of inner peripheries that we identified when classifying Functional Urban Areas. They are located in different geographic locations and have diverse configurations of deficits in various kinds of territorial capital. In the next chapter, we will focus on these areas and extend our analysis to a dynamic approach and the use of primary information sources to explore the barriers and interrelations in forming the territorial capital and the strategies to cope with the peripherality challenges in more detail.

References

Bański J, Czapiewski K, Mazur M (2012) Policentryczność rozwoju Mazowsza. Mazowsze. Studia Regionalne 10:71–88
Bański J, Mazur M (2016) Classification of rural areas in Poland as an instrument of territorial policy. Land Use Policy 54:1–17. https://doi.org/10.1016/j.landusepol.2016.02.005
Bartosiewicz B (2016) Obszary funkcjonalne małych i średnich miast w Polsce - koncepcja badawcza. Studia Ekonomiczne 279:234–244
Bartosiewicz B, Marcińczak S (2020) Policentryczność regionów miejskich w Polsce. Przegląd Geograficzny 92(4):455–474
Berry BJL (1973) Growth centers in the American Urban System. Ballinger, Cambridge
Bishop CM, Nasrabadi NM (2006) Pattern recognition and machine learning (Vol. 4, No. 4, p. 738). New York: Springer
Bożek J, Szewczyk J, Jaworska M (2021) Poziom rozwoju gospodarczego województw w ujęciu dynamicznym. Rozwój Regionalny i Polityka Regionalna 57:11–24. https://doi.org/10.14746/rrpr.2021.57.02
Bremer J (2004) Wprowadzenie do logiki. Wydawnictwo WAM. Karków
Challenges of Regional Policy Soc Indicat Res 164:31–54. https://doi.org/10.1007/s11205-021-02829-x
Chojnicki Z, Czyż T (1992) Region–regionalizacja–regionalizm. Ruch Prawniczy, Ekonomiczny i Socjologiczny 54(2):1–18
Chramiec A (1966) Dynamika rozwoju i przemiany strukturalne ludności w kształtowaniu aglomeracji miejskiej w Polsce. Instytut Urbanistyki i Architektury. Warszawa
Churski P, Perdał R (2022) Geographical differences in the quality of life in Poland: challenges of regional policy. Soc Indic Res 164:31–54. https://doi.org/10.1007/s11205-021-02829-x
Churski P, Herodowicz T, Konecka-Szydłowska B, Perdał R (2021) Spatial differentiation of the socio-economic development of Poland –"Invisible" historical heritage. Land 10:1247. https://doi.org/10.3390/land10111247
Churski P, Adamiak C, Szyda B, Dubownik A, Pietrzykowski M, Śleszyński P (2023) Nowa delimitacja miejskich obszarów funkcjonalnych w Polsce i jej zastosowanie w praktyce zintegrowanego podejścia terytorialnego (place based approach). Przegląd Geograficzny 95(1):29–55. https://doi.org/10.7163/PrzG.2023.1.2
Churski P, Herodowicz T, Konecka-Szydłowska B, Perdał R (2020) Teoretyczny i praktyczny wymiar polityki rozwoju zorientowanej terytorialnie. Studia KPZK. Cykl monografii. 9/201. Warszawa
Czapiewski K, Janc K (2009) Internal peripheries of socio-economic development in Poland. Kovács AD (ed) Old and new borderlines/frontiers/margins. Centre For Regional Studies of Hungarian Academy of Sciences. Discussion Papers, pp 109–121
Czyż T (2009) Koncepcje aglomeracji miejskiej i obszaru metropolitalnego w Polsce. Przegląd Geograficzny 81(4):445–459
Czyż T (2012) Poziom rozwoju społeczno-gospodarczego Polski w ujęciu subregionalnym. Przegląd Geograficzny 84(2):219–236. https://doi.org/10.7163/PrzG.2012.2.3
Davies N (2005a) God's playground. History of Poland. Volume I The Origin to 1795. Columbia Univeristy Press, New York
Davies N (2005b) God's playground. History of Poland. Volume II 1795 to Present. Columbia Univeristy Press, New York
Diemer A, Iammarino S, Rodríguez-Pose A, Storper M (2022) The regional development trap in Europe. Econ Geogr 98(5):487–509. https://doi.org/10.1080/00130095.2022.2080655
Domański B (2018) Zróznicowanie dynamiki wzrostu gospodarczego polskich regionów w latach 1995–2015. Studia KPZK PAN 183:248–262
Domański B (2021) Sprawiedliwość społeczna a przestrzenne zróżnicowanie rozwoju społeczno-gospodarczego Polski–kilka refleksji dyskusyjnych. Studia Regionalne i Lokalne 2(84):109–116. https://doi.org/10.7366/1509499528407

Drawsko Pomorskie (2025a) Urząd Miejski w Drawsku Pomorskim. (n.d) Archiwum - Drawsko Pomorskie. https://archiwum.drawsko.pl/asp/pl_start.asp?typ=14&sub=11&subsub=13&menu=69&strona=1. Accessed 15 Feb 2025
Drawsko Pomorskie (2025b) Studium uwarunkowań i kierunków zagospodarowania przestrzennego gminy Drawsko-Pomorskie. Część opisowa. https://archbip.drawsko.pl/public/zal.%20nr%202%20studium%20Drawsko1045.pdf?id=318769. Accessed 15 Feb 2025
Dziewoński K, Gawryszewski A, Iwanicka-Lyra E, Jelonek A, Jerczyński M, Węcławowicz G (1977) Rozmieszczenie i migracje ludności, a system osadniczy Polski Ludowej. IGiPZ PAN. Wrocław
Dziewoński K, Kosiński L (1964) Rozmieszczenie ludności w Polsce w XX wieku. Przegląd Geograficzny 36(1):3–36
Eberhardt P, Gontarski Z, Siemiński J (1973) Koncepcje delimitacji aglomeracji miejskich w Polsce. Biuletyn KPZK PAN 79:187–217
ESPON (2007) Study on Urban functions. ESPON Project 1.4.3. Final Report. ESPON Coordination Unit, Luxembourg. https://www.espon.eu/programme/projects/espon-2006/studies-and-scientific-support-projects/study-urban-functions
ESPON PROFECY (2017) Inner Peripheries: National Territories Facing Challenges of Access to Basic Services of General Interest. Luxembourg. https://archive.espon.eu/inner-peripheries
Faracik R (2017) Sandomierz–miasto historyczne jako ośrodek turystyki. Turyzm 27(2):25–44
Floerkemeier H, Spatafora N, Venables A (2022) Regional disparities. In: Cerra V, Eichengreen B, El-Ganainy A, Schindler M (eds) How to achieve inclusive growth. Oxford University Press, pp 613–644. https://doi.org/10.1093/oso/9780192846938.003.0017
Gontarski Z (1980) Obszary metropolitalne w Polsce. KPZK PAN. Warszawa
Gorzelak G (2021) Różnice regionalne–preferencje polityczne–sprawiedliwość społeczna. Studia Regionalne i Lokalne 2(84):117–127. https://doi.org/10.7366/1509499528408
Grochowski Z (2010) Historia Włocławka. 500 zadań i rozwiązań. Wydawnictwo Adam Marszałek. Toruń
Heffner K, Twardzik M (2022) Rural areas in Poland–changes since joining the European Union. Eur Countryside 14(2):420438. https://doi.org/10.2478/euco-2022-0021
Hołowiecka B, Szymańska D (2008) New changes in the functional uran region in the new socio-economic conditions in Poland: the case of Toruń. Bull Geogr Soc Econ Ser 9:63–78
Ilnicki D, Janc K (2021) Obszary intensywnych powiązań funkcjonalnych miast na prawach powiatu w Polsce–autorska metoda delimitacji. Przegląd Geograficzny 93(2):141–160. https://doi.org/10.7163/PrzG.2021.2.1
Iwanicka-Lyra E (1969) Delimitacja aglomeracji miejskich w Polsce. PWN, Warszawa
Jasiak M, Archutowska J, Konarski A, Morga B, Niewęgłowska M, Puławska-Obiedowska S, Pokora P, Pyzik M, Uryga B, Wielgo B, Zaremba J (2023) White book on railway development. How to create an effective railway system. Centralny Port Lotniczy Spółka z o.o. Warszawa
Kaufmann V (2002) Re-thinking mobility. Ashgate, New York
Kiryk F, Chyła R (2019) Sandomierz. Miasto fascynującej przeszłości. W 730. rocznicę lokacji miejskiej. Polska Akademia Umiejętności Kraków
Kochanowicz J, Sosnowska A (2011) Historia gospodarcza Polski przedrozbiorowej–porzucone terytorium? Roczniki Dziejów Społecznych i Gospodarczych 71
Kociuba D (2015) Miejskie obszary funkcjonalne–wyzwania planistyczne. Studia Miejskie 18:39–53. https://doi.org/10.25167/sm.2424
Komornicki T, Śleszyński P, Rosik P, Pomianowski W (2009) Dostępność przestrzenna jako przesłanka kształtowania polskiej polityki transportowej. KPZK PAN, Warszawa
Komornicki T, Korcelli P, Siłka P, Śleszyński P, Świątek D (2013) Powiązania funkcjonalne pomiędzy polskimi metropoliami. Wydawnictwo Akademickie SEDNO. Warszawa
Komornicki T, Rosik P, Stępniak M, Śleszyński P, Goliszek S, Pomianowski W, Kowalczyk K (2018) Ewaluacja i monitoring zmian dostępności transportowej w Polsce z wykorzystaniem wskaźnika WMDT. Ministry of Investment and Development, Warszawa

Komornicki T, Wiśniewski R, Miszczuk A (2019) Delimitacja przygranicznych obszarów problemowych. Przegląd Geograficzny 91(4):467–486. https://doi.org/10.7163/PrzG.2019.4.2
Kondracki J (1965) Geografia fizyczna Polski. PWN, Warszawa
Korcelli P (1973) Głos w dyskusji. Biuletyn KPZK PAN 79:157–159
Korcelli P (1976) Aglomeracje miejskie w systemach osadniczych, Wybrane hipotezy i perspektywy badawcze. Przegląd Geograficzny 48(4):589–600
Korenik S (2022) Kształtowanie rozwoju społeczno-ekonomicznegoze szczególnym uwzględnieniem polityki regionalnej. In: Pancer-Cybulska E, Baran B, Szostak E (ed) Procesy konwergencji i dywergencji w Europie. Monografia jubileuszowa dedykowana Profesorowi Janowi Borowcowi. Wydawnictwo Uniwersytetu Ekonomicznego, Wrocław, pp 15–36. https://doi.org/10.15611/2022.15.2.01
Kowalski H (1993) Dzieje Sandomierza, pod red. Henryka Samsonowicza, t. I: Średniowiecze, pod red. Stanisława Trawkowskiego, t. II: XVI—XVIII wiek, pod red. Feliksa Kiryka, cz. 1: W okresie świetności, cz. 2, W czasach stagnacji i upadku, cz. 3: Sztuka. Polskie Towarzystwo Historyczne, Warszawa
Leszczycki S, Eberhardt P, Heřman S (1971) Aglomeracje miejsko-przemysłowe w Polsce 1966–2000. KPZK PAN, Warszawa
Marcińczak S, Bartosiewicz B (2018) Commuting patterns and urban form: evidence from Poland. J Transp Geogr 70:31–39
Markowski T, Marszał T (2006) Metropolie, obszary metropolitalne, metropolizacja. Problemy i pojęcia podstawowe. KPZK PAN, Warszawa
Merta-Staszczak A (2020) Ochrona zabytków, przyrody i krajobrazu w warunkach reformy rolnej w Polsce w dwudziestoleciu międzywojennym. Ochrona Zabytków 1:111–129
Ministry of Regional Development (2011) Koncepcja Przestrzennego Zagospodarowania Kraju 2030. Warszawa
Muszyńska J, Wijaczka J (eds) (1996) Rzeczpospolita w latach potopu. WSP im. Jana Kochanowskiego
Nystuen JD, Dacey MF (1961) A graph theory interpretation of nodal regions. Pap Reg Sci Assoc 7(1):29–42
Okiński P (2016) Główne etapy formowania się miasta lokacyjnego w Sandomierzu w XIII-XIV wieku. Roczniki Historyczne 82:79–106
Parysek J (1982) Modele klasyfikacji w geografii. UAM. Poznań
Piekutowska A (2023) Nierówności na rynku pracy a zróżnicowanie regionalne w Polsce. Nierówności Społeczne a Wzrost Gospodarczy 73(1/2023):92–109. https://doi.org/10.15584/nsawg.2023.1.6
Pietrzykowski M (2024) Zróżnicowanie rozwoju społeczno-gospodarczego w Polsce w świetle europejskiej polityki spójności i celów zrównoważonego rozwoju. In: Cabańska J, Czyżewska-Misztal D, Mazur G (eds) Droga do zrównoważonej gospodarki światowej. Wydawnictwo Uniwersytetu Ekonomicznego w Poznaniu, pp 107–123. https://doi.org/10.18559/978-83-8211-245-0/8
Potrykowska A (1989) Funkcjonalne regiony miejskie w krajowym systemie osadniczym. Prace Geograficzne 152:55–76
Racoń-Leja K (2010) Wewnętrzne peryferia – przypadek miasta Oświęcim. 5ULAR: Między Miastem a Nie Miastem: Odnowa Krajobrazu Miejskiego: materiały Międzynarodowej Konferencji Naukowej Wydziału Architektury Politechniki Śląskiej, Gliwice 09–10 grudnia 2010. Wydział Architektury Politechniki Śląskiej i Komitet Architektury i Urbanistyki PAN, pp 387–394
Radzimski A (ed) (2017) Dzieje regionu kujawsko-pomorskiego. Towarzystwo Naukowe Organizacji i Kierownictwa „Dom Organizatora". Toruń
Rodríguez-Pose A, Dijkstra L, Poelman H (2024) The geography of EU discontent and the regional development trap. Econ Geogr 100(3):213–245. https://doi.org/10.1080/00130095.2024.2337657

Runge J (2007) Metody badań w geografii społeczno-ekonomicznej–elementy metodologii, wybrane narzędzia badawcze. Uniwersytetu Śląski. Katowice
Ruraż K, Panek M (2016) Rośliny inwazyjne wybranych siedlisk Wyżyny Sandomierskiej. In: Klich M, Kozłowski J (eds) Odnawialne źródła energii i gospodarka odpadami oraz ochrona przyrody i gospodarowanie zasobami przyrody–wybrane problemy w Polsce. Wyd. PWSZ w Tarnowie, Tarnów, pp 203–209
Rychłowski B (1967) Regionalizacja ekonomiczna (zagadnienia podstawowe). PWN, Warszawa
Rykiel Z (1978) Miejsce aglomeracji wielkomiejskich w przestrzeni społeczno-gospodarczej Polski. IGiPZ PAN, Wrocław
Śleszyński P (2012) Kierunki Dojazdów Do Pracy. Wiadomości Statystyczne 11:59–75
Śleszyński P (2013) Delimitacja Miejskich Obszarów Funkcjonalnych stolic województw. Przegląd Geograficzny 85(2):173–197
Śleszyński P (2014) Delimitation and typology of functional urban regions in Poland based on commuting, 2006. Geogr Pol 87(2):317–320. https://doi.org/10.7163/GPol.2014.20
Śleszyński P, Bański J, Degórski M, Komornicki T (2017) *D*elimitacja obszarów strategicznej interwencji Państwa: obszarów wzrostu i obszarów problemowych. IGiPZ PAN. Warszawa
Śleszyński P (2016) Delimitacja miast średnich tracących funkcje społeczno-gospodarcze. Opracowano dla potrzeb Strategii na rzecz Odpowiedzialnego Rozwoju. Instytut Geografii i Przestrzennego Zagospodarowania. Polska Akademia Nauk. Warszawa.
Śleszyński P (2021) Syntetyczny wskaźnik jakości życia w gminach polski na początku trzeciej dekady xxi wieku. Czasopismo geograficzne 92(2):325–352. https://doi.org/10.12657/czageo-92-14
Smętkowski M, Jałowiecki B, Gorzelak G (2009) Obszary metropolitalne w Polsce–diagnoza i rekomendacje. Studia Regionalne i Lokalne 1(35):52–73
Smętkowski M (2007) Delimitacja obszarów metropolitalnych w Polsce–nowe spojrzenie. In: Gorzelak G, Tucholska A (eds) Rozwój, region, przestrzeń. Ministry of Regional Devvelopment. Warszawa, pp 215–233
Sokołowski D, Szyda B (2015) Tourist function of the Gostynińsko-Włocławski Landscape Park (Funkcja turystyczna Gostynińsko-Włocławskiego Parku Krajobrazowego). Studia i materiały CEPL w Rogowie 17(45/4):216–223
Solon J, Borzyszkowski J, Bidłasik M, Richling A, Badora K, Balon J, Brzezińska-Wójcik T, Chabudziński Ł, Dobrowolski R, Grzegorczyk I, Jodłowski M, Kistowski M, Kot R, Krąż P, Lechnio J, Macias A, Majchrowska A, Malinowska E, Migoń P, Ziaja W (2018) Physico-geographical mesoregions of Poland: Verification and adjustment of boundaries on the basis of contemporary spatial data. Geogr Pol 91(2):143–170. https://doi.org/10.7163/GPol.0115
Spychała M, Spychała J (2022) Zróżnicowanie rozwoju regionalnego w Polsce, Czechach i na Słowacji. Przegląd ekonomiczny 3/2022:131–149. https://doi.org/10.31743/ppe.13420
Stanny M, Komorowski Ł, Rosner A (2021) The socio-economic heterogeneity of rural areas: towards a rural typology of Poland. Energies 14:5030. https://doi.org/10.3390/en14165030
Statistics Poland (2010) Dojazdy do pracy w Polsce. Terytorialna identyfikacja przepływów ludności związanych z zatrudnieniem. Poznań
Strategii rozwoju województwa kujawsko-pomorskiego do 2030 roku – Strategia Przyspieszenie 2030+. Załącznik do uchwały nr XXVIII/399/20 Sejmiku Województwa Kujawsko-Pomorskiego z dnia 21 grudnia 2020 r. Urząd Marszałkowski Województwa Kujawsko-Pomorskiego. 2020. Toruń
Strzyż M (2021) Wyżyna Kielecka (342.3). In: Richling A, Solon J, Macias A, Balon J, Borzyszkowski J, Kistowski M (eds) Regionalna geografia fizyczna Polski. Bogucki Wyd. Naukowe, Poznań, pp 415–422
Sudra P (2018) Ewolucja kryteriów delimitacji wielkomiejskich układów osadniczych w Polsce. Przegląd Geograficzny 90(2):181–208
Suszyna J (2000) Specyfika czynników klimatycznych Wyżyny Sandomierskiej. In: Puszkar T, Puszkar L (eds) Przyroda Obszarów Stykowych Ziemi Sandomierskiej i Polski Południowo-Wschodniej. Towarzystwo Naukowe Sandomierskie. Instytut Biologii i Ochrony Środowiska

WSP w Rzeszowie. Wyższa Szkoła Humanistyczno-Przyrodnicza w Sandomierzu. Sandomierz, pp 19–25
Swianiewicz P, Klimska U (2005) Społeczne i polityczne zróżnicowanie aglomeracji w Polsce–waniliowe centrum, mozaika przedmieść. Prace i Studia Geograficzne 35:45–70
Strategia Rozwoju Województwa Świętokrzyskiego 2030+, Załącznik do uchwały nr XXX/406/21 Sejmiku Województwa Świętokrzyskiego z dnia 29 marca 2021 r. Urząd Marszałkowski Województwa Świętokrzyskiego. 2021. Kielce
Szmytkie R (2013) W kwestii metropolii i obszarów metropolitalnych. Przegląd Administracji Publicznej 2:35–47
Wei Y (2015) Spatiality of regional inequality. Appl Geogr 6:1–10. https://doi.org/10.1016/j.apgeog.2015.03.013
Witwicki P (2021) Znikająca Polska. Wydawnictwo Zysk i spółka. Poznań
Wódz K, Nowalska-Kapuśniak D, Libora G (eds) (2016) Przestrzenie społeczne w teorii i badaniach empirycznych. Wyd. Uniwersytetu Śląskiego. Katowice
Zaborowski Ł (2016) Próba przebudowy układu województw z wykorzystaniem sieci ośrodków regionalnych. Przegląd Geograficzny 88(2):159–182. https://doi.org/10.7163/PrzG.2016.2.2
Zachodnie P (2023) Delimitacja Specjalnej Strefy Włączenia na obszarze Województwa Zachodniopomorskiego oraz planowane kierunki działań interwencyjnych, Zachodniopomorskie Regionalne Obserwatorium Terytorialne Urzędu Marszałkowskiego Województwa Zachodniopomorskiego. Szczecin
ZMP (2021) Strategia terytorialna. Partnerstwo Strefa Centralna. https://drawsko.pl/aktualnosci/f589ddd72aab54a59bd14fb8a83548db.html. Accessed 15 Feb 2025

Chapter 6
Development Processes of the Inner Periphery in the Period of Development Shocks

Abstract The chapter provides an analysis of the development trajectories of inner peripheries, supported by qualitative research conducted in four Functional Urban Areas (FUAs) classified as inner peripheries. It opens with a comparative examination of the dynamics of territorial capital formation in Poland during the SARS-CoV-2 pandemic, taking into account the differences between core, intermediate, and peripheral areas. This dynamic analysis enabled the identification of development trajectories of these areas in relation to more developed parts of the country and allowed for the specification of their distinctive characteristics. The chapter then presents a detailed description of the research methods employed, including content analysis of strategic documents, surveys among residents, and in-depth interviews and focus group interviews with local leaders. Subsequently discussed are the findings related to transport accessibility and its changes, this being a key development factor differentiating peripheries from central areas. Results concerning human, social, and institutional capital are presented in an integrated manner, allowing for the identification of their interrelations and overlapping deficits that constrain development potential. The analysis of financial and material capital in the studied FUAs concludes the presentation of the in-depth research results. The final section of the chapter highlights the importance of the identified factors in shaping the relational development environment of inner peripheries. It points to the relationships between these factors, the relational disconnection from developmental cores, the discrepancies between strategic planning and the actual needs of local communities, and the low resilience of these areas to developmental shocks.

Keywords Development trajectories · In-depth interviews · Local leaders · Relational disconnection · Low resilience to developmental shocks

P. Churski et al., *Inner Peripheries in the New Conditions of Cohesion Post-SARS-CoV-2*, Economic Geography,
https://doi.org/10.1007/978-3-032-10936-1_6

6.1 Developmental Trajectories

In Chap. 5, we presented the characteristics of inner peripheries as areas with a lower level of territorial capital than core and intermediate areas. The various kinds of territorial capital are described in Chap. 2 and include accessibility, human, social and institutional, financial, and material capital. The literature review in Chap. 3 suggests that the deficit of territorial capital is not only a static characteristic but may become increasingly apparent with time, particularly in a crisis situation such as the SARS-CoV-2 pandemic. The crisis is supposed to have specific effects that affect development in inner peripheries perhaps more adversely than in more resilient core areas. In this section, we attempt to verify the difference in the dynamics of territorial capital formation between Poland's core, intermediate, and peripheral areas (as delimited in Chap. 5) during the pandemic. We look at the changes in the metrics of various components of territorial capital. An in-depth look at the developmental trajectories of inner peripheries would help to understand the specific features of their vulnerability and resilience to the SARS-CoV-2 crisis and its socio-economic consequences. Methodologically, it will help us to suggest and test a system for monitoring socio-economic changes in inner periphery areas.

The dynamics of the development distance between central areas and peripheries is among the important research topics in regional studies. In regional economics, the concepts of economic convergence and economic divergence are used to describe two opposing dynamics: economic convergence appears when less affluent and economically productive regions catch up with more economically advanced regions and the development distance between them shrinks. Regional economic divergence, on the other hand, occurs when the distance between the economically advanced and disadvantaged regions grows as the more highly developed places escape from the poorly developed ones. An early theoretical explanation of regional convergence was based on the neoclassical growth model (Solow 1956), which suggests that diminishing returns on capital and technological advantages over time lead poorer economies gradually to catch up with richer ones. This assumption is challenged by later theoretical advantages of institutional and new economic geography. These suggest instead that institutional quality, innovation, human capital, and economies of scale may lead to the maintaining or even widening of the gap between economically advanced and disadvantaged regions (Chap. 2).

Empirical research on regional economic convergence usually refers to two ways of quantifying economic development dispersion across regions. Sigma-convergence is referred to as the reduction of disparities in income levels across economies (be it countries or regions). Conversely, sigma-divergence is the growth of such disparities. Beta-convergence, on the other hand, signifies a higher pace of growth in poorer economies than richer ones. Similarly, beta-divergence is the relatively quicker growth of already richer economies (Barro and Sala-i-Martin 1992). Both are mathematically and empirically interrelated but not equivalent (Furceri 2005; Young et al. 2008). The distance is usually measured by income or economic productivity metrics such as Gross Domestic Product.

In Europe, signs of convergence between countries are evident at the regional level; the trends are still diverse within individual countries. In Central and Eastern Europe, regional divergence occurred around the turn of the twenty-first century (Wójcik 2021), which is often explained by Williamson's (1965) regional adaptation of the Kuznets curve, as the disparities are expected to grow with economic development until it reaches a certain level and then start to fall (Monastririotis 2014). Therefore, even though the EU is moving toward regional convergence at the national level, it is debatable whether the same is true at the regional scale, and there is evidence of increasing disparities at the subregional level (European Commission 2024). The apparent convergence at the national scale is partially associated with the regional-level divergence. In Poland and other Central and Eastern European countries in recent decades, national economic indicators were elevated by values for the capital regions, as they were the fastest to approximate the average indexes for the EU, while peripheries lagged behind (Alcidi et al. 2018).

Our study conceptually refers to beta-convergence, as we look at the dynamics of the regions that we label "peripheral" and set them apart from the core and intermediate areas. If peripheral areas caught up with them, diminishing peripherality would be proven. However, two important assumptions differentiate our study from the general regional economic discourse. First, we do not solely analyze the levels of economic indicators but instead take additional account of broader territorial capital conceptualized after the redefinition of development factors (see Chap. 2). Traditional regional economic studies often use national or regional Gross Domestic Product, but GDP may not be the best metric to measure the quality of life or even the economic performance of the region. For example, despite the GDP divergence of Polish regions, no such trend in personal incomes has recently been noticed (Dubownik et al. 2022). We based the delineation of peripheral area on low levels of all kinds of territorial capital and compared these with areas with higher levels of capital. Therefore, we look at the general dynamics of the territorial capital within peripheries in relation to other areas. Secondly, in our study, we do not only look at the absolute change in indicators of territorial capital over a longer period of time. Instead, we consider their dynamics in the immediate temporal surrounding of the crisis impulse of the pandemic. We look at nonlinear changes as effects of disruptor events.

Our dynamic analysis has three major aims; firstly, to determine the developmental trajectories of inner periphery areas in relation to more developed areas in general. Secondly, by also looking at the specifics of the concrete case study areas that we selected for closer investigation, we aim to further confirm their representatives or detect specifics of their case in relation to the background. Thirdly, we tried to test and evaluate principles and criteria that future studies could use to identify the developmental trajectories of inner periphery areas.

We base the analysis of trajectories on the same set of variables as we used for classification into categories. We looked at the period immediately preceding the SARS-CoV-2 pandemic (since 2017), the period of pandemic emergency and aftermath, as long as data are available (until 2023). Due to the unavailability of a sufficiently long data series, we could not use all the variables that we took into account in FUA classification, for which we used various unconventional data sources other than easily accessible Statistics Poland data (Sect. 5.3). Our requests for data series of appropriate length were not able to be met by, for example, the Office of Electronic Communication, Polish police, of the Ministry of Finance. We could also not use time series for data produced explicitly for the purpose of the study based on real-time observations such as Google Maps Distance Matrix API data and data from the public transportation timetables. Finally, some data, such as Corine Land Cover, are produced once for a long time interval that exceeds our period of analysis. Due to these limitations, we only investigated the changes in 26 out of 47 variables that we considered in Sect. 5.3. These 26 variables represent all subfactors of territorial capitals, with the notable exception of accessibility. Accessibility factors were measured entirely by data created originally or obtained from previous studies based on novel methodologies and which could not be repeated for previous or consecutive years. Therefore, changes in accessibility of peripheral areas were excluded from the dynamic analysis in this section. In compensation, particular attention was given to accessibility in Sect. 6.3, where the results of document analysis, interviews, and surveys in four case study areas enabled us not only to describe the deficits of accessibility capital, but also to diagnose changes, challenges, and policy responses in peripheral regions.

Table 6.1 lists the indicators taken into account and a comparison of average general change in the value of these indicators between 2017 and 2023 in each group of FUAs. The subsequent graphs (Fig. 6.1) outline the trajectories of changes in the indicators over time in five categories of FUAs, complemented by those specific to the case study areas.

Out of human capital indicators, demographic and labor market characteristics are the most disadvantageous in peripheral areas compared to core and intermediate areas, while access to education positively distinguishes peripheries. The trends for most demographics are more negative in peripheral areas than in other kinds of areas. Most of all, the ratio of the post-working-age population (HD_POST on Fig. 6.1) to working-age population was around the national average in both types of peripheral areas in 2017. Yet, after much quicker growth than in other types of areas, peripheries and intermediate areas of larger cities became the most aged ones. In contrast, core areas were the only ones to resist the growth trend in the elderly population. The share of the pre-working age population (HD_PRE) and natural increase (HD_NI) was low in peripheral areas, and slight yet visible differences between types of FUAs were noticeable to the disadvantage of the peripheries. The natural increase trajectory was most affected by the SARS-CoV-2 pandemic, marking record-low values in 2021 due to the highest historic mortality since Second World War (519.5 thousand deaths). Yet, pandemic-related mortality was seemingly not correlated to the core–periphery division. Net migration (HD_MIGR) remained negative in both peripheral types of FUAs throughout the period, contrary to core areas. Healthcare measure (number of outpatients using healthcare advice per population, HH_PAT) was characterized by parallel trajectories in all types of FUAs: an increase that was interrupted by a pandemic-related drop in 2020–2021. Access to education, as measured by the places in kindergartens (HE_KIND) and sizes of classes in primary schools (HE_CLASS), is better in peripheries than in other types of FUAs, and the

Table 6.1 Changes in the characteristics of five types of Functional Urban Areas between 2017 and 2023

Capital/factor	Subfactor	Variable	Symbol in Fig. 6.1	Variable value in 2017 and percentage change between 2017 and 2023				
				Core	Intermediate larger cities	Intermediate smaller cities	2nd-order periphery	1st-order periphery
Human capital	Demographic situation	Population in post-working age per 100 inhabitants of working age	HD_POST	(35.1) 8.4%	(35.8) 21%	(30.6) 23.4%	(33.3) 28.6%	(33.3) 25.4%
		Share of population of pre-working age	HD_PRE	(18.2) 2.8%	(16.9) 1.6%	(19.4) 1.3%	(17.3) −1.2%	(17.6) −0.5%
		Rate of natural increase per 1,000 population	HD_NI	(1) −346.1%	(−1.5) 259.8%	(1.4) −279.6%	(−1.7) 248.4%	(−1.4) 245.2%
		Net registered migration increases per 1,000 inhabitants	HD_MIGR	(3.5) 7%	(−1.3) −9.4%	(−0.8) −6.6%	(−2.7) 0.2%	(−3.2) 3.3%
	Health and safety	Number of outpatients using healthcare advice per 10,000 inhabitants	HH_PAT	(83,072.3) 12.2%	(73,997) 9.6%	(65,354) 8.1%	(65,640.9) 8.4%	(62,164.2) 11.4%
	Education	Places in kindergartens per 1,000 children 3–6 years old	HE_KIND	(768.8) 10.6%	(789.9) 11.7%	(658.6) 12.6%	(624.2) 17.2%	(522) 18.7%

(continued)

Table 6.1 (continued)

Capital/factor	Subfactor	Variable	Symbol in Fig. 6.1	Variable value in 2017 and percentage change between 2017 and 2023				
				Core	Intermediate larger cities	Intermediate smaller cities	2nd-order periphery	1st-order periphery
		Number of pupils per class in primary schools	HE_ CLASS	(18.6) −1.8%	(18.2) −6.3%	(17.4) −5.7%	(16.8) −8.8%	(15.7) −8.1%
	Labor market	Share of unemployed in population of productive age	HL_ UNEMP	(3.3) −22%	(4.1) −19.5%	(3.4) 1.6%	(5) 7.3%	(5.4) 4.6%
		Share of population of productive age running personal businesses	HL_BUSI	(23.9) 21%	(16.2) 25.8%	(15) 28.2%	(14.5) 29.7%	(11.9) 32.7%
Social and institutional capital	Social activity	Graduates of courses run by cultural institutions per 10,000 inhabitants	SA_ COURS	(2.2) −4.4%	(1.4) −29.9%	(1.2) −8.5%	(1) −36.7%	(2.4) −61.4%
	Leisure	Sports and recreation institutions per 10,000 inhabitants	SL_REKR	(15.6) 10%	(12.8) 14%	(12.7) 14.6%	(12.9) 17.1%	(10.5) 15.4%

(continued)

Table 6.1 (continued)

Capital/factor	Subfactor	Variable	Symbol in Fig. 6.1	Variable value in 2017 and percentage change between 2017 and 2023				
				Core	Intermediate larger cities	Intermediate smaller cities	2nd-order periphery	1st-order periphery
		Attendees of sport and cultural events per 10,000 inhabitants	SL_ EVENT	(10,162.7) −1.1%	(7,378.4) 25.6%	(3,610.3) −10.5%	(2,693.4) −11.7%	(1,805.5) −2.1%
	Business infrastructure	Business environment institutions per 10,000 businesses (all NACE sections)	SB_ENVIR	(867.6) −4%	(928) −2.2%	(657.4) −2.1%	(1110.3) −2.7%	(548.5) −3.8%
		All-year-round hotel bed-places per 1,000 inhabitants	SB_ HOTEL	(10.2) 18%	(3.8) 22.1%	(4.4) 12%	(6.5) 23%	(3.7) 31.9%
	Economic structure and activity	Share of new economic entities in creative sector among all new entities (NACE J59, J60, M71, M73, M74, R90)	SS_CREAT	(0.7) −3%	(0.5) −12.9%	(0.5) −16.9%	(0.4) −16.1%	(0.5) −16.8%

(continued)

Table 6.1 (continued)

Capital/factor	Subfactor	Variable	Symbol in Fig. 6.1	Variable value in 2017 and percentage change between 2017 and 2023				
				Core	Intermediate larger cities	Intermediate smaller cities	2nd-order periphery	1st-order periphery
		Number of economic entities per 1,000 inhabitants of productive age	SS_ENTIT	(146.5) 19%	(99.4) 19.8%	(92.8) 22.2%	(90.2) 22%	(73.7) 25.8%
		Number of large (>49 employees) economic entities per 10,000 inhabitants	SS_ LARGE	(823.9) −18%	(192.9) −16.9%	(165.1) −21%	(146.2) −17%	(140.6) −17.6%
Financial capital	Financial situation of local governments	Own income of local governments per inhabitant	FL_OWN	(3,033) 41.7%	(2,265) 54.6%	(1,476.6) 62.3%	(1,325.9) 64.3%	(1,108.8) 74.5%
		Investment expenditures of local governments per inhabitant	FL_INW	(3,130,637.9) 201.5%	(530,878.8) 227.9%	(471,853.2) 246.4%	(316,806.2) 283.6%	(419,935.8) 333.2%
		Net operational surplus of local governments	FL_SURP	(−2,454.3) 857.1%	(−304.1) 1278.3%	(−337.7) 916.1%	(14.4) −20,141.9%	(−172.7) 1476.9%

(continued)

Table 6.1 (continued)

Capital/factor	Subfactor	Variable	Symbol in Fig. 6.1	Variable value in 2017 and percentage change between 2017 and 2023				
				Core	Intermediate larger cities	Intermediate smaller cities	2nd-order periphery	1st-order periphery
	Financial situation of inhabitants	Personal income per adult inhabitant	FP_INC	(4,518.6) 67.5%	(4,182.3) 66%	(3,786) 69.5%	(3,731.4) 70.8%	(3662.6) 70.2%
Material capital	Housing situation	Dwellings completed per 1,000 inhabitants	MS_NEW	(7.6) 13%	(2.5) 64.5%	(3.7) 37.7%	(2.2) 47.5%	(2.3) 38%
	Housing quality	Share of population in dwellings with central heating	MQ_CH	(87) 2.2%	(81.7) 3.4%	(79.9) 5.3%	(79) 5.3%	(71.7) 10.9%
		Share of housing buildings connected to sewerage system	MQ_ SEWER	(54.9) 6%	(55.6) 12.4%	(48.9) 9.1%	(51.1) 5.2%	(39.9) 10.1%
		Useable floor area per inhabitant	MQ_ SPACE	(30.1) 12%	(26.8) 14.7%	(26.8) 15%	(25.3) 15.1%	(27.1) 14%
	Ecosystem services	Share of protected areas in commune's area	ME_AREA	(24) −6.3%	(13.9) −0.5%	(24.8) −3.3%	(21.7) −1.3%	(28) −3%

Source Own elaboration

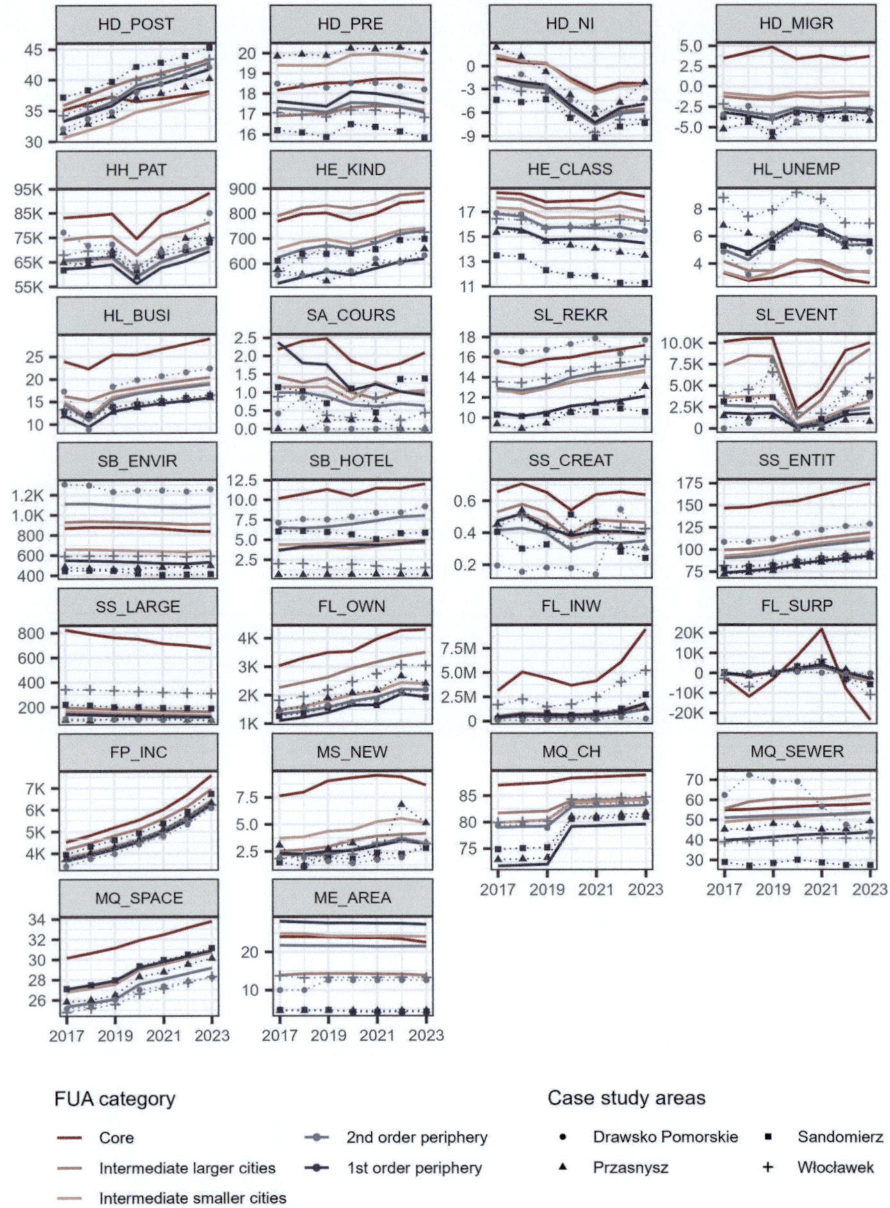

Fig. 6.1 Changes in the characteristics of five types of Functional Urban Areas between 2017 and 2023. *Source* Own elaboration

difference has grown over time. It is worth noting, however, that this divergence partially stems from more profound demographic depression of peripheral areas and measures of the quality of education, which are non-determinable due to the brevity of the time series. Labor market trajectories clearly differentiated peripheries from core and intermediate areas. Notably, the changes in the share of the unemployed population (HL_UNEMP) had opposite directions in peripheries (where it was high and growing over the pandemic period) and core or more urbanized intermediate areas (where values were low and decreasing). A closer look at the path of change shows a pandemic peak in 2020. These dynamics differ from those of changes in the share of the population running personal businesses (HL_BUSI): while this grew in all types of FUAs, even throughout the pandemic, the growth in peripheral areas outpaced that in core and intermediate areas. This is nonetheless an effect of low starting values and the convergence in the levels of entrepreneurship between core and peripheral areas is far from being achieved.

Among social and institutional characteristics, throughout the pandemic, the social activity indicator (graduates of courses run by cultural institutions per population, SA_COURS) collapsed in peripheral areas even more than the average for Poland. In 2017, the first-order peripheries had the same high level of such activity as core areas but leveled down to the values typical for intermediate areas in 2023. The dynamics of leisure infrastructure and activity indicators, such as sports and recreation institutions per population (SL_REKR) and participants of sports and cultural events per population (SL_EVENT), did not differ systematically between core, intermediate, and peripheral areas. Both, however, remained lowest in the first-order peripheries. Participation in events was understandably the variable that dropped most in 2020 and 2021, due to the direct limitations on this kind of activity imposed during lockdowns. Similarly to the leisure subfactor, business infrastructure characteristics did not present systematic differences between types of FUAs. The second-order peripheries boasted relatively high values of both, possibly due to group's inclusion of areas within poorly populated but touristically attractive areas in the mountains, lakelands, and coastal areas of Poland. Still, the capacity of tourist infrastructure (SB_HOTEL) did not grow as fast as in core areas. Among three variables describing the economic structure and activity, only one exhibited significant changes in dynamics between core and peripheral areas: the share of new economic activities in creative sectors (SS_CREAT) saw a fall (from the highest starting level) that was only slight (3%) in core areas, while it was much more evident in both intermediate and peripheral areas. The growth in overall number of economic entities per population (SS_ENTIT) and the decrease in number of large entities (SS_LARGE) occurred at a similar pace across all types of FUAs. Yet, the values were lowest in peripheral areas at the outset and remained so at the end.

The identified types of FUAs also differ in the dynamics of financial capital characteristics. Throughout the period in the study, peripheral municipal budgets were less wealthy than those of core and intermediate FUA municipalities as measured by own income (excluding state subsidies) per inhabitant (FL_OWN) and investment expenditures per inhabitant (FL_INW). By comparison, through the period, local budgets peripheral areas show higher increases in both indicators that are gradual yet convergent with one another. Operational surpluses (FL_SURP) of local budgets follow similar trajectories in all types of areas: positive values in the pandemic period due to the forced reduction in expenditures (including investments), followed by a negative trend afterward. Although we do not have the time series data for firm financial results, personal income (FP_INC) may indicate actual changes in residents' economic situation over time. The differences between core, intermediate, and peripheral areas are apparent, with values being highest in core areas and lowest in first-order peripheral areas. Still, the change over time does not vary between types of FUAs and fits within a narrow scope of between 67.5% growth in core areas and 70.8% growth in second-order peripheral areas. This invariability stems mainly from the fact that a high proportion of salaries are determined at the state level: it includes many public sector jobs on the one hand and low-paid jobs in the private sector, for which national minimum wages influence remunerations.

The dynamics of the material capital are measured by the dynamics of housing availability, quality of housing, and access to ecosystem services. The housing situation represents the dynamics of the housing market measured by the number of dwellings built per population (MS_NEW). The values of this indicator are highest in core areas, yet they tend to stabilize there while growing dynamically in all other types of areas, including peripheral ones. In housing, being equipped with central heating (MQ_CH) and sewerage systems (MQ_SEWER) is less diversified between types of areas, but the first-order parishes stand out negatively in both indicators. Differences in the rate of change suggest a slow catching up with other areas. Residents of peripheral areas do not have less floor space (MQ_SPACE) than residents of intermediate areas. Core areas have higher values there, yet the lower increase there will lead the difference to diminish in the future. The share of protected areas (ME_AREA) is relatively high in first-order peripheries and close to the national average in second-order inner peripheries. There is a national negative trend in the indicator's value, which is strongest in core areas where urbanization pressure is the most substantial.

Apart from the typical trajectories of the values of diagnostic indicators for types of Functional Urban Areas, Fig. 6.1 provides information about the four case study areas of Drawsko Pomorskie, Włocławek, Przasnysz, and Sandomierz on the background of their respective group of second-order (the former two) and first-order (the latter two) inner peripheries.

Drawsko Pomorskie FUA has a more favorable economic situation regarding income and business activity than other second-order peripheral areas. However, the share of creative sector firms only recently grew from a very low base level. While its demographic situation is better than the overall group average, it is also deteriorating rapidly. Migration trends align with the typical pattern for second-tier peripheral areas.

Meanwhile, the economic situation in Włocławek FUA is similar to the average for second-order peripheral areas. The number of large employers and size of local government budgets are more favorable than average for peripheral FUAs, probably due to the presence of a relatively large central city. The demographic situation has consistently been less favorable than the second-tier peripheral average, resembling that of large and medium-sized city regions, but has been worsening over time.

In Przasnysz FUA, income and business activity are at low levels typical for first-order peripheral areas. However, entrepreneurship is growing significantly, as is housing construction. The demographic situation has remained highly favorable throughout the period—comparable to transitional regions—but is showing clear signs of deterioration. Migration trends are typical for first-tier peripheral areas.

In Sandomierz FUA, income and entrepreneurship levels remain low, as typical of first-order peripheral areas. The demographic situation is highly unfavorable, worsening even faster than in all other regions of Poland, which is particularly visible in the share of the young population.

Summing up, the indicators in which peripheral regions stand out most negatively compared to core and transitional regions are those related to entrepreneurship, the labor market, local finances, infrastructure, and demographics. While there is apparent convergence in local finances and infrastructure indicators—meaning the differences between region types are decreasing—and wage disparities remain unchanged, a noticeable divergence is observed in unemployment rates, the structure of economic activity, and demographic aging. This divergence reflects a worsening relative situation for peripheral regions. However, in certain aspects of territorial capital, such as nature protection and access to educational services, peripheral regions maintain a lasting advantage over some core and transitional areas.

6.2 Case Study Research Methods

Field research conducted across the four case studies employed a combination of methodological approaches. The starting point was an analysis of strategic, programming, and planning documents, carried out using content analysis. Subsequently, to capture the perspectives of residents, entrepreneurs, NGO representatives, and local leaders, we employed survey methods, in-depth interviews, and focus group interviews. The empirical material gathered in this way constituted a unique information base concerning the conditions and course of development processes in inner peripheries, with particular emphasis on the consequences of external developmental shocks.

6.2.1 Analysis of Strategic, Programming, and Planning Documents

The analysis of strategic, programming, and planning documents was based on the method of content analysis. This approach is grounded in the examination of recorded human communication, both written (books, press articles, policy documents, websites) and oral, disseminated through radio, television, and the internet (Babbie 2005). The central premise of content analysis is the subjective interpretation of textual data through a systematic coding process and the identification of themes or patterns (Hsieh and Shannon 2005). Importantly, however, this method provides objective means for drawing valid inferences, with the goal of describing and quantifying specific phenomena (Bengtsson 2016). By reducing the content of the collected materials into categories based on a developed conceptual framework (Babbie 2005), the process reveals key trends, patterns, and relevant details contained within the analyzed documents.

In planning the study and establishing its assumptions, we followed a five-stage research procedure:

1. Defining the research objective;
2. Building the document base;
3. Developing a categorization system to structure the analysis of document content;
4. Coding: transforming raw data and performing qualitative analysis;
5. Interpreting results and formulating preliminary recommendations.

The primary aim of the study was to identify innovative proposals and solutions addressing the development challenges of inner peripheries, as articulated in the assumptions of intervention measures contained within strategic, programming, and planning documents. These findings are intended to form a foundation for recommendations regarding the design of development interventions targeting this category of lagging areas.

According to Polish legislation, "development policy is implemented on the basis of development strategies, programs, policy documents, and public policies" (Act of 6 December 2006 on the principles of conducting development policy, consolidated text: Journal of Laws 2024, item 324). Strategic programming documents, therefore, serve as a framework for effective, long-term governance by local government units. They contribute to shaping social, economic, regional, and spatial cohesion and foster the competitiveness of the economy based on available resources and tools at national, regional, and local levels (ibid.).

Broadly understood strategies function as auxiliary tools in the form of development programming documents. The most prominent strategic programming documents in the development policy system include national, regional (NUTS 2), and local (LAU) development strategies, as well as supra-local strategies (for groups of LAUs). The measures defined within these documents aim to implement the principles of regional policy, which constitute the foundation of the EU's Cohesion Policy, including:

- the principle of programming, which emphasizes the definition of long-term development priorities at the EU and member state levels and supports the financing of comprehensive, long-term development programs rather than isolated projects;

- the principle of concentration, both thematic and spatial, which highlights the need to target interventions at those areas and actions most in need of support, rather than dispersing efforts across all domains and regions;
- the principle of partnership, which ensures close cooperation between the European Commission and each member state and promotes the participation of regional and local authorities, other public institutions, civil society, and economic partners in the design and implementation of intervention measures;
- the principle of additionality (co-financing), which ensures that EU funds do not replace beneficiaries' own resources but rather complement them.

Ultimately, these measures aim to ensure the effective efficient, sustainable, and balanced development of the country, region, and local units. In addition to comprehensive development strategies, other strategic documents are developed that define goals and directions for sectoral, thematic, regional, or spatial development. All strategic programming documents at national, regional, and local levels are intended to be interconnected and coherent. Their mutual integration must apply to both content and objectives.

Our analysis was based on documents obtained through a systematic online search of public administration websites, supplemented where necessary through direct contact with local government representatives during the in-depth interviews. Due to the near-total absence of documents concerning Functional Urban Areas (with the exception of materials prepared for Włocławek FUA—including a development strategy, labor market analysis, transportation study, and territorial strategy), the analysis was conducted across two spatial scales. We examined documents prepared for regional-level units (NUTS 2) encompassing our case study FUAs, as well as for local-level units (LAU—counties and municipalities) forming part of these FUAs (see Table 6.2). The quality of the document base was influenced, first, by the completeness of the materials and, second, by the currency of the available documents. At both regional and local levels, the most commonly available documents were those mandated by legal requirements or necessary for accessing specific forms of investment support, such as EU funds. Additionally, some of the documents were clearly outdated, with their planning horizons having expired several years prior.

During the study, we employed a tool in the form of a document analysis questionnaire developed by our team. This instrument incorporated a content search scheme addressing the development factors as defined by our framework. For each Functional Urban Area (FUA) and every strategic programming document that was, to some extent, related to its territory, we conducted an analysis of the diagnostic information contained therein—identifying the development barriers and challenges, as well as the proposed changes along with their underlying assumptions. We utilized a predetermined coding key; the process of identifying phrases and, where appropriate, entire sentences were guided by the subfactors of development to which we assigned specific elements from the analyzed material. For instance, for the subfactor *inner transport mobility* we searched for phrases such as "private transport," "public transport efficiency," or "degree of connectivity," whereas for the subfactor *socialactivity* we looked for expressions like "social activation," "socially excluded persons," or

Table 6.2 Strategic, program, and planning documents selected for content analysis

Voivodeship	County	Municipality
– Regional development strategy – Regional innovation strategy – Social policy strategy of the province – Regional social economy development program – Regional transport plan/ transport infrastructure development program – Waste management plan – Spatial development plan of the voivodship – Strategy of strategic intervention area – Regional strategy for smart specialization – Regional economic development program	– County development strategy – Program for promotion of employment and activation of the local labor market – Strategy for solving social problems in the county – Program of cooperation with non-governmental organizations – Environmental protection program – Report on the state of the district	– Municipal development strategy with a diagnosis of the social, economic and spatial situation – Local development plan – Strategy for solving social problems – Plan for cooperation with non-governmental organizations – Local revitalization program – Low carbon economy plan – Environmental protection plan – Climate change adaptation plan – Program for counteracting domestic violence – Study of spatial development conditions and directions

Source Own elaboration

"civil society." The research incorporated a dual-aspect analysis. On one hand, it was crucial for us to understand the development barriers diagnosed in inner peripheries and the challenges these areas face—as elaborated by the preceding qualitative analysis. On the other hand, we also conducted a quantitative analysis, examining the frequency of occurrence of various findings and proposals, which, in our assessment, further underscores the significance of the described observations. The assumptions adopted, and the study's structure allows for future replication of the research, thereby enabling comparisons with previously obtained results.

6.2.2 Survey Research with Residents

Survey research is one of the most widely used methods in the social sciences. The essence of survey research lies in uncovering the opinions, attitudes, preferences, behaviors, or characteristics of individuals from a well-defined population through the use of a questionnaire (Babbie 2005; Visser et al. 2000). When conducted rigorously, survey research enables the formulation of nuanced descriptive statements. The data thus acquired may also be employed for explaining phenomena and presenting relationships between variables. The value and popularity of survey research compared to other methods is attributable to several factors: respondent anonymity, the ability to reach a large cohort, the relatively short time required to conduct the study, cost effectiveness relative to other methods, and the ease and extensive potential for analyzing results, particularly in relation to closed-ended questions.

In organizing the survey research, it was critical to adopt and implement guidelines that would influence the quality of the collected empirical material. These guidelines include:

1. Defining the sample size and selecting techniques: It is essential to determine the sample size and employ selection methods that adhere to the principles of randomness and representativeness. Random sampling ensures that each statistical unit has an equal chance of selection, whereas non-random sampling, based on the researcher's discretion, often fails to meet the criterion of representativeness.
2. Construction of the measurement tool: The questionnaire should be concise, clear, and well-aligned with the research objectives. Factors that affect the quality of the instrument include the number and types of questions, the logical sequencing of these questions, and the avoidance of bias.
3. Planning and application of a specific measurement technique: This may involve engaging an interviewer. Examples include Computer-Assisted Telephone Interview (CATI), Computer-Assisted Personal Interview (CAPI), Paper and Pencil Interview (PAPI). Alternatively, an interviewer-independent approach can be adopted, such as through Computer-Assisted Web Interviews (CAWI). Interviewer-administered surveys tend to generate fewer incomplete or misfiled questionnaires and reduce the risk of misinterpreting questions, resulting in higher response rates. Although CAWI methods are increasingly popular due to their relatively low cost and efficiency, they are also frequently associated with a higher dropout rate. The role of interviewers—as intermediaries between researcher and respondent—is critical, as their professionalism and commitment to ensuring data integrity significantly impact the quality of the research outcomes. Additionally, interviewer-administered techniques typically expedite the research process.

Our survey was designed to reach residents living in areas designated as inner peripheries. Its purpose was to gather their opinions as to what changes and threats stemmed from crisis shocks, as well as their assessments of potential opportunities for improving socio-economic conditions in their communities. The statistical population comprised residents of the four case study FUAs—more precisely, those residing in the central city and in its surrounding areas. For the study, a stratified random sample of $N = 1{,}524$ respondents was targeted, with strata corresponding to the individual case study FUAs. The sample was selected deliberately, with an emphasis on achieving a relatively high participation rate among young individuals (i.e., up to 35 years of age), whose life decisions and activities are likely to exert the greatest influence on the future functioning of peripheral areas. Consequently, the resulting sample does not fully reflect the actual age distribution of residents in inner peripheries.

The study was conducted on a sample of 1,576 respondents, with 19 interviews subsequently removed from the dataset due to missing key data. The final sample size was 1,557 respondents. The structure of the research sample is presented in Table 6.3.

The survey was conducted in each LAU unit forming the four case study FUAs. Among the respondents, 41% lived in the core cities of the FUAs, while the remaining 59% resided in the surrounding areas constituting the respective FUA—that is, outside the core city. Women slightly outnumbered men in the surveyed group (57%), reflecting the actual gender structure in the studied areas, where women constitute approximately 53% of the population. The largest share of respondents fell into the age categories of 35–49 and 18–24 years, each representing nearly 30% of the total sample. One in five respondents was aged 25–34, resulting in a relatively low average age of the sample. In terms of educational attainment, almost one third of the respondents had completed higher education. The largest group, however, comprised individuals with secondary education—41% had completed post-primary schooling. Another 35% declared a vocational education, while 29% reported the lowest levels of education—basic (6%) or vocational (23%).

In terms of marital status, the majority of respondents were either married (44%) or in informal partnerships (23%). One in four (25%) identified as single, while 8% were divorced or widowed. Single-person households accounted for only 14% of the sample; the dominant household type included at least two members. Notably, one in four surveyed households (25%) was multigenerational, with respondents most commonly living with both children and parents or in-laws, and less frequently with extended family members. Among households with children (43%), the prevailing family models were 2 + 1 (20%) and 2 + 2 (19%). A relatively high proportion of single individuals lived with parents or other family members. For nearly 56% of respondents, the main source of household income was salaried employment. Another 28% ran their own businesses, while only 7% reported deriving income from agricultural activity. The remaining 15% of respondents relied on other income sources such as pensions, disability benefits, or social assistance, which likely reflects the age structure of the sample. Moreover, 10% of respondents declared having no source of income, which may correspond largely to the youngest age group—students or those still in education.

Table 6.3 Structure of the survey sample

FUA	Number of LAUs (municipalities)	Population	Sample sizes
Włocławek	20	216,798	458 (30%)
Przasnysz	6	40,405	314 (20%)
Drawsko Pomorskie	2	23,517	315 (20%)
Sandomierz	9	71,112	470 (30%)
Total	37	351,832	**N = 1,557**

Source Own elaboration

As a research instrument, we used an original questionnaire specifically developed for the purpose of this study. Its structure followed a classic format. It opened with an introductory section (demographics) that captured basic characteristics of the respondent, including age, gender, education level, marital status, and place of residence, as well as household-level variables such as household size, income sources, and financial management practices. The main part of the questionnaire was divided into four thematic sections. This structure intentionally deviated from the analytical model of development factors comprising territorial capital, which we considered potentially unclear for respondents. However, the questions were formulated in such a way as to allow us to identify perceptions regarding the state and transformations of territorial capital in the context of developmental shocks within the studied FUAs.

The first section addressed changes experienced by individuals and households in the context of observed crises. These questions covered work conditions, material and housing situations, transport accessibility, service availability, and mental resilience. The second section focused on transformations in the labor market. Respondents were asked whether changes in employment were voluntary or forced, and about alterations in work modalities triggered by the SARS-CoV-2 pandemic. The third section referred to issues of identity and perceived safety. Questions concerned the duration of residence in a given locality and the willingness or motivations for potential relocation. Respondents were also asked about their place-based identity—locally, regionally, nationally, and in relation to Europe. Regarding safety, we asked about changes in general security as well as health-related security.

The final section pertained to transport accessibility. Questions were designed to gather information on the location of services used by respondents, the means of transportation employed, and the location of workplaces (both the respondents' and their household members'). Additional questions addressed commute times to workplaces and services and the quality of public transport services in the region. Respondents were also asked about changes in transport accessibility as a result of the pandemic. The final questionnaire contained 15 core questions and 10 demographic items. Thus, we met the criteria for developing a survey instrument well-aligned with the study's objectives.

We chose to conduct the survey using the CATI (Computer-Assisted Telephone Interviewing) method. To ensure representativeness in terms of place of residence and demographic characteristics, a small portion of the surveys was administered using the CAPI (Computer-Assisted Personal Interviewing) method. The survey was commissioned to a research agency equipped with technical infrastructure and an experienced research team. By combining CATI and CAPI techniques, we created conditions conducive to high effectiveness and smooth implementation of the survey.

6.2.3 In-depth Interviews with Local Government Leaders

The in-depth interview method, also called IDI (Individual In-depth Interview), is one of the key qualitative research techniques in the social sciences. It involves conducting detailed, one-on-one conversations with respondents to thoroughly understand their opinions, motivations, and experiences (Kvale 2012). In-depth interviews enable the exploration of difficult constructs, such as the subjective interpretations of reality formulated in our study by local government leaders. This method allows researchers to gain insight into the decision-making processes and values that drive the actions of individuals holding key positions in local government. It is commonly used in exploratory studies when a topic is poorly understood or requires in-depth analysis. Questions in this type of research address factual matters and the respondents' subjective evaluations. A key advantage of this method is its flexibility, allowing questions to be adapted during the interview, which enables the researcher to respond to new issues emerging during the conversation (Gaskell 2000). In conducting in-depth interviews, the researcher plays a central role, being responsible not only for asking questions but also for creating an atmosphere of trust that encourages openness and honesty on the part of the respondent. Essential researcher competencies include flexibility, empathy, and active listening skills, which enable effective exploration of the topics under study (Kvale 2012). An important ethical aspect of research involving human participants is the obtaining of informed consent. This means that before the interview begins, the respondent must be fully informed about the purpose of the study, its course, potential benefits and risks, and their right to withdraw at any stage without consequences. This information must be presented clearly and understandably, and participation in the study must be voluntary (Rosnow and Rosenthal 2013).

In research involving local government leaders, particular attention must be paid to confidentiality and anonymity. Due to the public nature of their work, the researcher should clearly specify how personal data will be protected and which information may be disclosed. Adhering to these principles is essential to maintaining ethical standards and building a relationship of trust between the researcher and the respondent.

Despite its many advantages, the in-depth interview method also comes with certain limitations. It is time-consuming and resource-intensive, both during the interview stage and in the analysis of the collected empirical material. Furthermore, due to the subjective nature of the data, the results may not be fully representative of the wider population. There is also a risk of interviewer effect, whereby the presence of the researcher and the way questions are posed can influence the respondents' answers, potentially distorting the study results (Babbie 2020; Brinkmann and Kvale 2018).

The aim of the in-depth interviews we conducted was to gather detailed opinions from local government representatives on key factors and barriers to the development of four selected Functional Urban Areas (FUA) that constitute inner peripheries from the perspective of the FUA's core city. The interviews focused on analyzing socio-economic processes occurring in Poland's inner peripheries, taking into account the impact of development shocks that emerged after 2020, such as the SARS-CoV-2 pandemic and the war in Ukraine. As part of the study's methodological framework, we empirically verified previously identified development factors by conducting in-depth interviews with local-level government leaders. Particular attention was paid to gathering opinions on the influence of the central city of the FUA on its surrounding areas. We expected respondents to identify key development barriers and propose strategies for overcoming them. Additionally, we assumed that conducting interviews during the late stage of the pandemic and amid the war in Ukraine would allow for an assessment of how such development shocks affected local development governance processes.

The research tool, in the form of an original in-depth interview script, consisted of five key sections corresponding to the identified development factors: accessibility, human capital, social and institutional capital, financial capital, and physical capital. The questions primarily focused on issues that could not be examined using statistical data employed in earlier project stages (Chaps. 4 and 5). In the area of accessibility, questions addressed the organization of public and individual transport, including the structure of transport operators, the range and frequency of services, fleet quality, and the level of use of available transport modes by FUA residents. We also analyzed changes in transport accessibility—inner, external, and digital—in the context of the SARS-CoV-2 pandemic's impact. Regarding human capital, the research questions concerned the local demographic situation and problems within the education system. Particular attention was paid to changes in perceptions of public and health security, as well as the dynamics and mobility of the local labor market, considering its transformation after 2020. Social and institutional capital was analyzed through the lens of local identity, which determines the level of civic engagement. Respondents were asked to assess the structure and economic activity of local businesses, as well as their cooperation with local government. In terms of financial capital, the research included an assessment of the financial condition of local governments, entrepreneurs, and residents. A key aspect involved analyzing the budgetary situation of local government units after 2020 and evaluating the effectiveness of government interventions aimed at mitigating the effects of the pandemic. In the analysis of physical capital, we focused on the availability and quality of housing resources, both municipal and market-based. In the context of improving residents' quality of life, we also inquired about the condition of the natural environment within the FUA and the dynamics of changes occurring there. One of the final topics discussed during the interviews was the possibility of identifying local interdependencies between the development factors examined in the study. Moreover, each interview included an assessment of the implementation level of activities set out in local development strategies. Respondents were also encouraged to discuss new development challenges arising from economic and social shocks. This approach allowed us to obtain a multidimensional picture of development processes and to identify key challenges facing the studied inner peripheries.

As part of the research, we conducted in-depth interviews with representatives of the local governments of the four main cities of the FUAs under study—mayors or persons designated by them. In the remainder of this chapter, we refer to the sources of quotations with the following codes: Włocławek (IDI_W), Drawsko Pomorskie (IDI_DP), Przasnysz (IDI_P), and Sandomierz (IDI_S). All interviews were conducted in the city hall offices—the respondents' workplaces. The average duration of the interviews was 90 min, and the conversations were recorded in audio format and subsequently transcribed. Respondents provided informed consent to participate, to be recorded, and to have their interviews transcribed, though the transcripts were not subject to respondent approval. The interviews were conducted directly by the authors. Subsequently, the interview transcripts were coded in order to assign analytical categories to text fragments. We then identified thematic patterns and the relationships between them in order to interpret the findings.

Among the respondents were a deputy mayor, two town mayors, and a city secretary. All interviewees held higher education degrees. Only one person declared affiliation with a political party, while the others had run for office in local elections on behalf of civic or local electoral committees. Among the respondents were three men and one woman, all aged between 35 and 55. All were residents of the studied FUAs and, by virtue of their official positions, possessed expert knowledge of the local conditions shaping socio-economic development processes. All respondents actively participated in and observed changes resulting from development shocks such as the SARS-CoV-2 pandemic. In our study, we applied the method of in-depth interviews, utilizing its flexible structure. The interview script remained open, allowing the direction of the discussion to be adjusted to the interviewee's insights. When a respondent wanted to elaborate on a particular topic, we were ready to explore it further, especially when it offered valuable insights from our research perspective. We are aware of the subjective nature of the data obtained through interviews, which prevents their direct generalization to all inner peripheries. We also recognize that respondents may have sought to present their cities in a more favorable light. To minimize the risk of reality distortion by participants, we employed methodological triangulation by complementing the in-depth interviews with a survey among residents and focus group interviews with local leaders.

6.2.4 Focus Group Interviews with Entrepreneurs, Local Government Officials and Local NGO Leaders

The focus group interview (FGI) is a widely used qualitative research method involving a structured discussion within a small group of participants, usually numbering between six and 12 individuals. The aim of this technique is to explore respondents' opinions, experiences, and attitudes toward a given issue, allowing for the collection of in-depth and multifaceted data. A key element in the effective use of this method is the careful selection of participants. Individuals with familial, business, or hierarchical relationships should be excluded to avoid potential distortions in group interaction dynamics (Babbie 2005). Focus group interviews are commonly employed in research on cross-sectoral collaboration, local development challenges, and the effectiveness of public policies. The interaction between participants allows for the identification of shared values, conflicts of interest, and potential areas of synergy, making this method particularly useful for studying social dynamics and group structures (Kvale 2012). However, the method is not without its limitations.

Group dynamics can affect the sincerity of responses—some participants may dominate the discussion while others remain passive, leading to an unequal representation of viewpoints. Moreover, the presence of individuals from different sectors or backgrounds may result in self-censorship or tension, especially in the context of prior conflicts. The limited number and non-representative nature of the sample further constrain the generalizability of the findings (Sim and Waterfield 2019).

Moderators play a crucial role in conducting focus group interviews. Their task is to guide the discussion in a way that ensures the collection of valuable research data, while also fostering participant comfort and engagement. The moderator should remain neutral, refrain from expressing personal opinions, and skillfully manage group dynamics by encouraging less-vocal participants and moderating dominant individuals (Kvale 2012). An essential part of the research procedure is the obtaining of informed consent from participants prior to the interview. Each individual must be fully informed about the study's purpose, the use of collected data, and their right to withdraw at any time without consequence. Ensuring anonymity and confidentiality is a key ethical requirement, particularly when discussing sensitive or controversial topics. This process contributes to building trust between researchers and participants and enhances the credibility of the results (Rosnow and Rosenthal 2013).

The aim of the focus group interviews conducted in our study was to collect detailed opinions from individuals associated with local governments (municipal councilors and public sector employees), as well as local entrepreneurs and leaders of non-governmental organizations, regarding key developmental barriers and factors fostering socio-economic growth in selected inner peripheries in Poland. The study also sought to identify recommendations for development interventions that reflect the needs of diverse stakeholder groups and highlight good practices worth broader implementation. The diversity of respondent perspectives enabled a comprehensive analysis of the impact of the SARS-CoV-2 pandemic on the development dynamics of the studied areas.

The focus group interview guide, similarly to that used for the in-depth interviews, encompassed discussion on five key factors shaping the development of inner peripheries. The first part of the interviews addressed transport accessibility within the Functional Urban Area (FUA), taking into account the perspective of local governments organizing public transport and entrepreneurs for whom employee mobility is a critical aspect of business operations. The discussion included an analysis of transport barriers and their impact on FUA development, especially in the context of the SARS-CoV-2 pandemic. The next section focused on the human capital of the studied areas, including depopulation processes, the outflow of working-age residents, and labor market changes resulting from ongoing economic and social crises. Another area of analysis was social and institutional capital, which included assessments of local identity, civic engagement, and the role of entrepreneurs and NGOs in shaping social and institutional capital. We also evaluated the functioning of business support institutions and the structure and activity of the local economy within the FUA. The financial standing of municipalities, local businesses, and residents' wealth were also examined, with particular attention paid to investment attractiveness and the ability of local governments to secure external funding. Discussions addressed local economies—how the pandemic and the war in Ukraine had affected them but also their capacity for adaptation and recovery. Regarding material capital, we assessed the importance of access to ecosystem goods and services for residents' quality of life, housing availability for new employees, and the quality of local spatial planning systems. The FGI scenario was designed to enable a comprehensive diagnosis of each FUA's situation, integrating the perspectives of public institutions as well as social and economic leaders. The interview structure facilitated the identification of development barriers and good practices that could inform regional development policy recommendations.

The focus group interviews were conducted in two respondent categories: (1) representatives of local governments and (2) representatives of entrepreneurs and NGOs. In total, eight moderated focus group interviews involving 60 participants were conducted across the studied inner periphery case studies. In each case study, two separate group interviews were conducted corresponding to the two respondent categories. Each group ranged in size from six to 12 participants. All respondents were invited to recorded sessions and provided informed consent for participation, audio recording, and transcription. Transcriptions were not subject to participant authorization. While the interviews were conducted by a professional research agency, the purpose, assumptions, and scenarios were developed by the research team. The transcripts were then coded, assigning analytical categories to relevant text segments. Subsequently, thematic patterns and interrelationships were identified to facilitate the interpretation of results. The first respondent group included local government representatives—councilors and municipal office staff from both core FUA municipalities and peripheral municipalities, in a 1:2 ratio. This group included 28 individuals. To ensure the open exchange of opinions and prevent the influence of hierarchical relationships, senior municipal officials were not invited to participate. All focus group interviews with local government representatives were coded FGI1 and assigned the following identifiers depending on the FUA core city: Drawsko Pomorskie—FGI1_D; Przasnysz—FGI1_P; Sandomierz—FGI1_S; Włocławek—FGI1_W. The second group consisted of entrepreneurs and NGO representatives from core cities and surrounding municipalities, considered local leaders within the studied FUAs. This group comprised 32 participants. Among the entrepreneurs were farmers and owners of micro and small enterprises employing up to nine people. NGO representatives came from village women's associations, volunteer fire brigades, health-promoting organizations, and residents' associations involved in preserving local history and cultural heritage. All focus group interviews with this group were coded FGI2 and labeled as follows: Drawsko Pomorskie—FGI2_D; Przasnysz—FGI2_P; Sandomierz—FGI2_S; Włocławek—FGI2_W.

The interview guides were adapted to the specific characteristics of each respondent group. The guide for local government representatives focused on legally defined tasks of municipalities and counties, enabling an in-depth analysis of their functioning. In contrast, the guide for entrepreneurs and NGO representatives covered cooperation with local governments, joint business initiatives, and social needs in the inner peripheries. Both versions of the interview guide included questions on the impact of the SARS-CoV-2 pandemic and the war in Ukraine on developmental factors, analyzed from the perspectives of public institutions and socio-economic actors. Special attention was paid to identifying inner relationships and feedback loops shaping the relational development environment of territorial capital in inner peripheries.

In designing the focus group interviews with local leaders in inner peripheries, we sought to capture social dynamics by analyzing shared stakeholder perspectives and potential conflicts of interest. Focus group interviews are well-suited for such analyses. However, the limited number of interviews per case study area means our findings pertain only to the specific areas studied and should not be generalized to all inner peripheries. We recognize that poorly moderated focus groups can lead to asymmetrical discussions, with one group dominating while others are marginalized. To mitigate this, we entrusted the interviews to experienced moderators from a specialized research agency. Aware of the risk of self-censorship arising from social or hierarchical relationships among participants, we formed groups of as equal status as possible. In the local government group, we invited councilors but excluded mayors and senior officials to avoid institutional dependencies. A similar approach was used in the other group, which was composed of entrepreneurs and NGO leaders. This design minimized the influence of hierarchical relationships and allowed for more open and authentic discussions.

The research methods described in this subsection, employed in the analysis of selected inner periphery case studies, constitute a key component of a prototype system for monitoring socio-economic changes in these areas.

6.3 Accessibility and Its Changes

The degree to which accessibility needs are met constitutes a key development factor differentiating peripheral areas from central ones. The concept of accessibility features in both geographic and relational understandings of peripherality. In the latter perspective, accessibility is the ability to engage in socio-economic interactions (e.g., cooperation, trade, knowledge exchange) between residents and institutions (including enterprises) located in different areas. Accessibility is shaped by various factors, including the state of transport infrastructure development (the extent, density, and quality of its components), traffic organization (including the types of transport modes and the number of public transport connections), as well as the state of digital infrastructure and services. In our approach, we conceptualize accessibility through three key subcomponents: external accessibility (the ability to establish interactions between the Functional Urban Area [FUA] and the rest of the country and Europe, particularly with major cities that act as socio-economic development cores); internal accessibility (the ability to form relationships within the FUA, including the connection of the entire area to the central city, as well as *horizontal* connections between localities within the FUA, contributing to its inner cohesion); and e-accessibility (access to digital communication services that enable interaction without physical movement; see Sect. 2.3.3).

The operationalization of accessibility is reflected in actual interactions between territories within FUAs, particularly between the core city and its surrounding area, as well as between different FUAs. Considering the directions and intensity of these connections, as well as the priorities expressed in strategic documents and by local stakeholders, it is possible to assess the level of satisfaction of accessibility needs and identify directions for desirable improvements. Based on the interviews and surveys conducted in the selected test areas, we identified the following categories of intra- and inter-FUA functional linkages:

1. Intra-FUA relationships—connections between the central city of the FUA and surrounding localities and between localities within the FUA, involving flows of people, goods, and services.
2. Relationships between FUAs and higher-order cities (regional capitals)—which concentrate administrative, economic, and educational functions and act as regional development centers.
3. Horizontal relationships between FUAs and cities of similar size to the FUA cores—including travel and business contacts between cities with comparable demographic and economic potential.
4. International relationships—encompassing international cooperation in the fields of economy, education, culture, and permanent or temporary migration related to employment or education.

According to our conceptual division into three subcomponents, external accessibility encompasses all types of functional linkages realized in the last three types of relationships. Internal accessibility refers to all types of linkages within the first type being realizable. E-accessibility refers to the possibility to use digital communication to conduct linkages related to work, education, and service use, as well as inter-firm collaboration, enabling the formation of functional linkages both within and beyond the FUA.

Functional linkages involving regular travel beyond the FUA most often refer to upward commuting within the settlement hierarchy—to larger cities such as regional capitals or Warsaw. This direction is determined by the attractive, absorptive, and diversified labor markets in larger cities situated within economic development cores. However, the long commuting time—usually exceeding one hour (the functional range threshold used in FUA delimitation)—limits the regularity of such commutes; only 5.8% of respondents reported commuting to regional capitals for work. Horizontal commuting to other cities of similar size to the FUA's central city also plays a significant role in local labor markets. For example, according to an entrepreneur from a medium-sized FUA in eastern Poland (Sandomierz), local employment is significantly influenced by commuting to three nearby cities of similar size (e.g., Tarnobrzeg), each forming their own functional area, located within a 35-min drive and exhibiting complementary labor market demands.

Work-related mobility is often linked to relocation, motivated by quality-of-life considerations, such as access to services and housing. A respondent from a medium-sized FUA in north-eastern Poland (Przasnysz) observed: *We see a trend of our residents buying homes in Ciechanów (the capital of a neighboring FUA)—even my employees at the city hall here. They prefer to live in Ciechanów and commute here* (IDI_S).

Such patterns of regular, permanent, or occasional relocation also occur at the international scale. On the one hand, this includes permanent or temporary labor-related emigration, including contractual work: *My son works in Germany. Also, a lot of people from here go abroad for work to other cities. They're gone for two weeks or more* (FGI1_W). On the other hand, peripheral areas in Poland—not just large cities—are now destinations for labor immigration, with foreign workers filling workforce gaps, especially in the industrial and construction sectors: *I naturally employ people from Ukraine and I'm very satisfied* (FGI2_D).

This also applies to the transport sector, where a large portion of services is export-oriented, and work involves constant mobility: *The trend has completely changed—seeing a Pole in a truck now is like winning the lottery. It's mostly dominated by Belarusians, Ukrainians, and Kazakhs. And yes, they live and are stationed here somewhere locally* (IDI_P). However, not all foreign workers adapt well to the work environment in peripheral areas: *The Kross company started bringing in workers from various parts of the world* [...] *Indians,* [...] *people from Ethiopia.* [...] *But something just didn't work out there.* [...] *At some point, they were let go* (FGI1_P).

Education, especially at the tertiary level, is also a key driver of functional linkages between peripheral FUAs and higher-order cities (regional capitals). Even if local universities exist, studying elsewhere is seen as fulfilling young people's aspirations and as an opportunity for both education and employment. *Completing a higher education degree here in Włocławek doesn't bring prestige or build self-worth. It's better to go to Nicolaus Copernicus University in Toruń-it's closer-or even to Poznań* (FGI2_W). Studying outside the area often marks the beginning of permanent migration, typically solidified by finding a job in a larger city: *Sure, there are colleges nearby, even in Sandomierz, but everyone wants to pursue the career they dream of. From my experience, my daughters studied in Kraków, and then life—and family—kept them there. They won't come back* (FGI1_S). Horizontal travel between FUAs also plays a significant role in some forms of mobility. A notable example is commuting to secondary schools, motivated by the availability and prestige of educational institutions in neighboring cities of similar size. *Our young people, for various reasons, decide to attend what's called "Krasiniak," the general secondary school, which has a fairly high reputation* (FGI1_P). Beyond regular commutes to school or work, transport accessibility to regional centers is important for incidental needs such as accessing specialized healthcare services unavailable within the FUA. From the perspective of passenger flows, access to regional capitals and larger cities outside the FUA is thus a latent yet significant asset.

Commercial relationships between FUA-based enterprises and their suppliers, clients, and partners outside the area may be regular or incidental. Regardless, the ability to ensure the easy, fast, and reliable transport of people and goods is a necessary condition for local economic development. The directions of commercial linkages include larger urban centers, horizontal relations with other functional areas of smaller towns, and international ties with partners connected to local businesses via ownership, client relations, or supply chains. Some highly specialized industrial plants in peripheral areas are part of the supply chains of large multinational corporations. One example is the ABB electrical equipment factory in Przasnysz: *This is a company that mainly specializes in exports. The transport comes, collects all the made-to-order items, and then* […] *they're sold in dollars* (FGI1_P). Another is the automotive glass manufacturer Pilkington Polska in Sandomierz: *It's one of over a dozen factories worldwide, and they produce windows for all cars globally* (FGI2_S).

Given the strength of functional linkages and the shape of Poland's strategic and spatial planning system—which emphasizes the role of regions—both strategic documents and respondents tend to prioritize accessibility to the regional capital. In many of the areas studied, this form of accessibility is assessed quite positively.

Most passenger and freight movements between FUAs and larger cities, as well as other areas in Poland and abroad, are carried out via road transport—using private cars and trucks. Given the significance of incidental travel, the fulfillment of external accessibility needs is measured not only by the frequency of regular journeys, but by the very possibility of undertaking such journeys or transporting goods as needed. In this context, both the quantity and quality of roads are crucial, especially high-capacity motorways and expressways, the network of which has been systematically expanded since Poland's accession to the European Union (increasing from under 1,000 km in 2004 to nearly 5,000 km in 2025, with a target of over 8,000 km). These roads primarily connect major cities and, by their nature, either bypass or marginally intersect peripheral areas. The only FUA in our study currently integrated into the high-capacity road network is the one centered on the large city of Włocławek in central Poland, through which the A1 motorway—linking Poland's southern border, the Katowice and Łódź agglomerations, and Gdańsk in the north—has passed since 2013. While the lack of expressways is perceived as an obstacle to attracting investment, it does not necessarily imply transport exclusion if the technical condition, capacity, and relatively short distance to the central city enable travel within a reasonably short and predictable time. Thus, for instance, in the small-town FUA of Drawsko Pomorskie in north-western Poland, informants assessed access to the regional capital by car positively: *Let's be honest—(national road no.) 20 to Szczecin takes about 50 min, which isn't much—it's roughly the same time it takes to get from one district to another in a big city.*

From the perspective of sustainable development—both environmental and social—it is desirable to achieve a high share of public transport (by rail, bus, and minibus) in travel, including outside FUAs. The structure of public transport networks is generally concentric, so passenger journeys by public transport are mostly to regional capital cities, although their share is relatively small. For example, only about 10% of commuters in regional capital cities use trains. It is worth noting, however, that only two central cities among the analyzed FUAs—Włocławek and Drawsko Pomorskie—have direct rail connections with their voivodeship capitals. These are assessed as sufficient: *looking at rail transport in Drawsko and the PKS (bus) transport to Szczecin, it's probably quite optimal in terms of frequency and occupancy* (FGI_D). The railway line also passes through Sandomierz (although the station is located outside the city) and along the outskirts of the Przasnysz FUA, but in both cases, getting to the voivodeship capitals is time-consuming and requires transfers. Przasnysz, halfway between the voivodeship cities of Warsaw and Olsztyn, has convenient bus connections with both cities: [...] *as for Warsaw, there is no problem, basically there's a departure every hour* [...] *as for the connection between our county and the voivodeship city, there's a 4 a.m. bus to Warsaw* [...] *and as for connections with Olsztyn, with Warsaw, there's no problem* (IDI_P). The situation is worse in the Sandomierz FUA, where public transport to the voivodeship capital Kielce is limited to private minibus services (IDI_S). The connectivity of neighboring FUAs with cities of similar size is less frequently mentioned in documents and interviews, and when it is, it is usually in a negative light. For example: *in Przasnysz we have problems with Ciechanów, because there really are very few buses, basically only school ones* [...] *And there's also a problem with the connection to Ostrołęka* (IDI_P).

In strategic documents and interviews, improving external accessibility is identified as particularly important in efforts to integrate the case study areas into economic development processes and improve quality of life. Attention is focused especially on expanding road and rail infrastructure. According to the National Road Construction Program for 2030, the construction of the S10 expressway connecting the Włocławek FUA with Warsaw is planned. Two other FUAs, which currently lack access to the expressway network, will receive such connections by the beginning of the 2030s. The S10 expressway is scheduled to pass along the outskirts of the Drawsko Pomorskie FUA by 2028, linking Szczecin with Bydgoszcz and eventually Warsaw. After 2030, the S74 road will pass through Sandomierz, connecting the FUA with its own voivodeship capital, two neighboring capitals, and central Poland. Planning documents also highlight the importance of improving the integration of expressways with the existing road network (e.g., strengthening access road capacity and constructing bypasses) to ensure that residents of the surrounding area can use the expressways safely and with minimal disturbance.

Concerning external public transport, attention is primarily focused on rail transport, which is considered the most environmentally friendly mode and one that can also compete with private car use in terms of travel time. Some of the studied FUAs could improve their external transport accessibility through the planned or proposed expansion of the railway network. The new *Northern Mazovia railway line*, currently in the project phase and part of the government's Kolej Plus program, is expected to connect Przasnysz with Warsaw by around 2040. This connection would offer a more comfortable and faster alternative to current bus and car travel: *today, it takes about 1 h and 30 min to drive to Warsaw, and up to 2 h during rush hour, whereas rail transport would probably cut that to around 1 h—1 h and 10 min* (IDI_P). Interestingly, local stakeholders are aware that improved external accessibility will bring not only benefits but also new challenges: *that would completely overturn the labor market in Przasnysz* [...] *and local employers would definitely face serious difficulties, because people would start commuting to Warsaw daily, taking advantage of the new prospects* (IDI_P). In the Włocławek FUA, a high-speed rail line is planned to connect the city with Warsaw and the new CPK airport. Long-term plans also include integrating smaller cities not currently served by rail into the network. Across these and other FUAs, strategic documents emphasize the need to improve regional rail services in areas with existing rail networks. This includes increasing train frequency, upgrading rolling stock to modern, low-emission vehicles, and integrating transportation systems by implementing a joint ticketing system for regional rail and bus transport organized at the voivodeship level.

For the two FUAs located along the Vistula River—Włocławek and Sandomierz—strategic documents also include ideas to improve external accessibility via inland waterway transport (e.g., construction of a multimodal port in Włocławek integrating river, rail, and road transport). However, the share of inland waterway transport in Poland's freight structure is negligible. Likewise, air transport receives little attention, as none of the studied FUAs has its own airport. Nevertheless, some strategic documents mention the need to improve public transport access to regional airports in larger cities.

The second subfactor of accessibility—internal accessibility—plays a key role in ensuring the cohesion of the functional area and enabling the spread of development processes from the central city to the outer zones of the FUA. A significant share of travel related to commuting, migration, and access to various services occurs within the FUAs. Among respondents, the majority (67.6%) work in the municipality where they live (Fig. 6.2). In central cities, this proportion is even higher (84.2%). Meanwhile, more than one third (37.9%) of employed residents of the outer zone commute to work, most often to the FUA's central city (16.7%) or to another locality within the same municipality. Only 5.8% of respondents work in the regional capital.

Work-related commuting is thus relatively rare, and, when it occurs, it typically happens within the FUA. These are often commutes to the central city; for example, *people from the surrounding municipalities look for work here in Przasnysz. […] A lot of these residents are here during the week, and […] on Sundays, the streets in Przasnysz are completely empty* (FGI1_P). However, horizontal commuting is

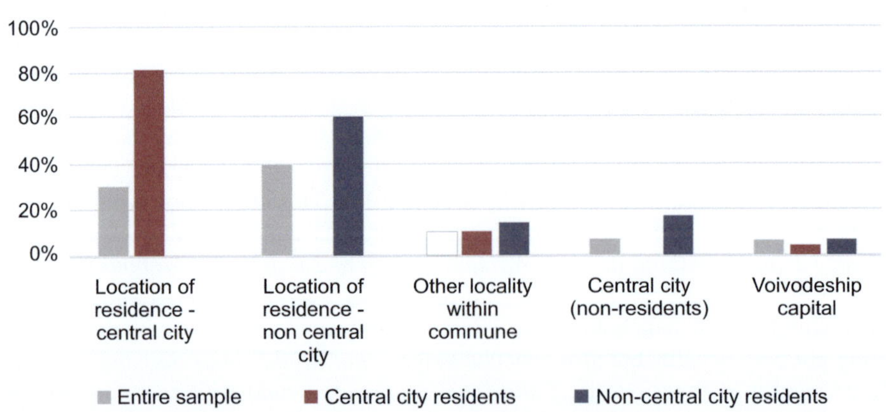

Fig. 6.2 Place of employment of survey respondents (residents in FUA central city and the rest of FUA). *Source* Own elaboration

also quite common, increasingly due to business relocations or uneven locational advantages within the MOF, stemming, for example, from differences in external accessibility that result in preferences for locating businesses in better-connected areas (FGI1_W, Włocławek).

This may be evidence of the diminishing importance of FUA central cities and the growing strength of surrounding areas—particularly those that are well-managed and feature accessible, well-equipped economic zones attracting investors.

Changes in place of residence also often occur within the FUA and reflect, among other things, suburbanization, which—outside Poland's largest cities—also takes place in the suburban zones of medium-sized towns. For their residents, the desire to improve housing conditions is the most important motivation for relocating, slightly surpassing the desire to find a better job (Fig. 6.3). Respondents emphasize, however, that, within the functional area, only the immediate surroundings of the largest city and areas attractive in terms of accessibility and landscape become targets of migration from the city to the countryside or to smaller towns. Areas further from the central city are less likely to be affected by suburbanization processes. A community activist from the Włocławek FUA asks: *do places like Choceń or Lubraniec* [small towns about 20 km from Włocławek] *stand a chance of becoming something like bedroom communities for Włocławek?* and answers: *Probably not.* [...] *A bit too far* (FGI2_W).

Migration to the outskirts is driven not only by the appeal of suburban areas but also by housing market conditions, personal resources, and family relationships. Residents of central cities return to the countryside because they cannot afford housing in the city. *If they have a plot of land at their parents' place, they get the land for free and can start building on it* (FGI1_S). Yet the same housing market dynamics, combined with a reluctance to invest outside of familiar and easily accessible areas, foster the formation of ownership ties within the FUA. This, in turn, triggers issues that are familiar to larger cities, such as the misalignment between the location of owned residential property and the actual or formal place of residence: *these apartments (in Przasnysz) are being bought largely by residents from neighbouring rural municipalities* [...] *with inflation running wild, people prefer to invest their savings in real estate* [...] *but they don't register here or become our official residents* (IDI_P). This results in a relative deterioration of local government finances, as municipalities incur the costs of providing public services to new residents without compensation in income tax shares.

Beyond commuting for work and permanent relocations, functional linkages within FUAs are also evident in travel for various services, including education, healthcare, culture, and retail. Residents of central cities meet 80–85% of their basic service needs within their own city. Residents of the surrounding areas are more likely to commute to access these services, most often to the FUA's central city. For instance, one in four respondents travel there for cultural and recreational services, and nearly one in four secondary school students commutes to school (Fig. 6.4). However, it is worth noting that commutes, for example to school, occur not only toward the central city. In the Włocławek FUA, *it used to be that Włocławek was the main center of secondary education. Now schools are closer—there's Chodecz, Choceń,*

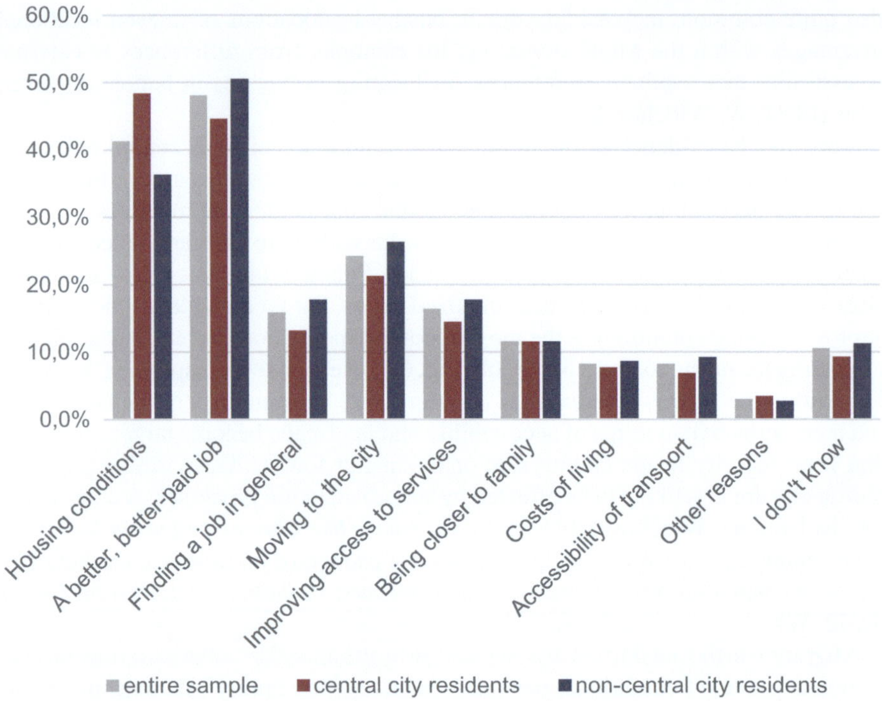

Fig. 6.3 Reasons for the willingness to move. *Source* Own elaboration

and Lubraniec—secondary schools, high schools (FGI1_W). This reflects, on the one hand, the declining relative position of the main city and, on the other, the improved educational offer in localities surrounding the FUA's central city. Functional linkages within the FUA also include cooperation between businesses.

Commuting to work, schools, and services within FUAs—as well as between FUAs and other areas—is primarily by private cars, and less frequently by public transport (Fig. 6.5). Due to the relatively short commuting distances, the majority of respondents (63.2%) spend no more than half an hour per day commuting. Only 8.3% of respondents commute for more than an hour daily—exceeding the commonly accepted threshold for a comfortable commuting time. These are mainly individuals who work in the central city of the FUA or in a regional capital. The most frequently used mode of transport is the private car, accounting for almost two thirds of work-related commutes and the majority of trips to various services. An exception is commuting to upper secondary schools, where public transport (bus or train) plays a relatively significant role, used by nearly 40% of respondents. One in eight respondents walks to work. The share of bicycles is surprisingly low (highest in commuting to cultural and recreational services, at 5.8%). The choice of transport mode depends on the destination. For instance, among residents of surrounding areas who work in the central cities, over 20% commute by bus, while the share exceeds

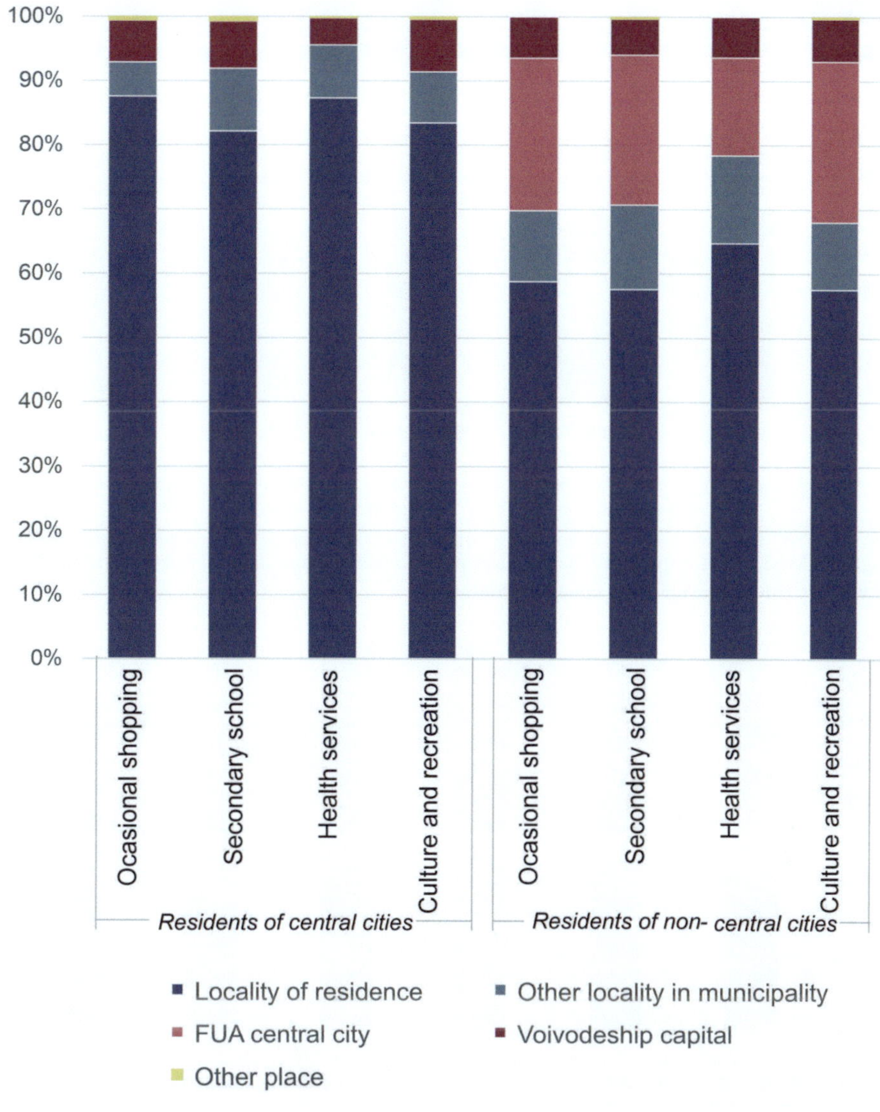

Fig. 6.4 Places of usage of selected services by respondents: residents of FUA central cities and the remaining parts of FUAs. *Source* Own elaboration

30% among students. The relatively high share of buses in these cases is due to radial networks linking surrounding settlements with central cities. However, in commuting to other locations (not central cities), residents often rely on their own cars due to the lack of public transport connections. Trains are most commonly used (10%) by those working in regional capitals, as rail routes typically connect to these cities. By contrast, trains are rarely used for commuting within FUAs (Fig. 6.6).

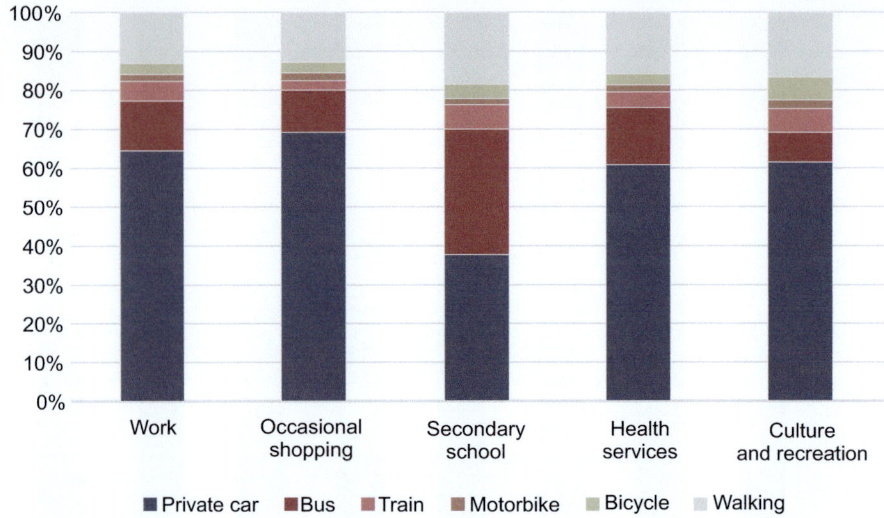

Fig. 6.5 Modes of transportation used to commute to work, school and services by the residents of the case study FUAs. *Source* Own elaboration

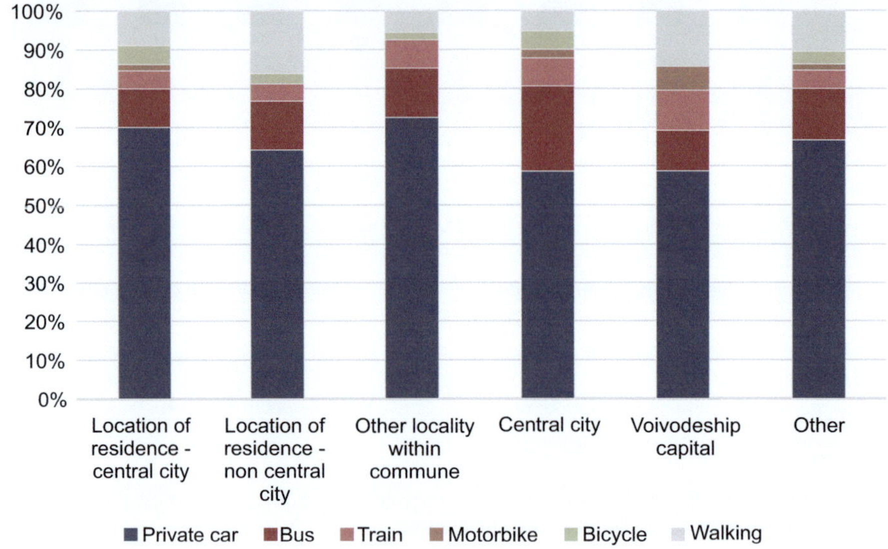

Fig. 6.6 Modes of transportation used to commute to work in relation to place of work. *Source* Own elaboration

Due to the dominance of individual car transport in commuting within FUAs, the density and quality of the road network are key elements of accessibility. The network of regional and local roads is relatively evenly developed across the country, and its quality has improved in recent years due to the increasing share of paved roads, construction of town bypasses, enhanced road safety measures such as sidewalks and bicycle paths, and the modernization of intersections. Nevertheless, planning documents still emphasize the need to improve the technical condition of the local road network, adapt it to current technical standards, and expand it to include, for example, access to new investment areas. Planners stress the importance of building bypasses to mitigate the nuisances of transit traffic and modernize connections between newly constructed expressways and the local road network.

Strategies and interviews emphasize local cycling transport as an environmentally friendly and healthy transport mode. The small size of central towns leads to the bicycle being identified as a mode of transport that should be used more frequently across FUAs. *I think that* [...] *building bike paths would help solve the* [*public transport*] *problem a bit. Especially in spring and summer. Around Przasnysz, students could easily ride their bikes* (FGI1_P). The main barriers to using bicycles as a means of transport include the shortage of safe (separated from car traffic) bike paths, the discontinuity of the network, their poor technical condition (narrowness, poor pavement quality), and maintenance (e.g., sand on the paths).

A crucial aspect of ensuring internal accessibility within FUAs is the organization of local public transport. This is most often operated by private or public bus companies or small private van operators. Its effective operation is hindered by a lack of coordination between private and public carriers and between services organized by public authorities at various administrative levels: *Today, a municipality near Włocławek, to have MPK buses run there, must sign an agreement with us and simply pay for each vehicle-kilometer. We used to pay* [...] *Not all municipalities agree to this* (FGI2_W). Integrating public transport services within FUAs (and at the county and voivodeship level) is a frequent postulate in strategic documents. Such integration includes several components: first, routes operated by different carriers should converge at common hubs (stations, stops); second, timetables should be coordinated to allow transfers; third (a more long-term goal), there should be a unified fare policy and mutually accepted ticketing system; and fourth, there should be a shared information and promotion policy, to help reverse the decline in public transport use that began with the rise in car ownership.

A major challenge in organizing public transport in FUAs is the limited demand, which officials noted as often being restricted to students and elderly people. These services are frequently unavailable or do not meet the needs of other residents. *In the Przasnysz FUA, it would be possible to use the school bus, but unfortunately one cannot.* [...] *A private person cannot get on, even if they pay for a ticket, because it operates under a different accounting system* (FGI2_P). In some areas, it has been possible to make school transport available to other groups: *In our municipality* [Włocławek FUA], *it's only school transport, but older people who don't have cars can also get on and pay* [local government officials]. Similarly, in Drawsko Pomorskie: *We also allowed other residents to board these buses. Previously it was school-only transport—no-one but students could use it. Now it's optimized and works jointly, giving us new possibilities* (FGI_D). However, the dependence of public transport on school schedules does not support its use for other purposes, such as recreation: *You can't get to an Anwil basketball game because there's no way back. The last bus is at 5:30 p.m. On Saturdays and Sundays—there's no service at all. To go to the cinema—youth have no options* (FGI1_W). Employers themselves often organize work commutes within FUAs. To enable public transport provision even under conditions of low demand, some strategies propose supporting the creation of flexible transport services—on-demand transport using small buses contracted by local governments. A good example of solving the problem of discontinued state-run services (PKS) is Lipno, where The County Office organized small minibuses and sent them out to the local villages. *Finally, children can go to school, and the elderly to doctors or government offices* (FGI2_W).

The third subfactor constituting the level of accessibility of a functional area is e-accessibility, that is, the quality of ICT infrastructure enabling work and contact with clients, suppliers, public offices, and service providers online. In today's world, e-accessibility can, to some extent, replace physical accessibility via transport. At the same time, the speed and quality of internet communication depend on connection quality rather than on distance. Therefore, while in the case of physical accessibility, the development of infrastructure and transport services can reduce the barrier of distance from centers of socio-economic development, a high level of development of telecommunication infrastructure can allow peripheral areas to match economic centers in terms of e-accessibility. This is illustrated by a quote from FGI2_P: *I'm very optimistic, because I believe that* […] *there is a gigantic development here. Technology helps because this closeness of technology means you no longer have to travel to big cities* […]. *You can order all the clothes from the internet.* […] *That pressure to seek adventures in big cities will be a bit different, especially in the context of rising costs, like housing or transport. I think that small towns* […] *will bounce back and likely move upward.* Nevertheless, the level of ICT network development—measured for example by the share of areas covered by fourth (and fifth) generation mobile networks and fiber-optic internet—is lower in peripheral areas than in large cities. In planning documents and interviews, problems with *e-accessibility* are not mentioned as a priority. Instead, there is a call for further digitalization of administration to allow residents to handle official matters without leaving home.

Spatial relations related to commuting to work, school, and services took new forms due to the pandemic, when remote work expanded the employment possibilities in places other than one's residence. Initially, even local government officials hoped that the spread of remote work would reduce the advantage of large cities as labor markets and encourage people to return to or settle in smaller centers. *People will look for smaller centers to relax. Now many people work remotely* […] *so in the future it won't matter so much where I live and where I work* (FGI1_W). *Do they have to be hammering on that keyboard in Warsaw? I mean, not just in Warsaw* (FGI2_P). However, reality has revised these expectations. From the pandemic outbreak to the study period, only 13.7% of working respondents changed their work mode to remote or hybrid. Some worked remotely during the pandemic for a time. Still, for the majority, the pandemic did not lead to a permanent change in the mode of work (Fig. 6.7). For example, the authorities in Przasnysz tried to orient the city's development policy toward remote work. Still, as they noted, it neither attracted nor retained residents: *while developing strategic documents* [...] *we wanted to create a vision for the development of our city as a remote work center* [...]. *From the perspective of two years after COVID, I must say it was only moderately successful* (FGI1_P). Mobility from large cities to peripheral areas does occur, but it often takes the form of a weekend or holiday trips to second homes rather than permanent relocation: [...] *where we have access to the lake and there are plots,* [...] *and the residents there are mainly from Łódź, Warsaw, large agglomerations* (FGI1_W). Thus, the changes in work conditions resulting from the pandemic did not improve the position of inner peripheries as a potential place for permanent residence and work. On the other hand, the popularization of internet communication not only for work but also for shopping or contact with clients and contractors has led to increased use of this communication channel instead of traditional ones. *The way residents communicate with entrepreneurs has changed. It used to be all about meetings or direct trade, now everything has moved online. Some things are ordered* [...] *just from platforms and not from stores, and that also has an impact* (FGI1_D).

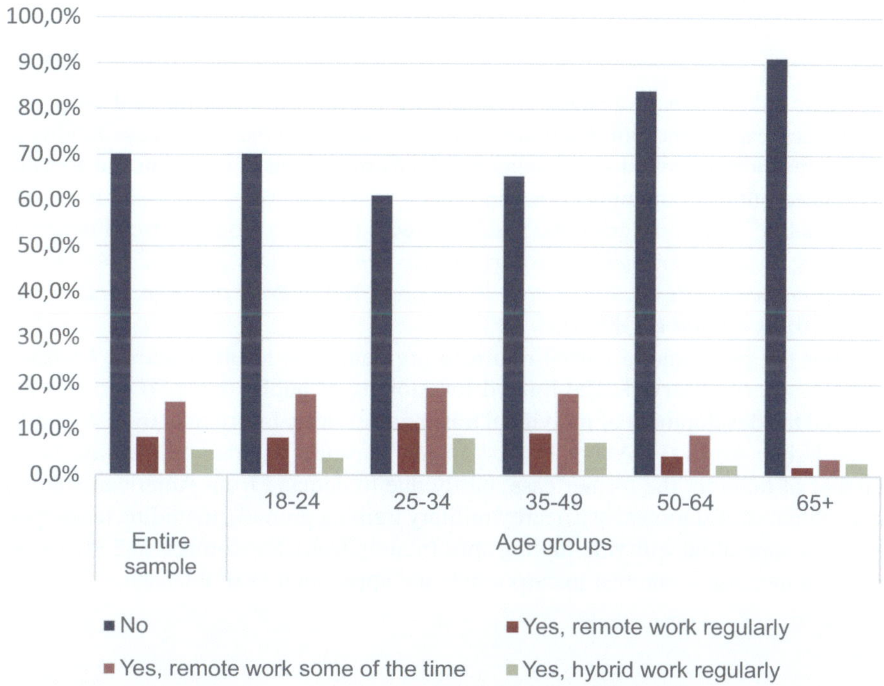

Fig. 6.7 Change in mode of work between 2020 and 2023. *Source* Own elaboration

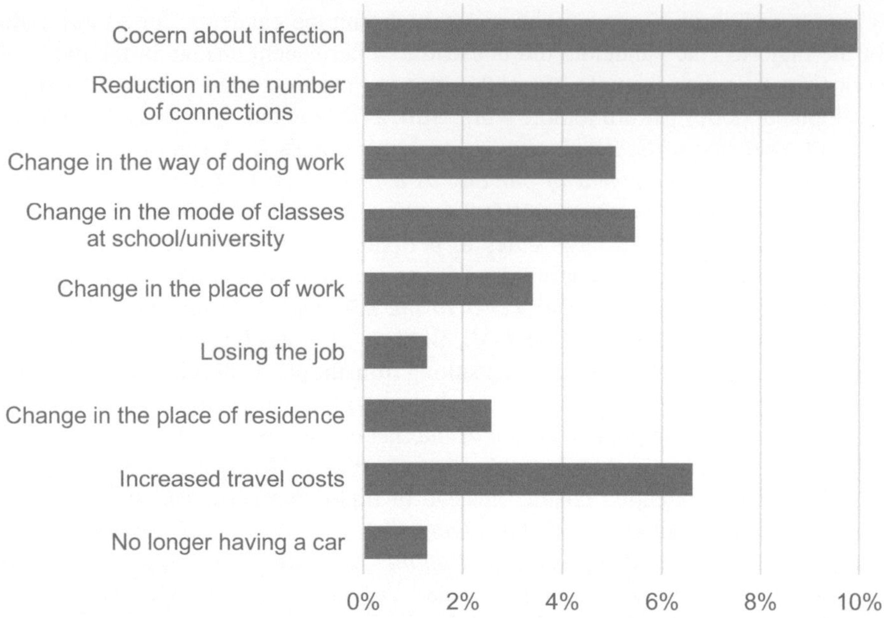

Fig. 6.8 Reasons for change of mode of transportation for commuting to work between 2020 and 2023. *Source* Own elaboration

When asked about changes in transport accessibility during and after the pandemic, respondents pointed to the continued decline in the use and availability of public transport. People stopped using it not only out of fear of infection, but also due to reduced numbers of connections and rising costs (Fig. 6.8). The reduction in the use of public transport during the pandemic was a continuation of a trend observed for years: *local government transport, public transport, was displaced because at some point we didn't want to choose it. We preferred to drive our own cars. We got a little drunk on that* (FGI2_S).

After the pandemic, a partial return to pre-pandemic levels of public transport usage has been observed. The limited functioning of public transport services has fostered the development of individual transport services. Informants from the Functional Urban Areas (FUAs) of Drawsko and Przasnysz reported an increase in the number of taxis (in the former case, partly due to demand from American soldiers and civilian staff stationed at a nearby military training ground), including taxis operating in cooperation with ride-hailing apps (mainly Bolt). Sandomierz and Przasnysz noted an increase in interest in carpooling and apps (such as Blablacar).

6.4 Human, Social, and Institutional Capital

Among all the factors influencing the development of inner peripheral areas, we assign key importance to the social sphere of territorial capital, as it is the inhabitants—by shaping the developmental environment of territorial capital—who form the foundation of economic and institutional processes. Human, social, and institutional capital are interconnected and demonstrate a deep interdependence, which determines their effective functioning through integration and mutual interaction.

We understand human capital as a resource of individual characteristics of residents, including their knowledge and creativity, skills and abilities, and health status. Its use translates into the emergence of unique functional linkages and influences territorial development dynamics. The analysis of human capital in inner peripheries included what we consider to be four key subfactors: demographic situation, health and safety, education level, and the labor market, which enables the evaluation of how effectively human capital is being utilized (Sect. 2.3.3).

An important complement to human capital is social and institutional capital, which were analyzed jointly due to their multidimensional nature. The two types of capital are strongly interrelated, forming a relational structure in which social capital includes a system of informal social institutions based on trust, shared values, and civic engagement. In contrast, institutional capital refers to formal administrative, economic, and organizational structures operating according to legal norms. The analysis of results regarding social and institutional capital as a development factor for inner peripheries in selected case studies considered social activity, ways of spending free time, business infrastructure, and the structure and economic activity of the population (Sect. 2.3.3).

In this subsection, the research results concerning the distinguished categories of capital are presented in an integrated manner, making it possible to show the interrelations and overlapping deficits that influence the development of the analyzed inner peripheries. The territorial approach (Sect. 2.1.1) constitutes the foundation of territorial capital, within which space is seen as a social construct. A strong sense of local identity promotes the formation of belonging to a specific place, which in turn can strengthen the process of individual anchoring in a given environment. Local identity is defined as the degree of identification of an individual or group with their place of residence, encompassing both the emotional bond with the area and its material and immaterial cultural aspects, including traditions, values, and social norms (Sects. 2.2.2, 2.3.2). At the same time, social identity—a key factor in forming social capital—has a collective character and is based on experienced and innerized tradition, a shared interpretation of the present, and a common vision of the future (Sect. 2.3.1). The territorial approach emphasizes the complex relations between identity, belonging, and rootedness, showing how local conditions influence not only individual identity but also social cohesion and its capacity for adaptation in the face of crises.

In analyzing the results of in-depth research concerning local human, social, and institutional capital, we included the varying degrees of place attachment among residents in our case studies. The analysis of the four selected FUAs reveals differences in this phenomenon depending on the historical context and the level of population mobility. In the FUA of a small town in north-western Poland (Drawsko Pomorskie), whose settlement structure was shaped after World War II, the sense of local identity is relatively low. This results from intensive migration to larger cities and abroad and the lack of multigenerational social ties. In contrast, the FUA of a medium-sized town near the country's capital (Przasnysz), based on a historic trading town later shaped by its military function, is characterized by a moderate resident identification with place. However, the significant outflow of population to Warsaw presents a major challenge. The FUA of a large city in central Poland (Włocławek), once a strong industrial hub, shows a high degree of social attachment to the local community despite the challenges of ongoing social stagnation. Finally, the FUA of a medium-sized town located in the eastern part of the country (Sandomierz), a city with a rich historical heritage, stands out for a very strong sense of local identity—in both the city itself and the surrounding area—due to limited population mobility and the dominance of tourism over other sectors of the economy.

The role of human capital in shaping the development of areas—not only peripheral ones—is exceptional, as it is the people—the residents of a given area, their knowledge, and accumulated experience—who constitute the potential of that territory and determine the quality of the territorial capital development environment. Our analysis of the components of this capital began with the demographic subfactor. In each of the four FUA case studies, in-depth research revealed ongoing processes of depopulation and population aging. These adverse changes are observed across the country; however, as demonstrated in Sect. 6.1, they are significantly more pronounced in peripheral areas. Our analysis of the content of strategic and programming documents of the entities forming the FUAs, at both regional and local levels, indicated that demographic problems are one of the dominant elements in diagnosing the current situation. Foremost among these is the depopulation of peripheral areas, including towns, caused by multiple interrelated factors. These include both natural population decline and, most notably, excessive outmigration, which together produce a highly unfavorable population age structure.

In the inner peripheries under study, natural population growth has been clearly negative for several years, mainly due to low fertility rates and decreasing births. Two fundamental elements underpin this situation. One is the abandonment of the large-family model, and even the traditional 2 + 2 replacement model is becoming increasingly rare. Young people often postpone the decision to start a family, are less likely to have children, and typically opt for just one child. The birth crisis in the case study areas is further deepened by the emigration of residents, especially young people, who frequently move to larger urban centers to pursue higher education or employment opportunities and do not return to their place of origin. As a result, inner peripheries lose young and qualified individuals; local identity is also weakening, understood as a sense of belonging to one's place of residence. Young people identify less with their local environment and community, making it easier to decide to leave the FUA and assimilate quickly in a new place. When confronted with their expectations and the actual offer of peripheral areas regarding education, labor market opportunities, or cultural amenities, they often decide to move to a large city even before it becomes formally possible (i.e., upon reaching legal adulthood). This becomes an additional motivation, especially in education (good exam results will allow them to apply to a good university). Their decisions are driven by a desire for self-development and a longing for a modern and dynamic lifestyle associated with greater social and cultural diversity and anonymity. However, this does not necessarily imply a complete renunciation of their attachment to the community they come from; an emotional bond may persist, though it is rarely strong enough to influence a decision to return permanently to their hometown. Emigration affects the youngest members of the peripheral population and working-age adults who often migrate abroad in search of better (implicitly, better-paid) employment. However, unlike younger people, these migrations are more frequently temporary and take the form of circular or shuttle migrations (seasonal departures and returns). This suggests that such individuals tend to have a high level of rootedness, a stable social and economic position, and a well-formed sense of belonging to the local community, significantly complicating the decision to leave a given area permanently. As a result of these highly adverse changes, the process of population aging is advancing at a pace that is bringing inner peripheries to the brink of demographic collapse.

Negative demographic changes pose a major developmental challenge. At a declarative level, strategic and programming documents of local government units formulate decisions and recommendations for implementing actions aimed at improving the quality of human capital. These efforts seek to enhance the attractiveness of inner peripheral areas as places to live, halt negative demographic trends, attract new residents, and strengthen local social identity. Among the proposed solutions, the strongest emphasis is placed on the development of pro-family policies. At the national level, solutions in operation include the "Family 800+" program, which provides financial support to help families cover child-rearing costs, and the "Active Parent" program, which offers additional financial assistance to working parents. A further important initiative is the reintroduction of the government-funded in vitro support program, which, for many families, represents the only opportunity to have children. However, in each case, the local governments forming the FUAs in the inner peripheries function merely as intermediaries in these programs, without significant influence over their design, scope, or delivery mechanisms. At the local government level, the recommended measures do not primarily involve direct financial support but aim to provide material and organizational assistance. A strong emphasis is placed on improving housing accessibility, particularly financially. In Sandomierz, plans include introducing housing incentives, such as the temporary provision of housing stock to young people. In Przasnysz, plans have been made to establish inter-municipal Social Housing Companies (TBS), offering an alternative to purchasing housing from private developers. (An additional merit of this concept is its supra-local cooperative character among LAU units.) A particularly distinctive initiative is the promotion of foster parenting in the FUA of a medium-sized town in north-eastern Poland (Przasnysz). Furthermore, the strategic documents for the FUA with a large city in central Poland (Włocławek) and the FUA of a medium-sized town in north-eastern Poland (Przasnysz) highlight the potential for improving the demographic situation through the influx of new residents from outside Poland: *recruiting personnel from abroad for the economy and public services* […] *and supporting their adaptation to the labor market* (Kujawsko-Pomorskie Region Development Strategy 2030+).

A much more pessimistic picture of the demographic situation emerges from our in-depth interviews with individuals responsible for managing administrative units. The authorities of all four case study FUAs agreed that low fertility rates and an aging society are fundamentally reducing the state of human capital—something that is difficult to influence effectively. Observing current realities and the directions of change, they draw particular attention to the interdependence between the demographic situation of peripheral areas and numerous aspects of social life, which in turn causes growing problems in both the economy and local governance. The mayor of Drawsko Pomorskie emphasized that […] *of course, this affects everything related to education, preschools, and so on* […] *but despite the authorities' efforts and the implementation of various solutions, no offer in peripheral areas is able to compete with the metropolitan lifestyle, and thus reverse the negative demographic trends* […] *so honestly, I see no chance of changing this process. And whether it's the first, second, or fifth social assistance package, it won't change anything* (IDI_D). Nevertheless, at the local government level, corrective actions are being taken that could potentially reverse these negative trends. In the FUA of a medium-sized city in the east of the country (Sandomierz), these actions include initiatives covering various dimensions, with a primary focus on improving access to infrastructure: *We recently built a new block of municipal flats, because there is always demand for that, and in fact, I think our municipality has a lot of such flats. Really a lot of municipal flats* […] *We also undertake preventive health actions. For example, we finance pneumococcal vaccinations for children. We opened a new nursery* (IDI_S).

The identified issues are also reflected in the opinions and declarations of the residents of the case study FUAs. In the survey, nearly 40% of respondents expressed an intention to relocate and leave the studied FUA area. This response rate was consistent among residents of both the core city and the surrounding areas. However, desire to relocate declined significantly among older age groups (Fig. 6.9). The propensity to move clearly decreased with age among young residents (under 30), and every second respondent declared the intention to leave their current place of residence. The main reasons given for such decisions were three strongly interrelated factors: a desire to live in a large city (30%), the wish to improve housing conditions (40%), and above all, the desire to improve employment opportunities (51%) or to find a job in general (18% of those planning to relocate) (Fig. 6.3). These decisions are driven by everyday life circumstances and reflect difficulties in the local labor market and the insufficient supply of housing, especially municipal and rental flats.

These tendencies intensify the process of population aging. The proportion of elderly residents is clearly increasing, unfortunately, at the expense of the youngest demographic group.

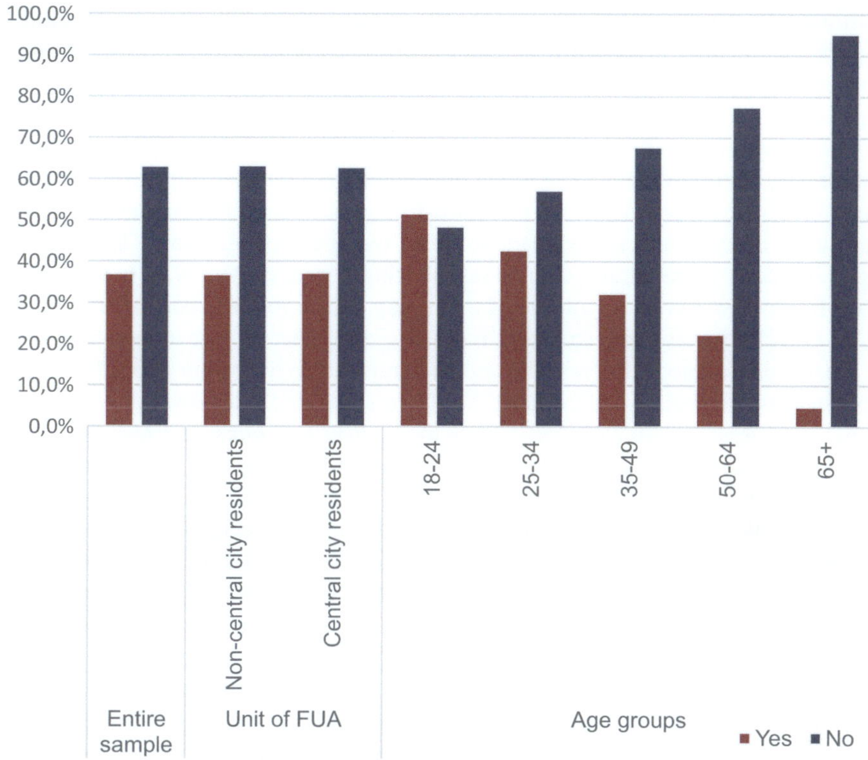

Fig. 6.9 Desire to move within the next two years. *Source* Own elaboration

Local leaders and entrepreneurs from the case study FUAs emphasized that, in light of negative demographic trends, the most pressing issue is the outflow of young and educated individuals, which depletes the labor force and disrupts the balance of the local labor market. This concern was frequently voiced in interviews: *they go to university and don't come back; young people leave; they don't stay: they don't see prospects for themselves* [FGI2_D]. In Sandomierz, participants noted that there are visible returns to rural areas among young people, mainly due to changing perceptions of the environmental dimension of material capital. Residents are beginning to appreciate life in smaller towns or rural areas—valuing peace, greater privacy, and contact with nature. The experience of living elsewhere, usually in larger cities, often leads to a renewed sense of local identity. Those who return often engage in activities related to tourism services. In the FUA of a medium-sized city in eastern Poland (Sandomierz), the city's rich history, supported by territorial marketing initiatives—particularly those leveraging the popularity of a television series about a priest detective—has attracted an increasing number of tourists each year and provided residents with a renewed perspective on the value of living in this location and the potential for developing new business ventures. Unfortunately, a less optimistic dimension of decisions not to settle in large cities also emerged. Economic factors significantly influence residential decisions, particularly access to housing. As noted by local leaders in Sandomierz, *young people return to the countryside because they cannot afford housing in the city. If they have a plot of land from their parents, they get the land for free and can start building* (FGI2_S). In Drawsko Pomorskie, interviewees pointed to the phenomenon of circular migration, although unrelated to the labor market. Recently, the FUA has seen cyclical inflows of people from large cities connected with the trend of owning second homes. The region's high landscape value attracts individuals seeking peace, contact with nature, and a place for leisure. Eventually, some of them may decide to settle permanently in the FUA, especially those who can work remotely, commute, or have already retired.

An influx of senior residents who decide to relocate to the countryside after retirement has also been observed in the FUAs of Przasnysz and Włocławek. Respondents from these areas noted: *they come back to us, they buy plots—retirees who have earned their pensions and just want to spend their final years peacefully* (FGI2_W). On the one hand, this generates a positive effect by bringing in new residents. On the other hand, it presents significant challenges related to developing the silver economy. It is worth emphasizing that the positive changes described still appear to be incidental rather than indicative of an emerging trend. Nonetheless, they draw attention to potential directions for local action aimed at mitigating the adverse demographic trends that are having a destructive impact on the state of human capital.

The state of human capital is significantly influenced by public and health security. This challenge received limited attention in the strategic and programmatic documents of the surveyed FUA units. The situation analysis revealed that residents of peripheral areas experience a high level of safety, and events that could disrupt or weaken this sense of security occur infrequently. In some areas, an improvement was even observed: with regard to road safety—a declining number of traffic accidents, and in terms of crime—a recorded decrease in thefts and robberies. A much stronger emphasis was placed on health security (understood as access to effective healthcare and a stable medical system) determining the conditions for maintaining a proper quality of human capital. The diagnosis presented in strategic documents highlights the issue of access to healthcare infrastructure, which is significantly limited by shortages in medical personnel and long waiting times for medical services. The assessment of the medical infrastructure itself is generally positive. The recommended and planned actions emphasize the necessity of strengthening human resources. However, fulfilling this need may prove difficult due to the growing shortage of medical staff in the Polish labor market. In Drawsko Pomorskie, authorities declared the provision of access to the so-called *basic service basket*, which includes, among others, public administration, culture, education, and healthcare, offered to residents through cooperation between administrative units based on existing functional connections. Similarly, in the case of the Sandomierz FUA, there was an expressed intention to establish inter-municipal associations tasked with delivering public services, including healthcare.

Adapting healthcare services to ongoing demographic changes, particularly population aging, is a significant challenge. Increasingly, solutions that consider the well-being of the elderly are being implemented. Strategic documents refer to the intensification of preventive programs and the development of services tailored to senior citizens. However, these measures do not specifically target inner peripheral areas and are designed for the entire region.

Interviews with representatives of local governments revealed a generally consistent perspective with the contents of strategic documents. Regarding public safety, interviewees emphasized the effectiveness of measures and safeguards developed, such as video surveillance in the Sandomierz FUA, and most importantly, the efficiency of emergency services and institutions responsible for public safety. In terms of health security, narratives varied depending on the specific FUA. In Sandomierz and Drawsko Pomorskie, interviewees stressed how crucial the presence of a hospital is for residents' sense of health security, particularly in providing specialized treatment and responding quickly in emergency situations (e.g., travel time). In the Włocławek and Przasnysz FUAs, hospitals also operate, but they struggle with financial, staffing, and investment-related issues, which direct affects the quality of care. As noted by a representative of the Włocławek FUA, *residents of Włocławek definitely complain about health security, and that comes up very often. Perhaps it's because we have a voivodeship hospital, which has its share of problems* [...] *even though it's not our unit, we still try to support it as much as we can* (IDI_W). Largely, the problems facing hospitals in Poland result from underfunding in the healthcare system. Although healthcare expenditure in Poland is increasing, it still remains below the EU average. This results in equipment shortages and limited access to innovative therapies. Many healthcare facilities also face debt issues resulting from the high operating costs of hospitals (including staff salaries and maintenance costs), which directly impacts the quality of services provided.

Residents of inner peripheries hold views on public and health security that contrast significantly with those of local government representatives. In the study, 57.5% of respondents reported a deterioration in their sense of public safety in recent years. The results varied across the analyzed Functional Urban Areas (FUAs). The lowest level of perceived safety was recorded in the Drawsko Pomorskie FUA, with similar values observed in the Sandomierz and Przasnysz FUAs. In all three cases, approximately 60% of respondents reported negative perceptions. A relatively higher level of perceived safety was recorded in the Włocławek FUA, although even there, as many as 50% of respondents assessed the situation negatively (Fig. 6.10).

A sense of local identity and belonging to one's place of residence plays a significant role in shaping subjective feelings of security within local communities. In areas surrounding the urban cores of FUAs, a higher overall sense of security was observed compared to their central parts. Respondents from these peripheral zones were less likely to report a decline in public safety, which may stem from stronger neighborhood ties and a greater sense of community. In every FUA, residents most frequently expressed fear of the war escalating onto Polish territory and perceived threats associated with the influx of migrants from abroad. These concerns are grounded in various events currently taking place in the country. Across different age groups,

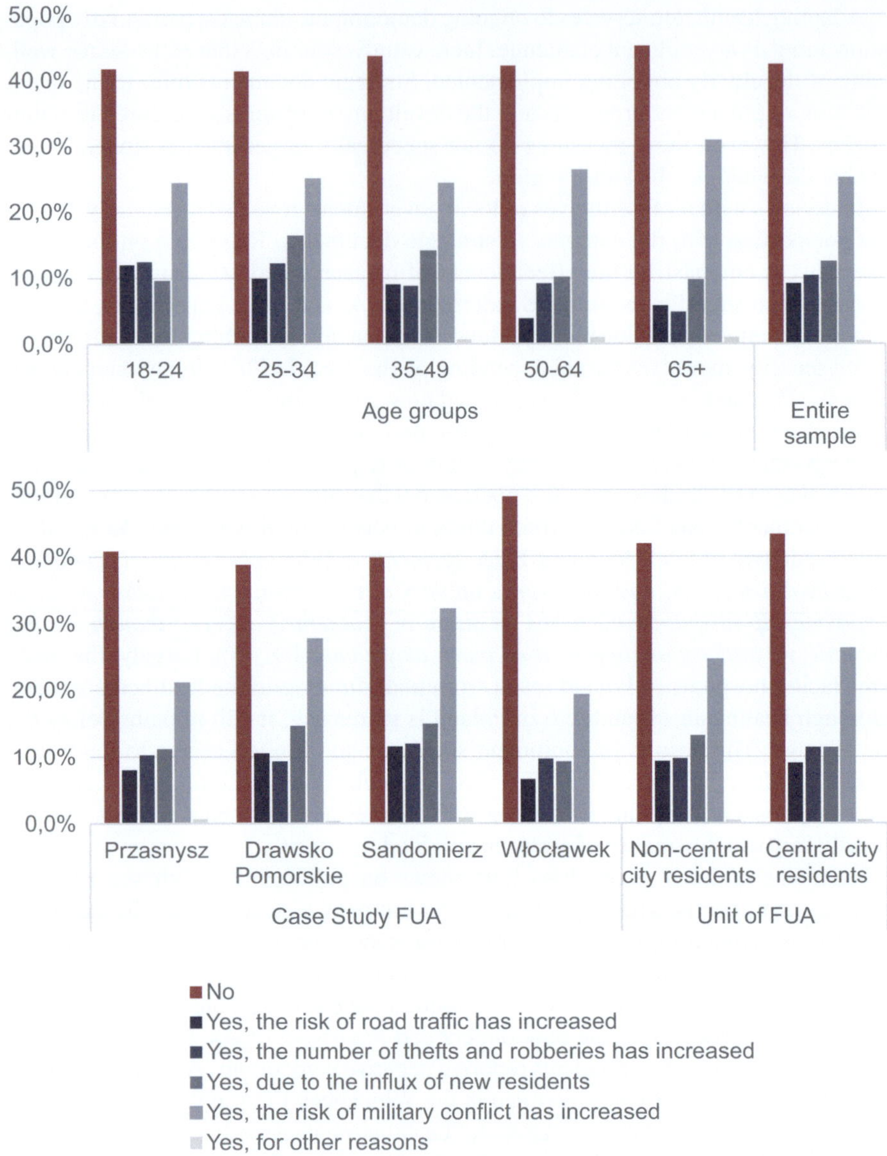

Fig. 6.10 Sense of security in FUAs. *Source* Own elaboration

the sources of fear varied. In every group, the most frequently mentioned concern was the potential escalation of war within Poland (25%). This fear was significantly more pronounced among older individuals, who possess life experience and often remember World War II and the martial law period in Poland. Respondents also voiced concern regarding the arrival of new residents from outside Poland

(12.5%). This aversion toward incoming populations is rooted in persistent stereotypes suggesting that migrants take jobs from locals and contribute to rising crime rates. Additional disparities emerged in responses regarding robbery and threats related to road traffic. In both cases, younger individuals dominated, which is naturally related to differences in daily activity patterns between younger people and seniors.

Regarding health security, residents of inner peripheries reported a decline in their sense of safety after 2020, with 55% of respondents expressing this view. This change was more pronounced among residents of areas surrounding central cities (58%) than in the cities themselves (52%). Differences between FUAs were minor, typically around 2–3 percentage points. However, with regard to age groups, the proportion of individuals declaring a decline in health security increases with age (reaching 69% among seniors) (Fig. 6.11).

The primary reason for this negative assessment, according to respondents, is the noticeable decline in the accessibility of healthcare facilities, which must be considered from a multidimensional perspective. One of the main issues is the shortage of medical personnel. The number of doctors and nurses reaching retirement age is increasing. At the same time, a growing proportion of young professionals choose to work abroad, where employment conditions are significantly better—and not only financially. As a result, waiting times for specialists are increasing, and access to diagnostic tests and treatment is becoming increasingly difficult, especially in smaller towns. Rising bureaucracy and administrative burdens on medical staff also have a negative effect. Doctors and nurses are spending more time on paperwork, significantly reducing patient availability.

The issue of healthcare accessibility is particularly acute for those who must travel to medical facilities. Poor access to public transportation forces patients to rely on private transport, which often poses a serious barrier—especially for elderly individuals (Sect. 6.3). All of this contributes to an alarmingly difficult healthcare situation. It is important to emphasize that the quality of healthcare has a substantial impact on overall quality of life. For instance, in the case of the Drawsko Pomorskie FUA, it may influence residents' migration decisions: *it's the same for us young people—if we have any serious health problems, we have to leave. If the healthcare system worked better, this wouldn't be a bad place to live. In fact, it would be a great place to live* [...] *We do have a hospital, but when it comes to primary healthcare, it's a disaster, it's a disaster* (FGI_D).

Access to high-quality education is an important factor in shaping the quality of human capital. Many local government units responsible for organizing education prioritize its quality and effectiveness. Diagnoses found in strategic documents of municipalities within the studied inner peripheries reveal three key conclusions regarding the functioning of the education system. Firstly, access to early childhood care and education facilities is rated poorly. Although inner peripheries experience strongly negative natural population growth, the access to childcare and preschool infrastructure remains very low—especially when it comes to public institutions, which are less financially burdensome for families. Almost every planning document highlights the need to increase the availability of nurseries and kindergartens to

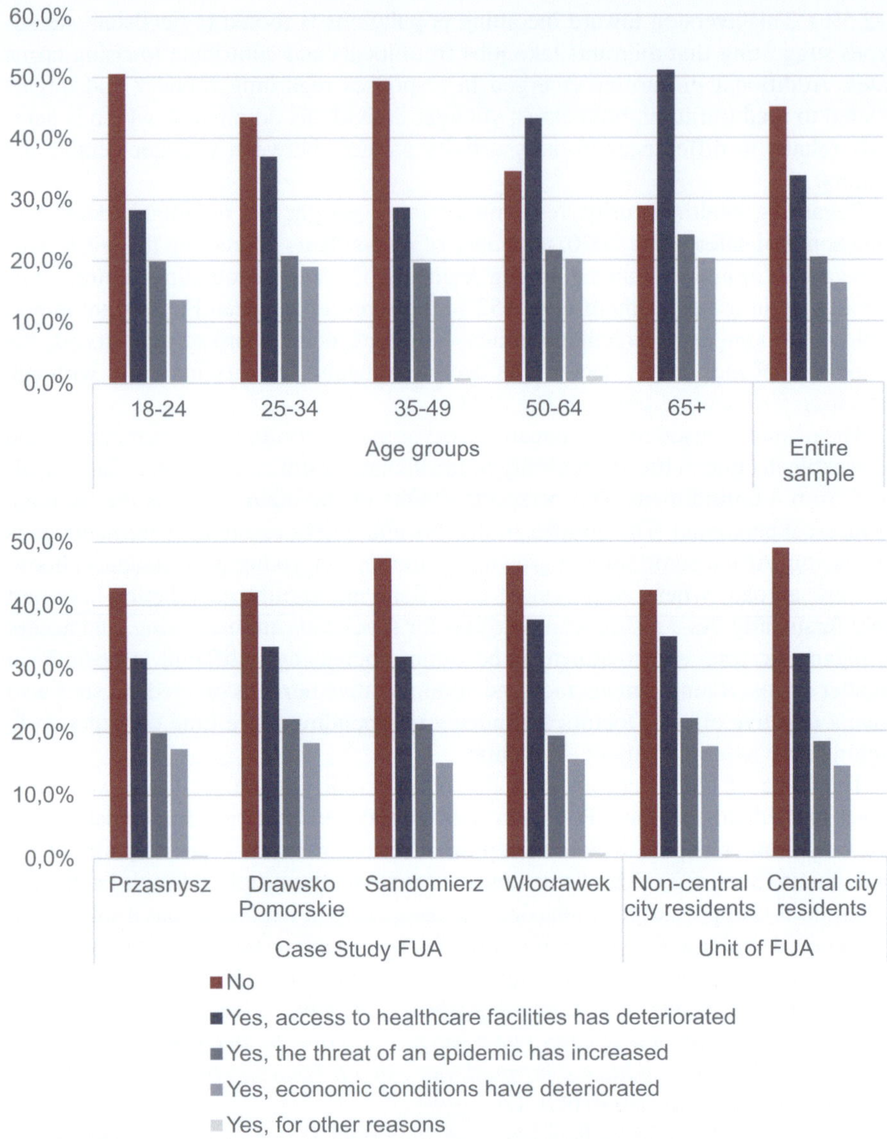

Fig. 6.11 Sense of health security in FUA. *Source* Own elaboration

improve residents' quality of life and enhance the attractiveness of peripheral areas for current and prospective inhabitants. Secondly, regarding primary education, good access to schools and small class sizes are seen as advantages. While staffing shortages are noted, they mostly affect vocational subject teachers in secondary schools. However, these shortages are not as severe at this stage as in Poland's larger urban areas. Thirdly, significant concerns relate to secondary education, particularly the mismatch between educational profiles and the needs of the labor market. Curricula are often misaligned with employer expectations, and there are major gaps in modern teaching infrastructure, especially in training and workshop facilities.

Local government representatives are aware of these challenges and recognize the pressing need to strengthen collaboration between vocational schools, higher education institutions, and local businesses in terms of curriculum development and internship opportunities. The current secondary education offer is often shaped more by competition for students and the popularity of trendy disciplines—such as psychological-social, artistic-theatrical, or e-sports tracks—than by labor market demand. These trends ignore local employer needs, although they may also reflect students' lack of interest in certain professions.

In this context, implementing vocational guidance for young people becomes particularly important, helping them make more informed decisions about their educational paths and career choices. According to local authorities, the mismatch between graduates' qualifications and labor market needs contributes to difficulties in the local job market. A local government representative from Drawsko Pomorskie explicitly referred to a crisis in vocational education, which should be preparing young people for specific professions. In response to these challenges, employers in Drawsko Pomorskie are increasingly engaging in the education sector—mirroring practices from the 1970s—by helping to train future employees.

I see a change in employers' attitudes [...] *most of them are now positively inclined to set up vocational classes within their companies. They're "producing" their future employees, training them from the ground up—and those people usually stay with them.* [...] *They're interested in opening such classes* [...] *and the local government provides vocational instructors and specialists. They sign agreements with schools and want trained staff—this model works. And maybe it's also thanks to good practices borrowed from Western Europe* [...] (IDI_D).

The growing collaboration between schools and the Polish Armed Forces is a noteworthy opportunity for young people. In both Drawsko Pomorskie and Przasnysz, military units cooperate with local schools to implement the *Certified Military Uniformed Classes* program run by the Ministry of National Defence. This initiative includes theoretical and practical classes, equipping graduates with basic preparation for military service and complementing the secondary education curriculum.

As a representative from Przasnysz noted: *an interesting aspect is the perspective this gives young people—because when they graduate high school at 19, a significant percentage* [...] *choose not to pursue higher education but instead opt for military service and ultimately secure employment in military units* (IDI_P).

Two schools in Przasnysz and one within the Drawsko Pomorskie FUA participate in this program. Włocławek presents a unique case among the studied FUAs, being the largest urban center classified as an inner periphery and hosting institutions of higher education—both public and private. A representative from the Włocławek municipal government highlighted that due to the city's industrial nature, where the ANWIL Chemical Company is the largest employer, there is a growing shortage of technically educated workers: *This is one of our major weaknesses and an increasingly serious issue in the job market. Many companies here in Włocławek—ones I stay in regular contact with—complain they simply don't have enough people to hire* […] *we're still educating too few people in technical disciplines* […] (IDI_W).

According to this official, the key to building strong human capital lies in further developing technical higher education. *The future of our city practically depends on the growth of the Włocławek Academy. I dream that one day it will evolve into a full-fledged technical university* […] *If we don't develop this academic sector, unfortunately we'll keep losing residents. Though of course, that's not the only cure for the disease* (IDI_W). It should be noted, however, that the Włocławek Academy was established on the foundations of a former State Vocational School, originally meant to compensate the city for the loss of its administrative functions after Poland's territorial reform. Unfortunately, this model was largely unsuccessful—mainly because these schools were not built around permanent academic staff relocating to the area but instead served as an additional income source for professors from larger universities (in Włocławek's case, the University of Lodz). This structural flaw still significantly limits these institutions' development potential and broader socio-economic impact.

According to residents of the studied areas, access to educational services has worsened since 2020. The average rating of this change was 2.9 (on a 1 to 5 scale, where 1 indicates significant deterioration and 5 significant improvement) (Fig. 6.12). The most negative responses came from individuals over the age of 35 and from seniors. Although these groups were not the direct users of educational services, they were the first to perceive a drop in educational quality. This perception stemmed mainly from difficulties supporting remote learning at home—especially in families with multiple school-age children. Another major reason for the low evaluation was the decline in school transportation access, due to reductions or complete cessation of public transport services.

Local leaders shared the views of local government representatives regarding the crisis in vocational education and the lack of practical job training. They also pointed to another crucial issue concerning primary education—namely, that small class sizes in schools were not a result of deliberate efforts to ensure high-quality teaching but rather a consequence of unfavorable demographic trends. This, in the future, could lead to further reductions in the number of classes or even school closures. One interviewee from Drawsko Pomorskie emphasized that *over the past several years no new schools have been opened, rather we've been closing them. More closures are coming.* […] *I'm not saying they have to be profitable, but maintaining five- or seven-student classes—well, it's just pointless* (FGI2_D). It's worth noting that the presence of both primary schools and teachers who originate from the local area can create opportunities to build a sense of social identity and belonging among students.

The challenges posed by the misalignment of the education system with labor market needs in peripheral areas impact unemployment levels and job opportunities, especially for young people. This leads to the underutilization of human capital in supra-local labor markets—on the one hand, employers struggle to find adequately skilled workers; on the other hand, residents without the required qualifications cannot secure employment. These issues are reflected in strategic and policy documents, which clearly emphasize that the difficulties in the labor market are closely

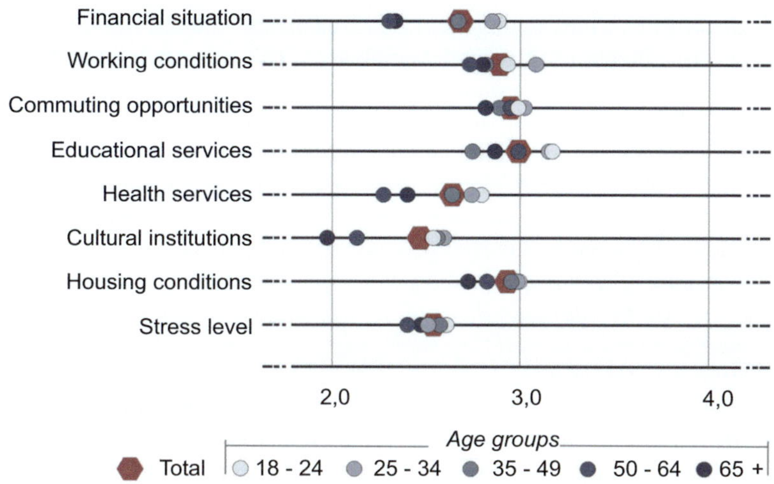

Fig. 6.12 Assessment of change in life situation due to SARS-CoV-2 pandemic (from 1 to 5). *Source* Own elaboration

linked to the demographic situation prevailing in peripheral regions. In this context, attention is drawn to the fact that young people are migrating to larger cities in search of better-paid jobs that match their qualifications and financial expectations. Simultaneously, local entrepreneurs are searching for skilled employees, particularly those with a technical or engineering education and a defined set of hard and soft skills, which are clearly lacking. Another significant issue lies in the inefficiency of labor offices, which are not effectively fulfilling their mission of supporting job seekers. What is evidently missing is the supra-local- or even regional-level collaboration between labor offices that could facilitate better information sharing and job matching, as well as expand the scope and quality of training and upskilling opportunities for job seekers. Among the studied FUAs, Przasnysz FUA reports the highest unemployment levels. Importantly, many long-term unemployed residents in this area do not express a desire to take up employment, even when job opportunities are available. The professional activation of unemployed residents is further hindered by poor access to childcare facilities for children under the age of three and to kindergartens. Improving this situation requires, among other measures, the involvement of employers in establishing such facilities or co-financing children's attendance in nurseries and preschools. Another important issue emphasized in strategic and programmatic documents is the need to develop educational offerings for adults and seniors. Given the increasing pace of population aging, such initiatives are seen as an opportunity to prolong the professional activity of older individuals, partially fill staffing gaps, and improve their quality of life. Extending the period of employment for seniors can result in higher earnings and, consequently, higher retirement benefits, helping secure a decent standard of living after retirement.

The underutilization of human capital has also been noted by local government representatives, who, during interviews, highlighted problems already identified in strategic documents. Chief among them is the mismatch between employer expectations and the qualifications of potential employees. This mismatch often drives residents of inner peripheries to permanently or temporarily relocate in search of better opportunities. Interviewees also agreed that developing childcare services is a major challenge—particularly in enabling young mothers to re-enter the labor market. As a representative of the Drawsko Pomorskie FUA local government pointed out, the problem is not just a lack of new investments but also the closure of existing childcare facilities, which places financial strain on municipal budgets: *We have very few places for the care of very small children. Now another preschool is closing, and once again, parents will be left out in the cold, so to speak, because they'll have nowhere to leave their little ones. Even if they want to go back to work after maternity leave, it's a real problem* (FGI1_D).

Since the beginning of 2020, nearly half (48%) of survey respondents have changed their professional qualifications. This rate was nearly identical between residents of central cities and those in surrounding areas (a difference of just 2 percentage points). Most changes occurred through completing professional development courses (30%), while another 15% completed university-level education, including postgraduate studies. These shifts in professional qualifications occurred almost exclusively among individuals under 50, with the youngest respondents being the most active—both in terms of completing university programs and participating in professional courses or training (Fig. 6.13).

Fewer than one in three respondents changed their place of employment after 2020, primarily due to a change in professional qualifications (8%), a change of residence (7%), or a voluntary decision (7%). Job loss during the pandemic affected only 5% of respondents, which should be interpreted positively (Fig. 6.14). Unfortunately, we could not assess what share of residents in the case study FUAs changed their residence by moving out of peripheral areas due to job loss and the inability to find new employment.

A particularly emphasized issue among local entrepreneurs and community leaders was the marked increase in labor and business operation costs, which often prevents entrepreneurs from undertaking developmental activities or expanding employment. This is a result of legal and fiscal regulations introduced at the national level; meanwhile, at the local level, there is a lack of instruments to offset these costs in other ways, such as through tax reliefs or incentive programs rewarding specific business behaviors. Additionally, there is a lack of substantial support for entrepreneurs who employ or wish to employ persons with disabilities. This refers not only to financial assistance but also to advisory support, legal consultation, or psychological counseling.

With regard to the local labor market, attention should be paid to the context of local identity, which influences both jobseekers' decisions and employers' actions. A sense of belonging to the local community can shape residents' attitudes and occupational choices, limiting the tendency to migrate, encouraging local entrepreneurship, and strengthening ties to the local work environment. From the employer's perspective, attachment to place may constitute a significant social asset, contributing to

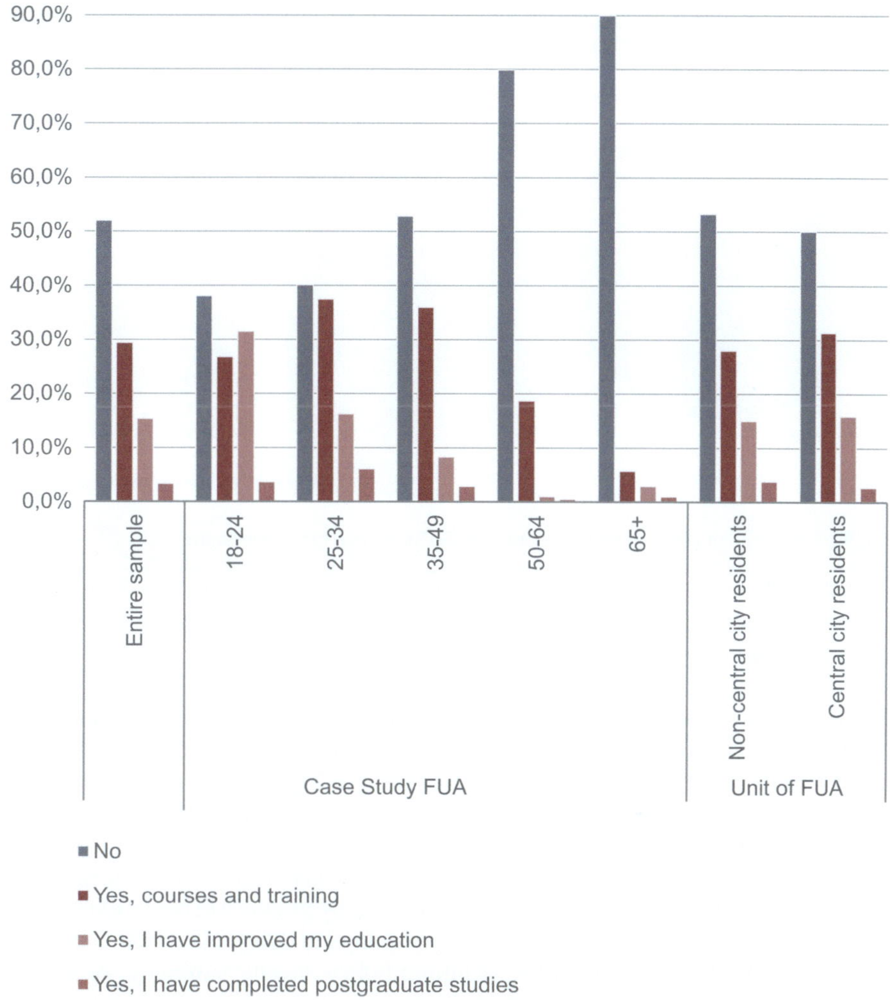

Fig. 6.13 Change in professional qualifications after 2020. *Source* Own elaboration

employment stability and fostering long-term investment in human capital development. This phenomenon may particularly emerge in inner peripheries, where limited availability—both in quantity and quality—of job offers intersects with strong place attachment among residents.

Our analysis of the state and transformation of social and institutional capital in the studied FUAs began with an assessment of the activity level of local communities. In all four case study FUAs selected for in-depth analysis—each representing inner peripheries—a low level of social activity was diagnosed. Content analysis of strategic documents at both regional and municipal levels revealed varied ways of verbalizing the same issue. References were made to *low participation in social*

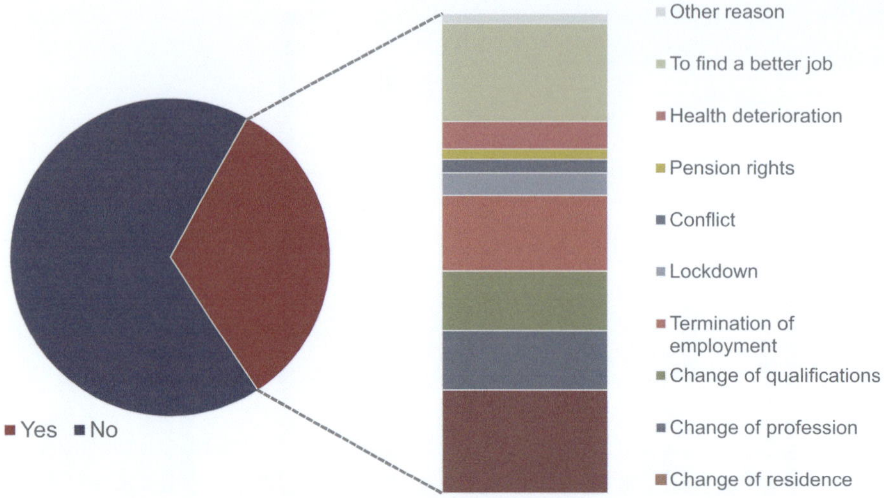

Fig. 6.14 Change of place of employment after 2020. *Source* Own elaboration

life, *low civic engagement*, *weak communal bonds*, or explicitly, *low levels of local identity among residents.*

Interviews with mayors of core cities in our case study areas generally confirmed the overarching trend of low social activity among local communities, although exceptions to this trend were also noted. Interviewees emphasized that the level of social activity in rural areas surrounding the cities was higher than in the central parts of Functional Urban Areas. This was attributed to the smaller population size of local communities, higher levels of trust, and a stronger sense of community and engagement in local matters. For example, a female interviewee from a medium-sized city FUA in eastern Poland (Sandomierz) explained this by pointing to the availability of funding for activities being greater in rural areas than in cities, often from resources associated with the EU Common Agricultural Policy already familiar to rural residents: *Sandomierz as an urban municipality is severely limited in terms of obtaining funds for activities; they don't have the same opportunities as rural areas. They don't have Rural Women's Circles or Volunteer Fire Departments—centers that bring people together. Because that's where people gather—in fire stations, community centers, churches—because these are small communities. Sandomierz is bigger and lacks those kinds of spaces where people could meet* (*IDI_S*). Another local female leader confirmed this phenomenon by offering a specific example of rural community centers: *I see how more and more Rural Women's Circles are being created in the Dwikozy municipality. Each village demands a community center, a place to meet* (*FGI2_S*). Thus, it can be concluded that in the studied areas, outside of the cities, there is a relatively high level of bonding social capital, rooted in closely integrated communities of like-minded individuals. At the same time, there is a noticeable deficit of bridging social capital, which entails the construction of an inclusive social structure based on openness to new actors and knowledge diffusion across diverse communities. Since bridging capital is key to development processes, this deficit points to an overall low level of social capital in inner peripheries. In the capitals of the studied FUAs—for example, in the largest city in the sample, Włocławek—there is a significant issue with social participation: *the challenge is engaging the residents of Włocławek in co-governance* [...] *It's a big problem. Successive rounds of the participatory budget show that interest is somewhere around one or two percent* (*IDI_W*). Meanwhile, in a similar situation, rural communities around a small-town FUA in north-western Poland (Drawsko Pomorskie) are more integrated and able to unite around common goals, which motivates their activity: *residents can come together quite strongly around certain causes, especially in the case of participatory budgets. If it turns out that two villages can come up with a joint project and lobby for it together, then it starts to show that they are indeed capable of cooperation* (*IDI_D*).

As indicated by the in-depth interviews, the same communities are not always active when offered, for instance, completely free pro-health campaigns for women, as in the FUA of Drawsko Pomorskie: *we're trying to implement health policy programs, but let me say this. There's always a huge outcry that there's no access. We recently organized a trip for women over 40 to the Oncology Center in Szczecin for mammograms. I went on the first bus myself—only 12 women came with me. We cancelled the second trip because the cost of the bus was simply too high* [...] *The women really didn't have to pay anything. They just had to come to the high school on Saturday. The bus was waiting* (*FGI1_D*). It thus appears that local communities in the studied inner peripheries are more likely to integrate around common goals within smaller groups in rural areas than in urban ones. The reason for this may lie in the already established awareness of the potential for local development co-financed by the European Union. The identified low level of social activity in the studied inner peripheries may result from a weak sense of identity and belonging to the local community. It is well-known that stronger place attachment fosters greater community engagement in shaping development conditions.

In the surveyed inner peripheries, a questionnaire asked respondents about which territorial unit they felt the strongest connection to. Participants were asked to rank the available categories in order of felt attachment—from strongest to weakest. Among the listed categories, responses such as *local community* and *city* were treated as indicators of local identity. In all four case studies, identification with the local community predominated, although this option was selected by an average of only 40% of respondents. The highest level of local identity was recorded among older individuals—nearly half of seniors (50%) declared the strongest bond with the local community, and a further 22% indicated a connection to the FUA's core city. Other age groups displayed similar distributions of responses, though the percentages were lower. Results show a clear trend—the younger the respondent, the weaker their declared local identity. Among the youngest respondents (aged 18–24), as many as 15% were unable to define their sense of belonging. Most respondents from Sandomierz primarily identified with the city. In the FUA of a large city in central Poland (Włocławek) and a medium-sized city in the east (Sandomierz), an average of 67% of respondents declared the strongest local identity, while in the FUA of a small town in the north-west (Drawsko Pomorskie) and a medium-sized town in the north-east (Przasnysz), this figure was nine percentage points lower (58%). This indicates a variation in the sense of local identity among the studied communities (Fig. 6.15).

The findings indicate that local government officials have developed strategies for stimulating civic engagement through fostering a sense of local identity. Strategies for addressing this deficit are reflected in strategic planning documents and municipal officials' statements. The development strategies of the municipalities comprising the studied FUAs reveal that local authorities recognize the need for initiatives to

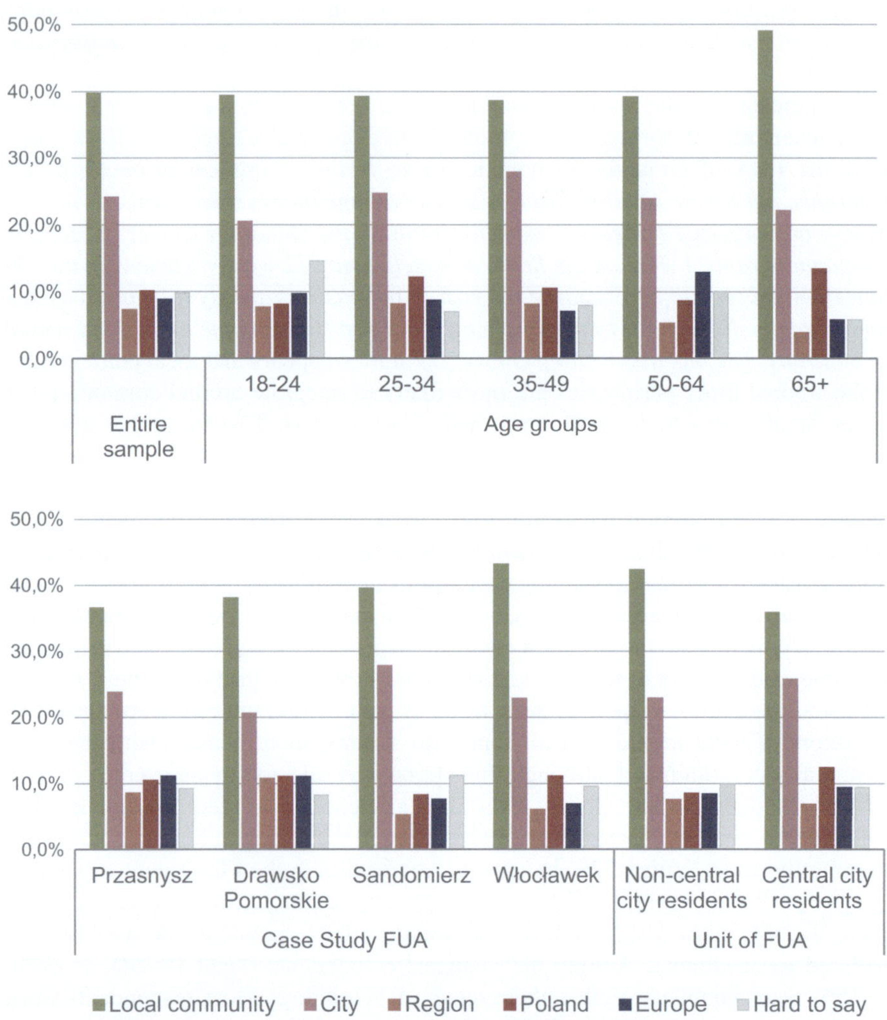

Fig. 6.15 Sense of identity among residents of FUAs. *Source* Own elaboration

strengthen identification with the region or place of residence. Among the proposed measures are cultural education initiatives and workshops focused on raising local awareness, primarily targeting young people. Respondents also emphasize the need to establish community activation centers and spaces for personal development, particularly in rural areas. Moreover, they point to the importance of intergenerational integration, facilitated by participatory decision-making processes such as participatory budgeting. The documents further show that one of the priorities in promoting social development is support for individuals at risk of social exclusion.

Building local identity is a long-term process and a considerable challenge for local governments. It is worth noting that the lack of attachment to their place of residence identified among residents often stems from historical conditions. Three of the four studied FUAs—Włocławek, Przasnysz, and Sandomierz—were under Russian rule following the collapse of the Polish state during the partition period (1795–1918). During this time, governance in these areas was chaotic, corruption was rampant, and there were extensive efforts to Russify the population. In the former Russian partition, social divisions between the elites (nobility and landowners) and peasants became entrenched, largely due to limited access to education. The land reform implemented for political reasons—intended to diminish the influence of Polish landowners—resulted in the fragmentation of land into small, unprofitable farms. As a consequence, the socio-economic development of these areas continues to lag behind that of, for instance, the former Prussian partition (Kowalski 2000; Zarycki 2015).

Our research shows that residents of Sandomierz and its surroundings are aware of and proud of their history as a *royal city*. However, according to officials, this is not a community inclined toward cooperation, but rather one that is highly critical of local government initiatives. *They identify with the place. Of course* [...] *but when something needs to be done for the city, there's no one.* [...] *They are proud, that's for sure, but this doesn't translate into collaboration. When we organize public consultations for various projects, the response is very low. The return rate on surveys is minimal. As for in-person meetings, two or three people show up* (IDI_S). Local leaders and residents of the Sandomierz FUA openly refer to a *post-partition identity* that shapes the region's social capital: *The partitions also played a role in all of this. It's a stigma, a historical legacy* [...]. *I think it's the inheritance of poverty or helplessness. Someone offers a helping hand, but the person says they don't want to try because their mother didn't work, their father didn't work, so why should they? And there always has to be some "Master" who says—get up and do it* (FGI2_S). In contrast, prior to 1945, the Drawsko Pomorskie FUA was part of Germany and its population was predominantly German. After World War II, due to border changes, many Germans left the region, leaving behind empty houses and farms. In 1947, during the "Vistula" operation, people from south-eastern Poland—particularly Ukrainians and Lemkos—were resettled there, resulting in an influx of communities with diverse cultural traditions unconnected to the region (Buczyło 2006). According to officials, processes of local community integration in the Drawsko Pomorskie FUA are ongoing, although gradually improving over time. [*A*]*ll the older generations were newcomers because there were no indigenous residents here* [...] *but I believe more and more young people are identifying with this location, treating it as their "little homeland"* (IDI_D). Local leaders confirmed this perspective when asked about identity: *the whole community was made up of newcomers, and a certain tradition, an identity, has not yet fully formed here* […] *I can confirm that there is still no clear sense of identity. I envy those who live in the historical borders of the Polish–Lithuanian Commonwealth, where families have lived for generations. We don't even have culinary traditions here, because this is a result of Operation Vistula—resettlers from Baranowicze, and other so-called "Zabugowcy", and from areas of Ukraine that were incorporated into the USSR* (FGI2_D). At the same time, some individuals believe that *we're now in the third or fourth post-war generation. And there is a growing sense of identification …*

In Włocławek, the rapid industrial development after 1975 attracted people from various parts of Poland. The city's growth created new development challenges due to the increasing influx of population. *I explain it to myself in this way—that our city doubled in size over the last few decades of the 20th century. This means we had an influx of population. Let's say the first generation, and even the second, doesn't identify with the place the way we would hope* (IDI_W).

At the same time, middle-generation local leaders note a growing desire to return to their roots: *many of my friends moved away, but they're coming back. Life is better there, but they miss their friends, family, and connections. Many people my age are returning, and I wouldn't leave this place for anything. This is my "piece of the floor," my family is here, this is where I pursue my passions, and I identify with this city. There was a trend of people leaving, but my generation sees it differently* (FGI2_W). A similar phenomenon was observed in Przasnysz, where a Polish Army garrison—the city's largest employer—has contributed to inward migration from other regions. When asked whether residents of Przasnysz feel a strong sense of local identity, the mayor responded: *No, I must sadly admit that they do not, although we've been working on it for quite some time* (IDI_P). The incoming populations in the studied inner peripheral areas generally exhibit a relatively weaker sense of local identity. In response, various strategies have been implemented to strengthen it, including references to local legends and the creation of regional tourism products such as thematic trails based on historical narratives, culinary traditions, or characteristic regional dishes—for example, *Przaśnik* bread in Przasnysz. Some respondents have identified a *renaissance of local identity* among younger generations, particularly in areas with high landscape and tourism value. This may be related to a gradual shift away from the negative connotations traditionally associated with rural origins in Poland, as well as a growing prestige associated with local traditions and cultural heritage. Our findings also suggest that residents of the areas surrounding the cores of the studied inner peripheries demonstrate a higher level of local identity, which may be attributed to lower levels of population mobility and stronger attachment to local traditions.

An important indicator for assessing the state of social and institutional capital is the analysis of how residents spend their leisure time, including their cultural activities. An examination of strategic documents from the studied inner peripheral areas shows that local governments recognize the low level of cultural engagement among local communities during their free time, primarily due to the limited cultural offerings. A particularly pressing issue highlighted in these documents is the insufficient number of initiatives aimed at young people. Local authorities see an opportunity to improve this situation through the development of cultural education and by raising awareness of the region's tourism potential, based on existing but underutilized resources. One proposed solution is intergenerational integration, involving the creation of diverse forms of cooperation between youth and seniors, enabling shared leisure experiences. The documents also include specific proposals for action. For example, Włocławek's development strategy includes the promotion of sailing through the creation of the "Kujawsko-Pomorska Żegluga Wiślana" river-sailing initiative. Additionally, two new cultural institutions are planned to enrich the city's leisure offerings: the Kuyavian Philharmonic and the "Energia Włocławka" Family Knowledge and Entertainment Center.

Interviews with local government officials and community members point to significant challenges related to organizing cultural and recreational activities in the inner peripheries. Despite numerous initiatives, levels of public engagement remain low. As noted by a representative of the cultural center in Sandomierz: *We, as the Cultural Center, really try to organize a lot. We want to do a lot for the residents, but they're just not that interested. We want to create something like a night museum, we have a gallery, events, we invite comedy groups for Women's Day, we organize concerts and so on, but there's always just the minimum number of participants* (FGI1_S).

Similar difficulties are observed in northern Mazovia, where the lack of recreational attractions and limited cultural infrastructure are perceived as key barriers to local development. In Przasnysz, efforts to develop tourism infrastructure were undertaken but not realized due to a breakdown in inter-municipal cooperation: *Maybe if we created more cultural and entertainment centers, including by developing the reservoir. There was even such an idea. An agreement had been signed with the district government, but it was never implemented. The district withdrew. Unfortunately, the municipality alone cannot afford it. Maybe one day we'll do it if there are some funds or national programs* (FGI1_P).

Concerns also arise regarding the potential expansion of cultural infrastructure without ensuring an adequate programmatic offering. As one respondent from Przasnysz remarked: *I'm a bit worried that a cultural center might be built, but it won't be the walls that determine the quality of the service—it's the program. Right now, we're in a place—the library—that has its own life. If that spirit could be transferred to those other buildings that were built for millions, then they too could come alive like this library does* (FGI2_P).

In Włocławek, despite rich natural and cultural resources, there is a lack of effective utilization: *Where the dam is located, we have the largest reservoir on the Vistula, which is also unused. All of Zawiśle plus one of the most beautiful boulevards in Poland. Highway access.* [...] *forests everywhere, maybe five or seven accessible lakes. And in terms of tourism, the city is basically failing. We should take advantage of the history of faience. We have the legend of "The Griffin," which hardly anyone talks about, but it could be turned into a fairytale, a gem, something that would draw crowds of visitors* (FGI2_W). The Drawsko Pomorskie FUA stands out from other areas in terms of residents' positive attitudes toward the available recreational opportunities: *If I want to go to Szczecin, to the theatre, or to the philharmonic, I can go whenever I like. We have a beautiful cinema. I enjoy cycling—new trails are being created, and the forest roads are beautiful because the State Forests take care of them. So, honestly, I really can't complain about the local offerings* (FGI2_D).

In summary, the interviews suggest that the challenge of strengthening social and institutional capital in the studied inner peripheries through the development of leisure opportunities lies not only in the lack of cultural and recreational infrastructure but also in the low level of resident engagement in such activities. At the same time, there is a recognized need to utilize local resources better and create new initiatives that address the needs of diverse social groups. In this regard, it is important to draw on the knowledge and experience of community members who identify strongly with the area. Of particular importance is the insufficient or ineffective supra-local cooperation among both formal and informal institutions. In certain cases, such as in the Drawsko Pomorskie FUA, local conditions are conducive to cultural and recreational activity, demonstrating that appropriate management of existing infrastructure can significantly improve the quality of life and effectively strengthen the social and institutional capital of a region.

To assess the functioning of institutional capital, we also examined the level of business infrastructure, understood as business support services. These services constitute a key component of the infrastructure that supports entrepreneurial development. They encompass a broad range of provisions offered by public institutions, including local governments, as well as private actors and non-governmental organizations, such as associations, foundations, or specialized companies, with the aim of facilitating business operations and stimulating the growth of enterprises, particularly within the small and medium-sized enterprise (SME) sector (Sect. 2.3.3). Within this framework, businesses may access strategic consulting, training, financial support, market information, and investment opportunities. Business support institutions also act as intermediaries in establishing economic relationships and facilitating access to business networks.

We consider local governments in inner peripheral areas to be a specific type of business support institution (Sect. 2.3.3). It is worth noting that the strategic documents of the studied inner peripheries do not explicitly refer to a *deficit in access to business support services*. Nonetheless, officials—for instance, in Sandomierz—acknowledge the lack of such institutions in their region, noting in their development goals a desire to support entrepreneurs implementing family-friendly policies for employees. Similar aspirations to strengthen business cooperation were articulated in Drawsko Pomorskie's strategy, with proposals to create inter-municipal partnerships for business support, and in Włocławek's development strategy, which called for the creation of network links with functional areas of development cores.

The development of business infrastructure and support services in the studied inner peripheral areas faces significant barriers. Officials in the analyzed municipalities attempt to attract investors, but their efforts do not always translate into tangible economic outcomes. One example is a municipality in the Włocławek FUA that received the "Golden Site" award for investment potential, yet: *No one has used that land. The medal hangs on the wall, but no one bought the land* (FGI1_W), highlighting the difficulties of translating symbolic achievements into practical investment results. In larger centers, business infrastructure is more developed, as in Włocławek, where a business incubator operates. Outside the core city, however, the only major investment spaces are economic zones such as the Brześć Economic Zone or the Industrial Zone in Lubień Kujawski. One of the key challenges for fostering economic development in inner peripheries lies in how entrepreneurs perceive local governments and public administration. In the Sandomierz FUA, business owners point to excessive bureaucracy, which significantly hampers business operations: *There are many inspectors who can literally stop production or block a shipment* (FGI2_S). Another recurring issue is the perceived attitude of officials toward entrepreneurs: *They treat entrepreneurs somewhat like a necessary evil* (FGI2_S), a sentiment that may drive away investors and weaken the local economy.

Another major issue is the availability of skilled labor and the recruitment of qualified professionals. In Przasnysz, which hosts one of Europe's largest bicycle factories—Kross—the lack of city attractiveness for highly skilled professionals poses a problem: *The city of Przasnysz is not attractive to specialists, to experts. They come here for very high salaries, but they don't want to live here* (FGI2_P). As a result, the company was forced to open a branch in Warsaw to successfully recruit and retain its expert staff.

The assessment of the quality of social and institutional capital also draws on findings related to the structure and economic activity observed in the inner peripheries. The structure and nature of economic activity in the studied inner peripheries vary depending on the relationship between the core city of a Functional Urban Area (FUA) and its surrounding area. In the case of the core city and its immediate vicinity, economic activity relies heavily on micro and small enterprises, whereas, in the surrounding (predominantly rural) areas, agricultural holdings prevail. The dominant sectors in the economic structure of the studied areas included agri-food processing and services, particularly tourism-related industries. Each core of the studied FUAs hosted a limited number of medium and large enterprises and public institutions employing a significant workforce. However, their scarcity meant that labor market stability depended primarily on the small-business sector.

Strategic documents from all four central cities of the FUAs include a diagnosis of a *low level of entrepreneurship*, which is attributed to the *emigration of young, potentially active individuals to larger urban centers*. Local leaders recognize this challenge: *Our economy is based on small and medium-sized enterprises. Small businesses, some service workshops, and a few medium-sized ones—and those are mostly tourism-related services* (FGI2_S). At the same time, changes in the employment structure are evident, particularly concerning the agricultural sector, which is gradually losing its significance: *In general, agricultural areas are beginning to die out here. People are cutting down a lot of gardens, eliminating them* (FGI2_S). Despite the establishment and operation of special economic zones in the analyzed areas, their utilization remains significantly insufficient. *We have an economic zone, but only a few percent of it is being used* (FGI1_P). While the availability of investment land is a key development factor, many areas remain underutilized. For example, in the Włocławek FUA, a designated economic development zone exists. Still, its limited size prevents the accommodation of large-scale investments: *The zone is too small for giants to be established there* (FGI1_W). Moreover, fragmented land ownership complicates economic development processes: *Our zone was owned by the city and was sold by the city. But when it comes to the Brzeska zone, most of the land is privately owned, and purchasing it required negotiations with 20 different owners* (FGI2_W). It is worth noting, however, that the economic zone dominated by private land ownership was more successful than the one with publicly owned property. This suggests that land ownership structures may influence management efficiency and, consequently, the quality of institutional capital.

In shaping favorable conditions for economic activity in inner peripheries, insufficient cooperation between formal and informal institutions is evident, along with significant administrative barriers. In this context, local governments should play a key role by stimulating economic activity and engaging local stakeholders in the development planning process. Meanwhile, entrepreneurs highlight excessive bureaucracy and administrative control, which hinder economic dynamism: *There's too much bureaucracy. And on the other hand* […] *it's so extensive that it interferes with running a business* (FGI2_S). Additionally, according to residents, labor offices are not fulfilling their function as employment support institutions, and their activities are perceived as ineffective. *The labor office itself, for me, is a strange entity. It doesn't help* (FGI2_P). The lack of effective economic promotion (territorial marketing) of Functional Urban Areas and their products leads to an underutilization of local resources. The Włocławek FUA, for example, is home to nationally recognized brands such as in ketchup production. Yet, this visibility does not translate into an effective marketing strategy: *We have products known throughout Poland, but no one talks about them* (FGI2_W). The absence of a comprehensive marketing strategy results in limited investor interest and underutilization of local products and the labor market, despite advantageous transport accessibility. Despite numerous institutional and administrative barriers, local entrepreneurs are taking steps to improve the economic situation and halt the outflow of young residents. For example, the company Anwil in Włocławek initiated scholarship programs that support young people pursuing education and ensure their employment in local businesses: *Anwil launched a scholarship program for young people going to university—provided that they choose chemistry-related majors* (FGI2_W). Such grassroots initiatives by entrepreneurs are more frequently observed in areas where a strong sense of local identity has been identified. These actions represent independent efforts to counteract the negative consequences of depopulation and the shortage of skilled labor—particularly pressing challenges in lagging regions such as inner peripheries.

The consecutive development shocks—namely the SARS-CoV-2 pandemic and the war in Ukraine—had a profound impact on the state of human, social, and institutional capital in the studied inner peripheries. Some changes proved to be temporary, while others are likely to persist and pose significant challenges to the development of these areas. Demographically, the pandemic acted as an accelerator of pre-existing depopulation trends. Increased mortality, especially among the elderly, directly contributed to population decline. Furthermore, uncertainty in employment, financial instability, and limited access to healthcare services led individuals to postpone parenthood plans, thereby lowering already low fertility rates. The authorities of Przasnysz have been grappling with this issue for several years: *We've had a very clear fertility problem since 2020, the beginning of the pandemic. Of course, before that, the natural increase was negative, but it was in single digits. Since 2020, however, we've seen some very worrying developments, and now the children born during that time are entering our nurseries—and for the first time in our history, we don't have a waiting list* (IDI_P).

During the SARS-CoV-2 pandemic, a predominantly temporary, and in some cases permanent, migration from large cities to inner peripheries was observed. Prolonged isolation and changes in the nature of work made living in a detached house and having regular contact with nature more desirable (IDI_S). Fears of infection and numerous restrictions prompted some urban residents to relocate—temporarily or permanently—to areas offering better living conditions and lower living costs. It should be noted, however, that this was a selective migration, primarily involving a relatively small group of middle-class individuals capable of maintaining their incomes outside urban centers. Another visible phenomenon during the pandemic was the return of young people to their hometowns, although this trend was mostly temporary. It was typically triggered by sudden changes in urban living conditions, such as job loss, transition to remote work, and the suspension or later transformation of academic instruction into online formats. For these individuals, returning to their family homes offered a form of security—both financial (lower living costs) and emotional (family support during times of uncertainty). Nonetheless, the pandemic did not reverse the long-term trend of youth migration to urban centers but merely slowed it temporarily. This phenomenon also underscored the importance of local and familial bonds as a buffer zone in times of crisis.

A direct consequence of the escalation of the war in Ukraine in February 2022 was a significant influx of Ukrainian nationals into Poland. Although migrants predominantly settled in larger cities—where they could rely on the support of family members or acquaintances who had previously arrived in Poland, mostly for work-related reasons—some also found their way to smaller towns and rural areas. As a result, the arrival of Ukrainian migrants led to a noticeable increase in population, even in peripheral areas. Only a portion of these individuals decided to remain permanently; others either returned to Ukraine or moved on to other European countries or to Canada, where Ukrainians constitute one of the largest national minorities. Those who chose to stay have already had—and are likely to continue having in the coming years—a positive impact on the demographic situation, as well as contributing to beneficial changes in the fields of education and local labor markets in inner peripheries.

A representative of the Przasnysz local government noted a positive demographic shift resulting from the influx of refugees: *At the peak, that absolutely critical moment of Russia's invasion of Ukraine, we had more than 200 Ukrainians in our town. Today, around 80 remain, and together with the Municipal Social Welfare Center, we estimate that these are people who will stay with us for good. They have found employment in the vicinity; their children have enrolled in our schools—there are about 20 Ukrainian children in each school. We also have Ukrainian children in kindergartens, and this has slightly improved our population statistics* (IDI_P).

When Russia invaded Ukraine, Poles responded with remarkable solidarity toward the refugees, offering help and support unconditionally. However, as the initial wave of empathy subsided, tensions began to emerge related to the total influx of approximately two million Ukrainians. As the war dragged on, factors such as language barriers, cultural differences, and competition for jobs, housing, and school places contributed to a gradual decline in the positive attitudes of Poles toward the Ukrainian newcomers. Public discourse also played a significant role in this process. Certain media outlets and political figures began to fuel hostility toward refugees by emphasizing issues such as the alleged abuse of social assistance, entitlement, or preferential treatment. This contributed to the perception of migrants as competitors, outsiders, and even a threat, which further intensified the growing social aversion.

The pandemic significantly impacted the functioning of the healthcare system, exposing its structural weaknesses. The dramatic surge in COVID-19 infections overloaded hospitals. Many hospital departments were converted into infectious disease units, and only patients suffering from COVID-19 or in urgent need of care were admitted. Numerous planned procedures were canceled or postponed, and patients with chronic conditions were often left without adequate support. This exacerbated the already limited availability of healthcare services and specialist doctors. Even once the epidemiological situation had stabilized, the healthcare system was still struggling with a "healthcare debt"—the need to catch up on missed diagnoses and treatments. These developments had particularly serious consequences for the perceived healthcare security of residents in peripheral areas. Respondents were unequivocal in stating that, since 2020, access to healthcare services has deteriorated. This change was rated, on average, at 2.6 points on a 5-point scale (where 1 indicates a significant deterioration, 3 is moderate deterioration, and 5 is a marked improvement) (Fig. 6.12). The decline affected older individuals most acutely, as they are the most frequent users of healthcare services. This deterioration was not only due to the suspension of in-person visits but also to reduced transportation accessibility. The reduction in or elimination of public transport routes made it more difficult to reach healthcare facilities providing both primary and specialist care.

The SARS-CoV-2 pandemic also posed considerable challenges for the education system at all levels, including in the studied inner peripheries. The format of teacher–student interaction and knowledge transmission changed significantly, and many individuals lacked sufficient digital competencies to adapt. Furthermore, access to the necessary digital infrastructure posed a major challenge, both in terms of stable internet connections and the availability of computers or other devices in households, particularly when multiple members were working or studying remotely simultaneously.

The war in Ukraine further disrupted the educational process. Practically overnight, refugee children from Ukraine—lacking Polish language skills, educated in a different system, and often traumatized by war—appeared in schools across the country. Creating the conditions for their successful integration into Polish schools posed a substantial challenge, especially in inner peripheries where access to qualified intercultural assistants was very limited. The integration of Ukrainian children also required considerable effort from teachers, who in peripheral areas often lacked institutional support. On the other hand, the presence of school-aged Ukrainian children contributed to increased class sizes, which in some cases helped justify the continued operation of classes that would otherwise have been deemed too small.

With regard to the labor market, the pandemic brought about significant changes. In peripheral areas, however, these changes were largely incidental. The most prominent challenge was the shift in how work was performed. In the studied Functional Urban Areas (FUAs), only one in three residents of peripheral areas changed the way they carried out their professional duties (Fig. 6.7). A similar pattern was observed in central cities and their surrounding areas. Most changes declared were temporary—16% of respondents worked remotely only during the strictest pandemic restrictions. Changes in work modalities affected younger individuals most significantly. Respondents under the age of 49, and those under 35 in particular, were more likely to report a shift to remote or hybrid work. In this youngest age group, 56% of working individuals reported the nature of their work changing, and 15% now work permanently in a remote or hybrid format. Nevertheless, these shifts in work modalities only minimally translated into lasting transformations. This is primarily due to the nature of the peripheral labor market, which relies heavily on jobs that require physical presence and, to a much lesser extent, includes positions conducive to remote work. Notably, in two of the studied FUAs—Sandomierz and Przasnysz—local government representatives attempted to leverage the pandemic-driven changes in work patterns to promote alternative development pathways. Unfortunately, their efforts to position these towns as remote work hubs and encourage both former and new residents to relocate did not yield the expected results. As the mayor of Przasnysz explained: *We wanted to create a vision for our town's development as a remote work center* [...] *we launched some marketing initiatives, including promotional videos, to highlight that life here is cheaper, housing is more affordable, everything is cheaper, and the pace of life is slower than in big cities. That's what we wanted to use to encourage people to move to Przasnysz and work elsewhere remotely* [...] *We succeeded in a few cases, but not on the scale we had hoped for* (IDI_P).

A different situation was observed in Drawsko Pomorskie, where the pandemic led to a reduction in economic migration, particularly to jobs abroad. Moreover, the possibility of remote work encouraged some individuals to return or remain in the Drawsko Pomorskie FUA. While this was a selective phenomenon, limited to a narrow range of professions, a representative of the local government emphasized its developmental potential: *There is a real opportunity for regions like ours, because these young people—or not only young people, but well-educated individuals—may want to escape the hustle and bustle of big cities. If they can work online, they'll do it from here, even for economic reasons: cheaper land, and so on* (IDI_D).

Residents of inner peripheries assessed that, as a result of the pandemic starting in 2020, working conditions in these areas had deteriorated. This change was rated on average at 2.9 points (where 1 indicates a significant deterioration, 3 a moderate deterioration, and 5 a marked improvement) (Fig. 6.12). The most critical evaluations came from older age groups, most likely because they found it more difficult to adapt to new work requirements and organizational changes, including acquiring digital skills necessary for remote work within a short timeframe. Some older employees, faced with uncertainty and health concerns, chose to retire early or were encouraged to do so by their employers.

The crisis resulting from the consequences of Russia's aggression against Ukraine also led to significant changes in the labor market. These were triggered by the influx of migrants and had different implications in 2022 than their current effects. Initially, uncertainty about the future (not knowing when the war would end and when returning home would be possible), combined with the language barrier, led Ukrainians to take jobs that did not require specific qualifications, often working below their skill levels. Over time, these barriers (including formal ones) disappeared, and the migrants became better integrated into the labor market. Unfortunately, due to the protracted nature of the conflict, many men who had previously lived in Poland returned to Ukraine, abandoning their jobs in Poland. This created disruptions in some sectors, such as construction and industry, and forced employers to seek alternative workers, often recruited in an organized manner from other continents.

Our respondents agreed that, during the pandemic, due to a sharp decline in the sense of health security and administrative restrictions, local communities ceased their activities. Family members and close friends helped one another—bonding forms of social capital came into play—while general social relations deteriorated. No meetings or events were held; people stayed at home and became accustomed to this behavior. As a result, social activity declined significantly across all studied areas during subsequent waves of infection. During a focus group interview with local leaders in the Włocławek FUA, we learned that, in 2021, in the midst of the pandemic, a social cooperative was established. The absence of active social economy entities in the city prompted this decision. The newly established cooperative faced significant organizational and financial challenges. It was initially intended to specialize in care services for the elderly and people with special needs. However, pandemic regulations posed a substantial barrier to providing such services. Legal provisions did not mandate compulsory vaccination against SARS-CoV-2 but still required it of those delivering care services, leading to a deadlock. Employees were not obliged to be vaccinated, but the lack of vaccination prevented them from working in, for instance, social care homes. As a result, staff were ready to work but unable to perform their duties, further exacerbating the cooperative's financial difficulties. In the following period, the cooperative's operations expanded to include additional responsibilities related to the outbreak of war in Ukraine. Although these activities represented an important element of social support, they were conducted on a voluntary basis and did not improve the organization's financial situation. The Social Cooperative EMPATIA in Włocławek continues to operate, securing external support from EU funds. In all studied areas, we observed a reversal of the negative trend of limited social activity, driven by the urgent need to help war refugees from Ukraine. This crisis may have contributed to the reintegration of communities and the initiation of collective support activities. In the urban functional area of Przasnysz, two organizations were established in the aftermath of the outbreak of war in Ukraine. They quickly gained recognition both locally and across northern Mazovia. These are the Civic Humanitarian Convoy and the Arena Dobra Association. Both entities, operating as associations, have been involved in numerous relief and social initiatives since the beginning of the conflict. The scale of their activities is described as impressive. These organizations consistently undertake new projects, and various social and charitable events help integrate the local community: *in all this misfortune that is the war, two organizations were created here* [...] *[S]ince the war started, they have been initiating beautiful actions, and the number of these is really overwhelming, because something good happens here every weekend* (IDI_P).

The SARS-CoV-2 pandemic led to major changes in how local communities in inner peripheries spent their leisure time. First, initial travel restrictions meant that residents stayed home more often, using digital technologies for cultural and entertainment purposes. Second, communities in areas with high environmental value, such as the FUAs of Drawsko Pomorskie, Sandomierz, and Włocławek, began to rediscover the potential of their local surroundings, which was reflected in a growing emphasis on outdoor physical activity.

The pandemic significantly impacted business operations, even among larger enterprises. In Przasnysz, the company Kross *laid off over 100 workers during COVID-19 but has now stabilized its economic situation* (FGI1_P), indicating the scale of temporary difficulties but also the adaptability of large enterprises. In the case of Włocławek, effective policies by local authorities played an important role in mitigating the crisis: *A wise policy by the office and the protection of entrepreneurs from the detrimental effects of the pandemic meant that the companies managed to recover* (FGI1_W). A positive development is the growth of social and environmental engagement among large companies, such as Anwil in Włocławek, which *is throwing millions at the environment or for organizations and the community* (FGI2_W), suggesting an evolution in the role of enterprises in the local socio-economic ecosystem. The pandemic also highlighted the importance of private business support networks, which became more active. In Włocławek, for example, *the sphere of private entrepreneurship is developing dynamically—entrepreneurs are gathering in clubs or various councils* (FGI2_W), indicating a rise in grassroots, locally rooted business initiatives as an alternative to institutional support.

An analysis of business infrastructure and business support services in inner peripheries reveals numerous challenges, such as difficulties in attracting investors, bureaucratic barriers, and the migration-driven loss of specialists. In many cases, the pandemic exacerbated these problems but also accelerated the development of alternative forms of business support, including local government initiatives and grassroots entrepreneurial actions.

The SARS-CoV-2 pandemic significantly affected the economic situation in the studied regions, leading to business closures and reduced investments. *Business activity is still stagnant. The stagnation stems from the fact that people simply don't know where they stand* (FGI1_W). Many local enterprises could not survive the crisis: *Entrepreneurs began to suspend and close their businesses en masse, and this had nothing to do with the number of new firms being set up* (FGI2_D). This resulted in increased investment caution and a slowdown in local economic development.

The findings indicate that economic activity and structure, which influence the quality of social and institutional capital in inner peripheries, are based primarily on micro and small enterprises, while the underutilization of special economic zones often limits large investments. A lack of constructive communication among local stakeholders, non-transparent administrative procedures, and insufficient trust hinder effective cooperation and constitute a significant development barrier. At the same time, local entrepreneurs are taking steps to improve the situation. Still, their efforts are being undermined by external factors, such as the SARS-CoV-2 pandemic and the economic consequences of Russia's aggression against Ukraine. A key challenge remains the intensification of supra-local cooperation and more effective economic promotion of the studied urban functional areas. Actions in this regard require enhanced cooperation between local governments and private-sector entities. An important mechanism supporting development is integrated territorial investments, which enable the effective acquisition and distribution of EU funds for joint projects. So far, *this cooperation has been based solely on money* (FGI1_W). Such partnerships may form the basis for a long-term supra-local development strategy that includes both economic investments and social and educational initiatives.

6.5 Financial and Material Capital

Socio-economic development requires solid material foundations. Among all dimensions of territorial capital, financial and material capital gain particular importance, because they constitute the essential basis for the functioning of public institutions, the local economy, and the everyday lives of residents. Deficits in these resources within peripheral areas can significantly constrain development opportunities, deepening spatial disparities in access to services between core areas and peripheries. Conversely, the effective use of both types of capital can mitigate the negative consequences of geographical peripherality by creating conditions conducive to activating other forms of capital (Sect. 2.2.1).

While differing in origin and form, financial and material capital are closely interrelated. The relationship between the availability and allocation of financial resources and the state of infrastructure, environmental assets, and residential space quality largely determines a territory's ability to sustain its social and economic functions in the face of external volatility and economic shocks.

Financial capital is the pool of financial resources (both own and external) available to public institutions (especially local governments), private enterprises, and individual residents. Our analysis of financial capital in inner peripheries focused on three key subfactors: the financial standing of local governments, enterprises, and residents (Sect. 2.3.4). Each of these components influences the others, collectively shaping the overall capacity of an area to mobilize developmental resources and its resilience to economic disruptions. Material capital, in turn, refers to physical structures and assets—both man-made and natural—that enable or constrain local development. It includes developed spaces (municipal and social infrastructure), natural environmental assets, and housing conditions. The analysis included three subfactors: spatial development, the quality of the housing stock, and environmental assets—all of which directly affect the quality of life, investment attractiveness, and a territory's adaptive capacity (Sect. 2.3.3).

In this subsection, findings concerning financial and material capital are presented in an integrated and coherent manner, which allowed for the identification of overlapping developmental barriers and an assessment of the adaptive capacities of specific FUAs. The decision to examine both types of capital within a single analytical block reflects their observed interdependencies—specifically, the tight link between financial resource availability and the quality and scope of local infrastructure. Limited financial resources constrain the full utilization of material potential, while infrastructure deficits and environmental degradation require high levels of financial investment—creating a feedback loop that further entrenches peripherality.

When interpreted through the lens of territorial justice (Sect. 2.2.4), the state of financial and material capital translates into lower material living conditions for residents of inner peripheries compared to those in core areas. The territorial perspective adopted in our analysis identifies spatial inequalities in access to key resources and services—both between urban cores and their surrounding areas and among different types of Functional Urban Areas. These disparities are particularly evident in the context of disruptions triggered by the pandemic and the war in Ukraine; areas with stronger material foundations and more stable budgets demonstrate a higher capacity for response and continuity. At the same time, those with limited capital are considerably more vulnerable to the loss of functions, deepening marginalization, and the outflow of human resources.

In the following sections of this chapter, we present a synthesis of our findings based on selected case studies of inner peripheries. Each identified subfactor is addressed in turn, with reference to insights derived from strategic documents and data obtained through IDI, FGI, and CATI interviews. Additionally, we highlight changes observed in the wake of the pandemic and the war in Ukraine.

The significance of financial capital held by local government units (LGUs) for the socio-economic development of inner peripheries stems primarily from its role in enabling the delivery of basic public services, maintenance of local infrastructure, and co-financing of investments to improve residents' quality of life. Despite the complex interplay of factors shaping the financial condition of LGUs, the analysis of four selected FUAs confirmed that budgetary shortages were a systematic issue, particularly in areas outside core zones. LGUs operating in the vicinity of central hubs possessed limited revenues, were more reliant on external transfers, and exhibited low credit capacity. These constraints translated into difficulties securing co-financing for investment projects, thereby effectively impeding the utilization of available external funding sources.

Strategic and planning documents at regional and local levels rarely identified LGUs' financial condition as a standalone development barrier. In most cases, financial issues were mentioned only indirectly—for example, in the context of low investment capacity, the need for coordinated inter-municipal action, or the imperative to seek external funding actively. References to financial stability were generally vague and did not translate into concrete strategic measures. While strategic documents for the analyzed FUAs included commitments to support economic growth, activate local labor markets, or improve quality of life—all of which could potentially strengthen LGUs' financial foundations—there was a lack of instruments directly targeting the improvement of local government finances.

In the FUA centered around a large city in Central Poland (Włocławek), development strategies emphasized the diversification of the economic base and investment in "future-oriented sectors" such as chemical and food processing, logistics, and water-related tourism. These actions aimed to shift the city from an industrial development path and create new revenue sources. Nonetheless, they included no specific references to improving the LGU's financial position, apart from calls for the decentralization of finances and competencies as a condition for strengthening local budgets.

In other cases—primarily in regional strategies and functional partnership documents (the latter available only for Włocławek)—the emphasis was placed on maintaining access to public services, including in peripheral areas. However, this was not accompanied by any dedicated financial support mechanisms—neither in the form of equalizing transfers nor instruments targeted at low-income municipalities. In the FUA centered around a medium-sized town in north-eastern Poland (Przasnysz), development strategies highlighted the potential of an economic zone located on the site of a former military airfield. This zone was intended to serve as a developmental catalyst. However, in practice, its development progressed slowly, revealing a gap between strategic aspirations and the actual ability to attract investment.

Across all analyzed FUAs, strategic documents expressed aspirations for socio-economic growth, yet issues of LGU financial stability and resilience to external shocks were largely overlooked. Proposed measures failed to reflect local financial conditions or offer tools that could meaningfully improve the fiscal standing of peripheral municipalities. As a result, a clear gap emerged between strategic planning and the everyday realities faced by local governments—a gap that significantly undermines their capacity to respond to economic shocks and build long-term, strategic resilience.

The financial condition of local governments constitutes a major development constraint for inner peripheries—a conclusion strongly supported by qualitative and quantitative research findings. The CATI survey results suggest that, while residents did not explicitly comment on LGUs' financial health, they frequently expressed dissatisfaction with the availability and quality of public services—an indirect indicator of fiscal limitations at the local level.

A more explicit picture of LGU financial conditions emerges from qualitative research (FGI and IDI). Interviews with local government representatives highlighted the challenges associated with low own revenues and the rising costs of day-to-day operations. Particularly frequent were references to increasing expenditures related to the maintenance of municipal infrastructure, energy, and employee salaries—all of which exceeded the budgetary capacity of many peripheral municipalities.

In the FGI studies conducted with local leaders and municipal officials, the issue of limited capacity to secure own contributions for projects co-financed with external funds was also strongly emphasized. Respondents pointed to the lack of equalizing instruments allowing peripheral municipalities to compete for development funding on equal terms with units possessing higher own revenues. One participant highlighted the absence of realistic economic activation opportunities: *We somehow can't attract investors, we fail to convince them to come here, or they have requirements we simply can't meet. There's a lack of investment support structures, and public–private partnerships practically don't exist in our area* (FGI1_P).

In other cases—namely the FUAs of small and medium-sized towns—representatives of local administration also noted increasing difficulty in attracting and retaining qualified administrative staff, which directly stemmed from financial constraints and further limited the municipalities' capacity to absorb funds and implement development projects. As a result, the units surrounding the cores of the studied FUAs (rural municipalities) remained particularly financially weak compared to the central towns, operating on modest budgets and struggling to secure funds for basic public services. For instance, joint municipal transportation initiatives often failed to materialize due to lack of funding: *So yes, we need that transport, but when it comes time to calculate the costs, that's when* [...] *we pull out, because it's just too hard* (FGI1_P).

The overall picture of local government finances in peripheral areas, as revealed by the research, is thus unequivocally negative. Local government units faced chronic budgetary shortfalls, dependence on external transfers, and limited ability to implement their own development projects. Consequently, the financial marginalization of municipalities deepened, exacerbating the disparities between urban cores and their surrounding peripheral areas.

The significance of enterprise-level financial capital for socio-economic development primarily stems from the key role businesses play in generating local income, creating jobs, and supporting innovation and investment. Despite the diversity and volatility of factors affecting firms' financial conditions, the analysis of the four FUAs revealed clear financial constraints being faced by enterprises, particularly those operating outside the core areas. Firms located in the peripheries of major cities often had less access to diverse funding sources. They exhibited limited creditworthiness, which hindered their ability to grow or invest in necessary technological or market transformations. This phenomenon particularly affected micro and small enterprises, which dominate peripheral zones and possess limited capital reserves for weathering periods of economic hardship.

Strategic documents for the analyzed areas included numerous declarations of support for entrepreneurship, the development of local economic zones, or investment promotion. However, these initiatives were primarily oriented toward attracting new investors or creating business infrastructure—rarely did they address the real financial challenges faced by firms already operating in external zones. There was a lack of policy measures aimed at improving credit access, supporting financial liquidity, establishing revolving funds, or creating credit guarantee institutions. As such, the financial situation of enterprises was not identified as an area requiring targeted intervention.

The financial landscape of local enterprises in the studied areas was characterized by low diversification and dependence on a few key economic actors or sectors. Strategic documents frequently noted the dominance of traditional industries: agriculture in the peripheries of the medium-sized city FUA in eastern Poland (Sandomierz) and the small-town FUA in north-eastern Mazovia (Przasnysz), the chemical industry in Włocławek, the military sector in Drawsko Pomorskie—and the associated need for modernization and diversification. Regional and local strategies proposed support for non-agricultural entrepreneurship in rural areas, the development of leisure industries (such as tourism), and the use of regional specializations in medium-sized cities. In the strategy for the large city FUA in Central Poland (Włocławek), as mentioned earlier, particular emphasis was placed on developing *future-oriented sectors*. This approach reflects an attempt to reduce the vulnerability of local economies to external shocks by diversifying income sources, though without a clear focus on improving the financial condition of businesses already operating in the local market. In the case of other studied areas (e.g., the medium-sized city FUA in north-eastern Poland—Przasnysz), strategic documents pointed to the potential of a special economic zone intended to stimulate local economic activity. However, the development of this zone proceeded very slowly, highlighting the absence of effective mechanisms for attracting investors that could strengthen the financial standing of local entrepreneurs.

A more direct picture of enterprise finances emerges from the qualitative research (FGI and IDI). A frequently mentioned concern was the dependence on a single large employer and the lack of sectoral diversification—a structure that renders the entire area vulnerable in the event of sector-specific shocks. The financial condition of firms thus largely reflects the specificity of the local economic structure: where a strong public employer (e.g., the military) or a large entity in the wood-processing sector was present, income stabilizers were in place. One interviewee observed: *The military is a completely closed topic—neither the war nor the pandemic had any impact on them* (IDI_D). This statement aptly illustrates the inequality between firms covered by systemic public contracts and small local businesses, which operated under significantly more volatile and cost-pressured conditions. As another respondent put it: *Current revenues have not increased—they've even gone down, and operating costs have increased several times over* (IDI_D). A frequently raised issue was the challenge of maintaining profitability under sustained wage pressure: *The cost of running a business here is incomparably higher than in big cities, and the wages we can offer workers are lower. That makes it hard for us to keep good specialists* (FGI1_S). Elsewhere, a participant noted: *Entrepreneurs have the same problem when it comes to the national minimum wage. They simply can't afford it* […] *It's holding back the economic development of our country* (IDI_P). Another statement added: *The level of support for employing people with disabilities has remained* [...] *but, in real market terms, that support has clearly decreased* (FGI1_P). One respondent also highlighted that growing *entitlement mentalities*—in terms of expectations of state support—were demotivating segments of the local labor force: *Social welfare patterns mean there's little desire to take up work* (FGI1_P).

The financial capital of residents plays a crucial role in shaping socio-economic development—household financial resources determine their ability to meet basic needs, participate in social life, engage in employment, and invest in education, health, and housing. Stable household incomes feed into local consumption, the tax base, and investments in human capital (e.g., children's education). Conversely, financial deficits among residents suppress demand and may lead to deepening spatial disparities—particularly between the FUA core and its surrounding areas. In this way, household finances form one of the key foundations of quality of life, social resilience, and the long-term sustainability of the areas in which people live.

In each of the four FUAs analyzed, we identified significant economic disparities depending on location (core vs. periphery). Financial hardship was especially evident in the peripheral zones among older individuals and households dependent on microentrepreneurship.

Regional and local development strategies of the studied FUAs rarely addressed residents' financial situation or the fiscal condition of local governments as distinct challenges. These documents typically focused on stimulating economic growth, improving infrastructure, and enhancing the quality of life—assuming indirectly that these efforts would translate into improved incomes for residents and stronger municipal budgets. Content analysis of regional and local strategies revealed that the financial stability of households and municipalities appeared only marginally—often in contexts unrelated to direct financial support. The dominant themes were economic development and the maintenance of public services, with little mention of compensatory mechanisms. Although regional strategies did highlight the need to ensure access to services in peripheral municipalities, they lacked proposals for dedicated financial transfers for low-income units. The proposed lines of action did not consider the weaker financial base of peripheral areas (e.g., lower resident incomes, smaller tax base), nor did they offer tools to improve their budgetary situation genuinely.

The only notable exception found in all the analyzed FUAs were planned interventions aimed at activating unemployed individuals and improving the living conditions of groups at risk of long-term marginalization (e.g., social inclusion programs funded by the ESF). As a result, a gap becomes evident between strategic planning and the functioning of local governments and communities. While nearly all strategic documents of the studied FUAs formally declared their pursuit of economic growth, issues of financial stability and resilience to external shocks remained merely implied. This approach may reflect the assumption that the market will resolve issues related to residents' incomes—an assumption that proves overly optimistic in light of further research findings.

Strategic plans do not explicitly address challenges associated with difficult income situations; no programs aim to strengthen local budgets or improve the material conditions of the most disadvantaged groups. Although the need for economic diversification and job creation is recognized, there is a lack of protective measures for existing problems (e.g., debt, poverty, financial exclusion).

The results of the CATI survey conducted in the four FUAs shed direct light on the material situation of households and the differences between residents of central cities and their surrounding areas. One particularly revealing question concerned sources of income. Wage labor is the dominant form of income across all FUAs (approximately 60%). Still, self-employment is notably more prevalent in larger centers (Sandomierz 31.1% and Włocławek 21.8% vs. Drawsko Pomorskie 16.4% and Przasnysz 15.5%) (Fig. 6.16).

The higher percentage for Sandomierz may result from the area's more tourism- and agriculture-oriented (particularly fruit-growing) character, which may also be reflected in the observed variation between the core and its surroundings—to the latter's advantage (24.1% vs. 19.1%). In smaller FUAs, pensions and retirement benefits more frequently constitute the primary source of livelihood. In comparison, in the larger FUAs, a significantly higher proportion of respondents did not indicate any income source (Sandomierz 13.0%, Włocławek 11.8%, and only 7.3% in Przasnysz and 5.4% in Drawsko Pomorskie). Subjective assessments of material well-being were captured through a question on how respondents manage financially (i.e., whether they have enough money to meet current needs or experience financial shortfalls). The results indicate significant variation both spatially and within individual FUAs (Fig. 6.17).

The level of financial well-being—measured by the proportion of households that *have enough for everything* (including some that are even able to save) was found to be highest in FUAs centered around larger towns (Włocławek and Sandomierz) and significantly lower in smaller urban centers (Przasnysz and Drawsko Pomorskie). Only approximately 22% of residents in the smaller FUAs declared that they *have enough for everything and more*, whereas this figure reached nearly 40% in the FUAs of larger cities. In other words, in the two smaller FUAs, fewer than one in four respondents reported feeling financially comfortable, while in the other two, this share rose to roughly one in two. These differences are likely attributable to divergent economic structures: Włocławek (a major industrial center) and Sandomierz (a subregional center with growing tourism) offer more and better-paid employment opportunities than more peripheral locations such as Przasnysz (dominated by agriculture and small business) or Drawsko Pomorskie (where the economy is based primarily on the military unit, the timber industry, and basic services). On the other hand, financial hardship was more commonly reported among residents of rural areas surrounding the urban cores. The proportion of households struggling to meet current obligations (e.g., paying bills or loan installments) or even to cover basic needs averaged approximately 7% in rural municipalities, compared to around 5–6% in the central towns (according to respondents' declarations). In Przasnysz FUA, which was the most problematic, as many as 10% of surveyed households reported arrears or an inability to cover essential expenditures. For comparison, in the FUA of the

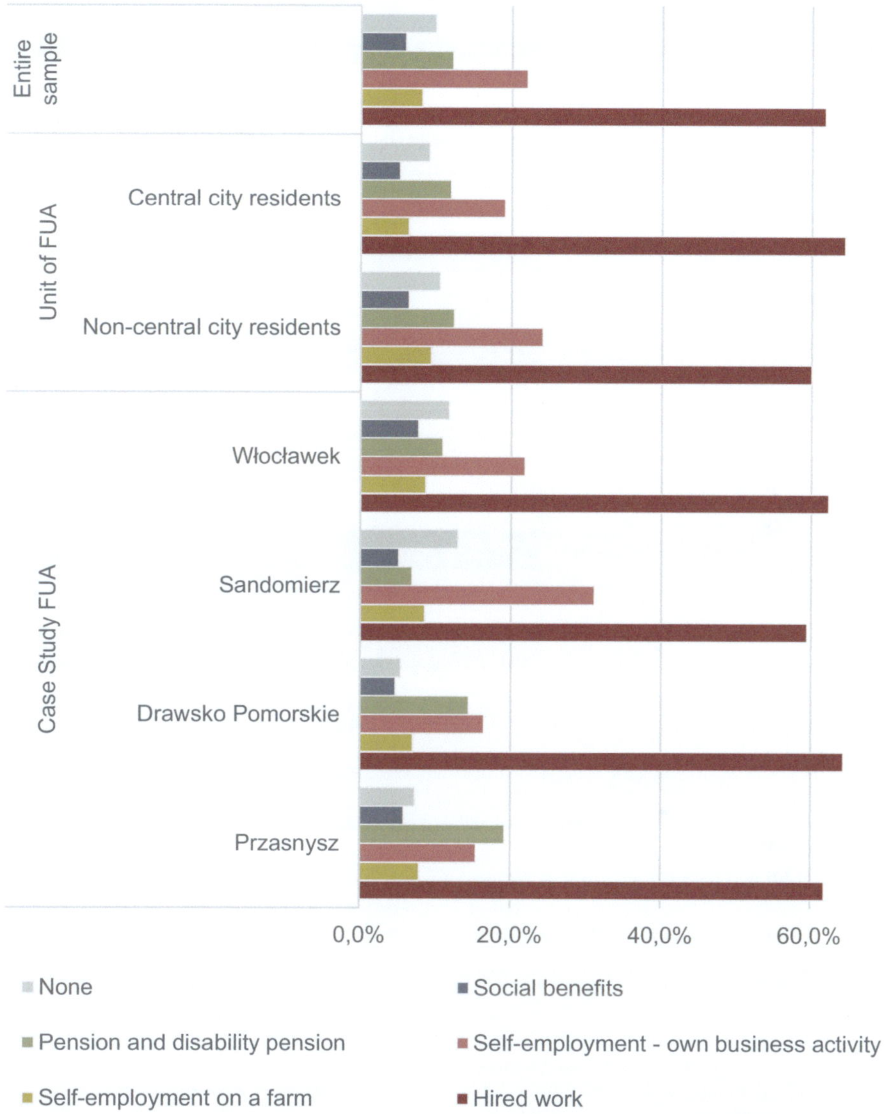

Fig. 6.16 Source of income in households of FUA residents. *Source* Own elaboration

larger city of Włocławek, the proportion was around 4%—still indicating a non-negligible number of families facing extreme financial hardship. More than half of respondents (approximately 50% in surrounding areas and 54% in cores) described their living conditions as *relatively stable*—meaning they had enough for comfortable day-to-day living, even if saving was not always possible. At the same time,

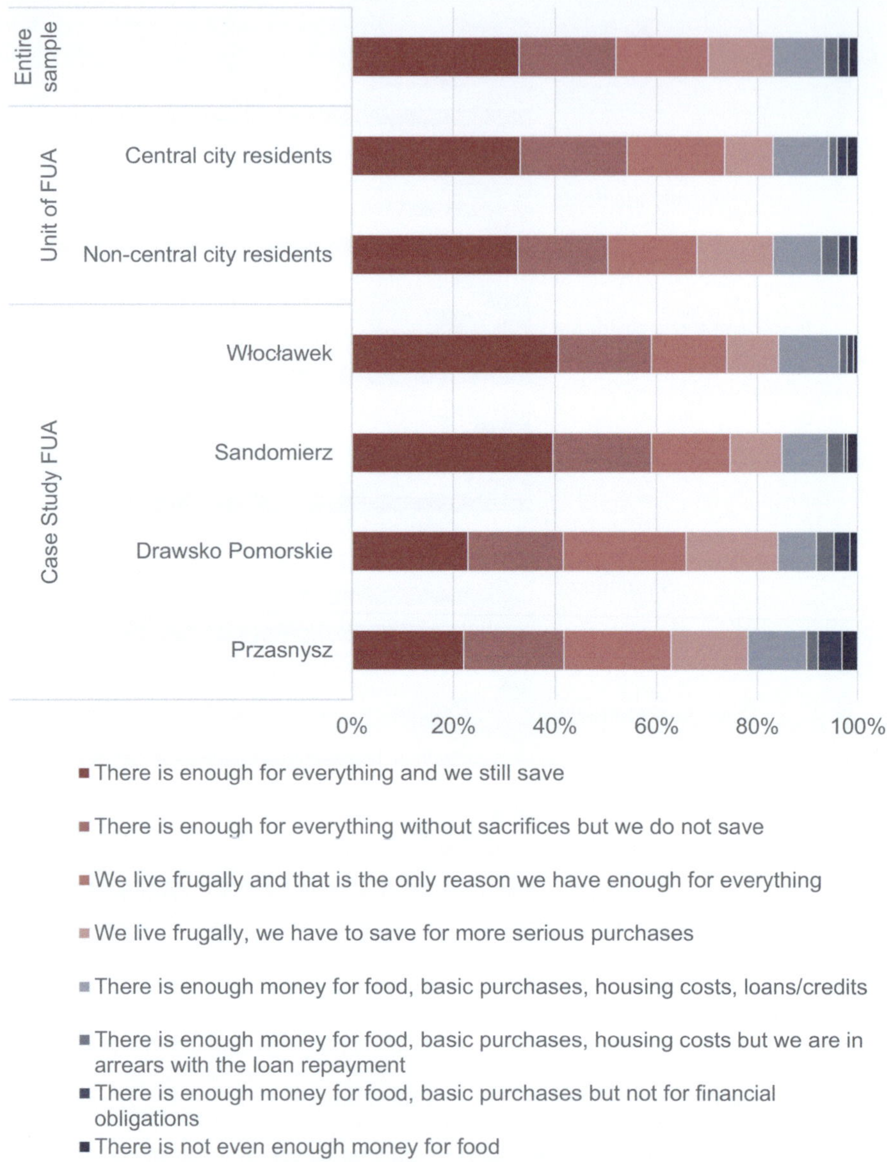

Fig. 6.17 Assessment of financial situation in households of FUA. *Source* Own elaboration

a significant portion reported the need to live *very frugally* to make ends meet—about 33% in the surrounding municipalities versus 29% in the central towns, many of whom could only afford basic necessities and had to postpone larger purchases. These data suggest that suburban and rural residents are more likely to experience financial constraints than their urban counterparts.

The CATI results also provided more comprehensive insights into how households adapt to financial difficulty. Some respondents pointed to the need to cut consumption, forgo certain expenses, or seek additional sources of income. Importantly, a substantial number of residents considered labor migration as a strategy to improve their financial situation. When asked about potential reasons for relocating, *a better or higher-paying job* was the most frequently mentioned motive—cited by nearly 50% of respondents on average (Fig. 6.3).

This tendency was slightly more pronounced in the municipalities surrounding the core towns, where roughly half of those considering relocation cited employment-related reasons, compared to around 44% in the cores. These findings confirm that economic factors—wage levels and job availability—constitute the strongest migration driver in the studied areas. This implies that, if the local labor market fails to offer adequate income, residents—particularly the young and more mobile—are inclined to leave for larger urban centers or abroad in search of better opportunities. This is a critical signal for local governments: the outflow of human capital to more developed areas will likely continue unless employment conditions and wage levels improve locally.

Qualitative research—including focus group interviews (FGIs) with residents and local leaders, as well as in-depth interviews (IDIs) with representatives of local authorities—complemented the survey data by illustrating the everyday financial struggles of communities in inner peripheries while also highlighting good practices and local coping strategies.

The relatively poor financial situation of residents in smaller towns stems from a lack of well-paid jobs and limited development prospects. As the mayor of a small-town FUA (Przasnysz) explained: *We just can't seem to attract investors or convince them to come here, or they have requirements we simply cannot meet. There's a lack of investment support structures, and public–private partnerships practically don't exist here* (IDI_P). Meanwhile, local entrepreneurs noted during interviews that the local market is shallow (i.e., dependent on a handful of larger firms or sectors), making it vulnerable to economic fluctuations. In one of the smaller FUAs (Drawsko Pomorskie), respondents pointed to the dominance of the military sector and administration linked to the training grounds; in another, to the prevalence of agriculture and traditional sectors (Przasnysz). This employment monoculture means many families have no alternative income sources, undermining their financial security and preventing the development of long-term economic resilience. Residents' narratives often confirmed the statistical findings. Many FGI participants described the need to save and limit their needs. For instance, young women emphasized the cost of raising children and employment insecurity: *There's this constant fear—will I be able to raise this child, can I afford it, what if I lose my job? Because let's be honest, not every employer is willing to tolerate maternity or parental leaves* (FGI2_D). These explicit dilemmas directly influence reproductive decisions and demographic trends (see Sect. 6.4), driven by a lack of resources or perceived long-term financial instability.

Respondents' views clearly reflected these challenges: *We basically only have cheap stores—Dino, Biedronka, or second-hand shops. That tells you how poor we really are* (FGI2_P), remarked one participant, underscoring the limited purchasing power of the local community. In the smaller-town FUAs, there was a notable absence of modern retail and service facilities—partly because large retail chains did not perceive these markets as sufficiently profitable. In rural areas, the situation was even more difficult—households relying exclusively on small-scale agriculture were weaker, and interviews revealed patterns of intergenerational poverty: *It's like poverty or helplessness being passed down. Someone gets offered a hand to pull them out of it, but they say they don't want to, because their mom didn't work, their dad didn't work—so why should they?* (FGI1_S).

The material situation of residents shapes migration patterns. In areas closer to economically vibrant urban centers (such as Włocławek or Przasnysz), people tend to commute or relocate to find better-paying jobs (e.g., in Łódź or Warsaw). In more remote FUAs (like Drawsko Pomorskie), emigration or accepting lower local wages is the alternative. The low purchasing power of residents translates into weak demand for local services and products, which hinders entrepreneurship and bottom-up initiatives. Crucially, lower purchasing power and limited creditworthiness may contribute to the deepening of energy poverty: rising energy costs disproportionately affect low-income households, forcing them to cut other expenses or rely on outdated, inefficient heating systems.

Another factor analyzed in the development of inner peripheries is material capital, which is closely linked to financial capital, as it provides the foundation for the functioning of public institutions and the local economy. The first subfactor considered is spatial development, which allows for assessing the quality of territorial capital generated by residents through the lens of the spatial organization of socio-economic activities.

High-quality public space, efficient transportation networks, and well-maintained surroundings contribute to quality of life and social integration, creating an inclusive environment for all resident groups—from families with children to seniors. In FUAs with features of inner peripherality, spatial quality becomes particularly important—it can either mitigate the consequences of remoteness from main centers or, conversely, deepen developmental disparities. Unfortunately, the analysis reveals clear spatial inequalities between urban cores and their surrounding municipalities—especially regarding access to infrastructure, public services, and developed areas.

Strategic documents from the units comprising the studied functional areas generally address spatial development from the perspective of territorial accessibility, infrastructure, and spatial revitalization. Investments in the expansion of infrastructure such as roads, bypasses, and bicycle paths or in the revitalization of existing space aim to enhance the quality of life in these areas. Thus, development challenges are primarily focused on improving territorial cohesion. Key projects include improving mobility (construction of a multimodal hub based on the railway station), decarbonization and revitalization (FUA with a large city in Central Poland—Włocławek), the development of low-emission transport, the expansion of road and transport networks (FUA of a medium-sized city in Eastern Poland—Sandomierz), a railway line and city bypass (FUA of a medium-sized city in Northeastern Poland—Przasnysz), and the modernization of roads and rail transport (FUA of a small town in north-western Poland—Drawsko Pomorskie). Insufficient utility infrastructure and communal infrastructure deficits are highlighted as limitations to local development. Therefore, strategic plans include preparing additional investment and residential land and securing funding for basic amenities (water supply, sewage systems, street lighting) in dispersed villages.

The CATI survey results also provide (indirectly through reasons for relocation) insight into subjective assessments of spatial development. The findings confirm the general image emerging from strategic documents: FUA cores are rated more positively in terms of spatial quality due to the concentration of investments and services in cities. Inner peripheries, by contrast, face a shortage of high-quality public spaces (e.g., lack of developed squares or parks), poorer technical infrastructure, and lower institutional accessibility. These objective differences translate into subjective experiences of residents—often feelings of dissatisfaction or frustration with living conditions outside of cities, as well as lower expectations regarding standards of the local environment. (*You can't expect the same amenities in a small town as in a city.*)

Interestingly, a recurring theme was residents' attachment to place: despite noticeable deficiencies, some peripheral respondents emphasized their appreciation of peace, proximity to nature, and the sense of community in their surroundings—factors that constitute a potential to be preserved while improving infrastructure.

Qualitative research participants' statements highlight specific spatial problems. Many peripheral area respondents pointed to deficiencies in basic road infrastructure, such as sidewalks and cycle paths, the absence of which poses safety risks to pedestrians and cyclists. As one participant vividly described: *we don't have sidewalks, so there's zero safety, there are lots of accidents, collisions, and it's just a tragedy* (FGI2_P). Improving pedestrian and bicycle infrastructure therefore appears to be an urgent need. In discussions, the demand for a network of bike paths was raised, which would make cycling safer and more appealing: *I think that* [...] *building these bike paths would help solve this issue a bit* [...] *children could use them to get to school from nearby villages* [...] *around Przasnysz they could easily ride their bikes* (FGI2_P).

There are also positive mentions in interviews of newly developed infrastructure—like a recently built road: *before it was built, hardly anyone cycled due to fear of car traffic, but now lots of people use the path, they jog in the afternoon* [...] *people were simply afraid* [*before*] [...] *it was a danger not only for cyclists but also for drivers* (FGI2_P). These accounts show that even relatively small investments (a sidewalk or cycle path) can significantly improve the functionality and safety of spaces in inner peripheries, influencing residents' everyday habits.

Respondents—especially local residents such as business owners and community activists—also drew attention to the upkeep of public areas. Some mentioned the neglect of cleanliness or order in rural public spaces: *unfortunately, roads are rarely cleaned. The amount of sand left on the shoulders after winter* [...] *it's a small thing, but* [*it's annoying*] (FGI2_P). Other interviewees emphasized the lack of investment in recreational infrastructure in villages: no sidewalks for walking, street lighting, insufficient playgrounds, or maintained green spaces.

The contrast between the well-maintained urban spaces and the neglected rural surroundings was a recurring theme—residents of municipalities around FUA cores often felt their communities were "off the radar" regarding attention from local authorities. Importantly, local government representatives did not dispute these perceptions. Instead, they pointed to limited financial resources that require gradual, staged efforts to catch up on communal infrastructure development.

Statements from local government representatives (especially from the Sandomierz and Przasnysz FUAs) underscored the problem of insufficient land-use planning and lack of prepared investment land. The absence of local spatial development plans in peripheral municipalities often serves as a major barrier for investors and for meeting infrastructural needs. One local official vividly illustrated this issue through the example of searching for a site for a large processing plant: *Let's be honest—entrepreneurs won't come to municipalities that aren't prepared with land and infrastructure—*[*there are no*] *local development plans* […] *This company spent five years looking for a location in Sandomierz County* […] *and couldn't build anywhere because there were no plans* (FGI2_S). This quote illustrates the problem well—many municipalities cannot accommodate investors due to the lack of plans and designated areas.

Another important subfactor of material capital is housing conditions. These play a key role in ensuring residents' quality of life and determining urban areas' development potential. The availability of adequate housing is essential for meeting basic living needs, influences health and comfort, and can help attract and retain residents—especially young families and professionals—in a given region.

For FUAs identified here as inner peripheries, housing issues are particularly significant. Housing infrastructure deficits in such areas can severely limit development opportunities, exacerbating spatial disparities between the core (central city) and surrounding areas. On the other hand, improvements in housing conditions can help offset the negative effects of peripheral location, increasing overall living standards and community resilience during crises. Housing conditions and housing policy are recognized in strategic documents at both the regional and local (municipal and functional area) levels. Most of these documents emphasize the importance of improving housing availability and the quality of existing housing stock to ensure a high quality of life and counteract negative demographic trends.

At the regional level, housing issues appear in spatial development plans and regional development strategies. For example, the Spatial Development Plan of the West Pomeranian Voivodeship (relevant to the Drawsko Pomorskie FUA) includes the objective of limiting excessive building expansion into undeveloped suburban areas—that is, counteracting the sprawl of single-family housing into rural zones. It emphasizes the need for sustainable spatial planning—to allow for housing development, but in a controlled manner that avoids spatial chaos. Likewise, the Kujawsko-Pomorskie Region Development Strategy 2030+ (relevant to the Włocławek FUA) highlights the importance of ensuring a high quality of life across the region, which includes access to housing infrastructure. The regional strategy identifies functional areas as key platforms for implementing infrastructure investments, pointing to the need for an integrated approach beyond the local scale. Although it does not explicitly address housing, it emphasizes territorial and infrastructural cohesion—to which settlement networks and living conditions are integral.

The development strategy for the FUA of a smaller city (Sandomierz) introduces a dedicated priority 1.3 titled "*Developed housing economy*", under a broader strategic goal concerning quality of life. This strategy outlines concrete actions, including preparing investment areas for housing construction (both multi and single family), developing municipal infrastructure for residential purposes, expanding social housing (municipal units, housing associations, etc.), and supporting access to housing for young people. This comprehensive framing indicates that local authorities in a medium-sized city FUA in eastern Poland (Sandomierz) perceive the insufficient housing offer as a developmental barrier and are planning to actively stimulate the housing market—for instance, by equipping plots for housing development or creating support programs for young families.

Housing-related issues also feature in the *Włocławek City Development Strategy 2030+* . The diagnostic section notes that Włocławek's favorable location creates economic and residential development opportunities. The city is perceived as the core of its FUA, providing surrounding municipalities access to the labor market and higher-order services. In this context, the city's strategy emphasizes strengthening the city's role as a settlement center—implying the need to expand the housing offer for new residents migrating from the surrounding areas.

The *Przasnysz City Development Strategy 2022–2030* and its accompanying *Local Development Plan* also identify certain challenges related to residents' quality of life. Diagnostic studies indicate that access to housing in Przasnysz is rated relatively low—although the main barriers to quality of life are limited job opportunities and business development prospects. Locally, it is also observed that new apartments in Przasnysz are often bought as investment properties for rental purposes, and rental prices remain high. Municipal documents of this medium-sized FUA in north-eastern Poland (Przasnysz) thus advocate, among other things, for creating land plots for multifamily housing development and the containment of uncontrolled single-family housing sprawl onto agricultural land.

Strategic documents highlight several common issues across the analyzed FUAs: the need to expand the housing offer (especially affordable rental housing), improve the standard of existing housing stock (e.g., through thermal upgrades and estate revitalization), and promote spatial planning that limits suburbanization and dispersed development. Planned actions include both local government investments (e.g., construction of municipal and cooperative housing) and creating better conditions for private developers (e.g., plot servicing and simplified planning documentation). Some strategies also emphasize the need for targeted solutions for specific groups, such as young families (programs facilitating housing independence) or low-income individuals (expanding municipal and social rental housing stock).

The results of a CATI survey conducted among residents of the four analyzed FUAs provided valuable insights into subjective assessments of housing availability and quality and the housing needs of these communities. Respondents were asked, among other things, whether they plan to change their residence in the coming years and the main reasons for a potential move. These responses help to understand to what extent current housing conditions are perceived as unsatisfactory and whether they are a push factor in migration decisions.

Respondents' declarations show that many residents are considering changing their residence, with housing conditions being a key motivation. Depending on the FUA, the share of people planning to relocate within the next two years ranged from 36.0% (Przasnysz) to as much as 47.7% (Sandomierz). Interestingly, a desire to move due to housing reasons was much more frequently expressed by residents of core cities (48.3%) than those living in surrounding areas (36.3%). For comparison, the most commonly cited reason for moving was the need for better employment or higher earnings (about 45–50% of responses)—highlighting the importance of economic factors—but housing-related reasons came right after. This suggests that urban residents experience a greater dissatisfaction with the available housing offer (e.g., they want a larger apartment, their own house, or a better-quality building). In contrast, residents of surrounding areas are somewhat less concerned with housing issues. The latter more often emphasized economic and transportation-related motives—43% of peripheral residents indicated *better job opportunities* as a migration reason (compared to 44% in cities), and 26% from the periphery expressed a desire to move closer to a larger city (a voivodeship or central city) in search of better services and opportunities.

CATI results also point to generational differences in housing needs and perceptions. Younger adults are more likely to perceive inadequate housing conditions: in the 25–34 age group, nearly 47% of those planning to move cited the desire to improve their housing situation as the main reason, whereas in the 50+ group, this share was significantly lower (older individuals are less likely to plan relocation, and if they do, it is more often for health or family reasons). Young families and single adults transitioning into independent adulthood thus appear to be the most affected by the shortage of available housing—coinciding with life stages when forming a household and achieving housing independence become important.

Qualitative research—focus group interviews (FGIs) with residents, entrepreneurs, and local leaders and in-depth individual interviews (IDIs) with city officials—provided detailed and vivid examples of how housing conditions are perceived in the analyzed FUAs. Respondents' comments offer insights into specific problems (shortages, barriers) and initiatives recognized as good practices in housing.

Housing availability (supply of units) was a challenge in smaller towns and their surroundings. *Finding a place, a rental apartment in Drawsko is next to impossible (IDI_D)*—stated one of the local government representatives, emphasizing that the rental market in this FUA is fully saturated. The shortage of available housing applies to both private and public units. Similar concerns were voiced in other areas—housing is also in high demand in the eastern FUA (Sandomierz) and its vicinity, although the local authorities there indicate they have managed to build a relatively large municipal housing stock. The waiting list for social or municipal housing can be very long—participants in one of the FGIs mentioned that over 250 people had submitted applications for social housing (FGI2_S). Such a queue implies a wait of several years for the lowest-income groups to obtain housing, which creates serious social problems (such as renting expensive private quarters or living in overcrowded, substandard housing).

Local-specific factors sometimes exacerbate the housing shortage. In the FUA of the small town of Drawsko Pomorskie, one significant *co mpetitor* for housing is the stationed military (the Drawsko training ground is used by NATO forces, among others). Local hotels and rented accommodations are occupied by military personnel, reducing the pool of available housing for residents and driving up prices. This means that external workers (in this case, soldiers) increase demand for housing, making it even more difficult and costly for local residents to find a place to live. A similar situation was reported in another smaller FUA (Przasnysz), where some new apartments were bought by outside investors for rental purposes—making it more difficult for young families to purchase a home and causing rental prices to rise to levels that are relatively high compared to local incomes.

The cost of purchasing or renting a home is another theme that emerged in interviews. Generally, property prices in inner peripheries are lower than in metropolitan areas, but even these values can be a barrier for local residents with moderate incomes. Respondents noted that many people *fall through the cracks* of the housing market—unable to meet banks' credit requirements or secure sufficient down payments—meaning that a certain group is *too rich for municipal housing but too poor for a mortgage*. Włocławek's authorities have identified this very group as requiring support: *we have something that I think sets us apart—rental housing. This has helped those people who are too poor for a mortgage but a bit too wealthy to qualify for municipal housing (IDI_W).*

The example of a large central Polish FUA (Włocławek) illustrates how investment in housing can benefit the entire area. As part of a housing program, Włocławek built 288 rental apartments (with an option for future ownership), for which more than 1,000 people applied. Importantly, the city also introduced an innovative mechanism using some of these units for staffing needs: 20 apartments were designated as *staff housing*—rented, for example, to the local hospital for doctors, to attract much-needed specialists (many of whom came from Ukraine). FGI respondents also praised these efforts. This good practice shows that local governments can intervene actively and thoughtfully in the housing market, combining social goals (meeting families' housing needs) with developmental ones (attracting skilled workers and addressing local shortages, such as in healthcare).

Other FUAs also see local governments undertaking remedial actions, albeit on a smaller scale. In peripheral areas less attractive for residential settlement, investments in servicing large tracts of land for new housing developments may be risky. As a representative of a small-town government (Drawsko Pomorskie) stated, *it makes more sense for us to invest* [*in*] *15 km of transport for residents than to sink 20 million into land that may or may not be developed (FGI1_D).* This financial pragmatism reflects the limitations faced by smaller municipalities: even if land for housing exists, there is a lack of capital—public or private—to quickly convert it into ready-to-occupy dwellings.

Residents and authorities in inner peripheries consistently point to the housing shortage—especially housing affordable for the average citizen—as a real development barrier.

The final subcomponent of material capital examined in this report concerns ecosystem goods and services—the benefits society derives from nature. These include clean air and water, natural resource availability, green spaces, biodiversity, and the regulatory functions of ecosystems (e.g., pollution absorption and climate mitigation). Their availability and condition directly affect residents' quality of life, public health, and the potential for sustainable local development. In urban areas and their surroundings, ecosystem services influence living conditions—from opportunities for outdoor recreation and landscape aesthetics to climate resilience (e.g., mitigating the urban heat island effect or improving water retention). Across the four analyzed FUAs, natural resources play a diverse role. These areas differ considerably in character, ranging from a larger city (Włocławek) to smaller centers such as Przasnysz, Sandomierz (a historic city), and Drawsko Pomorskie. Each demonstrates a different spatial relationship between the urban core and surrounding areas. Despite these differences, ecosystem services are consistently recognized as essential: they underpin residents' well-being and serve as a factor attracting (or retaining) both population and investment in the studied inner peripheries.

Analyzing local and regional strategic documents for the four FUAs reveals that environmental issues and ecosystem services are important development elements. However, the emphasis placed on them varies. In most cases, the term "*ecosystem services*" is not used explicitly; instead, the concept is addressed through objectives related to environmental protection, green space development, or natural resource management.

The strategic documents of the FUA centered on the larger city (Włocławek) prioritize air-quality protection and climate-change adaptation, emphasizing developing green infrastructure and investing in low-emission transport. The strategy for the smaller urban FUA in Mazovia (Przasnysz) highlights the need to restore green areas within the city and protect surrounding agricultural land. The strategic documents of the Sandomierz FUA emphasize the exceptional landscape conditions of the loess hills and the Vistula River Valley, advocating for sustainable tourism and improved air quality in the city. Meanwhile, the FUA of the small town in northwestern Poland (Drawsko Pomorskie), per local and regional planning documents, focuses on protecting numerous lakes and forest complexes, aiming to balance the development of nature-based tourism with biodiversity conservation.

In summary, all four FUAs reflect an awareness of the role of the environment and ecosystem services in their development strategies. However, each local government sets priorities tailored to local resources and challenges. Urban documents focus on improving access to green areas and environmental quality in city cores. In contrast, rural and county-level documents emphasize the protection and economical use of natural resources in peripheral areas (e.g., tourism, agriculture, renewable energy). It is worth noting that urban greening initiatives (as in Przasnysz and Włocławek) indicate efforts to improve living standards in central areas. In contrast, investments in recreational infrastructure in peripheral zones (e.g., trails around Drawsko's lakes or neighborhood parks in Sandomierz) aim to reduce disparities in access to ecosystem services between the core and periphery.

CATI survey results show that respondents positively evaluate the state of the natural environment in their localities, especially in peripheral areas. Clean air, peace and quiet, and proximity to nature were frequently mentioned (in open-ended questions and spontaneous comments) as advantages of living outside large cities. Environmental assets were not explicitly cited as a reason for relocation. This may suggest that residents do not perceive the environment as a major concern—it is not typically a source of dissatisfaction leading to migration. However, when asked about the advantages of living in a given area, many respondents (particularly those from FUA outskirts) referred to the natural environment—indicating that good environmental quality is perceived as a given or a "bonus" of living in these FUAs, valued more implicitly. In other words, residents of inner peripheries are accustomed to the everyday availability of ecosystem services (tranquillity, greenery) and, unless these are lost, they are not perceived as problematic.

Qualitative research provided a deeper understanding of the significance of ecosystem services for residents of inner peripheries, offering insight into their own voices—rich with emotion, everyday observations, and expectations. Urban and rural residents emphasized that the natural environment strongly influences their daily lives and overall well-being. Rural respondents, in particular, cited nature as a reason they valued their place of residence: *I moved here from a big city precisely for peace and clean air. We have a forest outside our window, our own garden—you can't put a price on that* (FGI2_S). *Our area is beautiful and green, in the summer it feels like we're on holiday. People come here to relax by the lakes, but for us, it's everyday life* (FGI2_D).

Such statements indicate that healthy ecosystems (forests, water bodies, clean air) contribute to a high subjective quality of life for residents of inner peripheries. Even in the more urbanized context of a larger city (Włocławek), respondents appreciated elements of urban nature: the proximity of the Vistula River, city parks, and green squares. One interviewee (IDI_W) remarked that *Włocławek is a city surrounded by greenery—we have a large park, Zawiśle, with its forests—if someone wants to be close to nature, they'll find a place here.* Of course, these are individual opinions—some Włocławek residents complained that greenery in the city center is declining—but overall, respondents strongly attached to access to nature as a core part of local identity and attractiveness.

Despite many positive aspects, residents pointed to several environmental challenges:

- Neglect of green areas: Some respondents noted insufficient parks and residential greenery maintenance. *The grass is uncut, weeds up to the waist—that's what our green spaces look like* (FGI2_P). This suggests that the presence of green areas and their upkeep is essential.
- Air pollution and smog: This issue was frequently raised by residents of larger FUAs (Włocławek and Sandomierz), especially during the heating season. *In winter you can't breathe, chimney smoke drifts into the neighborhoods*

(IDI_W). In another FUA, respondents also mentioned smog: *It's a beautiful city, but in winter there's a fog of smog hanging over it—this urgently needs to be addressed* (FGI1_S). Smog was less often mentioned in rural areas (with less dense development). However, some homes in the municipalities surrounding Włocławek also contributed to the problem and the local residents' frustration.

- Uncontrolled urbanization and loss of natural areas: Some respondents expressed concern that increasing suburbanization is harming the environment. *New houses are springing up like mushrooms around the city—forests and orchards are being cleared for them; the periphery is losing its greenery, everywhere you look there are new building plots* (IDI_S). These developments raised concerns about lost ecosystem services and underscored the need for better spatial planning to protect ecologically valuable areas from haphazard development.
- Illegal dumping and waste: A problem specific to peripheral areas was illegal waste disposal in forests and roadside areas. In every FUA, peripheral residents cited examples of illegal dumping: *People bring trash bags to the forest because they don't want to pay for waste collection—it breaks your heart to see it* (FGI2_D). These accounts point to insufficient environmental education and/or weak enforcement. Residents expressed frustration at witnessing the degradation of their surroundings.

There were also examples of good practices in ecosystem stewardship:

- Grassroots environmental initiatives: Every spring we organize a clean-up. It used to be full of bottles and plastic, but now it's much cleaner—people see that we care, and they litter less (FGI1_W).
- Investment in green and recreational infrastructure: Many respondents appreciated recent local government projects that improved access to green spaces. Now we have a beautiful park built with EU funds—there are walkways, a playground, an outdoor gym. In the evenings, it's full of people walking (FGI1_S).
- Cooperation between local authorities and residents: Several focus group participants mentioned that authorities have started consulting with residents on environmental projects. There was a meeting about revitalizing the square—we could share our ideas. Some were even included—it's great, because we feel this is our place (FGI2_W). Such participatory approaches support the tailoring of solutions to community needs and foster a sense of shared responsibility for public space.

Overall, it can be concluded that residents of the four FUAs strongly identify with their local environments, viewing them as assets and extensions of home. At the same time, they recognize threats stemming from neglect and new pressures, such as smog and urban sprawl. While they welcome improvements in green infrastructure where these occur, they also expect further action. The inner peripheries, as described in these voices, are places where the potential of ecosystems remains underutilized, and their protection requires conscious, locally attuned policy.

The SARS-CoV-2 pandemic and the war in Ukraine have significantly affected the financial and material capital of the analyzed areas. Most notably, they exposed the chronic budgetary shortages local governments face, particularly in municipalities outside of core urban centers. As a result, many development initiatives could not be implemented due to a lack of local funding. The pandemic imposed additional expenditures related to healthcare infrastructure and access, remote education, support for local services, and emergency aid for the most vulnerable amid declining revenue streams. The war in Ukraine and the ensuing energy crisis triggered a wave of inflation and surging costs for energy, materials, and wages, drastically increasing the day-to-day operating expenses of local administrations and municipal units. As one local official observed, *The operating costs have multiplied. A simple example: in the last two years, debt servicing cost us about 280,000–300,000 annually, and now it is 1.8 million* (IDI_D).

Many municipalities were forced to cut non-essential spending, abandon planned investments, or seek savings to balance their budgets in the face of rising energy and service costs. An additional burden for some local governments was the urgent need to provide support for war refugees—housing, food, and schooling for children—which, given their limited resources, posed a significant challenge. Importantly, the central cities of the studied FUAs managed these shocks relatively better. With larger budgets, financial reserves, and greater borrowing capacity, they could respond more swiftly, whereas poorer rural municipalities were far more vulnerable to functional destabilization. Peripheral municipalities are chronically underfunded, while central cities—despite facing financial pressure—enjoy greater access to revenue sources (e.g., local business and property taxes, larger PIT/CIT due to higher population) and easier access to credit. For instance, in 2021, the *per capita* budget revenue in Włocławek was several dozen percent higher than in the surrounding rural municipalities (based on local government budget reports). Even basic services and infrastructure maintenance were underfunded in many of the studied peripheral municipalities. Interviews often emphasized that inter-municipal initiatives (e.g., public transport) failed to materialize due to a lack of funding. Local leaders repeatedly stressed the problem of limited own-source revenues and rising fixed costs. They also pointed to the outflow of qualified administrative staff to wealthier municipalities or the private sector, resulting from the inability to offer competitive wages. Consequently, lower financial capital in local governments translates into lower human capital in public administration, reinforcing a vicious cycle of weak fund absorption and deepening developmental lag. Meanwhile, the scope of responsibilities continued to expand, with the pandemic and war introducing additional administrative burdens.

The pandemic particularly affected small businesses (retail, services, gastronomy). Periodic lockdowns and a drop in demand led to revenue losses, and some small firms were forced to suspend operations. While national anti-crisis support schemes softened the blow, many businesses in peripheral areas—already financially fragile—incurred debt or used up their last reserves to survive. The war in Ukraine, in turn, triggered a cost shock: fuel, energy, and raw material prices surged, becoming an unbearable burden for low-margin enterprises. As one local entrepreneur noted, *Companies' current revenues did not increase—some even declined—while operating costs multiplied* (FGI2_D). High inflation forced many businesses to raise wages to retain staff, which not all could afford. As a result, many of the smallest businesses scaled down their operations or disappeared from the market entirely between 2020 and 2022.

Conversely, some periphery sectors proved resilient—or benefitted—from the crises. For example, firms linked to the defense sector or those with stable public contracts (e.g., supplies for the military or healthcare) operated with relative stability. *The military is a completely closed-off subject—neither the war nor the pandemic had any effect on them,* remarked one respondent, referring to the presence of a large military unit in the small-town FUA of Drawsko Pomorskie, which ensured a steady flow of orders and financial liquidity for associated local entities (IDI_D). This statement underscores the growing disparity between firms with stable contracts and local microenterprises left to fend for themselves in conditions of extreme uncertainty.

Business conditions in central cities were somewhat more favorable, with a greater presence of medium-sized enterprises, better client access, and better-developed financial infrastructure (banks, business support institutions). In contrast, peripheral areas are dominated by family-run microenterprises and farms that are often undercapitalized. For example, in the large-city FUA of Włocławek, major industrial plants operate (e.g., chemical producer Anwil, a fertilizer manufacturer) with access to capital and credit. Surrounding rural municipalities are characterized by small-scale farmers and service providers with limited working capital. Similarly, in the smaller-city FUA of Sandomierz, the urban core hosts a few larger processing companies (e.g., food and fruit processing). At the same time, most surrounding villages are made up of small orchards dependent on procurement prices in agricultural markets. Large investors are absent in another small FUA in Mazovia (Przasnysz). Despite establishing an economic zone aimed at stimulating entrepreneurship, development has been slow, highlighting the lack of effective mechanisms for attracting external investment and supporting local business growth, especially in small municipalities. In areas with a monocultural economic structure, sectoral shocks affect the entire region. For example, financial problems at a large chemical plant or an agricultural downturn have indirect consequences for all residents (e.g., reduced tax revenues, fewer orders for local subcontractors, lower local consumption).

In qualitative interviews, local entrepreneurs identified several key financial barriers. First, labor costs: *Running a business here is incomparably more expensive than in large cities, and the wages we can offer are lower. This makes it difficult to retain good specialists* (FGI1_S). Low wage attractiveness in peripheral areas leads to the outflow of skilled workers to larger centers or discourages skills development. Moreover, the rising minimum wage increases labor costs, which are particularly difficult for microenterprises—and smaller municipalities—to offset, especially under the pressures of the pandemic.

The crises of 2020–2022 also degraded the financial situation of households, though the effects varied over time and across social groups. Initially, the pandemic reduced income for some families, particularly those working in sectors shut down during lockdowns (e.g., hospitality, services, small trade). People employed under civil-law contracts or running sole proprietorships abruptly lost their income. Many households bridged the gap with government support or by dipping into savings.

Meanwhile, some groups—pensioners and public sector employees—maintained stable incomes, which, in the context of reduced spending (e.g., travel, entertainment), even allowed for saving. However, the war in Ukraine and the subsequent inflation surge affected nearly all residents, especially the most impoverished. Rising food, heating, and fuel prices, combined with static incomes, resulted in a real decline in purchasing power.

CATI survey results show that residents unanimously perceived a deterioration in their material situation since 2020—the average self-assessment score was just under 3 on a 1–5 scale (where 3 = no change, below 3 = worsening). Rising living costs forced many families to cut spending, draw on savings, or go into debt. The winter of 2022/2023 was particularly difficult: high heating and energy prices proved burdensome, with some households relying on government subsidies (e.g., coal allowance) or turning to cheaper fuels (e.g., wood, lower-quality coal), at the expense of comfort and health. Paradoxically, in some rural areas, farmers' incomes increased in 2022 due to high prices in agri-food markets. However, this benefit mainly applied to larger commercial farms; smallholders more often faced rising prices of fertilizers and production inputs—many imported from Ukraine or Russia—eroding their margins.

The pandemic and inflation disproportionately affected groups already in difficult situations: low-skilled individuals, large families dependent on benefits, or persons with disabilities. Those living on the financial edge often fell into debt or required support (e.g., from social welfare centers or food banks). In contrast, young and mobile individuals often responded to the crises by migrating for work—some moved to larger cities or abroad in search of better wages, thereby depleting the local human capital base. The study also revealed that hidden unemployment remains a reality in peripheral communities. Local employers observed that many individuals do not seek legal employment, relying on social support or working informally. This attitude, described during FGIs as *"social entitlement,"* became more visible during the mass distribution of aid (e.g., crisis shields, allowances). Some residents expected continued support rather than actively seeking employment.

In interviews with residents (IDIs) and local leaders, concerns about inflation and declining purchasing power were a recurring theme. People complained about rising electricity, heating, and fuel bills. The fuel price hike hit commuting workers particularly hard in rural municipalities, where distances are greater and public transport is lacking. Seniors worried about access to healthcare during the pandemic and about whether they could make ends meet amidst rapidly rising medication and food prices. In the survey, nearly half of those planning to leave their hometown cited a desire to improve their housing or financial situation—48% mentioned better job opportunities and around 41% better housing. This indicates that a significant portion of peripheral residents see their future outside their current locality unless their living conditions improve. On the other hand, some statements reflected a more positive theme—social solidarity and neighborly support during the crisis. The pandemic brought communities closer together, with people helping one another (e.g., shopping for elderly neighbors), and the war in Ukraine prompted many grassroots aid initiatives for refugees, even in small towns and villages. Such forms of social capital may indirectly translate into financial capital—for instance, through the creation of local purchasing cooperatives or mutual support networks exchanging services.

The pandemic and war also had a considerable impact on material capital. The pandemic temporarily reduced spatial mobility, highlighting both the disadvantages and, in some respects, the advantages of peripheral spatial arrangements. On the one hand, the lockdown exposed problems faced by residents living far from urban centers—lack of nearby shops, health centers, or pharmacies; limited remote work capacity due to poor digital infrastructure in some villages; and the need to travel long distances to access hospitals. On the other hand, during social distancing, there was an appreciation for the dispersed settlement pattern and proximity to nature. Some urban dwellers temporarily relocated to rural areas or recreational plots in peripheral regions, seeking space and tranquility. The remote work trend diminished the significance of geographic location—some professionals were able to settle outside metropolitan areas while continuing to work. However, for this trend to yield lasting benefits for peripheral areas, robust digital and service infrastructure is essential, though often still lacking. While the war in Ukraine did not directly alter spatial arrangements, it indirectly affected the housing market. The influx of people seeking refuge in Poland increased housing demand in larger cities and temporarily raised population density in certain areas. In the analyzed FUAs (e.g., Sandomierz and Włocławek), the arrival of refugees may have led to better use of previously vacant housing or facilities (e.g., sanatoria, dormitories) as temporary accommodation. This phenomenon, however, was limited in scale and primarily short-term.

The SARS-CoV-2 pandemic highlighted the importance of housing quality. As people were forced to spend most of their time at home, cramped or uncomfortable living conditions became a source of frustration. The lack of a separate space for remote work or children's schooling proved particularly troublesome. This prompted some families to accelerate their decisions to move to larger dwellings or to relocate from urban apartment blocks to rural areas in search of homes with gardens. During 2020–2021, demand for building plots and houses outside of major Polish cities increased—including in the attractively located peripheries of the FUAs under study (e.g., the surroundings of Sandomierz or the Drawskie Lakeland may have attracted some Warsaw or Tricity residents seeking remote refuge). However, this phenomenon mainly involved wealthier immigrants rather than the local population.

For many residents of peripheral areas, the pandemic led to a deterioration in housing conditions in a social sense. For example, adult children who lost their jobs in cities returned to their family homes, increasing residential density under one roof. The war in Ukraine also significantly impacted the housing situation—Poland received approximately 1–1.5 million refugees, who occupied available housing stock. Further, it increased demand for rental units, which were already under pressure due to the economic crisis and inflation. In large cities, rents rose sharply, and available housing was quickly exhausted. Although the situation was less strained in smaller centers, increased competition for rental units was still noticeable. For instance, in the FUA of a larger city (Włocławek), many apartments were rented to refugees—beneficial for property owners but problematic for local residents seeking housing.

Furthermore, the war triggered a drastic increase in the prices of construction materials (steel, bricks, insulation) and reduced their availability, which stalled or raised the cost of housing investments. Developers suspended some new projects, while individuals building homes had to expand their budgets or halt construction altogether. These developments exacerbated the housing gap—fewer new dwellings entered the market in 2022–2023, which may be felt in subsequent years.

In the CATI survey, respondents reported relatively few changes in their housing situation since 2020 (average score ~3.0, indicating no change). This stems from the fact that housing stock changes slowly—short-term crises neither improved nor drastically worsened housing conditions. Nevertheless, in qualitative interviews, housing emerged as a developmental barrier. Local government officials noted that their municipalities lacked the resources to construct social or subsidized housing (e.g., TBS units). At the same time, the private sector showed little interest in investing in areas with low-income populations due to perceived demand risks. *The market will not solve this on its own, and we cannot afford it*, stated one mayor about the housing shortage for seniors and young couples (IDI_D). In rural municipalities, there were recurring references to eligibility for government programs—some families wished to access subsidized mortgage schemes but did not meet creditworthiness criteria. This points to the issue of low incomes translating into limited housing access.

Finally, FGI discussions with residents included demands for improvements to the immediate residential environment, such as the renovation of old apartment blocks, the creation of more playgrounds, and the revitalization of neglected estates. This reflects the understanding that housing conditions are not limited to a roof over one's head but also encompass the quality of the neighborhood or village as a place to live.

The pandemic period brought about rather atypical and ambiguous environmental effects. Lockdowns and reduced traffic temporarily improved air quality and lowered noise levels, even in urban centers and small towns. Restrictions on international travel shifted public interest toward local attractions. Accordingly, many individuals began discovering nearby forests, parks, and cycling routes, using more intensive recreational ecosystem services. For example, the Drawskie Lakeland—part of a small-town FUA (Drawsko Pomorskie)—recorded increased domestic tourism in the summer of 2020 as people sought remote, safe places within Poland instead of crowded foreign resorts. This demonstrated the natural potential of peripheral areas as safe havens during crises.

Simultaneously, pressure on certain resources increased. For example, intensively used recreational areas required more maintenance, and a surge in grass burning in early 2020 (possibly due to reduced enforcement) resulted in numerous wildfires across the country. The war in Ukraine triggered an energy crisis that also impacted the environment: a return to more polluting fuels (coal, wood) worsened air quality during the 2022/2023 heating season, particularly in rural areas. The government temporarily relaxed emission standards (permitting the sale of lower-quality coal), which led to increased smog—e.g., in January 2023, particulate matter levels in many localities registered record exceedances. On the other hand, the high cost of energy prompted some residents to invest in renewable energy sources (e.g., photovoltaic systems and heat pumps) to reduce dependence on traditional fuels—a development beneficial for the environment in the long term due to reduced emissions. It is also worth noting that the war heightened the importance of local food production—agricultural areas in the peripheries became critical for food security, which also meant agricultural intensification (increased fertilizer use, expanded crop areas), potentially at the expense of biodiversity.

Poland's inner peripheries' financial and material capital was severely tested in the face of the pandemic and war. Financial challenges (municipal poverty, weak enterprises, low household incomes) intersected with material shortcomings (infrastructure gaps, housing shortages, underutilized spatial assets). Areas with better infrastructure and more stable budgets (such as core cities) demonstrated greater resilience—they could maintain service continuity and return to development paths more quickly. In contrast, weaker peripheral municipalities experienced more severe disruption and faced the risk of further marginalization. Reducing spatial disparities requires a comprehensive approach: simultaneous strengthening of financial foundations (through reforms and local initiatives) and investment in material capital, so that natural and infrastructural resources can be fully leveraged for development. The implementation of these recommendations by local governments, with central support and the involvement of local communities, can enhance the adaptive capacity of inner peripheries in the face of future shocks and improve the quality of life for their inhabitants.

6.6 The Examined Factors' Significance in Shaping Inner Peripheries' Relational Developmental Environment

The concept of a relational development environment refers to how residents, by utilizing the resources available within a given territory, assign them functional value through their interconnection and embedding in local social practices. In other words, the development of inner peripheries largely depends on networks of social, economic, institutional, and informational relations that connect these areas to development cores. This section explores the interrelations among the analyzed development factors and their significance for shaping a relational development environment. We illustrate how deficits in individual types of capital (human, social, institutional, financial, material) and accessibility—as well as the lack of synergies between them—limit the formation of an optimal relational environment for territorial capital, creating feedback loops that reinforce the peripheral status of the studied regions.

- Accessibility (both transport and digital) constitutes a fundamental condition for leveraging the remaining development factors in inner peripheries. It enables both physical and virtual access to public, educational, and health services and to labor markets (thus supporting the development of human capital) and fosters the creation of social and institutional networks. Furthermore, it affects the functionality of material capital by improving access to infrastructure and increasing the investment attractiveness of a given area.

- Human capital acts as a catalyst within the system. Its level is contingent upon investment in education, healthcare, and professional development (i.e., financial capital availability), while at the same time driving demand for improved accessibility and infrastructure. It also enhances social and institutional capital through active engagement, trust, and shared responsibility for local development.
- Social and institutional capital underpins the effectiveness of local governance, the quality of service provision, and the readiness for collaboration—factors that directly affect infrastructure quality, accessibility, and the overall attractiveness of a territory. Its quality depends on the presence of qualified residents and on the cohesion of communities with a strong local identity. At the same time, this form of capital influences the capacity to absorb external funding—that is, the level of financial capital.
- Financial capital is a transferable factor that determines the developmental potential of all other forms of capital. It provides the necessary resources for investments in people, infrastructure, institutions, and services. A lack of financial capital exacerbates existing deficits, which can mutually reinforce one another and lead to persistent marginalization and entrenchment of peripheral status.
- Material capital, in turn, shapes the quality of life and settlement opportunities, which directly influence migration decisions and residents' lifestyles (human capital). It is also a key factor in fostering trust in institutions and the quality of social relations. Its development is mutually dependent on financial capital—public and private investments determine the quality of space, services, and infrastructure.

The studied inner peripheral areas largely remain relationally disconnected from the country's development cores. This disconnection is multidimensional—beyond the mere geographical distance from major urban centers (such as regional capitals)—and involves weak social, economic, digital, and institutional linkages with broadly defined development networks.

Inner peripheries are characterized by limited access to resources and institutions crucial for human capital development. This is reflected, for example, in more difficult access to healthcare services—particularly specialized care—as well as to high-quality education and in the limited scope of local labor markets. As a result, residents of these areas have significantly fewer opportunities to develop their skills and competencies, negatively impacting their competitiveness in the labor market and their social and economic activity levels. These deficits also intensify migration toward central areas, further weakening human capital in peripheral regions and accelerating depopulation trends. Particularly evident are the deficits in social and institutional integration. Insufficient inclusion in national and regional development networks results in the relative isolation of local communities, limiting the flow of knowledge, the transfer of innovation, and the adaptation of effective governance practices. Residents of inner peripheries face difficulties accessing cultural, educational, and developmental events, which are typically concentrated in larger urban centers. Digital exclusion further intensifies this isolation, restricting access to remote education, e-government services, and telework opportunities. The economic linkages of these areas with the external environment are also weakened—local enterprises often remain disconnected from supra-local value chains and operate in isolation from key markets and national sources of innovation. Consequently, local economies develop in a fragmented, low-dynamic manner that is disconnected from prevailing development trends.

This multidimensional relational disconnection prevents inner peripheries from fully benefiting from the developmental impulses generated by the cores. In turn, this contributes to the reproduction of their peripheral status and the deepening of territorial disparities. The findings confirm the need to pay greater attention to strengthening the diffusion effects of development processes and to moderating the course and consequences of spatial polarization.

Unfortunately, our in-depth research highlights a clear gap between strategic planning and the actual needs of local communities inhabiting inner peripheries. Strategic documents insufficiently address the necessity of shaping a relational development environment and counteracting the disconnection of these areas from the country's development cores. Inner peripheries are also characterized by low resilience to development shocks, making them particularly vulnerable to destabilization. In the studied areas, residents rarely participate in creating strategic documents—a phenomenon largely rooted in a long-standing sense of powerlessness and limited influence over development decisions. As a result, the perspectives of key local actors' (residents and representatives of the entrepreneurial sector) are marginalized, further limiting the relevance and effectiveness of public policies implemented in these regions. A lack of effective tools for facilitating communication between local governments and stakeholders was also identified. Social consultations are often superficial, and beyond obligatory mechanisms (such as participatory budgeting), there is a lack of permanent dialogue platforms with residents and non-governmental organizations. All of this weakens the integration of local communities around development efforts and limits the potential for building social capital based on participation. In short, the relational perspective, which is focused on building linkages and collaboration between different actors, weakly represents the strategic planning of the studied peripheries. A particularly critical deficit from the perspective of entrepreneurs is the lack of consideration of business support services in strategic planning despite their essential role in fostering economic development. Low resilience to development shocks, such as the pandemic and the war in Ukraine, is another consequence of insufficient attention to potential risks in strategic documents.

Strategic planning deficiencies also translate into the low resilience of inner peripheries to development shocks. Local government strategic documents lacked consideration of potential threats and crisis scenarios, leaving these areas ill-prepared for sudden external disruptions. The SARS-Cov-2 pandemic, followed by the war in Ukraine and the accompanying energy crisis and inflation surge, starkly exposed peripheral regions' vulnerability to crises. Local labor markets and enterprises were more severely affected than those in more affluent centers—many small service and retail businesses failed, further waves of outmigration occurred, and local government revenues dropped sharply. These shocks intensified existing socio-economic problems (unemployment, exclusion, municipal debt) and deepened the marginalization of inner peripheries. In other words, the lack of prior resilience-building actions, such as economic diversification, financial reserves, or crisis planning, widened the developmental gap between peripheral areas and development cores in the face of these crises. This underscores the importance of long-term resilience-building and proactive support for the adaptive capacities of territories exposed to economic shocks.

Despite the outlined challenges, certain positive exceptions and development potentials were also identified in the studied areas. As previously mentioned, an asset of inner peripheries is their natural environment, which contributes to a high quality of life—clean air, proximity to nature, and lower noise and stress levels compared to large urban agglomerations. These attributes create the potential for the development of tourism, recreation, and leisure industries and for attracting individuals seeking a quieter place to live—a trend gaining momentum in the era of widespread remote work. Moreover, in some municipalities, grassroots local initiatives and committed community leaders are evident. These manifest in associations, rural women's circles, local foundations, and informal groups that initiate development projects, ranging from small cultural events and regional promotional campaigns to the establishing of local civic activity centers. Although isolated, these initiatives reveal the latent potential for community self-organization and cooperation.

An essential resource is the existing educational and social infrastructure: even in peripheral areas, schools (including secondary ones), libraries, cultural centers, and local vocational colleges operate. With adequate support, these institutions could become centers of social integration and incubators of new ideas (e.g., through third-age universities, youth clubs, or upskilling courses tailored to local labor markets). All these positive elements—the natural environment, cultural heritage, civic activity, and local institutions—constitute potential foundations for building the developmental advantages of inner peripheries. However, these assets alone appear insufficient to bring about a lasting shift in current developmental trajectories and require well-directed support. Harnessing these advantages on a broader scale necessitates a deliberate, relationally oriented, territorially anchored development policy that integrates bottom-up initiatives with external support and embeds peripheral areas into wider cooperation networks. Under such conditions, entrenched negative feedback loops can be gradually broken, and a more development-conducive relational environment can be fostered in the inner peripheries.

References

Alcidi C, Ferrer JN, Di Salvo M, Musmeci R, Pilati M (2018) Income convergence in the EU: a tale of two speeds. CEPS. https://www.ceps.eu/system/files/ConvergencePDF.pdf

Babbie E (2005) The basics of social research (3rd ed.). Wadsworth Publishing. https://doi.org/10.3112/erdkunde.2015.02.02

Babbie E (2020) The practice of social research. Cengage AU

Barro RJ, Sala-i-Martin X (1992) Convergence. J Polit Econ 100(2):223–251. https://doi.org/10.1086/261816
Bengtsson M (2016) How to plan and perform a qualitative study using content analysis. NursingPlus Open 2:8–14. https://doi.org/10.1016/j.npls.2016.01.001
Brinkmann S, Kvale S (2018) Doing interviews. SAGE
Buczyło M (2006) Akcja Wisła. Wypędzić, rozproszyć. Karta 49:32–63
Downe-Wamboldt B (1992) Content analysis: method, applications, and issues. Health Care Women Int 13(3):313–321. https://doi.org/10.1080/07399339209516006
Dubownik A, Churski P, Adamiak C, Szyda B (2022) Cohesion Policy in the struggle against the marginalization of the inner peripheries: polish experience and recommendations. In: Pradhan PKW, Leimgruber W (eds) Nature, society, and marginality: perspectives on geographical marginality. Springer, pp 271–290. https://doi.org/10.1007/978-3-031-21325-0_18
European Commission: Directorate-General for Regional and Urban Policy (2024) Forging a sustainable future together—cohesion for a competitive and inclusive Europe—report of the high-level group on the future of Cohesion Policy. Publications Office of the European Union. https://doi.org/10.2776/974536
Furceri D (2005) β and σ-convergence: a mathematical relation of causality. Econ Lett 89(2):212–215. https://doi.org/10.1016/j.econlet.2005.05.026
Gaskell G (2000) Individual and group interviewing. In: Bauer MW, Gaskell G (eds) Qualitative researching with text, image and sound. SAGE, pp 38–56
Hsieh H-F, Shannon SE (2005) Three approaches to qualitative content analysis. Qual Health Res 15(9):1277–1288. https://doi.org/10.1177/1049732305276687
Kowalski M (2000) Geografia wyborcza Polski: Przestrzenne zróżnicowanie zachowań wyborczych Polaków w latach 1989–1998 [Electoral geography of Poland: Spatial differentiation of Polish electoral behavior in 1989–1998]. IGiPZ PAN
Kvale S (2012) Doing interviews. SAGE
Monastiriotis V (2014) Regional growth and national development: transition in Central and Eastern Europe and the regional Kuznets curve in the east and the west. Spat Econ Anal 9(2):142–161. https://doi.org/10.1080/17421772.2014.891156
Rosnow RL, Rosenthal R (2013) Quantitative methods and ethics. In: Little TD (ed) The Oxford handbook of quantitative methods. Oxford University Press, vol 1, pp 32–54
Sim J, Waterfield J (2019) Focus group methodology: some ethical challenges. Qual Quant 53(6):3003–3022. https://doi.org/10.1007/s11135-019-00914-5
Solow RM (1956) A contribution to the theory of economic growth. Q J Econ 70(1):65–94. https://doi.org/10.2307/1884513
Visser P, Krosnick J, Lavrakas P (2000) Survey research. In: Reis H, Judd C (eds) Handbook of research methods in social and personality psychology. Cambridge University Press, pp 223–252
Williamson JG (1965) Regional inequality and the process of national development: a description of the patterns. Econ Dev Cult Chang 13(4, Part 2):1–84. https://doi.org/10.1086/450136
Wójcik P (2021) Parallel regional convergence in Poland before and after EU accession. Misc Geogr 25(2):83–92. https://doi.org/10.2478/mgrsd-2020-0050
Young AT, Higgins MJ, Levy D (2008) Sigma convergence versus beta convergence: evidence from U.S. county-level data. J Money Credit Bank 40(5):1083–1093. https://doi.org/10.1111/j.1538-4616.2008.00148.x
Zarycki T (2015) The electoral geography of Poland: between stable spatial structures and their changing interpretations. Erdkunde 69(2):107–124. https://doi.org/10.3112/erdkunde.2015.02.02

Chapter 7
Conclusions and Recommendations

Abstract In this chapter, we present final conclusions and operational recommendations based on the research findings outlined in this book. We begin by drawing the reader's attention to the findings that serve as the foundation for verifying the research hypotheses adopted in the study. We aim to demonstrate how the achievement of the empirical, methodological, and applicative objectives outlined in the book has led us to formulate these findings, which in turn formed the basis for confirming or falsifying the hypotheses. Furthermore, we offer our assessment of the impact that the identified facts and regularities may have on the potential incorporation of priority directions of contemporary Cohesion Policy transformations into development processes in inner peripheries. As part of our operational recommendations, we formulate both general and specific guidelines. The former category includes priority actions that, in our view, should be considered when shaping conditions conducive to the development of inner peripheral areas. These actions are organized according to the main contemporary transformations of Cohesion Policy, in response to the challenges of Cohesion Policy 2027+ .

In the latter category, we propose two specific operational changes related to the principles of programming and implementing development interventions financed under Cohesion Policy. The first concerns the use of emerging functional areas at the supra-local/subregional level as the primary actors and beneficiaries of these interventions. The second highlights the need to give greater consideration to the pursuit of territorial social justice in the programming and implementation of EU Cohesion Policy interventions. This includes the full application of integrated development planning and the use of an operational definition of the spatial social minimum. This approach entails a proposal to develop, at the European level, criteria for defining a spatial social minimum that, on the one hand, would ensure necessary and expected access to basic needs for every EU citizen while maintaining the required economic efficiency of intervention measures, and on the other, would provide a framework for a socially acceptable scale of spatial development disparities.

Keywords Final conclusions · Operational recommendations · Cohesion Policy 2027 + · Functional areas · Territorial justice · Spatial social minimum

P. Churski et al., *Inner Peripheries in the New Conditions of Cohesion Post-SARS-CoV-2*, Economic Geography,
https://doi.org/10.1007/978-3-032-10936-1_7

7.1 The Role of Inner Peripheries in Addressing Economic, Social, and Political Challenges in the Post-2027 Phase of the European Cohesion Policy, Following the SARS-CoV-2 Pandemic

We started our analysis with the country-wide study, which aimed to achieve the first empirical research goal (EG1), which was the delimitation and typology of spatial development disparities within the framework of FUAs, leading to the identification of inner peripheries. Considering our decision to use FUAs as basic units of analysis and the lack of a previous exhaustive delimitation of such areal units in Poland, we decided to produce our delimitation based on commuting and migration flows. We distinguished 413 FUAs, which we later used in the classification procedure to identify those that fall under the definition of inner peripheries. In the classification, we measured five types of territorial capital (development factors) by 47 metrics derived from public statistics and non-published statistical data, which we used to classify using cluster analysis and Gaussian mixture models. The construction of this framework of spatial units of analysis was the first step to achieving the methodological goal MG1—preparing a model indicator system to identify inner periphery areas and monitor the developmental processes occurring in these areas. FUAs delimited using our methodical approach are inherently volatile—a change in commuting flows may extend the functional area of some cities, shrink those of others or create new central cities around which zones of influence will be formed. However, the procedure may be re-applied with updated data to create more up-to-date delimitations. Also, other metrics of functional relations can be easily added to the algorithm or exchanged for the ones we applied (e.g., commuting to schools), which also makes the study procedure applicable in other national contexts. The classification of the FUAs into five types—core areas, two types of intermediate areas, and two categories of inner peripheries—added the next step to MG1. It also fulfilled MG2—the development of principles and criteria for the delimitation and typology of inner periphery areas. Again, future studies or ones conducted in another country may not bring entirely comparative results, due to the differences in source databases and the Gaussian mixture modeling method that renders differences in classifications due to even a minor change in input variable values. However, the principles and procedure of the classification are so general that it may be applied even if a different set of variables is available (provided that they represent five theoretical factors within territorial capital) or the researcher decides on another number of resulting clusters. The number of identified FUAs is similar to the number of second-level administrative division units of Poland, *poviats*. Many FUA boundaries, in fact, overlap with those of *poviats*. However, the key distinction between FUAs and *poviats* is that administrative units have roughly similar sizes, measured by the number of municipalities and area. In contrast, the sizes of FUAs vary between 2 and 92 municipalities, between 14 and 8,800 km^2, and between 7.5 thousand and 3.3 million inhabitants. This variability is highly correlated to the size of the central city. For example, the Warsaw FUA is the largest one in all regards. Such a trend is explained by relatively strong functional relations generated by larger cities, which reach further away from that city than in the case of smaller central towns. Also, transport infrastructure and services are more developed and centered around the central city in areas surrounding major cities. This explains why places located further away from them maintain functional links to larger cities rather than forming smaller FUAs around smaller towns. Major cities—centers of core cities—tend to have large areas of influence. In the case of intermediate areas, FUAs are often relatively small. In the peripheries, their sizes vary. Rare larger cities have extensive areas of influence, as they often do not neighbor zones of influence of other large areas, which exacerbates their peripherality. Smaller peripheral FUAs are organized around small towns, particularly the most remote ones beyond the zone of influence of large cities. It all corroborates hypothesis (H1), stating that the inner peripheries identified within the system of functional areas are characterized by varying sizes of both core cities and their surrounding areas. Moreover, the identified FUA arrangement—distinctive from the current administrative division into *poviats* yet supported by the functional connections—provides grounds for making adjustments to the administrative division at the subregional level.

After selecting case study areas, we looked at the recent dynamics of socio-economic indicators in the different types of areas, particularly in case study areas. In this, we aimed mainly to carry out EG2 to determine the developmental trajectories of inner periphery areas in relation to more-developed areas, with particular emphasis on their vulnerability and resilience to the SARS-CoV-2 crisis and its socio-economic consequences. Based on the literature review, we assumed (H2) that the development trajectories of inner peripheries differ from those of more-developed areas and exhibit greater sensitivity to the socio-economic consequences of the SARS-CoV-2 crisis. We designed a novel procedure for identifying development trajectories in order to achieve the methodological aim MG3 "Development of principles and criteria for identifying the developmental trajectories of inner periphery areas." This part of the research also aimed to contribute to the methodological aim of MG1 in monitoring development processes occurring in inner peripheries. A comparison of recent changes across 26 of 47 indicators related to four dimensions of territorial capital (excluding accessibility) leads to the general conclusion that peripheral regions not only continue to lag behind core and transitional regions in terms of human, socio-institutional, financial, and certain aspects of material capital, but that the gap between them is widening in some of these aspects. This is particularly evident in subcategories such as entrepreneurship, the labor market, local public finances, infrastructure, and demographic trends. Unemployment rates, the structure of economic activity, and population aging are three key variables in which the disparity between core and peripheral regions has significantly increased in recent years. In particular, unemployment demonstrates a clear divergence in response to the pandemic, while core regions have seen rates fall below pre-pandemic levels, peripheral regions continue to experience higher levels of unemployment. This trend appears to confirm hypothesis H2 partially. On the other hand, some indicators suggest a stabilization or even a narrowing of the gap between cores and peripheries. For instance, local finances and infrastructure metrics show signs of convergence, and wage disparities have remained largely unchanged. Therefore, divergence is not observed uniformly across all dimensions.

When performing in-depth studies on case studies with the use of document analysis and survey and interview data, we intended, among other goals, to (EG3) determine the intensity of the diffusion of developmental processes in spatial terms between inner periphery areas and urban centers serving as subregional hubs of functional areas, as well as the factors facilitating and hindering these processes. In particular, we assumed that (H3) the spatial spread and diffusion of development processes between inner peripheral areas and urban centers serving as subregional hubs of functional areas are constrained by factors unfavorable to these processes, which have gained prominence as a result of the SARS-CoV-2 crisis. To achieve this aim and verify the hypothesis, we measured the diffusion of development processes both within and between FUAs, identifying the following categories of flows as potential vehicles of this diffusion: commuting to work, permanent and temporary migration, commuting for education and services, goods delivery, and business cooperation. In terms of spatial direction, we distinguished intra-FUA relations, "horizontal" connections between FUAs and cities of similar size and economic potential, relations with higher-order cities, and international linkages. The majority of these flows—including commuting to work and school, as well as migration occur within FUAs. Besides the concentric flows between the central city and its surrounding areas, lateral and reverse flows (e.g., commuting from central city to a rural industrial park) are also emerging, reshaping but maintaining strong internal FUA linkages. External functional connections bring socio-economic benefits, such as capital inflow, opportunities through international trade and supply chains, and return migration of individuals with education and professional experience. They also bring challenges, including the outflow of human capital by student migration that often results in permanent emigration. Most travel within and out of the FUA is car-dependent, partly due to insufficient public transportation infrastructure and services. Where public transport exists, it primarily serves vertical (hierarchical) connections to FUA centers or regional capitals, but rarely supports horizontal links, which constrains the development of broader functional relationships. The pandemic temporarily restricted this access even further. Hopes that remote work might open new opportunities, halt emigration, or attract new residents ultimately were not realized. All this contributes to the positive verification of H3.

As part of an in-depth study of selected case studies, we aimed to determine (EG4) the vulnerability levels of individual industries and economic sectors to the shocks induced by the SARS-CoV-2 pandemic crisis and to identify both the observed and projected economic and social consequences of this developmental shock. We formulated the hypothesis (H4) that inner peripheries are characterized by a specific vulnerability of their economic structures to pandemic-related shocks, which results in a different course and intensity of crisis effects compared to more-developed areas. To address this objective and test the proposed hypothesis, we conducted an analysis of changes in economic activity and structural dynamics within the studied inner peripheries. For this purpose, we developed a monitoring system for socio-economic changes in inner peripheral areas (MG4) (Fig. 7.1). The system's design emphasizes the necessity of conducting both quantitative analyses of territorial capital features and qualitative analyses of documents, opinions, and assessments using tools such as CATI (Computer-Assisted Telephone Interviewing), IDI (In-Depth Interviews), and FGI (Focus Group Interviews). In our view, the two analytical stages (quantitative and qualitative) complement each other and collectively enable the development of a comprehensive geographic information base. We also emphasize that the commonly occurring lack of data at the quantitative analysis stage, particularly at the local and subregional levels, should be addressed through the implementation of well-planned and meticulously executed primary research efforts, which provide unique source material on the territorial specificities of the areas under study. We propose structuring the acquired source information on the state and transformation of socio-economic conditions in inner peripheries according to the dimensions of territorial capital adopted in our framework: accessibility, human capital, social and institutional capital, financial capital, and material capital.

We established that the local economies of peripheral areas are predominantly composed of micro- and small-sized enterprises, operating mainly in processing industries (primarily agri-food) and in services that require physical presence (e.g.,

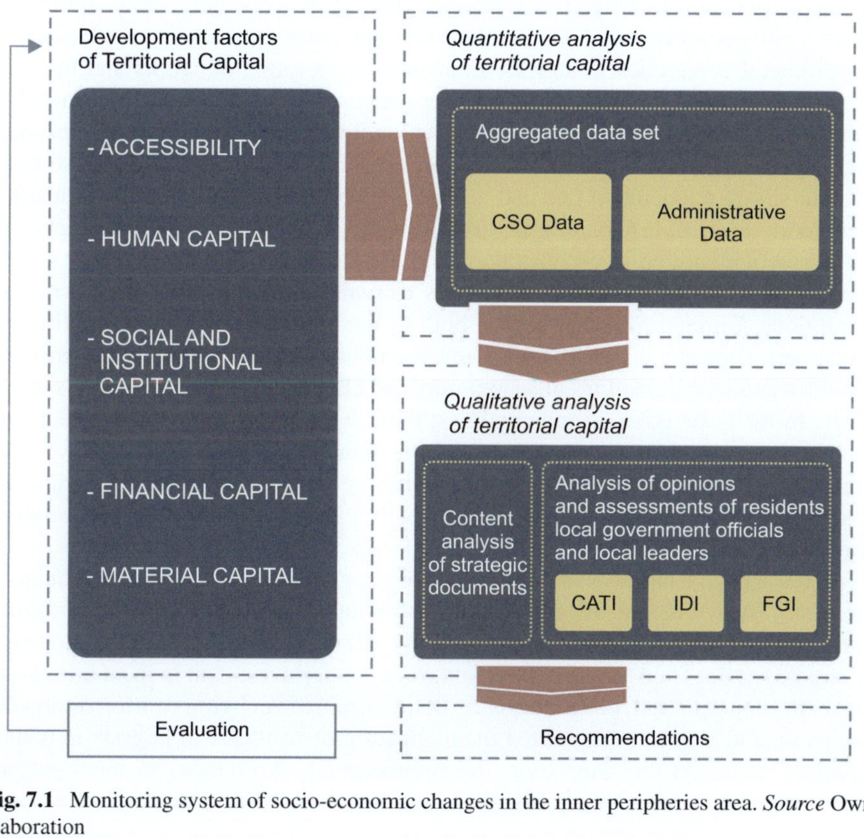

Fig. 7.1 Monitoring system of socio-economic changes in the inner peripheries area. *Source* Own elaboration

transport, tourism). These sectors proved to be the least resilient to the pandemic shock. Many entities were forced to temporarily suspend or permanently cease operations as a result of lockdown measures, leading to a sharp increase in unemployment and a decline in local economic activity. In contrast, sectors not requiring on-site labor poorly represented in peripheral areas and those with stable public funding (e.g., public administration) experienced a significantly milder impact. Overall, however, the economic crisis had a more severe course in inner peripheries than in developed urban centers, confirming the specific vulnerability of these areas to the pandemic.

The socio-economic consequences of the pandemic revealed several weaknesses in the territorial capital of inner peripheries. A key differentiating factor in resilience was human capital. The studied peripheral areas have long experienced an outflow of young and educated residents, and the local educational offer particularly in technical and vocational training does not fully align with labor market needs. This resulted in a shortage of specialized competencies. As a consequence, adaptation to new conditions during the pandemic (e.g., transitioning to remote work or changing business profiles) was limited. The lack of qualified personnel forced some firms to relocate parts of their operations to larger cities or reduce employment. The low level of human capital in the peripheries also constrained the capacity to absorb innovation and implement new technologies, further weakening their competitiveness during the crisis. Deficiencies in social and institutional capital also increased the vulnerability of peripheral economies to shocks. Economic cooperation networks and business support institutions (e.g., clusters, chambers of commerce, local development agencies) are underdeveloped in these areas, which limits knowledge transfer, resource exchange, and mutual support among enterprises. During the crisis, local businesses without strong cooperative ties had to face demand collapse on their own, making them more susceptible to bankruptcy. Furthermore, deficiencies in labor market institutions were identified in peripheral areas. Poor coordination between labor supply and demand hindered effective responses to rising unemployment during market disruptions. Limited financial capital both in the private sector (low liquidity and investment capacity of small firms) and the public sector (low local government revenues and constrained fiscal space) impeded efforts to mitigate the effects of the crisis. In many peripheral municipalities, there were insufficient funds to actively support the local economy, further depressing demand and investment levels. At the same time, local households' reduced purchasing power weakened their ability to absorb the shock through consumption, which widened the financial gap between core and peripheral regions.

Material capital and accessibility factors also significantly influenced the course of the crisis. Weaker technical and social infrastructure in peripheral areas (e.g., slower internet, underdeveloped transport, and utility infrastructure) hindered the provision of remote services and the shift to remote work. Underinvestment in local infrastructure both economic and related to quality of life meant lower living comfort during the pandemic and may have increased the inclination of residents to migrate to better-equipped areas. At the same time, the environmental advantages of inner peripheries' clean air, access to green spaces, and lower population density contributed to an improved subjective quality of life during lockdowns. However, the potential of these assets for local development remained underutilized due to the lack of complementary tourism infrastructure and weak promotional efforts.

The pandemic thus highlighted the distinct trajectory and intensity of crisis impacts in inner peripheries compared to more-developed regions, confirming the validity of hypothesis H4 on the specific vulnerability of the economic structures of inner peripheries to pandemic shocks. The findings of this analysis underscore the need to develop dedicated instruments aimed at enhancing the resilience of peripheral economies to similar shocks in the future.

The results of in-depth research conducted in accordance with the monitoring system for socio-economic changes in inner peripheral areas enabled us to identify the developmental factors and barriers shaping socio-economic transformations that determine the living standards and conditions of residents in selected inner peripheries, particularly in the context of the consequences of the SARS-CoV-2 crisis shock (EG5). We hypothesized (H5) that social and institutional capital is the most significant factor influencing socio-economic transformations that determine the level and quality of life in inner peripheral areas, and that its importance increases under the impact of the SARS-CoV-2 crisis shock. To pursue this objective and verify the hypothesis, we applied the socio-economic change monitoring system (MG4) developed in our research and conducted in-depth analyses of local and regional strategic documents, survey research with residents, and interviews with local leaders. Based on this approach, we examined the status and evolution of development factors in inner peripheries in the face of the developmental shock caused by the SARS-CoV-2 pandemic.

The crisis triggered by the SARS-CoV-2 pandemic significantly exposed and deepened existing developmental barriers in inner peripheries, revealing their particular vulnerability to socio-economic disruptions. In many respects, the pandemic not only intensified pre-existing issues but also effectively halted development opportunities in these areas. One of the barriers that was exacerbated during the pandemic was limited transport accessibility. The implementation of public health restrictions led to the reduction of public transport services, including a decreased number of routes, suspension of connections, and the discontinuation of unprofitable lines. For residents without alternative means of transportation, this resulted in heightened transport exclusion and social isolation. Consequently, population mobility was further diminished, limiting access to the labor market, education, and public services. Another identified developmental barrier was insufficient digital infrastructure (e-accessibility). The pandemic exposed the extent of digital exclusion affecting both residents and local entrepreneurs. The lack of access to broadband internet and appropriate technological resources prevented many individuals from participating in remote work, education, and e-services. Digital exclusion also impacted local businesses, preventing them from adapting to the new reality, weakening their competitiveness, and obstructing the development of innovative entrepreneurship.

One of the most severe consequences of the pandemic was the weakening of local labor markets, leading to the underutilization of human capital. The relatively lower quality of this capital, along with limited flexibility in adapting to rapidly changing conditions, became starkly apparent. A decline in demand for goods and services led to job losses and the closure of many local businesses, directly increasing unemployment and income insecurity among residents. An additional developmental barrier was the employment structure. The dominance of occupations requiring physical presence at the workplace meant that only a relatively small portion of the workforce could transition to remote work. The lack of labor market flexibility and the low diversification of employment significantly limited the capacity of peripheral areas to adapt under crisis conditions. The pandemic also deepened pre-existing problems in healthcare accessibility, further weakening the quality of human capital. The restriction of medical services, suspension of planned procedures, interruption of treatment continuity, and chronic staff and infrastructure shortages in healthcare facilities revealed the long-standing underinvestment in healthcare systems in peripheral regions. Moreover, the SARS-CoV-2 pandemic exposed the low adaptive capacity of public institutions operating in peripheral areas. Difficulties in organizing healthcare, adapting to new legal regulations, and communicating with residents pointed to insufficient preparedness of local administrations to operate under non-standard conditions. The resulting outcomes included a decline in the health security of residents and a deterioration of the overall population health status even among individuals with non-COVID-related conditions, which may contribute to a long-term decrease in quality of life. Attention should also be drawn to the worsening demographic situation. The pandemic accelerated depopulation and population aging—two of the most significant barriers to development in inner peripheries. Economic and health insecurity led many young people to delay starting a family, while increased mortality among older adults further deepened population decline—an effect that was not offset by the temporary inflow of urban residents. As a result, the available human capital resources in these areas became even more limited.

The results of the conducted research confirmed our initial assumption regarding the priority role of social and institutional capital in shaping the development processes of inner peripheries. For this reason, we paid particular attention to the social resources of territorial capital in the studied inner peripheral areas. It is important to emphasize that these resources are fundamentally built upon human capital, which, as previously indicated, exhibits various dysfunctions in these areas—conditions that significantly influence the functioning of social and institutional capital.

Our findings confirm that this type of social capital is strongly conditioned by local identity and sense of place attachment. The data indicate that inner peripheries are predominantly characterized by bonding social capital, which plays a key role in reinforcing local communities during times of crisis. However, this primarily applies to pre-existing groups connected through shared values and interpersonal ties.

Unfortunately, our research also identified that inner peripheries suffered a deficit of bridging social capital—the type of capital that enables cooperative capacity beyond the local level. We consider this form of capital essential for overcoming developmental barriers and establishing durable relationships with external stakeholders. Meanwhile, the findings revealed high levels of trust and solidarity within tightly knit social groups based on familial or neighborhood ties. While this intra-group trust is helpful in crises (e.g., neighborly assistance during lockdowns), it does not translate into the creation of broader coalitions or the development of new economic cooperation networks. We also observed a low level of collaboration with external partners, including entrepreneurs and civil society organizations from outside the municipality or region. This limits access to know-how and restricts the diffusion of innovation, which is especially crucial in the context of rapidly changing market conditions induced by the pandemic. In the context of the co-occurring war crisis in Ukraine, we noted positive relief actions directed at refugees. However, inner peripheries did not stand out in this regard compared to communities located in more-developed core regions. A fundamental barrier in the domain of institutional capital appears to be the lack of trust and constructive communication between local government authorities and entrepreneurs or other local leaders. The absence of transparent, jointly developed procedures significantly restricts the potential for cooperation, thereby posing a major obstacle to the development of inner peripheries. A key challenge remains the intensification of intermunicipal cooperation and the implementation of professional territorial marketing for FUA, which we have classified as inner peripheries in our analysis. Our interpretation of the findings allows us to confirm the validity of hypothesis H5.

The obtained results allow us to formulate recommendations regarding the direction and implementation methods (including the selection of instruments) for development interventions under Cohesion Policy in inner peripheral areas, with particular emphasis on the role of their urban centers in creating conditions for networking and the diffusion of development processes (AG1). These recommendations are organized according to the proposed directions of change in EU Cohesion Policy, which we derived from the ongoing discourse in the academic literature and on political platforms at various governance levels, as presented in Sect. 3.2.3. We place particular emphasis on the degree of alignment between the observed developmental patterns in inner peripheries and the rationale behind these proposed policy directions, highlighting both the areas of convergence and the domains where implementation remains incomplete, as well as the barriers encountered in inner peripheral areas.

The results of the conducted research emphasize the need to tailor regional development policy to the specific characteristics of each territory, including inner peripheries. As a priority, efforts should focus on strengthening the long-term resilience of each area. However, given the limited internal resources of inner peripheries, achieving this objective can be particularly challenging. Findings from the analysis indicate that, without targeted measures aimed at enhancing the resilience of these territories, they will be unable to adequately benefit from broader growth trends, and their developmental lag relative to core regions may continue to widen.

In this context, core cities within FUAs should play a key role. As local development poles, these cities should actively stimulate the diffusion of development processes into their surrounding hinterlands. This implies strengthening functional linkages including transport, economic, and social connections with smaller centers in the region, as well as serving as coordination hubs for pro-development actions. Core cities, equipped with relatively stronger resources (such as institutions, labor markets, and human capital), can make a tangible contribution to economic diversification and enhanced resilience in peripheral areas. They should therefore initiate and support initiatives that integrate the combined potentials of the entire functional area. Such an integrated approach supports overcoming development barriers in areas overshadowed by stronger centers and helps to mitigate the negative effects of excessive spatial polarization.

The promotion of inclusive development opportunities for all, combating economic exclusion and poverty requires integrated planning of both transport infrastructure and services. Efficient transport connections support commercial relations and flexible work arrangements that allow individuals who seek jobs outside of peripheral areas to maintain ties to their home regions. Yet, this planning should ensure connectivity not only between peripheral FUAs and economic centers but also within individual FUAs, particularly between the central city of a peripheral FUA and its surrounding areas. Such a connection enables the central city to take an active role in integrating the functional area both functionally and in terms of development policy, stimulating its economic growth and safeguarding the provision of services including public services such as education and healthcare for the residents' entire area. While car travel remains the preferred mode of transportation, greater emphasis must be placed on public transportation, which is often economically unsustainable in less populated areas. Innovative solutions, such as on-demand public transport, could be introduced to address this challenge. Within smaller FUAs, bicycle transport could also be a practical solution for short-distance travel. However, its potential is currently limited by the underdevelopment of safe infrastructure—particularly dedicated bike lanes. Mobility systems adapted to dispersed settlement patterns can help overcome isolation by enabling older adults to access healthcare services, youth to participate in educational activities, unemployed individuals to seek employment, and workers to commute to service and employment locations. This can effectively enhance the utilization of local human capital and potentially reduce its outmigration. From the perspective of social and institutional capital, transport is not merely a physical connection it is, above all, a tool for relationship-building. The design of transport systems for functional areas can stimulate dialogue and create sustainable mechanisms for long-term collaboration. Initiatives that involve employers or civil society organizations in organizing transport (e.g., employee shuttles, school buses managed by local cooperatives) help mobilize social capital and strengthen reciprocity networks. However, financial and material deficits limit the capacity of municipalities and FUAs to invest in transport and service infrastructure. Even when external funding is available, constraints related to local co-financing hinder both the implementation and maintenance of services. Existing investment backlogs in road infrastructure, public transport fleets, and digital networks remain significant. Therefore, it is essential to integrate physical and digital accessibility solutions, provide adequate financing mechanisms for rural transport, and promote cross-sectoral planning that links mobility with education, healthcare, and labor markets.

Strengthening the adaptive capacity and agency of communities residing in inner peripheries is crucial for their development. This process should be based on a comprehensive identification of local developmental deficits and the construction of new, durable systems of functional relationships. It must be initiated and implemented from the bottom up, relying on the social engagement of all key local and supra-local development stakeholders, while simultaneously fostering a sense of co-responsibility for both the directions and outcomes of undertaken actions. Equally important is the creation and expansion of access not only in physical terms, but also in informational and institutional dimensions. A lack of platforms for cooperation and participation often leads to a sense of exclusion and social passivity. Therefore, the openness of institutions and support for building a civil society are essential preconditions for local mobilization and the development of a shared sense of agency. Peripheral communities are often characterized by strong bonding social capital, yet they lack bridging capital i.e., external relationships and openness to change. The loss of young, educated residents weakens the base of local leadership, while deficits in civic and digital education hinder citizen engagement in local life. To reverse this trend, it is necessary to support grassroots initiatives, create conditions for cooperation, and strengthen institutions as social partners. Financial and material shortages, especially in social infrastructure (libraries, cultural centers, public spaces), mean that even active communities lack the means to pursue their plans or realize their initiatives. Support for grassroots initiatives should be systematic, easily accessible, and understandable free from bureaucratic barriers. Investments in social infrastructure, combined with skills development and micro-financing, can rapidly result in increased civic engagement and a tangible influence of residents on local development directions. In the implementation of these efforts, the core cities within FUAs play a key role, as they are perceived as regional growth poles. Their active involvement in shaping development across the FUA, as well as in building and implementing territorial development strategies, requires not only sufficient endogenous capacity, but also the ability to generate and effectively transfer development impulses to peripheral areas, and to initiate collaboration among local governments and other stakeholders. Cities serving as FUA cores should act as leaders of regional development, initiating and coordinating institutional, economic, and social cooperation processes. A key aspect of this role involves the development of functional network connections, including collaboration through public–private partnerships or innovation platforms. An essential component of this function is also the capacity to transfer technology or know-how to cooperating areas, thereby enhancing the competitiveness of the entire FUA and strengthening its integration with the broader regional context.

From the perspective of inner peripheries identified within FUAs, a key priority is to overcome barriers that limit the diffusion of growth and development impulses from core cities to their surrounding areas. Central urban centers possess a relatively high concentration of territorial capital, including infrastructure, public services, a dynamic labor market, and business support institutions, which enables them to stimulate development in neighboring peripheral municipalities. Main cities function as network hubs through which resources, knowledge, and innovation flow; their influence is thus crucial for the diffusion of development processes into less developed peripheral areas. Currently, the dynamic growth of major urban centers does not always translate into improvements in the conditions of adjacent areas, thereby intensifying spatial disparities. At the same time, it is essential to mitigate the negative effects of agglomeration, which are increasingly evident in both core areas (e.g., congestion, rising operating costs) and surrounding zones (e.g., the vicious cycle of stagnation and decline). A viable solution lies in the broader use of functional areas as the main entities for planning and implementing supra-local interventions. Planning interventions at the functional area level fosters stronger linkages between cores and their hinterlands, improving the effectiveness of development efforts and enabling wider dissemination of growth impulses to peripheral regions. The new approach in EU Cohesion Policy, which positions FUAs as supra-local, empowered structures, fully supports our findings. On the one hand, FUAs possess the capacity to integrate human, financial, and institutional capital and to diffuse innovation; on the other hand, they face a chronic lack of formal mechanisms for intermunicipal cooperation, shared strategies, and coordination platforms. To effectively reduce inequalities and strengthen economies of scale, functional areas should become the main entities responsible for programming and implementing supra-local interventions. Financial support should be directed toward the creation of legal and institutional frameworks that enable FUAs to plan, co-finance, and manage infrastructure investments and public services in an integrated and complementary manner. Core cities can serve as connectors, integrating peripheral areas into broader socio-economic processes and providing residents of remote areas with improved access to markets, specialized services, education, and employment opportunities.

Decentralization of development interventions and the strengthening of local institutions have become an urgent necessity, as particularly evidenced by the situation of inner peripheries identified in our research. The decentralization of development efforts requires not only an expansion of legal competencies, but also the systematic reinforcement of local institutions. The objective of these efforts should be to build institutions capable of efficient, multi-level communication with all stakeholder groups both within the governance system and in their interactions with local communities. It is essential to develop and enhance staff competencies, foster trust in institutions, and create the conditions for a civil society that is aware of and engaged with the functional area context. Given the financial and infrastructural limitations of peripheral areas, fiscal strengthening and operational support are critical. Without adequate resources, newly assigned competencies to local governments risk remaining purely declarative and not implementable in practice. Therefore, decentralization must be accompanied by sustained financial support, simplified administrative procedures, and capacity-building at the local level, including through partnership-based training and advisory programs.

To improve the effectiveness and efficiency of the future EU Cohesion Policy, it is essential to strengthen supra-local cooperation, which should be particularly promoted within FUAs. The deficits we identified in this regard in inner peripheries must be recognized as one of the key barriers to their development. The development of relational capital at the supra-local level including across administrative boundaries is necessary to enhance the position of areas historically marginalized due to their peripheral location relative to regional cores. Local governments in peripheral areas often act in isolation, focusing on their relationship with larger cities instead of developing their own support networks. Joint projects could strengthen these areas' bargaining power for resources and support the development of functional niches (e.g., clusters, shared service networks), but they require coordination capacity and the ability to overcome institutional barriers.

Another barrier to cooperation is the insufficient transport infrastructure and public transport services within FUAs, particularly between peripheral and intermediate FUAs. Current efforts to improve transport accessibility in peripheral FUAs are primarily focused on their connection to large cities key economic growth centers. However, it is equally important to enhance the potential for development impulse diffusion between FUA core cities and their outer zones, as well as between neighboring functional areas. Strengthening these interconnections can foster cooperation, knowledge exchange, and more effective use of complementary resources and infrastructure. However, financial and material capital in these areas remains fragmented, reducing the efficiency and impact of development interventions. There is a lack of financial instruments to support intermunicipal cooperation among peripheral municipalities, and existing mechanisms still favor vertical integration (periphery–center) at the expense of horizontal collaboration. It is therefore necessary to create funding instruments that incentivize territorial cooperation, promote joint infrastructure investment planning, and support the development of shared intermunicipal services a task for the next phase of Cohesion Policy. In our assessment, such measures would enable more effective use of public funds allocated for development interventions. Overcoming the isolation of peripheral areas should not result in stronger subordination to central cities; instead, it should involve the creation of alternative, network-based development models built on reciprocity and complementarity.

The conclusions outlined above lead us to the finding that ensuring the effectiveness and efficiency of cohesion policy interventions targeted at inner peripheral areas requires, to a greater extent, changes in implementation instruments rather than in the programming principles of the policy itself (hypothesis H6), which allows us to positively verify this hypothesis. In this context, the final subsection of our book presents two proposals for operational changes related to the programming and implementation of development interventions financed under the cohesion policy framework.

7.2 Toward a More Effective and Efficient Cohesion Policy: Recommendations for Operational Changes

The results obtained, along with their alignment with the assumptions of Cohesion Policy 2027+ , provide a solid basis for formulating recommendations concerning the operational dimension of actions undertaken within this policy framework. Based on this, we argue that ensuring the effectiveness and efficiency of cohesion policy interventions aimed at inner peripheral areas more greatly requires that implementation instruments be modified than that the fundamental principles of policy programming be altered. This conclusion leads to a positive verification of hypothesis H6 (Chapter 4). In our view, the findings support the formulation of several general recommendations regarding the direction and implementation of interventions targeting lagging areas, including internal peripheries. We also consider it necessary to introduce two key changes that could significantly enhance the effectiveness and efficiency of development efforts aimed at reducing spatial development disparities while fostering competitiveness and territorial cohesion across the European Union. These changes concern: (1) the greater integration of the supra-local level in particular, the use of FUA as key entities in the programming and implementation of cohesion policy and (2) the enhanced incorporation of territorial social justice objectives into the programming and implementation of EU Cohesion Policy interventions, achieved through full use of integrated development planning and the application of an operational definition of the spatial social minimum. All of the proposed changes aim to improve the implementation of cohesion policy within the framework of the prevailing place-based policy paradigm, which we regard as the most effective and efficient approach for reducing spatial disparities in development.

As part of the general recommendations supported by the findings of our research and aligned with conclusions drawn by other authors in the relevant literature, we propose the adoption of a dedicated set of measures for peripheral areas. These measures are intended to better orient cohesion policy interventions toward the following strategic priorities:

- Enhancing the resilience of peripheral areas through comprehensive investments in transport and digital infrastructure. These areas require not only improved physical accessibility, but also high-quality digital infrastructure and the development of residents' digital competencies, which are crucial for the transition to a knowledge-based economy and for building long-term resilience (Bo et al. 2021; Gnap et al. 2021). Complementing technical infrastructure with mobility solutions such as on-demand transport can help reduce both geographical and economic isolation (Lestari and Dewi 2022; Yuliana 2021).
- Supporting economic diversification to reduce dependency on a single sector, thereby creating opportunities for income diversification and economic mobility among residents (Adegbie et al. 2022; Creutzig et al. 2022; Sotiroski 2023). The creation of investment incentive schemes, such as tax relief or dedicated support funds, can attract entrepreneurs and increase the competitiveness of local economies. In particular, the development of local value chains based on regional resources should be promoted to strengthen their long-term competitive advantages.
- Prioritizing investments in human capital, especially in education and vocational training. Programs aimed at counteracting youth outmigration, such as local scholarships or support for youth entrepreneurship, are essential to mitigating depopulation trends in peripheral areas. Education systems should be tailored to the

specific needs of regional and local labor markets to retain skilled workers and reduce demographic pressures (Podgorskaya 2021).

- Improving access to high-quality public services in peripheral areas. The development of telemedicine, mobile health services, and digital administrative tools can reduce barriers to healthcare and administrative services. These innovations lower operational costs while enhancing quality of life for residents (Haefner and Sternberg 2020).
- Strengthening institutional capacity at the local level in managing projects and cohesion policy funds. Institutional deficits are among the main barriers to effective implementation of development actions (European Commission: Directorate-General for Regional and Urban Policy 2024). Capacity-building programs for local leaders and local government units should be an integral component of interventions.
- Promoting interregional and transnational cooperation, which generally increases the effectiveness of development efforts (Sarin et al. 2020). Supporting functional areas as platforms for interregional projects will enhance spatial and functional integration. In particular, mechanisms for cross-border collaboration and the exchange of best practices should be developed.
- Placing greater emphasis on green transition in peripheral areas. Investments in renewlong-term resilienceable energy, the circular economy, and green technologies can not only improve local environmental conditions but also generate employment and strengthen local economies—thereby contributing to while unlocking often dormant regional potential (Bo et al. 2021).

In our view, the implementation of the above recommendations in the operational layer of programming and delivering Cohesion Policy interventions would benefit significantly from the broader application of the functional area concept. From the perspective of the development needs of all European regions, it is essential to recognize the importance of strengthening urban–rural linkages within the development paradigm for less developed areas, and to overcome the developmental barriers of regions caught in a path-dependent development trap. Cities, as local and supra-local growth poles, should reinforce their connections with both their immediate and extended hinterlands, consolidating and expanding the scope of functional relationships. They should also serve as coordination centers for actions aimed at diversifying local economic structures, thereby contributing to the resilience of supra-local socio-economic systems. Actions in this area must be programmed with a focus on reducing the negative effects of excessive spatial polarization and creating conditions for the diffusion of growth. All of these objectives can be achieved through the full alignment of intervention activities with the territorially oriented policy approach, based on objectively identified endogenous resources and the promotion of supra-local cooperation within functional areas. Strengthening supra-local cooperation, including interregional collaboration (beyond administrative boundaries), should be based on the development of supra-local strategies, the implementation of which must be embedded in revised legal frameworks that confer appropriate competencies and guarantee funding sources for tasks carried out within functional areas. This would establish a lasting foundation for improving the functioning of these areas across the administrative boundaries that often divide them. Actions in this regard should also create the conditions for the institutionalized development of Functional Urban Areas involving cities of various sizes. This proposed modification is supported by the following arguments:

- Increased effectiveness of cohesion policy through:

- integration of actions around shared development potentials and opportunities (e.g., economic specialization), as well as around common development problems and challenges (e.g., border regions along the external frontiers of the EU adjacent to the Russian Federation) of local administrative units (LAUs);
- creating conditions within functional areas for generating synergies between the resources and territorial capital potentials of local units (LAUs). This can both overcome development barriers and limitations that individual units are unable to address alone, and enable the implementation of initiatives that exceed the financial and institutional capacity of single LAUs;
- facilitating the diffusion of development processes by stimulating the positive influence of central nodes within functional areas, addressing their local development barriers, mitigating their negative externalities (e.g., resource extraction or "drain" effects), and enhancing the adaptive capacity of surrounding areas to absorb and benefit from development impulses over the long term;
- integration of local socio-economic systems at the supra-local level, including labor markets, housing markets, educational services, healthcare services, etc., to optimize their functioning both economically and organizationally;
- operationalizing the place-based policy approach through the decentralization of competencies and financial resources, with an emphasis on supra-local cooperation within functional areas. This would create the conditions for planning and implementing development interventions at the supra-local level, with guaranteed funding;
- preparation and implementation of supra-local development strategies aligned with the principles of territorial development plans for functional areas;
- strengthening results-oriented approaches at the supra-local level through the preparation and execution of evaluation studies specifically tailored to functional areas.
- Enhancing the effectiveness of cohesion policy through:
- creating opportunities to implement development-oriented actions dedicated to a broader range of potential beneficiaries;
- reducing the number of isolated, small-scale projects with very limited impact;
- increasing the number of supra-local, networked initiatives, both in terms of infrastructure (e.g., bicycle paths) and in terms of operational integration (e.g., the coordination of public transport systems);
- enabling the formation of purchasing groups, which create favorable conditions for negotiating prices of raw materials, energy, and services (e.g., electricity or

natural gas purchasing cooperatives, supra-local thermal modernization programs, etc.);
- allowing for differentiated internal development support within functional areas, taking into account variations in available resources and territorial capital potentials, in line with the place-based policy approach and using instruments of internal territorialization of actions applicable within supra-local development strategies;
- counteracting exclusion, peripheralization, and marginalization of areas which, lacking independent development capacity outside of cooperation within the functional area, risk becoming dependent on solidarity-based support and permanently losing their ability to function and develop autonomously.

The implementation of the proposed modification to the current practice of cohesion policy delivery requires the introduction of legal and organizational regulations that enable supra-local cooperation aligned with the spatial scope of emerging functional relationships, without the need for fundamental changes to the territorial division or governance structures of EU Member States. A key requirement in this regard is the establishment, within the legal systems of each Member State, of formal mechanisms allowing the pooling of local government budgets, and the ability to allocate those funds to initiatives carried out not only within their own administrative boundaries, but across a supra-local functional area. These legal changes must also include provisions allowing for the implementation of local tasks at the supra-local level. Ensuring such conditions would enable the programming and implementation of supra-local development strategies, with guaranteed funding. Supra-local development strategies should serve as the territorial equivalent of development plans for functional areas and should be prepared and implemented in full alignment with the principles of the place-based policy approach. We view these changes as an opportunity for more effective and efficient intervention in addressing spatial development disparities, potentially leading to improvements in the situation of inner peripheral areas, and in optimizing the programming and implementation of development interventions.

We maintain that the current practice of programming and implementing EU Cohesion Policy should evolve to place greater emphasis on territorial social justice within its intervention logic, fully utilizing the potential of integrated development planning and applying an operational definition of the spatial social minimum. To effectively integrate the principles of territorial justice into more efficient and impactful cohesion policy interventions aimed at strengthening economic, social, and territorial cohesion, three key conditions must be met:

- The first condition involves the need to perceive space not merely as a Cartesian physical substrate for development processes, but also as their outcome—a product of users' engagement and capacity to shape their environment. This leads to the conclusion that justice, as a result of social relations, is deeply intertwined with the spatial dimension that conditions those relations across places (Dikeç 2001), linking directly to Lefebvre's (1968) reflections on the "Right to the City" and spatial inequalities in development. In this view, space is co-created by its users, who organize and exercise control over it—an idea aligned with the concept of territory not as "space-as-place" but as "space-as-actor" (see Chapter 2). Space is thus produced through relational territorial capital, shaped by locally available assets that can offset structural deficiencies (Camagni 2008). Within this context, a territorially oriented cohesion policy, in which the concept of relational territorial capital plays a central role (Churski et al. 2021), becomes particularly suited to fostering territorial justice. This is especially relevant for functional areas, where the goal should be to enhance positive interactions both within core urban areas and between these cores and their wider hinterlands, thereby reducing the exclusion, marginalization, and peripheralization of people and territories.

- The second condition directly relates to the previously mentioned dilemma between equalizing outcomes versus equalizing opportunities (see Chapter 2). Addressing this dilemma requires recognizing that development-oriented actions aimed at territorial justice should not merely focus on leveling economic results (e.g., income indicators) (Rauhut 2017). More important is the creation of capabilities, which entails ensuring freedom of choice and access to basic goods and services for all individuals. Cohesion policy interventions should therefore establish conditions for people to change their life situations and access essential services—a prerequisite for equal opportunity and the foundation of territorial justice. Without access to fundamental resources, individuals suffer from capability deprivation (Sen 1999). Overcoming this condition cannot rely solely on market forces; it requires targeted interventions (Rauhut & Hatti 2005) that support the developmental potential of every resident. Opportunity inequalities reduce social mobility across generations and perpetuate long-term spatial disparities. In peripheral and marginalized areas, this can result in entrenched poverty, a cycle of marginalization, and deteriorating socio-economic conditions—including depopulation, reduced local revenues, falling investment levels, and declining public service provision. Enhancing supra-local development effects by removing barriers in cities of all sizes, stimulating their influence on surrounding areas, and preparing economically weaker territories to benefit from these effects through engagement of their endogenous territorial capital can help reverse these negative trends. It may also prevent "sunk-cost" effects, which drive away existing and potential investors, further limiting development prospects. The territorially integrated approach of cohesion policy, based on the multidimensional identification of local resources and development potentials, appears optimal for designing interventions that level the playing field for lagging regions, including inner peripheries. This approach supports targeted responses to local deficits by ensuring access to core services such as education and healthcare, fulfilling the basic needs of residents. These efforts align with Rawls' (1971) idea of a social minimum, Dworkin's (1981) theory of justice, and Smith's (1994, 2000) geographic interpretations of social justice and serve as a foundation for operationalizing the concept of the spatial social minimum.
- The third condition concerns the need to establish a legal basis for the full implementation of integrated development planning, as well as for the programming and execution of interventions at the supra-local level, including across administrative

boundaries such as NUTS-2 regions (especially relevant for inner peripheries situated along such borders). This requires granting legal institutional and financial empowerment, which would enable actions such as:

- the full integration of socio-economic and spatial planning, along with the institutional empowerment necessary for the comprehensive implementation of development projects planned under such an integrated framework;
- the planning and implementation of supra-local cooperation within functional areas, extending beyond regional capital areas to also include functional areas of medium-sized and small towns. This requires legal reforms that guarantee the competences and financial means for implementing jointly agreed development initiatives by groups of relationally connected local administrative units, which would be granted the right to pool their tasks and budgets.

In our view, it is essential to create the conditions for utilizing the supra-local level as both the architect and executor of intervention measures. This is not feasible without further decentralization and the provision of the technical assistance necessary to build institutional capacities at the local level (European Commission: Directorate-General for Regional and Urban Policy 2024). Current capacity deficits hinder full participation in interventions, prevent responsible multi-level governance, and ultimately impede the integration of processes, documents, and outcomes in development planning across territories that differ significantly in terms of available resources (Pinheiro et al. 2022).

The implementation of the proposed modifications should be reinforced by introducing the concept of a spatial social minimum into the operational practice of EU Cohesion Policy. As previously mentioned, this idea directly refers to Rawls' (1971) and Dworkin's (1981) concepts of a social minimum, Smith's (1994, 2000) geographical interpretation of social justice, and, in the Polish context, Gorzelak's (1985) notion of a spatial social minimum. As early as the mid-1980s, Gorzelak (1985) advocated for the establishment of a minimum threshold of socially recognized accessibility standards to key elements of social infrastructure including housing, retail services, and environmental quality indicators as a means of identifying areas requiring urgent intervention. In our view, this spatial minimum should be understood more as a social rather than merely welfare-based concept, and should serve as a reference point for territorial development policy to identify areas in particular need of intervention to ensure social acceptance among residents of diverse socio-economic spaces. While defining criteria for such a minimum is undoubtedly more complex than for an income threshold (Gorzelak 2021), it is essential to recognize that it must go beyond technical measures and include the empowerment and participation of the residents whose development opportunities it aims to shape (Weckroth and Moisio 2020). We advocate for a deepened reflection on this concept, acknowledging that ensuring development opportunities beyond the spatial social minimum may not be possible in all areas and that cohesion policy cannot radically transform the development status of every peripheral and marginalized territory (Halamska 2018; Wilczyński 2017). In such cases, non-standard, adaptive measures should be implemented, including ad-hoc transport solutions in place of regular public transport, mobile retail services, mobile healthcare units, or optimizing school networks by providing transportation to well-equipped, high-quality education centers. These are instruments of proactive governance, needed to address inevitable but unfavorable changes, and should replace reactive (and thus less effective) responses taken after such changes have already occurred (Gorzelak 2021). Theoretically, the spatial social minimum should define the expected target state toward which regional development interventions should strive. However, an open dilemma remains regarding the scope of public services that should be guaranteed to ensure equality of opportunity, and above all, the means to ensure these services are available in every spatial unit. This scope should be determined based on dedicated research. Importantly, the resulting set of criteria should be applied flexibly, in line with the place-based policy approach. This does not imply imposing a uniform standard of the spatial social minimum across the EU, but rather treating its definition as an ex-ante conditionality, to be established individually in each country and/or region.

The recommended solutions are expected to increase the effectiveness of Cohesion Policy interventions by moving away from idealistic assumptions such as the belief that development and innovation can be generated in every area regardless of constraints on territorial capital. They should also help define the type and scope of development interventions needed in territories that are not only relatively under-resourced, but also small in size and/or located at significant physical or relational distances from core areas. Our research findings indicate that these are also the areas most vulnerable to the compounded impacts of major contemporary challenges including climate change, demographic shifts, geopolitical instability, and technological transformation—all of which have been significantly reshaped by the financial crisis, the COVID-19 pandemic, and Russia's invasion of Ukraine, resulting in a resurgence of inflation and heightened concerns over food, energy, and resource security. Under these new conditions, development interventions in such areas must focus on building long-term resilience, leveraging locally available resources to develop new competitive advantages (Lange et al. 2021), and reorienting global and local linkages (European commission 2024; MacKinnon et al. 2022; OECD 2023). On the one hand, this necessitates a shift in how development actions are programmed and implemented, making the process more complex and integration-intensive (Pike et al. 2023). On the other hand, it allows for the strengthening of local capacities to respond appropriately to development challenges, which is particularly critical for peripheral areas lacking territorial capital and the critical mass for local development (Diemer et al. 2022; OECD 2023). To realize this effectively, a comprehensive understanding of the underlying causes of developmental challenges and the possibilities for overcoming them is essential. This requires mobilizing both local resources and the development of new functional relationships, with the broadest possible engagement of all stakeholders (Rodríguez-Pose et al. 2023, 2024).

Our analysis confirms that inner peripheries are a widespread component of spatial development disparities. We assume that the findings presented in this book will contribute to optimizing their identification and serve as a basis for reforming the programming and implementation of development interventions under Cohesion

Policy, thereby improving the living standards and quality of life for residents in these areas.

References

Adegbie FF, Ajayi A, Aguguom TA, Otitolaiye ED (2022) Diversification of the economy, tax revenue and sustainable growth in Nigeria. Int J Innov Res Sci Stud 6(1):115–127. https://doi.org/10.53894/ijirss.v6i1.1114

Bo L, Chan TCY, Saxe S (2021) The impact of COVID-19 cycling infrastructure on low-stress cycling accessibility: a case study in the city of Toronto. Findings. https://doi.org/10.32866/001c.19069

Camagni R (2008) Regional competitiveness: towards a concept of territorial capital. In: Capello R, Camagni R, Chizzolini B, Fratesi U (eds) Modelling regional scenarios for the enlarged Europe: European competitiveness and global strategies. Springer, pp 33–48

Churski P, Herodowicz T, Konecka-Szydłowska B, Perdał R (2021) European regional development and socio-economic changes—contemporary perspectives. Springer Nature

Creutzig F, Lohrey S, Franza MV (2022) Shifting urban mobility patterns due to COVID-19: comparative analysis of implemented urban policies and travel behaviour changes with an assessment of overall GHG emissions implications. Environ Res: Infrastruct Sustain 2(4):041003. https://doi.org/10.1088/2634-4505/ac949b

Diemer A, Iammarino S, Rodríguez-Pose A, Storper M (2022) The regional development trap in Europe. Econ Geogr 98(5):487–509. https://doi.org/10.1080/00130095.2022.2080655

Dikeç M (2001) Justice and the spatial imagination. Environ Plan a: Econ Space 33(10):1785–1805. https://doi.org/10.1068/a3467

Dworkin R (1981) What is equality? Part I: equality of welfare; Part II: equality of resources. Philos Public Aff 10(3/4):185–246. https://www.jstor.org/stable/2264894

European Commission (2024) Ninth report on economic, social and territorial cohesion. Directorate-general for regional and urban policy. https://ec.europa.eu/regional_policy/information-sources/cohesion-report_en

European Commission: Directorate-General for Regional and Urban Policy (2024) Forging a sustainable future together—Cohesion for a competitive and inclusive Europe—report of the high-level group on the future of cohesion policy. Publications office of the European Union. https://doi.org/10.2776/974536

Gnap J, Senko Š, Kostrzewski M, Brídziková M, Cződörová R, Říha Z (2021) Research on the relationship between transport infrastructure and performance in rail and road freight transport—a case study of Japan and selected European countries. Sustainability 13(12):6654. https://doi.org/10.3390/su13126654

Gorzelak G (1985) Przestrzenne aspekty jakości życia i sprawiedliwości społecznej [Spatial aspects of quality of life and social justice]. In: Kukliński A (ed) Gospodarka przestrzenna Polski wczoraj, dziś i jutro II. Wydział Pracy Posłów i Radnych Stowarzyszenia PAX, pp 69–83

Gorzelak G (2021) Różnice regionalne—preferencje polityczne—sprawiedliwość społeczna [Regional differences—Political preferences—Social justice]. Studia Regionalne i Lokalne 2(84):119–127

Haefner L, Sternberg R (2020) Spatial implications of digitisation: State of the field and research agenda. Geogr Compass 14(12):e12544. https://doi.org/10.1111/gec3.12544

Halamska M (2018) Wspierać czy zalesiać? Dylematy rozwoju wiejskich obszarów problemowych [To support or afforest? Dilemmas of rural problem areas development]. Wieś i Rolnictwo 3(180):69–91

Lange B, Hülz M, Schmid, B, Schulz C (eds) (2021) Post-growth geographies: spatial relations of diverse and alternative economies. Columbia University Press

Lefebvre H (1968) The right to the city. The Anarchist Library
Lestari PW, Dewi GK (2022) Availability of infrastructure and COVID-19 prevention behavior in public places. J Kesehat Masy 18(1):107–113. https://doi.org/10.15294/kemas.v18i1.33478
MacKinnon D, Kempton L, O'Brien P, Ormerod E, Pike A, Tomaney J (2022) Reframing urban and regional 'development' for 'left behind' places. Camb J Reg Econ Soc 15(1):39–56. https://doi.org/10.1093/cjres/rsab028
OECD (2023) OECD regional outlook 2023: the longstanding geography of inequalities. OECD Publishing. https://doi.org/10.1787/92c3f028-en
Pike A, Béal V, Cauchi-Duval N, Franklin R, Kinossian N, Lang T, Leibert T, MacKinnon D, Rousseau M, Royer J, Servillo L, Tomaney J, Velthuis S (2023) 'Left behind' places: a geographical etymology. Reg Stud 57(5):1–15. https://doi.org/10.1080/00343404.2023.2167972
Pinheiro FL, Balland P-A, Boschma R, Hartmann D (2022) The dark side of the geography of innovation: relatedness, complexity and regional inequality in Europe. Reg Stud 56(12):2109–2124. https://doi.org/10.1080/00343404.2022.2071422
Podgorskaya S (2021) Methodological aspects of rural economy diversification in the context of modern civilisational transformations. E3S Web Conf 273:08041. https://doi.org/10.1051/e3sconf/202127308041
Rauhut D (2017) A note on territorial justice [Conference paper]. In: 57th conference of the European regional science association, Groningen, Netherlands. https://www.researchgate.net/publication/319443258_A_Note_on_Territorial_Justice
Rawls J (1971) A theory of justice. Oxford University Press
Rodríguez-Pose A, Dijkstra L, Poleman H (2023) The geography of EU discontent and the regional development trap (Working Paper No. 03/2023). European Commission
Rodríguez-Pose A, Bartalucci F, Lozano-Gracia N, Davalos M (2024) Overcoming left-behindness: moving beyond the efficiency versus equity debate in territorial development (Policy Research Working Paper No. 10734). World Bank Group
Sarin V, Mahapatra SK, Sood N (2020) Export diversification and economic growth: a review and future research agenda. J Public Aff 22(3):e2524. https://doi.org/10.1002/pa.2524
Sen A (1999) Development as freedom. Oxford University Press
Smith DM (1994) Geography and social justice. Blackwell
Smith DM (2000) Moral geographies: ethics in a world of difference. Edinburgh University Press
Sotiroski L (2023) Consolidation and crisis: The evolution of the European Union's legal framework in pursuing peace, stability, and unity amidst global challenges. Soc Leg Stud 6(3):171–177. https://doi.org/10.32518/sals3.2023.171
Weckroth M, Moisio S (2020) Territorial cohesion of what and why? The challenge of spatial justice for EU's cohesion policy. Soc Incl 8(4):183–193. https://doi.org/10.17645/si.v8i4.3241
Wilczyński R (2017) Odnowa wsi a paradygmat dopasowania—czyli co robić, gdy wsie zanikają [Village renewal and the matching paradigm—what to do when villages disappear]. Paper presented at the XXXIII seminar on rural geography, Poland
Yuliana Y (2021) Infrastructure building adjustment in the COVID-19 pandemic. Border 3(2):81–86. https://doi.org/10.33005/border.v3i2.86

Index

P. Churski et al., *Inner Peripheries in the New Conditions of Cohesion Post-SARS-CoV-2*, Economic Geography,
https://doi.org/10.1007/978-3-032-10936-1

MIX
Papier aus verantwortungsvollen Quellen
Paper from responsible sources
FSC® C105338

If you have any concerns about our products,
you can contact us on
ProductSafety@springernature.com

In case Publisher is established outside the EU,
the EU authorized representative is:
Springer Nature Customer Service Center GmbH
Europaplatz 3, 69115 Heidelberg, Germany

Printed by Libri Plureos GmbH
in Hamburg, Germany